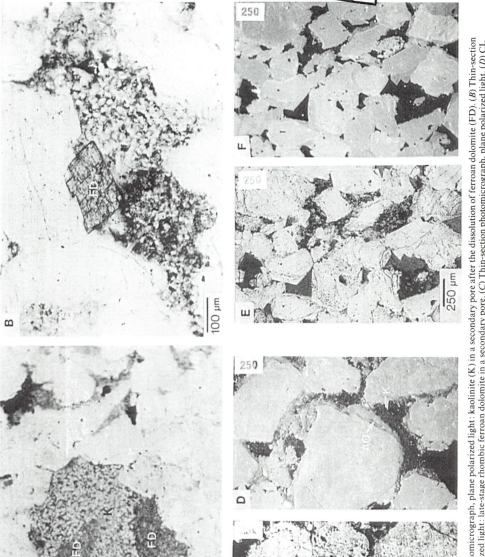

PLATE 1. (*A*) Thin-section photomicrograph, plane polarized light: kaolinite (K) in a secondary pore after the dissolution of ferroan dolomite (FD). (*B*) Thin-section photomicrograph, plane polarized light: late-stage rhombic ferroan dolomite in a secondary pore. (*C*) Thin-section photomicrograph, plane polarized light. (*D*) CL micrograph of the same field: quartz overgrowths (AQ) show turquoise emission colours and rugose boundaries with a kaolinite-filled pore (K); kaolinite shows strong blue emission colours. (*E*) Thin-section photomicrograph, plane polarized light. (*F*) CL micrograph of the same field. Thin-section micrographs show apparently highly compacted rock, whereas the CL micrograph shows that 'sutured' grain contacts are caused by interlocking quartz overgrowths. From paper by Cowan, page 62.

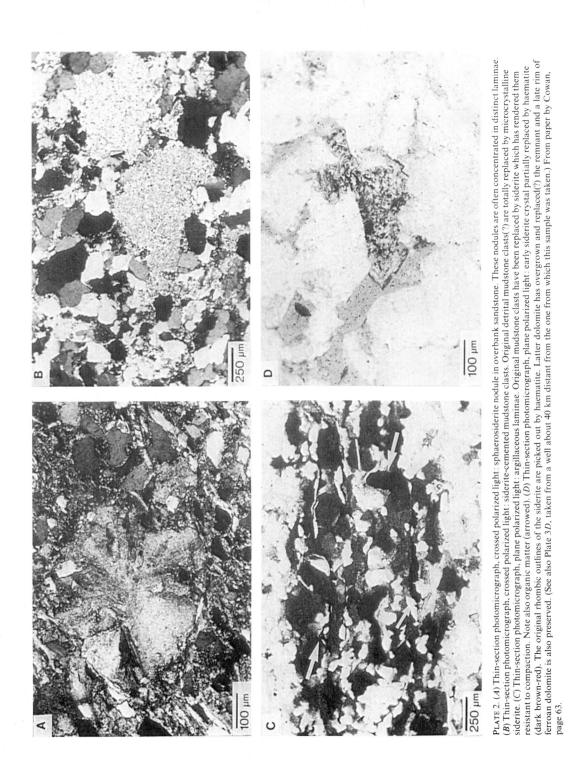

PLATE 2. (*A*) Thin-section photomicrograph, crossed polarized light: sphaerosiderite nodule in overbank sandstone. These nodules are often concentrated in distinct laminae. (*B*) Thin-section photomicrograph, crossed polarized light: siderite-cemented mudstone clasts. Original detrital mudstone clasts(?) are totally replaced by microcrystalline siderite. (*C*) Thin-section photomicrograph, plane polarized light: argillaceous laminae. Original mudstone clasts have been replaced by siderite which has rendered them resistant to compaction. Note also organic matter (arrowed). (*D*) Thin-section photomicrograph, plane polarized light: early siderite crystal partially replaced by haematite (dark brown-red). The original rhombic outlines of the siderite are picked out by haematite. Latter dolomite has overgrown and replaced(?) the remnant and a late rim of ferroan dolomite is also preserved. (See also Plate 3*D*, taken from a well about 40 km distant from the one from which this sample was taken.) From paper by Cowan, page 63.

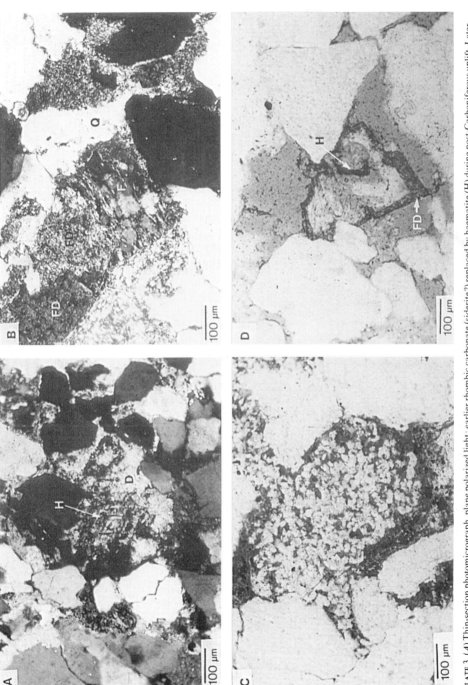

PLATE 3. (A) Thin-section photomicrograph, plane polarized light: earlier rhombic carbonate (siderite?) replaced by haematite (H) during post-Carboniferous uplift. Later, non-ferroan dolomite cement (D) infills the remaining porosity. (B) Thin-section photomicrograph, cross polarized light: ferroan dolomite cements (FD) partially replacing feldspar (F) and showing embayed replacive boundaries with quartz (Q). This non-rhombic form of ferroan dolomite was precipitated prior to the generation of secondary porosity at depth. (C) Thin-section photomicrograph, plane polarized light: secondary dissolution pore infilled with kaolinite. In this example it is not clear whether the kaolinite formed as a direct replacement of a feldspar or precipitated in an open secondary pore after feldspar dissolution. (D) Thin-section photomicrograph, plane polarized light: multiphased carbonate precipitation and dissolution. The earliest phase of carbonate has been replaced by haematite (H) during post-Carboniferous uplift and infiltration of meteoric fluids. With renewed burial non-ferroan dolomite overgrew the haematite–carbonate remnant and suffered dissolution during deep burial. The outer rim of rhombic ferroan dolomite (FD) is complete, shows no evidence of dissolution and has clearly precipitated after the dissolution event (see Plate 1B). From paper by Cowan, page 63.

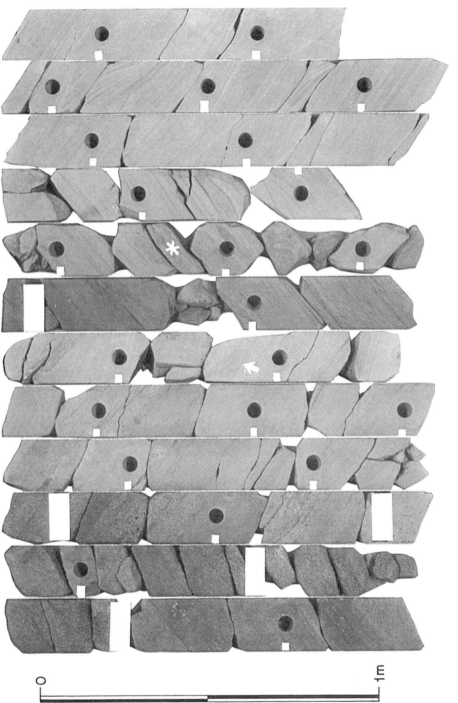

PLATE 1. Core photographs of the main coarsening-upward sand body of unit 3.1 taken from well B of Fig. 7. The lower part of the photograph is dominated by amalgamated fine-grained hummocky cross-stratified sandstone. The upper section contains thin decimetre-graded beds capped by wave-generated ripple cross-lamination (arrow) overlain by cross-stratified medium- and coarse-grained sandstone. The micaceous horizon discussed in the text is indicated with an asterisk. The core is taken from a deviated well and horizontal bedding is at an angle of 45° to the axis of the core. The scale is corrected for this deviation. From paper by Livera, page 277.

Deltas
SITES AND TRAPS FOR
FOSSIL FUELS

GEOLOGICAL SOCIETY SPECIAL PUBLICATION CLASSICS

Deltas

SITES AND TRAPS FOR

FOSSIL FUELS

EDITED BY

M. K. G. WHATELEY & K. T. PICKERING

Department of Geology

The University

Leicester LE1 7RH

Published by

The Geological Society

London

THE GEOLOGICAL SOCIETY

The Society was founded in 1807 as The Geological Society of London and is the oldest geological society in the world. It received its Royal Charter in 1825 for the purpose of 'investigating the mineral structure of the Earth'. The Society is Britain's national learned society for geology with a membership of 7374 (1992). It has countrywide coverage and approximately 1000 members reside overseas. The Society is responsible for all aspects of the geological sciences including professional matters. The Society has its own publishing house which produces the Society's international journals, books and maps, and which acts as the European distributor for publications of the American Association of Petroleum Geologists and the Geological Society of America.

Fellowship is open to those holding a recognized honours degree in geology or a cognate subject and who have at least two years relevant postgraduate experience, or who have not less than six years relevant experience in geology or a cognate subject. A Fellow who has not less than five years relevant postgraduate experience in the practice of geology may apply for validation and, subject to approval, may be able to use the designatory letters C. Geol (Chartered Geologist).

Further information about the Society is available from the Membership Manager, The Geological Society, Burlington House, Piccadilly, London W1V 0JU, UK.

Published by The Geological Society from:
The Geological Society Publishing House
Unit 7
Brassmill Enterprise Centre
Brassmill Lane
Bath BA1 3JN
UK

(*Orders*: Tel. 0225 445046
  Fax 0225 442836)

First published 1989
Paperback edition 1993

British Library Cataloguing in Publication Data
A catalogue record for this book is available from the British Library

ISBN 0-903317-98-2

Printed in Great Britain by
The Alden Press, Oxford

Distributors

USA
  AAPG Bookstore
  PO Box 979
  Tulsa
  Oklahoma 74101-0979
  USA
  (*Orders*: Tel: (918) 584-2555
    Fax (918) 584-0469)

Australia
  Australian Mineral Foundation
  63 Conyngham Street
  Glenside
  South Australia 5065
  Australia
  (*Orders*: Tel: (08) 379-0444
    Fax (08) 379-4634)

India
  Affiliated East-West Press PVT Ltd
  G-1/16 Ansari Road
  New Delhi 110 002
  India
  (*Orders*: Tel: (11) 327-9113
    Fax (11) 326-0538)

Japan
  Kanda Book Trading Co.
  Tanikawa Building
  3-2 Kand Surugadai
  Chiyoda-Ku
  Tokyo 101
  Japan
  (*Orders*: Tel: (03) 3255-3497
    Fax (03) 3255-3495)

Contents

Coal-related Case Histories

Preface

Deltas, with their economic, political and scientific importance, have long fascinated and attracted Man's attention. The term 'delta' was coined by Herodotus in approximately 450 BC for the triangular-shaped sedimentary body at the mouth of the River Nile. Today, the concept of a delta, its morphology and its controlling processes, are almost as numerous as the people who work on such systems.

This volume, of 23 papers on many aspects of modern and ancient deltaic sedimentary systems, will be useful to researchers as well as teachers and students alike. The well-balanced content of the book should prove particularly attractive to those who seek a detailed state-of-the-science overview of this large and ever-expanding subject area.

The international meeting from which this volume resulted was convened at the Geological Society of London, Burlington House, Piccadilly, from 21 to 22 April 1987, in order to fill the gap in publications on deltaic systems since the flurry of conferences on deltaic sedimentation took place in the early and mid-1970s. We felt that significant advances have been made in our understanding of deltas and yet not collated into a useful reference source, and a timely forum was long overdue to enable workers to disseminate this knowledge. The two-day meeting included 27 papers and eight posters covering modern and ancient deltas in marine to freshwater environments. Topics included processes, facies models, petroleum-, gas- and coal-related deltaic environments, together with general case studies. The meeting attracted about 180 speakers and delegates from Germany, France, Norway, Canada, Italy, Denmark, the USA and Britain. The delegates were warmly welcomed by Professor Alec Smith in his role as Vice-President of the Geological Society. Dr John Hudson stepped in at the eleventh hour to provide a superb after-dinner speech at the conference dinner.

Following the two days of plenary sessions, there were two successful two-day field trips, one led by Professor Trevor Elliot to the Westphalian of N Devon and the other led by Professor John Collinson to the Namurian of the Pennines. At the weekend following the field trips, there was a one-day core workshop in Edinburgh, organized by Stuart Brown and Philip Richards (British Geological Survey, Edinburgh), to examine the Jurassic Brent cores from the northern North Sea. This was so popular that an additional workshop had to be organized for the following day (Sunday) to cope with demand.

We would like to thank the following companies for their generous sponsorship of the meeting and this publication: Britoil plc, BP Petroleum Development Ltd, Chevron Exploration North Sea Ltd, Clyde Petroleum plc, Enterprise Oil plc, Esso, GAPS Geological Consultants, LASMO Exploration & Production UK Ltd, Neste Oy (Exploration), Norsk Hydro, North Sea Sun Oil Co. Ltd, Occidental International Oil Inc., Occidental Petroleum (Caledonia) Ltd, Robertson Research plc, Teredo Oils Ltd, Texas Eastern North Sea Inc., Total Oil Marine, Tricentrol Oil Corp Ltd and UNOCAL Ltd.

We also wish to thank the Royal Society and the Geological Society of London for their financial support of the meeting. The abstract volume was published by British Gas plc and we thank them for their generosity.

Our thanks to Barbara Afergan, Cherry Walker and Mark Wilkinson for helping us in the many essential operational/administrative tasks on the days of the conference. Sue Button is thanked for last minute drafting of diagrams, and the efficient secretarial staff at the Geological Society and at Leicester University are thanked for their help. Naturally, we thank the authors and the referees of the manuscripts for providing us with the material for this book. Finally we thank our families for their support during the production of this volume.

M. K. G. WHATELEY & K. T. PICKERING, Department of Geology, University of Leicester, Leicester LE1 7RH, UK.

Deltas: sites and traps for fossil fuels

M. K. G. Whateley & K. T. Pickering

This volume does not aim to review deltas, nor to provide a comprehensive reference source, but merely to highlight some of the recent advances in our knowledge of deltas and their economic significance. The discursive papers by Elliott and Alexander consider the definition of a delta and emphasize the wide spectrum of deltaic and associated deposits. Syvitski and Farrow describe fan deltas in fjords, with their rugged topography and dynamic history of basin infilling as analogues for small hydrocarbon-bearing submarine fans. Ito adds a further dimension to our knowledge of fan deltas and how they relate to water depth in the receiving basin.

A section of the book is devoted to subsurface and geophysical techniques. Cowan considers the porosity evolution of Carboniferous fluvio-deltaic sandstones, and illustrates the importance of understanding poroperm development in these economically attractive reservoirs. The Bristow and Myers combination papers describe how detailed gamma-ray logging, using a potassium, thorium, uranium scintillometer, can be used to distinguish lithofacies associations in a Namurian deltaic succession. This offers the possibility of adding to our already expanding knowledge on facies associations and brings the detailed correlation of surface with subsurface exposure using petrophysical logs. In this respect, Selley's contribution on reservoir prediction using rotational dipmeter patterns complements the previous authors' contributions.

The third section of the book is based around general case studies of various deltaic systems. Sestini provides a review of the sedimentary and tectonic history of the Nile Delta, and Okazaki and Masuda describe Quaternary deltas from palaeo-Tokyo Bay. Pedersen describes the fine-grained sediments in a fluvially dominated lacustrine delta in a volcanic province of W Greenland and highlights the need for provenance studies to understand poroperm and source rock potential better. Siedlecka, Pickering and Edwards document the oldest deltaic system described in this book, from the late Precambrian, N Norway, where wet-sediment deformation is also common. Harris discusses the sedimentology of a Middle Jurassic lagoonal delta system in the Hebrides off NW Scotland. Papers on Irish Carboniferous deltaic deposits are presented by Martinsen and Pulham. The former deals with styles of wet- or soft-sediment deformation on a Namurian delta slope, whilst Pulham considers the internal structure and architecture of sandstone bodies within the Upper Carboniferous deltaics of County Clare, also containing superb examples of sediment liquefaction and fluidization.

Petroleum- and gas-related case studies form the fourth part of this volume. Flint, Stewart and van Reissen describe the reservoir geology of a lacustrine delta in a Tertiary intermontane basin of the Sirikit oilfield in Thailand and their work shows how important relatively areally restricted basins can be in considerations of oil resource potential. Helland-Hansen, Steel, Nakayama and Kendall investigate sandstone geometry by computer modelling of the Brent Group. This review of the Brent Group is complemented by the papers by Brown and Richards, and Livera, who describe facies associations of formations within the Brent Group. Brown and Richards mainly describe the northern limit of progradation of the Brent Delta, whilst Livera concentrates on sandbody geometry of the Ness Formation.

The final section covers the coal-related case histories in deltaic settings. Haszeldine describes the relationship between depositional controls of coal and coal-bearing strata and climate and structural setting, and points out the need for integrated sedimentological and palaeobotanical studies to enable us to understand fully the relationships between coal and its enclosing sediments. Scott makes similar points, but also shows that modern peat-forming ecosystems require more detailed investigation in order to provide better modern analogues for many ancient coal-related delta systems. Whateley and Jordan consider a thick Tertiary coal deposit in Sumatra, Indonesia, whose geometry appears to be strongly influenced by fan-delta lobes which built out into a lake, the coal forming in interlobe areas. The final paper by Read describes the relationship between subsidence and deposition of fluvio-deltaic environments in the Namurian, Midland Valley, Scotland.

M. K. G. WHATELEY & K. T. PICKERING, Department of Geology, University of Leicester, Leicester LE1 7RH, UK.

From WHATELEY, M. K. G. & PICKERING, K. T. (eds), 1989, *Deltas: Sites and Traps for Fossil Fuels,* Geological Society Special Publication No. 41, p. ix.

Deltaic Systems and General Models

Deltaic systems and their contribution to an understanding of basin-fill successions

T. Elliott

SUMMARY: Deltaic systems are prominent features of many clastic basin-fill successions throughout the geological record and are also important as sites of fossil fuel reserves. The process-based classification of deltaic systems into fluvial-, wave- and tide-dominated types adopted in recent years remains valid, but requires additional information on, for example, the sediment load of the system, its position in the basin and the extent of synsedimentary deformation in order to be a meaningful characterization of a delta. The range of delta types is further extended by research into ancient deltaic systems which can result in the definition of a non-actualistic delta which differs from modern systems. This is an important trend which should develop in the future as well-constrained ancient systems are characterized to their full extent. A further avenue of future research lies in examining the nature of links of deltaic systems to shelf, slope and submarine-fan systems, with particular reference to depositional analysis of seismic stratigraphic information.

Deltaic systems are important elements of many clastic basin-fill successions; they are important in their own right and have significant links with alluvial, shelf, slope and deep-sea fan systems. Interpretations of sediment dispersal and deposition in clastic basins are often greatly enhanced if the position, character and behaviour of basin-margin delta systems are constrained. Deltaic successions are also important economically; organic-rich shales and coals can be important as hydrocarbon source rocks, sandstone bodies serve as reservoirs in numerous hydrocarbon provinces and significant coal reserves are hosted in deltaic successions. Modern and ancient deltaic systems have been extensively researched for many decades, with many of the basic principles of sedimentation such as cyclicity, facies and sequence analysis being developed and tested in these successions. However, despite the longevity and high quality of this research, important problems still remain. The aim of this paper is to review recent progress in deltaic studies and to identify outstanding problems which are currently attracting attention or might do so in the near future. The paper commences by considering various aspects of deltas as discrete depositional systems and continues by discussing the links between deltas and other depositional systems in clastic basin-fill successions.

Present status of deltaic models

The notion that deltas are highly variable depositional systems which cannot be summarized in terms of a single facies model is now well established. The variability of modern deltas has, for some time, been reviewed by means of a ternary diagram with fluvial-, wave- and tide-dominated end-members (Galloway 1975). This process-based scheme is directly applicable to ancient deltaic successions and has been readily adopted in numerous studies. Once the subenvironments of an ancient delta have been recognized on general facies criteria, it is possible to examine the nature of the physical processes which operated in various subenvironments in more detail and hence to reconstruct the character of the ancient delta. The nature of the delta-front sequence and, more particularly, the river-mouth part of the delta front is critical in this regard since it is here that the interaction between the sediment-laden river waters and those of the receiving basin takes place. The identification and thorough examination of this subenvironment at an early stage in an investigation can yield significant insights into the nature of the delta system by providing a reliable first-order interpretation of the type of delta. Fluvial- and wave-dominated deltaic systems have become increasingly well understood in both modern and ancient settings in recent years, but progress in understanding tide-dominated deltas is, by comparison, limited. The addition of the tide-dominated Mahakam Delta in Indonesia to our suite of well-documented modern deltas is a significant development in this respect (Allen *et al*. 1979), but further developments are required to advance our understanding of modern and ancient tide-dominated deltaic systems. What is the medium- to long-term behaviour of distributary channels in tide-dominated deltas? Do tide-dominated deltas develop lobes or lobe complexes, or is this tendency suppressed by the confined gulf-like basin settings in which these deltas often form? How are tide-dominated deltas

From WHATELEY, M. K. G. & PICKERING, K. T. (eds), 1989, *Deltas: Sites and Traps for Fossil Fuels,* Geological Society Special Publication No. 41, pp. 3–10.

modified during abandonment and how efficient are the reworking processes? These and other questions are central to an understanding of tide-dominated deltaic systems, particularly in terms of predictive facies modelling in hydrocarbon-bearing successions.

Despite the success of the process-based classification of delta systems it is clear that depositional systems as complex as deltas cannot be adequately summarized in terms of a single parameter, no matter how central to the formation of the delta this parameter is in our view. In any comprehensive assessment of a modern or ancient delta a number of other factors should be included in the characterization of the system. For example, the amount and calibre of the sediment load directly influences the nature of the delta and the resultant facies pattern. Variations in sediment load caused by climatic or tectonic events in the hinterland may induce a change in the nature of the delta which, to a degree, is independent of the basin regime. With regard to modern deltas, information on present sediment loads and recent fluctuations in supply is extremely limited. As a result our understanding of this important parameter is rather vague.

The basin type and the position of the delta in the basin are both important in determining factors such as the salinity of the basin waters, the water depth into which the delta is prograding, the nature of the basinal processes which are operative and the prevailing subsidence rate. The extent of synsedimentary deformation processes such as slumping, diapirism and growth faulting is also important. In many deltaic systems features related to synsedimentary deformation appear scarce, but in others they are abundant and the correct identification and interpretation of the various styles of deformation are essential to a full understanding of the facies and sand-body characteristics of the system (see below). Viewing the modern Mississippi Delta in this regard, it is undeniably a fluvial-dominated delta, but it is also an extremely fine-grained highly unstable delta system located in a relatively deep-water setting within 30 km of the present shelf edge. Key features of the modern delta such as the fixing of distributary channel courses due to their incision into previously deposited cohesive muds and the distinctive bar-finger sand-body pattern stem from the fine-grained nature of the system rather than its fluvial-dominated character. The logical conclusion of emphasizing a wide range of factors may be to render each delta unique, which is probably the truth of the matter. The way forward seems to be to use information on all the major factors controlling the delta system within a process-based framework.

River-mouth processes in deltaic systems

Rather surprisingly one of the key problems which often faces sedimentologists concerned with interpreting ancient deltaic successions is the manner in which fluvial currents decelerate and deposit sediment in the vicinity of the river mouth. Despite the pioneering work of Bates (1953) and more recent research it remains difficult to generalize on precisely what happens to fluvial currents as they leave the river mouth and enter the basin. In the case of deltas forming in low energy marine basins the density difference between freshwater and saline water gives rise to a buoyantly supported plume of sediment which is dispersed into the basin as the plume mixes with the basin water (Wright & Coleman 1974; Wright 1977). During low river stage saline basin waters can enter the lower reaches of the channel (the salt wedge) and cause the fluvial currents to detach from the channel floor. As river stage rises during flood periods, the salt wedge is flushed out of the channel and the fluvial currents transport bed-load sediment over the floor of the channel and the shallow crestal regions of the mouth bar. But how far and to what depth do these fluvial currents extend into the basin? Information on this point is surprisingly sparse and this creates problems in the interpretation of fluvial-dominated delta-front sequences in ancient sequences. The middle to upper parts of these sequences are often characterized by numerous thin erosive-based beds of sandstone deposited by basinward-directed waning currents. These currents are considered to have issued directly from the river mouth, but to judge from their position in the delta-front sequence the lowest examples of these beds must have been deposited in moderate water depths beyond the immediate river mouth. Evidence from modern delta systems is insufficient to determine whether this is feasible. In the case of relatively small delta-filled cratonic basins which prevailed, for example, during the Upper Carboniferous in northern Europe it is often debated whether fully marine salinities were maintained permanently in the basin. The lowering of basin salinity, coupled with high concentrations of suspended load, could reduce the density difference between the fluvial and basin waters and may have permitted traction currents to extend down the delta front for greater distances. This notion has been extended in some cases to include the idea that density-driven currents may be initiated at the river mouth, in a similar manner to sediment-laden fluvial waters entering a freshwater lake (*eg* the Rhone river entering Lake Geneva).

Support for the notion of density-driven under-flows operating on delta fronts in marine basins has come from the Yellow River Delta in China (Wright *et al.* 1988). The sediment load of this system is predominantly loess silt which is supplied in high suspended-load concentrations during flood periods. The flows are dispersed as underflows 1–4 m deep which decelerate rapidly across the delta front, depositing most of their sediment load between the 5 and 10 m isobaths, 1–5 km from the river mouth. The origin of these flows lies mainly in the exceptionally high concentrations of suspended load during flood period. Traction transport does not appear to be associated with these underflows, although this may be a function of the fine-grained nature of the sediment supply.

Shelf deltas versus shelf-margin deltas

The distinction drawn by early researchers in the Gulf of Mexico between pre-modern shoal-water deltas and the modern deep-water Mississippi Delta has been revived in subsurface studies of Tertiary and Pleistocene deltas in the region (Edwards 1981; Winker 1982; Winker & Edwards 1983). These workers use the terms shelf deltas and shelf-margin deltas to discriminate between deltas which prograded across a shallow water shelf and those which were relatively fixed in position in the vicinity of the shelf–slope break. Shelf deltas are characterized by laterally extensive delta-front sands, large progradation distances and a lack of growth faulting. In contrast, deeper-water shelf-margin deltas are characterized by short progradation distances and are extensively growth faulted. The stratigraphical successions of shelf-margin deltas are thick and localized within major growth fault structures. The successions are characterized by repeated vertical stacking of delta-front sequences, pronounced basinward increases in sediment thickness and rapid basinward facies changes. In view of the large displacements associated with the growth faults in these systems correlation of stratigraphic units is difficult and there are marked deviations in dip from the regional. Seismic facies mapping in Pleistocene shelf-edge deltas reveals steep basinward clinoform stratification and, in some cases, chaotic patterns of seismic reflectors which are considered to reflect synsedimentary sliding and slumping (Winker & Edwards 1983). In plan view the deltas appear lobate rather than elongate as in the case of the modern shelf-margin Mississippi Delta. In the Tertiary Wilcox Group Edwards

(1981) considers that this is due to a relatively high sand load which prevailed at the time and was being partially reworked along-shore by basin-wave processes. However, in Pleistocene shelf-margin deltas Winker & Edwards (1983) interpret the strike-oriented nature of the sediment body as a response to the geometry of the growth fault structure which defined the depocentre of the delta rather than the process regime of the delta.

Synsedimentary deformation in modern and ancient deltas

It has been appreciated for some time that certain delta systems are characterized by synsedimentary deformational processes such as slumping, mud diapirism and growth faulting which operate during delta progradation. However, the full range and scale of these processes has only really been appreciated in recent years following a major research programme in the Mississippi Delta, initiated after the failure of offshore petroleum platforms sited on the delta-front regions of the delta (Coleman *et al.* 1974; Coleman 1981). New maps of distributary mouth bars of the Mississippi Delta reveal a high diversity and density of synsedimentary deformational features which are active during deposition. Mud diapirs, rotational slides and slump sheets, collapse depressions, mudflow gullies and lobes, and normal faults proliferate on the surface of the mouth bars. It is estimated that 50% of sediment deposited in distributary mouth bars of the Mississippi Delta is involved in down-slope mass movements which translate upper mouth-bar facies into the deeper parts of the basin, often beyond the limits of the delta itself. The recurrence interval of slumps is high and mouth-bar progradation is therefore erratic with the slopes acting as transfer or bypass routes for sediment for much of the time (Lindsay *et al.* 1984). Slumping is initiated aseismically by a combination of processes including high pore-water and methane pressures in the sediment, wave-induced cyclic loading or pounding of the sediments during storms and hurricanes and oversteepening of the bar front due to differential sedimentation rates on the bar front during flood periods and jacking-up of the bar-front slope by mud diapir activity beneath the mouth bar. Recent research in the Gulf of Mexico has also demonstrated that major canyons in the outer shelf/continental slope in front of the Mississippi Delta were formed by very-large-scale slope failures which removed a substantial amount of sediment (12 000 km^3) from the upper part of the slope (Shepard 1981;

Coleman *et al.* 1983). The slump material is estimated to account for 15% of the deposits of the Mississippi fan. The slump scar formed a subbasin on the slope which was infilled mainly by the distal parts of shelf-edge delta lobes.

In view of the range, density and frequency of operation of these synsedimentary deformational processes it is necessary critically to re-evaluate our view of the facies pattern which is likely to be finally preserved. At present our view is still dominated by our understanding of the depositional processes active on the delta front, with only a limited role for deformational processes having been recognized. The work of Fisk and the early work of Coleman and co-workers in describing the facies patterns of the Mississippi Delta includes very few features attributable to synsedimentary deformation and is now in conflict with the new information from the delta. The deformational features referred to above are by no means confined to the Mississippi Delta; research programmes as intensive as that recently undertaken in the Mississippi Delta may reveal the same range and density of deformational features in other deltas.

In recent years a number of studies of ancient deltaic successions have recognized features which reflect synsedimentary deformation processes analogous to those of modern delta systems (Edwards 1976; Chisholm 1977; Rider 1978; Elliott & Ladipo 1981; Nemec *et al.* 1988; Pulham, this volume). Slump sheets, collapse structures and growth faults are the most widely described features. The examples of growth faults clearly demonstrate the role of these structures in defining localized sandstone-dominated depositional centres and suggest that delta systems characterized by extensive growth faulting should be regarded as a distinct category as the facies patterns may be determined principally by the history of growth faulting (as in Tertiary successions of the Niger Delta; Weber 1971; Evamy *et al.* 1978). Growth faults recognized in exposed ancient deltaic successions are directly analogous to structures identified in subsurface studies in all aspects except scale. All examples of growth faults described to date from surface exposures are substantially smaller than subsurface equivalents (tens of metres scale as opposed to kilometres scale). Large-scale growth faults have still to be recognized, or inferred, in exposed terrains and, conversely, smaller-scale growth faults beyond the resolution of seismic data await discovery in subsurface investigations. Further research in well-exposed deltaic successions and greater refinement in subsurface interpretation techniques may address the range of scales of growth faults in the near future.

Non-actualistic deltaic models

One consequence of stressing the wide range of controls which influence deltas is, as noted above, the tendency for each delta to be regarded as unique. It follows from this that ancient delta systems might differ substantially from the known range of modern deltas, giving rise to the notion of non-actualistic delta models. In the early stages of an investigation or where the available data are strictly limited it is perhaps inevitable that comparisons with modern deltas will be made and that the system may be interpreted with reference to a modern system (*eg* as Rhone-like or Niger-like). However, in the latter stages of study of a well-exposed or extensively cored delta system it should be possible to produce a model which honours all the available data and thereby produces a distinctive delta type which may differ in several important respects from modern deltas. To date, there are very few examples of this kind of interpretation. The Carboniferous Kinderscout Delta system provides an example of this point (Collinson 1969; McCabe 1978). Large high-discharge river systems characterized by coarse-grained bedforms several tens of metres in height supplied copious amounts of sediment to the Central Pennine Basin. The basin was a small confined intracratonic basin in which the basin salinity was appreciably diluted by the large amounts of freshwater discharge entering the basin. As a result, most of the sand-grade sediment bypassed the shoreline and was transported to the deeper parts of the basin via turbidity currents. Delta-front slumping may also have assisted in the transfer of sand-grade sediment to the deeper part of the basin (McCabe 1978). This delta system is characterized by major sand-rich fluvial distributary channels, a fine-dominated delta front with incised delta-front channels and a sand-rich submarine fan at the foot of the delta front. It is quite unlike any modern delta system.

In a less striking although still significant way, the oil-bearing Brent Delta system of the northern North Sea may provide another example of a non-actualistic delta system which can be interpreted from the large data base which is now available. There is a widespread consensus that the Brent Delta system was wave dominated in view of the predominance of storm-deposited facies in the coarsening-upward delta-front sequence and the laterally extensive sheet-like nature of the sandstone unit at the top of this sequence (the Rannoch Formation; Bowen 1975; Brown *et al.* 1987). The most detailed published account of the system makes a comparison with the Nile Delta system, in terms of both delta type and scale of the system (Johnson & Stewart 1985). In keeping with other modern wave-dominated delta systems the Nile Delta has few active distributary channels (two, in fact), with the result that the channel sandstones form relatively narrow bodies which are locally incised into the more widespread sheet sandstone of the upper delta front. In contrast, channel sandstones are extremely common in the lower part of the Etive Formation which directly overlies the delta-front sequence of the Rannoch Formation, seeming to form an extensive multistorey multilateral channel-belt sandstone body. This may suggest that the channels were more numerous and/or more laterally unstable than in modern wave-dominated deltas. An important control on this distinctive feature of the Brent Delta system was that the delta was supplied with a high sediment load which included a very significant proportion of sand-grade material. The high sediment load enabled the delta to prograde at a rapid rate despite a concurrent rise in sea level, and the abundance of sand may have caused the delta plain to resemble a braid plain of highly mobile shallow sand-dominated channels which coalesced vertically and laterally to produce the extensive channel-belt sandstone body found in the underlying Etive Formation. Once again, this view of the delta system renders it different from modern delta systems.

Delta–shelf interactions

The supply of sand-grade sediment to continental shelves often results from interactions between deltaic systems and the adjacent shelf. An understanding of sandstone-dominated shelf successions in the geological record is therefore aided by considering the nature, position and history of coeval deltaic systems along the inner margin of the shelf in view of their importance as a source of sediment. This may occur in response to low stands in sea level when deltas and their fluvial trunk streams deposit sand-grade sediment which is reworked during the ensuing transgression and by the physical regime of the shelf established by the transgression (*eg* the E coast, USA). However, this mechanism will produce only relatively thin shelf sandstone units (a few tens of metres thick?). Thicker sand-dominated shelf successions require either a more direct link between the delta and the adjacent shelf which operates during delta progradation or a repeated history of shoreline progradation and transgression.

The transfer of sand-grade sediment to the shelf requires a set of processes which are capable of transporting sand beyond the nearshore wave-induced 'littoral fence' which tends to retain sand in the nearshore area. Storm-driven processes and tidal currents are capable of transporting substantial volumes of sand from the delta front and transporting it to the adjacent shelf. These processes can be effective during delta progradation and may be enhanced by abandonment of the delta. Aerial photograph or map views of modern wave-dominated deltas are characterized by parallel beach ridges which are broadly aligned with the trend of the delta front (Psuty 1967; Curray *et al.* 1969). The ridges record the incremental growth of the delta during progradation; in effect, they are the growth rings of the delta. Often the ridges occur in groups or clusters separated by erosional discontinuities which truncate the earlier formed ridges and establish the trend for a new cluster of parallel ridges. The discontinuities reflect temporary abandonment phases in the history of the delta which are generally caused either by the avulsion of a distributary channel or by the waning of discharge in a channel. With a reduction in clastic input to part of the delta wave processes rework that part of the delta front and create a discontinuity surface whilst at the same time progradation continues elsewhere (as in the present-day Rhone Delta). Wave reworking is most significant during storm periods and therefore a large proportion of the sand eroded from the upper delta front during the temporary abandonment phases may be redeposited on the adjacent shelf by geostrophic currents.

Tidal currents can provide continuity of process between the delta front and the shelf and can override the retaining ability of the wave-induced littoral fence. In view of the persistent nature of tidal sand transport as opposed to the intermittent nature of storm-induced sand transport it is likely that tide-influenced deltas provide the most efficient means of transferring sand-grade sediment to the shelf. This mechanism has been used to explain thick sand-rich shelf successions in the geological record which lack evidence for substantial fluctuations in sea level. For example, the late Precambrian of northern Norway includes a 1500 m succession of mature cross-bedded sandstones which are interpreted as tide-dominated shallow-marine deposits (Levell 1980). The sand-grade sediment is considered to have been supplied to the shelf during the transgressive abandonment phase reworking of a series of tide-dominated deltas located at the basin margin.

Thus, both storm-driven and tidal current processes can be effective in transporting sand from deltas to the adjacent shelf during delta progradation and, more particularly, during delta abandonment. Cases where storm-driven processes enhance tidal current transport may provide the greatest potential in this respect. Consideration of the nature, position and history of deltaic systems along the inner margin of a shelf is crucial to understanding the genesis of sandstone-dominated shelf successions.

Delta–slope–submarine-fan links

Numerous modern and ancient delta systems are associated with submarine-fan systems in the deeper parts of basins. A key point in these systems is the extent to which the delta, slope and submarine fan are a linked set of coeval systems or conversely whether they are somewhat independent, with sediment transfer between the systems occurring primarily when there is a reduction in sea level (see below). The geological record provides several examples of deltas and submarine fans which are considered to be linked coeval systems and this has led to the view that delta-related submarine fans should be considered separately from other submarine fans (Galloway & Brown 1973; Heller & Dickinson 1985; Collinson 1986). Distinctive features of delta-related submarine fans include (i) the lack of a shelf separating the delta and submarine-fan systems, and commonly a small-scale slope system by comparison with continental slopes on passive margins, (ii) the presence of numerous channels incised into the basin slope rather than a single major-canyon feeder route and (iii) a comparative lack of organization in the fan deposits arising from the numerous slope feeder channels and the fact that the avulsion of distributary channels in the delta system can cause appreciable fluctuations in the locus of sediment supply to the fan. As with selected shelf systems, a comprehensive understanding of certain submarine-fan systems will require information on the nature and behaviour of the adjacent delta system. Discussions on the spectrum of delta-related submarine fans, the precise nature of links between the delta and the fan, and the facies patterns of the fan deposits are at an early stage and are likely to be a topical research area in the future.

Deltaic systems and seismic stratigraphy

Deltaic systems are often a prominent component of passive margin settings, forming preferentially during the post-rift phase during which regional thermally induced subsidence prevails. The dep-

ositional histories of passive margins have been exhaustively studied in recent years using the seismic stratigraphic approach and it is therefore appropriate in this review to consider the significance of deltaic systems in this approach.

In terms of seismic sequence and depositional systems tracts analysis, deltas are regarded as prominent features of high-stand periods. At times of high sea level wide shelves prevail and deltas prograde and aggrade on the shelf, producing a regressive facies sequence across the shelf. Shelf deltas or shoal-water deltas should therefore form extensively during these periods. In contrast, during periods of lowered sea level the shelf may be partially or entirely exposed, giving rise to type 1 and type 2 unconformities respectively. During these periods former high-stand deltas are eroded and active deltaic systems are confined to the outer part of the shelf where they form either after type 2 unconformities which exposed the inner part of the shelf or during the early stages of onlap which follow type 1 unconformities. In the latter case, deltas form at the shelf–slope break when sea level is beginning to rise and the heads of submarine canyons cut into the upper slope during the low stand in sea level are being filled rather than acting as transfer routes to the deeper parts of the basin.

Following this reasoning there is a tendency to equate widely developed shelf deltas with periods of high stand in sea level and more localized shelf-margin deltas with periods of low stand in sea level. However, consideration of the Gulf of Mexico region in this regard suggests that a degree of caution is required. In this region the development of shelf-margin deltas during the Tertiary and through to the present day has been controlled by sediment supply rather than eustatic sea-level fluctuations (Edwards 1981; Winker & Edwards 1983). The distribution of shelf-margin deltas is related to variations in sediment supply which are not basin-wide and are controlled by events in the hinterland source area. In the Gulf of Mexico, the main cause seems to be tectonic events in the Rocky Mountains and consequent changes in the continental drainage basin. Large and highly organized drainage basins are an essential feature of any major long-lived deltaic system and tectonic events which influence the drainage basin will directly influence the delta system, irrespective of how remote the tectonics are from the depositional site. Periods of shelf-margin progradation involving delta systems in the Gulf of Mexico deltaic province correspond with changes in the sand-to-shale ratio and sand provenance which can be related to discrete phases in the tectonic evolution of the Rocky Mountains (Winker 1982).

In terms of seismic facies analysis, Mitchum *et al.* (1977) interpret certain basinward-dipping clinoforms as successive phases in the progradation of a delta system, with the production of clinoforms presumably relating to lobe or delta switching events. Berg (1982) argues that fluvial- and wave-dominated deltas can be distinguished using details of the seismic reflector patterns. Fluvial-dominated deltas are often characterized by oblique tangential reflectors which are terminated up-dip. This pattern is considered to represent the deposits of deltas which accumulated under low subsidence rates and are dominated by delta-front facies. Alternatively, fluvial-dominated deltas can exhibit complex sigmoid-oblique reflectors which include an interval of flat reflectors at the top of the package. These are considered to have accumulated under conditions of higher subsidence and therefore include an interval of aggradational delta-plain facies. Wave-dominated deltas are considered to be characterized by relatively simple, shingled and oblique (parallel) reflectors, although the problem of discriminating between deltaic and non-deltaic prograding shorelines is not addressed by Berg (1982).

ACKNOWLEDGMENTS: Dr K. Pickering and Dr H. Williams are thanked for their critical comments on an earlier version of the manuscript.

References

ALLEN, G. P., LAURIER, D. & THOURENIN, J. 1979. Etude sédimentologique du delta de la Mahakam. *Notes et Memoires* **15**. Compagnie Francaise des Petroles, Paris.

BATES, C. C. 1953. Rational theory of delta formation. *Bulletin, American Association of Petroleum Geologists* **37**, 2119–2162.

BERG, D. R. 1982. Seismic detection and evaluation of delta and turbidite sequences: their application to exploration for the subtle trap. *Bulletin, American Association of Petroleum Geologists* **66**, 1271–1288.

BOWEN, J. M. 1975. The Brent oilfield. *In:* WOODLAND, A. W. (ed.) *Petroleum and the Continental Shelf of Northwest Europe.* Applied Science Publishers, London, 353–360.

BROWN, S., RICHARDS, P. C. & THOMSON, A. R. 1987. Patterns in the deposition of the Brent Group (Middle Jurassic) U.K. North Sea. *In:* BROOKS, J. & GLENNIE, K. W. (eds) *Petroleum Geology of North-west Europe.* Graham & Trotman, London, 899–913.

CHISHOLM, I. C. 1977. Growth faulting and sandstone deposition in the Namurian of the Stainton syncline, Derbyshire. *Proceedings of the Yorkshire Geological Society* **41**, 305–323.

COLEMAN, J. M. 1981. *Deltas: Processes and Models of Deposition for Exploration.* Burgess, CEPCO Division, Minneapolis, 124 pp.

——, PRIOR, D. B. & LINDSAY, J. F. 1983. Deltaic influences on shelf edge instability processes. *In:* STANLEY, D. J. & MOORE, G. T. (eds) *The Shelfbreak: Critical Interface on Continental Margins.* Society of Economic Paleontologists and Mineralogists Special Publication **33**, 121–127.

——, SUHAYDA, J. N., WHELAN, T. & WRIGHT, L. D. 1974. Mass movement of Mississippi river delta sediments. *Transactions, Gulf Coast Association of Geological Societies* **24**, 49–68.

COLLINSON, J. D. 1969. The sedimentology of the Grindslow Shales and the Kinderscout Grit: a deltaic complex in the Namurian of northern England. *Journal of Sedimentary Petrology* **39**, 194–221.

—— 1986. Submarine ramp facies model for the delta-fed, sand-rich turbidite systems: discussion. *Bulletin, American Association of Petroleum Geologists* **70**, 1742–1743.

CURRAY, J. R., EMMEL, F. J. & CRAMPTON, P. J. S. 1969. Holocene history of a strand plain, lagoonal coast, Nayarit, Mexico. *In:* CASTANARES, A. A. & PHLEGER, F. B. (eds) *Coastal Lagoons—a Symposium.* Universidad Nacional Autonoma, Mexico, 63–100.

EDWARDS, M. B. 1976. Growth faults in Upper Triassic deltaic sediments, Svalbard. *Bulletin, American Association of Petroleum Geologists* **60**, 341–355.

—— 1981. Upper Wilcox Rosita delta system of South Texas: growth-faulted shelf-edge deltas. *Bulletin, American Association of Petroleum Geologists* **65**, 54–73.

ELLIOTT, T. & LAPIDO, K. O. 1981. Syn-sedimentary gravity slides (growth faults) in the Coal Measures of South Wales. *Nature* **291**, 220–222.

EVAMY, B. D., HAREMBOURE, J., KAMERLING, P., KNAAPP, W. A., MOLLOY, F. A. & ROWLAND, P. H. 1978. Hydrocarbon habitat of Tertiary Niger delta. *Bulletin, American Association of Petroleum Geologists* **62**, 1–39.

GALLOWAY, W. E. 1975. Process framework for describing the morphologic and stratigraphic evolution of deltaic depositional systems. *In:* BROUSSARD, M. L. (ed.) *Deltas, Models of Exploration.* Houston Geological Society, Houston, 87–98.

GALLOWAY, W. E. & BROWN, L. F. JR. 1973. Depositional systems and shelf-slope relations on cratonic basin margin, Uppermost Pennsylvanian of north-central Texas. *Bulletin, American Association of Petroleum Geologists* **57**, 1185–1218.

HELLER, P. L. & DICKINSON, W. R. 1985. Submarine ramp facies model for delta-fed, sand-rich turbidite systems. *Bulletin, American Association of Petroleum Geologists* **69**, 960–976.

JOHNSON, H. D. & STEWART, D. J. 1985. Role of clastic sedimentology in the exploration and production of oil and gas in the North Sea. *In:* BRENCHLEY, P. J. & WILLIAMS, B. P. J. (eds) *Sedimentology: Recent*

Developments and Applied Aspects. Blackwells Scientific, Oxford, 249–310.

LEVELL, B. 1980. A late Precambrian tidal shelf deposit, the Lower Sandfjord Formation, Finnmark, north Norway. *Sedimentology* **27**, 539–557.

LINDSAY, J. F., PRIOR, D. B. & COLEMAN, J. M. 1984. Distributary-mouth bar development and role of submarine landslides in delta growth, South Pass, Mississippi delta. *Bulletin, American Association of Petroleum Geologists* **68**, 1732–1743.

MCCABE, P. J. 1978. The Kinderscout delta (Carboniferous) of northern England: a slope influenced by turbidity currents. *In:* STANLEY, D. J. & KELLING, G. (eds) *Sedimentation in Submarine Canyons, Fans and Trenches.* Dowden, Hutchinson & Ross, Stroudsberg, PA, 116–126.

MITCHUM, R. M. JR., VAIL, P. R. & SANGREE, J. B. 1977. Seismic stratigraphy and global changes of sea level, Part 6: Stratigraphic interpretation of seismic reflection patterns in depositional sequences. In: PAYTON, C. E. (ed.) *Seismic Stratigraphy—Applications to Hydrocarbon Exploration.* American Association of Petroleum Geologists Memoir **26**, 117–133.

NEMEC, W., STEEL, R. J., GJELBERG, J., COLLINSON, J. D., PRESTHOLM, E. & OXNEVAD, I. E. 1988. Anatomy of collapsed and re-established delta front in Lower Cretaceous of eastern Spitsbergen: gravitational sliding and sedimentation processes. *Bulletin, American Association of Petroleum Geologists* **72**, 454–476.

PSUTY, N. P. 1967. The geomorphology of beach ridges in Tabasco, Mexico. *Louisiana State University Coastal Studies Series* **18**, 51 pp.

RIDER, M. H. 1978. Growth faults in the Carboniferous of Western Ireland. *Bulletin, American Association of Petroleum Geologists* **62**, 2191–2213.

SHEPARD, F. P. 1981. Submarine canyons: multiple causes and long time persistence. *Bulletin, American Association of Petroleum Geologists* **65**, 1062–1077.

WEBER, K. J. 1971. Sedimentological aspects of oilfields of the Niger delta. *Geologie en Mijnbouw* **50**, 559–576.

WINKER, C. D. 1982. Cenozoic shelf margins, northwestern Gulf of Mexico. *Transactions, Gulf Coast Association of Geological Societies* **32**, 427–448.

—— & EDWARDS, M. B. 1983. Unstable progradational clastic shelf margins. *In:* STANLEY, D. J. & MOORE, G. T. (eds) *The Shelfbreak: Critical Interface on Continental Margins.* Society of Economic Paleontologists and Mineralogists Special Publication **33**, 139–157.

WRIGHT, L. D. 1977. Sediment transport and deposition at river mouths: a synthesis. *Bulletin of the Geological Society of America* **88**, 857–868.

—— & COLEMAN, J. M. 1974. Mississippi river mouth processes: effluent dynamics and morphologic development. *Journal of Geology* **82**, 751–778.

——, WISEMAN, W. J., BORNHOLD, B. D., PRIOR, D. B., SUHAYDA, J. N., KELLER, G. H., YANG, Z.-S. & FAN, Y. B. 1988. Marine dispersal and deposition of Yellow River silts by gravity-driven underflows. *Nature* **331**, 508–511.

T. ELLIOTT, Department of Earth Sciences, The Jane Herdman Laboratories, University of Liverpool, Brownlow Street, Liverpool L69 3BX, UK.

Delta or coastal plain? With an example of the controversy from the Middle Jurassic of Yorkshire

J. Alexander

SUMMARY: Deltaic and coastal plain deposits can be distinguished from each other in the rock record where enough data are available to interpret the three-dimensional distribution of facies and unconformities over a wide area. This distinction is based on the distribution of facies rather than purely on a marine-to-alluvial transition as has been the case in the past. The distinction between deposits of the two environments is often very difficult or impossible, and if based on limited data may lead to inaccurate and even incorrect models that can cause problems with prediction of reservoir rocks or coal location in the subsurface. Models of the gross facies distribution for tectonically, eustatically and autocyclically controlled coastal plains are seen as the end-members of a continuum between coastline depositional styles.

The Middle Jurassic sedimentary rocks of the Yorkshire Basin are used to illustrate the controversy that can arise from imprecise definitions and the application of incorrect facies models. The Middle Jurassic Ravenscar Group is interpreted here as the deposits of a tectonically and eustatically influenced coastal plain. Repeated fluctuations between shallow-marine and non-marine conditions are interpreted as mainly caused by vertical tectonic movements and sea-level changes rather than by autocyclic processes.

There is considerable controversy over the interpretation of the depositional environment of many sedimentary successions. Much of the controversy arises because of differing interpretations of specific words and probably fundamentally different perceptions. A survey of over 100 geologists and geomorphologists from British and European universities and oil companies shows that the use of the term 'deltaic' varies considerably. This wide variation of definition is particularly disquieting as large volumes of fossil fuels are found in sediments that are attributed to deltaic environments of deposition. Also, deltas and other coastal localities are sites for the deposition of exceptionally thick sediment piles.

It has been the tendency in recent years to apply the term 'deltaic' to any sedimentary succession that involves a transition from a marine to an alluvial environment of deposition. It is clear that in many cases these deposits will not strictly be deltaic in origin although they may contain similar facies or even be made up in part of deltaic deposits.

Barrell (1912, 1914) proposed criteria for the recognition of deltaic deposits in the rock record. These criteria were based on Gilbert's description of deltas with topset, foreset and bottomset beds. Barrell stressed that not all deltas would exhibit this structure. The distinction rested largely on the recognition of large coarsening-upward cycles reflecting progradation and the passage from marine to alluvial conditions at any one site. However, these are not sufficient criteria to distinguish between deltaic and other types of coastal plain deposits.

The main reason for the ubiquitous use of the word 'deltaic' for any sediments containing a marine-to-alluvial transition is the variation in the definition of 'deltaic' used by geologists when applied to the rock record. The survey revealed that there was no commonly accepted definition of either delta or deltaics.

Where a number of deltas coalesce laterally, the resulting area of land can be called a coastal plain although this may also result from other mechanisms. For example, distinction should be made between passive progradation of deltas or coastal plains resulting from sedimentary processes and an area of alluvial conditions created as a result of tectonic uplift or sea-level fall.

In this paper geometrical facies models are presented for the end-members of prograding coastline systems before models applicable to the Middle Jurassic sedimentary rocks of the Yorkshire Basin, NE England, are discussed briefly.

Results of the survey of deltaic definitions

An informal survey of geologists and geomorphologists was undertaken primarily to clarify the author's own ideas on deltaic deposits. This survey was carried out by direct questions and discussion with people at the meeting *Deltas: Sites and Traps for Fossil Fuels*, the British

From WHATELEY, M. K. G. & PICKERING, K. T. (eds), 1989, *Deltas: Sites and Traps for Fossil Fuels,* Geological Society Special Publication No. 41, pp. 11–19.

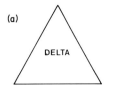

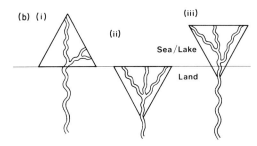

FIG. 1. The delta of Herodotus: (*a*) the fourth letter of the Greek alphabet; (*b*) three possible plan-form geometries that fit into this definition.

Sedimentological Research Group Meeting, 1986, and assorted other occasions. The results obtained from this informal survey fell into three main categories.

(a) A delta is any sedimentary succession that has a transition from marine to non-marine, or any marine–fluvial or lacustrine–fluvial interface.

(b) A delta is a crudely triangular lobe of sediment deposited where a fluvial channel enters a sea, ocean or lake.

(c) A deltaic deposit is the rapidly deposited sediment that progrades into an ocean, sea, lagoon or lake and originates from a single river system. The river supplied sediment more rapidly to its mouth than could be dispersed by basinal processes although the sediment may be reworked *in situ* by wave or tidal activity.

Definition (b) includes some of the most succinct definitions given (Fig. 1*a*) and relates most nearly to the original definition of Herodotus. This prompted the comment: 'Herodotus knew what he was talking about. Something appears to have been lost in the sedimentological translation! Most "things" we geologists call deltas or deltaic deposits probably/certainly are not'. The triangle definition also introduces some variation in the land forms that would be included in this definition (Fig. 1*b*).

50% of those questioned agreed essentially with the third definition, and this is the one that

is followed in this paper. Many people emphasized that deltaic deposits were primarily the result of progradation of a single river system. Several people observed that distributary systems were not ubiquitous in deltas although they were common, and that interaction of alluvial, tidal and wave activity allowed a wide range of morphological forms, thus allowing a wide range of facies patterns to come under the same general delta definition. In such cases the landforms near the mouth of single river systems resulting from deposition of the river load, but without the triangular shape, can be called deltas.

It is important to stress the range of scale of deltas from metres to kilometres depending on the system under consideration. The classic Gilbert-type delta is probably developed in relatively rare conditions. It is common for deltaics not to include topset, foreset and bottomset beds as stressed by Barrell (1912, 1914).

Deltaic versus coastal plain models

It is possible to erect four theoretical end-member models for constructive coastline depositional systems (Fig. 2): delta; autocyclically controlled coastal plain; tectonically controlled coastal plain; eustatically controlled coastal plain. These types can be subdivided on the basis of climatic conditions, sediment grade and the extent of tidal and wave activity. Major problems arise when defining the boundary conditions between these classes and when interpreting the nature of deposits where the predominant control changed through time or several controls were of equal importance, as appears to have been the case in the Middle Jurassic Yorkshire Basin as discussed below. Coastal plain and delta complexes as defined here have many sedimentary processes in common and consequently have similar facies, although the facies architecture is different.

Distinguishing features of the models and discussion

Deltaic deposits

The definition of 'deltaic' used in this paper follows (c) above. Significant features of this model are as follows: (i) they are the deposits of a single alluvial system forming near the mouth of that system; (ii) the deposits prograde by autocyclic processes into a marine or lacustrine basin, and this progradation leads to a vertical marine-to-alluvial transition and a distinct shore-

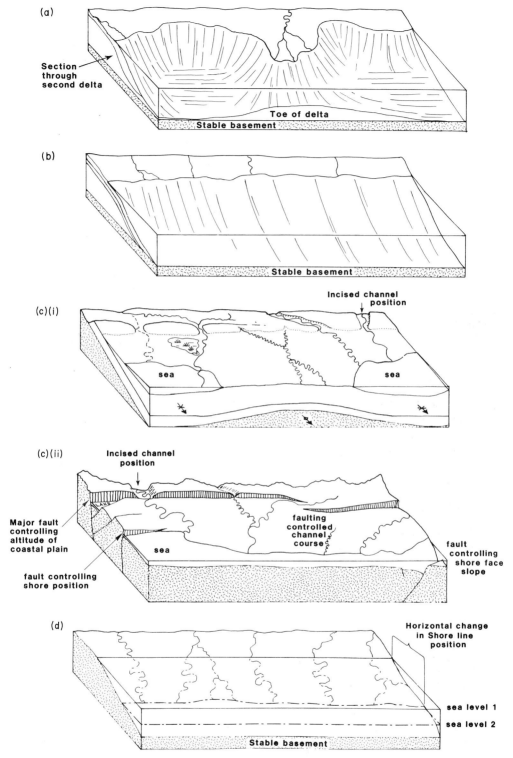

FIG. 2. Block diagrams representing the large-scale facies distribution in the four end-member models of aggrading coastlines: (a) deltaic model, (b) autocyclically controlled coastal plain, (c) tectonically controlled coastal plain ((i) coastal plain controlled by folding; (ii) coastal plain controlled by faulting) and (d) eustatically controlled coastal plain.

line perturbation; (iii) a coarsening-upward sequence may be developed as a result of lobe progradation, and this sequence may subsequently be removed by distributary channel erosion; (iv) distributary channel systems may have been present, producing a reduction in channel sandstone body size down-palaeocurrent; (v) a fanning pattern of channel-belt deposits is developed as a result of the point-sourced alluvial system, and this fanning sandstone body architecture develops even if distributary systems were not present; (vi) lobe switching is an inherent process in many forms of deltas and may lead to repeated lobe progradation over the same site to produce a rhythmical succession.

Deposition on the upper delta plain (area of alluvial deposition away from major marine influence) may result in facies patterns that are very similar to those formed on coastal plains.

Autocyclically controlled coastal plain

This depositional model results from progradation of a prism of terrestrially derived sediment into a marine or lacustrine basin. In general, an autocyclically controlled coastal plain comprises a number of deltas that have coalesced laterally and/or prograded at the same time. This configuration of laterally interfingering alluvial systems results in a sedimentary body with a more regular rectangular distribution of facies than is seen in isolated deltas. In most cases it would be impossible to distinguish the deposits of individual alluvial systems in the rock record unless (i) there is considerable difference in the load characteristics of the rivers, (ii) there is considerable spacing between river systems of similar size, as in the Cenozoic Gulf Coast sediments (Galloway 1981) and/or (iii) there is a persistent topographic control of channel positions resulting in vertical stacking of channel sandstone bodies owing to differential compaction of the substrata, minor tectonic deformation or upstream control of the site of river entry into the depositional basin. Even in this case, there would still be considerable problems in distinguishing the overbank deposits for the different systems.

Tectonically controlled coastal plain

In this model the change from marine to non-marine depositional conditions is controlled by synsedimentary tectonic deformation. This is seen on a small scale in several neotectonic areas, notably in Greece (Jackson *et al.* 1982) and along the shore of Hebgen Lake, SW Montana (Alexander & Leeder 1987). In both these modern

examples the horizontal distance of near-instantaneous shoreline movement, resulting from single seismic events, was small owing to the high relief of the areas concerned. This effect is most pronounced in areas of low coastal relief where only small amounts of vertical movement will cause the shoreline to move by a considerable distance across the gently sloping surface (Leeder & Gawthorpe 1987).

In the deposits of tectonically controlled coastal plains, evidence of synsedimentary tectonic activity will be preserved in the rock record. This evidence will be in the following form.

(1) Little, if any, evidence of a prograding wedge, although this may occur locally; there may be sudden switches between marine and non-marine conditions.

(2) Variability in formation thickness relating to folds or faults in the sedimentary succession, with such thickness variations frequently being associated with facies variations.

(3) Facies variations relating to persistent topographic features rather than vertically changing facies due to progradation or random distributions. In particular, examples include stacked sandstone bodies formed as a result of topographic control of the drainage pattern. Changes in the extent of marine influence along strike, associated with tectonic features, may be common, as observed in the area around Long Nab in the Middle Jurassic of the Yorkshire Basin. This effect can be seen to a lesser extent in other models.

(4) A non-random pattern of soft-sediment deformation superimposed on an inherent random pattern. Of particular diagnostic value are beds deformed by load and founder structures that can be traced over a considerable lateral extent (Mayall 1983; Alexander & Williams, in prep.) and concentrations of deformation structures near synsedimentary tectonic features (Alexander 1987; Leeder 1987).

(5) Abrupt vertical changes of marine influence, notably localized flood deposits containing marine microfossils.

Eustatically controlled coastal plain

This facies model is very similar to the tectonically controlled coastal plain model. In the eustatically controlled plain, vertical changes from marine to non-marine conditions result from fluctuations of sea level. These changes may be rapid or slow and superimposed on other local changes. Intuitively, it seems that hybrid models with a eustatic influence would be more common than examples where major changes in coastline position were controlled solely by sea-level change. However, many examples of major changes in shoreline position are recorded as a result of the major sea-level changes that occurred in the Quaternary associated with fluctuating climatic and glacial conditions (Curray 1960; Nelson & Bray 1970).

Distinguishing eustatic control from other effects in the rock record is currently very difficult since there appear to be few features that are unique to this case. Eustatic changes may be more widespread than tectonic effects, but this may not always be the case.

Conclusions for different coastline models

In the same climatic conditions, the facies distribution in a cross-section perpendicular to the shoreline through deltaic and autocyclically controlled coastal plain deposits will be similar, although the facies distribution parallel to the coastline will be significantly different. Similarly, tectonically and eustatically controlled coastal plains may have similar facies architecture but the mechanisms responsible for the pattern will be significantly different. This difference can be seen in the nature of channel-belt sandstone body stacking patterns and in the distribution of soft-sediment deformation. Individual deltas may not conform, *sensu stricto*, with the delta model throughout their history and may temporarily display facies relationships more consistent with other coastal plain models. In cases where there are insufficient data to make a distinction between the environments, a more general term such as 'paralic' or 'marine–alluvial transition' should be used in preference to a model that has a large probability of being inaccurate, thus avoiding basing any hydrocarbon or coal exploration on an incorrect facies distribution model.

These models are only end-members in a continuum and it is thought that hybrid coastline depositional systems will be the rule rather than the exception, as is the case with the Middle Jurassic of Yorkshire.

Introduction to the Ravenscar Group

The Middle Jurassic Ravenscar Group is well exposed in the cliffs between Yons Nab and Staithes (Fig. 3). These very well exposed outcrops are frequently used to demonstrate the nature and pattern of alluvial and deltaic facies to students, professional geologists and reservoir engineers working on time-equivalent rocks in the North Sea oil province. Care should be taken to acknowledge the differences between these two areas. In both these areas, local regression occurred in the Middle Jurassic despite rising sea levels.

The Ravenscar Group consists of alternating 'marine' and 'non-marine' formations, and demonstrates lateral and vertical variations in sandstone architectural style (Alexander 1986a). The pattern of formation thickness and facies variation is shown schematically in Fig. 4.

There has been considerable debate over the extent of marine influence in the alluvial formations of the Ravenscar Group and discussion as to the most applicable sedimentary models (Leeder & Nami 1979; Hancock & Fisher 1981; Livera & Leeder 1981; Fisher & Hancock 1985; Alexander 1986a,b). The continuing disagree-

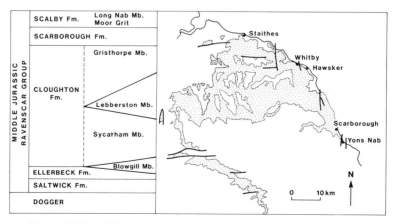

FIG. 3. Location map of the Yorkshire Middle Jurassic and the nomenclature used in this paper. (After Hemingway & Knox 1973.)

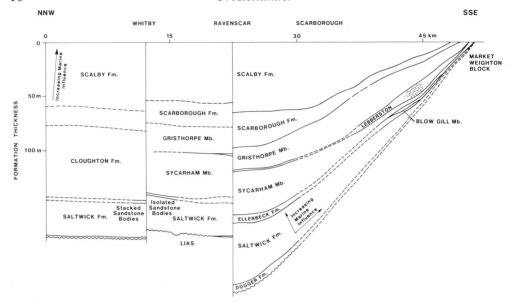

NNW   SSE

WHITBY   RAVENSCAR   SCARBOROUGH

FIG. 4. Schematic diagram of the major thickness variations in the Middle Jurassic across the Yorkshire Basin. The Dogger, Ellerbeck and Scarborough Formations and the Lebberston Member are marine.

ment has been caused, in part, by people working on limited data and on variations in the criteria thought to indicate deltaic conditions, as well as by scale problems. A brief discussion is presented below to justify the use of an alluvial-dominated tectonically and eustatically influenced coastal plain model for these rocks.

The onset of non-marine conditions in the Yorkshire Basin was sudden, with the main break in sedimentation occurring between the open marine Liassic and the shallow-marine Dogger Formation (Figs 3 and 5). This break in sedimentation and change in depositional environment is represented by an unconformity with a slight angular discordance, and demonstrates a period of uplift and folding prior to the Middle Jurassic that appears to have continued, albeit at a slowed rate, throughout that time. The Dogger Formation contains clasts of phosphatized Liassic shale and locally derived conglomerates. At many sites the top of the Dogger Formation contains rootcasts and evidence of pedogenesis, indicating superposition of subaerial conditions on marine sediments. Locally, the Dogger Formation is overlain by a thin coal.

The base of the Saltwick Formation, the first non-marine formation in the basin (Fig. 3), is also a plain of non-deposition or erosion. This is particularly well displayed where there has been incision of Saltwick channels into the Dogger Formation and underlying shales, as seen for example between Whitby and Robin Hood's Bay,

indicating a lowering of river base level rather than simple progradation.

There is considerable evidence throughout the Ravenscar Group that the depositional surface

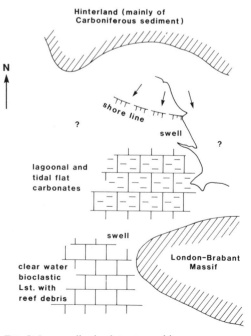

FIG. 5. A generalized palaeogeographic reconstruction. (After Livera 1981.)

was never far removed from marine influence. This evidence includes intercalation of marine formations and members, early diagenetic features and microfossils characteristic of brackish marine conditions within sediments with a dominantly alluvial character.

At the transition from the Lebberston Member to the Gristhorpe Member (Fig. 3), the Yons Nab Beds represent a rapid progradation of a wave–tide–fluvial influenced shoreline (Livera 1981). This is the only part of the succession where shoreline deposits are recognized. In the Saltwick and Scalby Formations there is evidence for a gradual upward increase in marine influence, shown by the occurrence of brackish marine microfossils and changes in palaeosol characteristics. This increased marine influence is interpreted as the result of a slow rise in sea level and subsidence of the area owing to sediment consolidation (Alexander 1986*a*).

One of the main differences between the Ravenscar Group and many other successions is that, in this case, the alluvial formations are relatively much thicker than the marine formations. In other successions described as deltaic, such as the Yordales of northern England, the converse is true. In the Ravenscar Group there is only locally preserved evidence of progradation into standing water (Livera 1981).

Palaeogeography of NE England during the Middle Jurassic

During the deposition of the Ravenscar Group, the Yorkshire Basin slowly and periodically subsided relative to the Pennine and Mid-North Sea Highs to the N (Kent 1974; Hemingway & Riddler 1982) and the Market Weighton Block to the S (Bott *et al.* 1978). The main sediment supply was from the N with rivers apparently converging into the Yorkshire Basin from the Pennine and Mid-North Sea Highs, and a shallow shelf sea lay to the S over the Market Weighton Block. S of the Market Weighton Block, the Bajocian is represented by a variety of shallow-marine lagoonal clastic and carbonate sediments of the Inferior Oolite Series. There was therefore a restricted marine connection to this area throughout the Middle Jurassic.

The area generally had a gentle southerly slope that allowed the position of the shoreline to change drastically in response to relatively minor tectonic movements and/or changes in sea level. One of the effects of this gentle slope is seen near the base of the Saltwick Formation where there is evidence of an increase in marine influence to the S through a small increase in the appearance of brackish marine microfossils and a slight change in the types of plant fossils (Harris 1953). There is no evidence, however, of 'deltaic' deposits in the Sole Pit Trough (Kent 1980) and little clastic input to the carbonate shelf to the S.

Coastal plain model for the Ravenscar Group

The interpretation of an alluvial-dominated tectonically and eustatically controlled coastal plain model is the result of considering a large volume of data collected over the last two decades (Knox 1969; Nami 1976; Livera 1981; Alexander 1986*a*). The main factors that lead to this model are as follows.

(1) A sudden onset of marine conditions at the base of the Saltwick Formation. The initial onset of alluvial conditions is interpreted to have resulted from local tectonic uplift rather than progradation of an alluvial system by autocyclic processes from the N or a sea-level change.

(2) The palaeogeography with inferred 'highs' to the N and S of the basin (Hemingway & Riddler 1982).

(3) The limited connection of the basin to open marine conditions, with a carbonate shelf to the S, so that small amounts of vertical movement would result in abrupt changes in depositional conditions. The slow subsidence of the Yorkshire Basin produced a clastic sediment trap that prevented dispersal farther S and so allowed the development of the carbonates to the S.

(4) No evidence of fanning or dividing of channel sandstone bodies southward has been found. Convergence of flow in the northern part of the basin has been suggested from provenance studies (Hemingway 1974; Hallam 1975).

(5) Tectonic influence on the Middle Jurassic Yorkshire Basin coastal plain is demonstrated by abrupt thickness changes across structural features (Fig. 4) and dramatic lateral variations in alluvial architecture (Alexander 1986*a*, *b*).

(6) The rising sea level throughout the Middle Jurassic is interpreted to be the cause of the gradual vertical increase in marine influence through the Scalby and Saltwick Formations.

Conclusions

Many depositional styles are possible on a constructive coastline. These can be classified by a four-end-member system into delta, autocyclically controlled coastal plain, tectonically controlled coastal plain and eustatically controlled coastal plain. Generally, a hybrid model will be most applicable. In the example from the Middle Jurassic of Yorkshire, a tectonically and eustatically influenced coastal plain model was found to fit the data best, rather than a delta environment as defined at the beginning of this paper.

ACKNOWLEDGMENTS: Heartfelt thanks are due to all those people who, in good spirit, answered my questions on definitions and generally put up with my persistent interruptions of civilized conversations at the bars of various conferences. I would like to thank Mike Leeder for his much-valued supervision during my 3 years in Leeds, British Petroleum for their financial support of my PhD, the French Institute of Petroleum for collaboration in the summer of 1986, and J. F. Freeman Ltd for donating the money to pay my wages at University College, Cardiff. Thanks are also due to Julie Jones, Liz Jolly and Mike Fisher for comments on the manuscript.

References

ALEXANDER, J. 1986a. *Sedimentary and tectonic controls of alluvial facies distribution: Middle Jurassic, Yorkshire and the North Sea Basins and the Holocene, SW Montana, USA.* Unpublished PhD thesis, University of Leeds.
—— 1986b. Idealised flow models to predict alluvial sandstone distribution in the Middle Jurassic Yorkshire Basin. *Marine and Petroleum Geology* 3, 298–305.
—— 1987. Syn-sedimentary and burial related deformation in the Middle Jurassic non-marine formations of the Yorkshire Basin. *In*: JONES, M. & PRESTON, R. M. F. (eds) *Deformation of Sediments and Sedimentary Rocks.* Geological Society of London Special Publication 29, 315–324.
—— & LEEDER, M. R. 1987. Active tectonic control of alluvial architecture. *In*: ETHRIDGE, F. G., FLORES, R. M. & HARVEY, M. D. (eds) *Recent Developments in Fluvial Sedimentology.* Society of Economic Paleontologists and Mineralogists Special Publication 39, 243–252.
—— & WILLIAMS, G. D. *Seismically induced soft-sediment deformation in Triassic marginal lacustrine deposits.* In preparation.
BARRELL, J. 1912. Criteria for the recognition of ancient delta deposits. *Bulletin of the Geological Society of America* 23, 377–446.
—— 1914. The Upper Devonian delta of the Appalachian geosyncline. *American Journal of Science* 37, 225–253.
BOTT, M. H. P., ROBINSON, J. & KOHNSTAMM, M. A. 1978. Granite beneath Market Weighton, east Yorkshire. *Journal of the Geological Society, London* 135, 535–543.
CURRAY, J. R. 1960. Sediments and history of Holocene transgression, continental shelf, north west Gulf of Mexico. *In*: SHEPARD, F. P., PHTEGER, F. B. & VAN ANDEN, T. H. (eds) *Recent Sediments, North West Gulf of Mexico.* American Association of Petroleum Geologists, 221–226.
FISHER, M. J. & HANCOCK, N. J. 1985. The Scalby Formation (Middle Jurassic, Ravenscar Group) of Yorkshire: reassessment of age and depositional environment. *Proceedings of the Yorkshire Geological Society* 45, 293–298.
GALLOWAY, W. E. 1981. *Depositional Architecture of Cenozoic Gulf Coastal Plain Fluvial Systems.* Society of Economic Paleontologists and Mineralogists Special Publication 31, 127–155.

HALLAM, A. 1975. *Jurassic Environments.* Cambridge University Press, Cambridge, 269 pp.
HANCOCK, N. J. & FISHER, M. J. 1981. Middle Jurassic North Sea deltas with particular reference to Yorkshire. *In*: ILLING, L. V. & HOBSON, G. D. (eds) *Petroleum Geology of the Continental Shelf of N.W. Europe.* Institute of Petroleum, London, 186–195.
HARRIS, T. M. 1953. The geology of the Yorkshire Jurassic Flora. *Proceedings of the Yorkshire Geological Society* 29, 63–71.
HEMINGWAY, J. E. 1974. The Jurassic. *In*: RAYNER, D. & HEMINGWAY, J. E. (eds) *The Geology and Mineral Resources of Yorkshire.* Yorkshire Geological Society, 161–223.
—— & KNOX, R. W. O'B. 1973. Lithostratigraphical nomenclature of the Middle Jurassic strata of the Yorkshire Basin of north-east England. *Proceedings of the Yorkshire Geological Society* 39, 527–537.
—— & RIDDLER, G. P. 1982. Basin inversion in North Yorkshire. *Transactions of the Institute of Mineralogy and Metallurgy* 91, 175–186.
JACKSON, J. A., GAGNEPAIN, J., HOUSEMAN, G., KING, G. C. P., PAPADIMITRIOU, P., SOUFLERIS, C. & VIRIEUX, J. 1982. Seismicity, normal faulting and the geomorphological development of the Gulf of Corinth (Greece): The Corinth earthquakes of February and March 1981. *Earth and Planetary Science Letters* 57, 377–397.
KENT, P. E. 1974. Structural history. *In*: RAYNER, D. & HEMINGWAY, J. E. (eds) *The Geology and Mineral Resources of Yorkshire.* Yorkshire Geological Society, 13–28.
—— 1980. Subsidence and uplift in East Yorkshire and Lincolnshire: a double inversion. *Proceedings of the Yorkshire Geological Society* 42, 505–524.
KNOX, R. W. O'B. 1969. *Sedimentary Studies of the Eller Beck Beds and Lower Deltaic Series in North-east Yorkshire.* Unpublished PhD thesis, University of Newcastle.
LEEDER, M. R. 1987. Sediment deformation structures and the palaeotectonic analysis of extensional sedimentary basins, with special reference to the Carboniferous of N. England. *In*: JONES, M. & PRESTON, R. M. F. (eds) *Deformation of Sediments and Sedimentary Rocks.* Geological Society of London Special Publication 29.

—— & GAWTHORPE, R. 1987. Sedimentary models for extensional tilt block/half-graben basins. *In*: HANCOCK, P. (ed.) *Extensional Tectonics*. Geological Society of London Special Publication.

—— & NAMI, M. 1979. Sedimentary models for the non-marine Scalby Formation (Middle Jurassic) and evidence for Late Bajocian/Bathonian uplift of the Yorkshire Basin. *Proceedings of the Yorkshire Geological Society* **42**, 461–482.

LIVERA, S. A. 1981. *Sedimentology of Bajocian Rocks from the Ravenscar Group of Yorkshire*. Unpublished PhD thesis, University of Leeds.

—— & LEEDER, M. R. 1981. The Middle Jurassic Ravenscar Group ('Deltaic Series') of Yorkshire: recent sedimentological studies as demonstrated during a field meeting 2–3 May 1980. *Proceedings of the Geological Association* **92**, 241–250.

MAYALL, M. 1983. Earthquake induced synsedimentary deformation in late Triassic (Rhaetian) lagoonal sequence, South West Britain. *Geology Magazine* **120**, 613–622.

NAMI, M. 1976. *Sedimentology of the Scalby Formation (Upper Deltaic Series) in Yorkshire*. Unpublished PhD thesis, University of Leeds.

NELSON, H. F. & BRAY, E. E. 1970. Stratigraphy and history of Holocene sediments in the Sabine–High Island area, Gulf of Mexico. *In*: MORGAN, J. P. (ed.) *Deltaic Sedimentation Modern and Ancient*. Society of Economic Paleontologists and Mineralogists Special Publication **15**, 48–77.

JAN ALEXANDER, Department of Geology, University College, PO Box 78, Cardiff CF1 1XL, UK.

Fjord sedimentation as an analogue for small hydrocarbon-bearing fan deltas

J. P. M. Syvitski & G. E. Farrow

SUMMARY: Fjord basins have sediment accumulating at rates between 0.1 and more than 1000 mm a^{-1} with sea-level fluctuations up to 50 mm a^{-1}, lithologies varying from poorly sorted conglomerates to unimodal sands and muds, carbon contents from less than 1.0% to more than 10% for methane-producing sediments found in some anoxic basins, and variably consolidated and sensitive sediment. Four major mechanisms affect the rate and style of basin infilling: (i) hemipelagic sedimentation of particles carried seaward by river plumes; (ii) delta-front progradation, particularly as affected by bed-load deposition at river mouths; (iii) proximal slope bypassing, primarily by turbidity currents and cohesionless debris flows; (iv) down-slope diffusive processes, mainly creep and slides, that work to smear previously deposited sediment into deep water.

Fjords, with their rugged topography and dynamic history of basin infilling, are prone to slope failures. Release mechanisms include sediment loading, ground motion from earthquakes, giant waves, sea-level fluctuations and changes in the mass physical properties of basin sediments. Slope failures occur on a semicontinuous basis and with displaced volumes of 10^3–10^6 m^3, or with much less frequency (1000 year intervals) but affecting larger volumes of sediment (10^6–10^9 m^3). Although slope failures are common to the delta front, they may occur anywhere along the prograding prodelta surface, along the basin margins and over basement highs.

Sediment accumulations in fjords have many similarities to small hydrocarbon-bearing basins, such as those of the North Sea, that once received submarine fans, often experienced anoxic conditions and contained high rates of sedimentation. Fjords typically contain more sand and conglomerate than other commonly employed modern analogue environments such as the Indus, Mississippi or Amazon fans.

Two stratigraphic plays are suggested on the basis of fjord-infilling processes. The first involves the semicontinuous failure of foreset sands where, through progradation, gravity flow sands form an areally extensive reservoir (up to 10 km wide, 100 m thick and 100 km long) sandwiched between basinal muds and prodelta muds. The second play involves the catastrophic failure of a major portion of the prodelta where reservoir-quality sands (from either the basin-head delta or marginal delta-beach complexes) infill a slide-formed megachannel (up to 500 m wide, 50 m thick and tens of kilometres long). These megachannel sands would be underlain by faulted and disturbed mud blocks, and capped by conformable hemipelagic muds.

The newer deep petroleum plays in mature basins, like those of the North Sea or Newfoundland's Hibernia, present a real challenge to today's explorer. Many of these plays are associated with synsedimentary faults that bound narrow depositional basins (Fig. 1). Cores from reservoir horizons commonly show massive ungraded sands that seem, from seismo-stratigraphic, tectonic and palaeo-ecological evidence, to have been deposited in a basinal setting (Stoker & Brown 1986). Previous predictive models have dominantly employed facies analysis from large-scale submarine fans on aseismic or seismically active ocean margins (Walker 1978) or have employed turbidite facies associations from the rock record whose mapped distribution cannot be matched on modern fans (Shanmugam et al. 1985). However, there is mounting evidence from the rock record, rather than from studies of Recent marine geology, of much smaller and more dynamic basinal settings. The Late Jurassic graben-margin play in the North Sea, for example, has been compared with Middle Volgian sequences exposed in E Greenland, where Surlyk (1978, p. 89) has described the palaeogeography: 'the sea was restricted to narrow N–S running fjords'. Furthermore, the bottom waters were 'stagnant' and normal benthic biota were absent, thus strengthening comparisons with modern-day fjords which we wish to explore further in this paper. Postma (1984) has recently noted the relevance of modern fjord processes to Tertiary fan-delta sequences in southeastern Spain.

Fans and deltas that prograde into basins normally experience frequent failures along their prodelta slopes. For instance, Lindsay et al. (1984) have estimated that at least 50% of sediment deposited in Mississippi River mouth bars ultimately finds its way into deep water via a variety of slope-failure mechanisms. Fjords are also

From WHATELEY, M. K. G. & PICKERING, K. T. (eds), 1989, *Deltas: Sites and Traps for Fossil Fuels,* Geological Society Special Publication No. 41, pp. 21–43.

21

22 *J. P. M. Syvitski & G. E. Farrow*

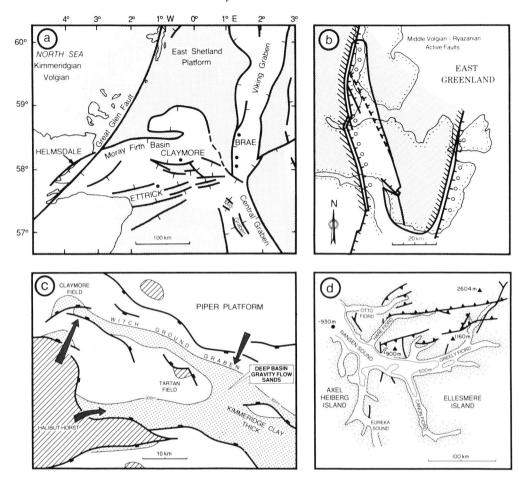

FIG. 1. Comparison of scale and tectonic elements in different basins: (*a*) Upper Jurassic of the North Sea including Helmsdale, the Moray Firth and the Brae fields (Ziegler 1975); (*b*) Upper Jurassic of E Greenland (from Surlyk 1978) (note the relationship to the modern fjordic coast); (*c*) an enlarged view of the Claymore field (from Turner *et al.* 1984) (*cf* (*a*)) for scale comparison with Knight Inlet, British Columbia, highlighted in Fig. 3; (*d*) Nansen Sound–Greely Fiord system of Ellesmere Island, North West Territories.

regarded as sediment traps dominated by gravity flow deposits (Holtedahl 1965). The purpose of our paper is to describe the variety of basin-infilling styles typical of the modern fjord environment.

Fjords may provide a better modern analogue to incipient rift-related palaeo-basin infilling than studies of open ocean continental slopes. In shape, fjords closely approximate the narrow and elongated grabens developed during the early stages of continental rupture, *eg* the Upper Jurassic of Greenland and the North Sea (Surlyk 1978; Ziegler 1975, 1981). A tectonic origin for modern fjords has been restated recently (Dowdeswell & Andrews 1985; England 1985). Fjords often contain a high proportion (20%–60%) of

relatively immature sand as the sediment-transport pathway is short. This appropriately compares with the areally limited 'highs' from which Brae sediment must have been derived (Fig. 1a). Many fjordic basins accumulate sediment very rapidly (*cf* certain wells W of Shetland, where less than 1 km of sand was deposited during one stage—the Albian—in the Cretaceous (Ridd 1981, p. 422)). As a consequence of the Earth's recent glacial history, sea-level fluctuations in modern fjords range from -10 to $+40$ mm a^{-1}, depending on the position of the fjord relative to the landward-retreating crustal peripheral bulge (Syvitski *et al.* 1987). Holocene prodelta muds are often uplifted. (Although fjords are dominated by isostatic uplift and grabens are more influenced by basin down-warping, we suggest that the net results, in terms of sedimentary deposits, are very similar.) Finally, the sedimentary sequences resulting from failure that we highlight in this paper have occurred in the past few thousand years and are thus synsedimentary. This is of great interest to the petroleum geologist, because a close association exists between synsedimentary faulting and major hydrocarbon accumulations (Price 1980).

Fjord characteristics and study area

Fjords are defined as deep elongate arms of the sea that characterize the glaciated margins of tectonically uplifted coasts. Some 3500 modern fjords lie within a belt N of 43°N and a belt S of 42°S. Fjords are immature non-steady-state systems, evolving and changing over relatively short time-scales. They may contain one or more submarine sills and thus exist as effective sediment traps (*cf* California Borderland basins, Field & Edwards 1980). The main sediment input is usually found at the landward head of the basin, although sediment input may occur as point sources anywhere along the flanks, and in this respect there is an important difference between borderland basins and fjords. Table 1 provides a brief review of the main geological parameters affecting fjord sediment processes.

Fjords by definition (Syvitski *et al.* 1987) have been (or are presently being) excavated or modified by land-based glaciers. During the deglaciation phase, basin infill may be strongly influenced by ice-contact sedimentation processes (Syvitski, in press). In this paper we are not interested in these initial deposits, except where they provide sources of sediment to be reworked and delivered to the post-glacial delta. In many fjords, Holocene basin fill is so substantial (hundreds of metres) that Pleistocene deposits cannot even be observed on high resolution seismic records.

With the vast number of fjords available for study and the near continuum of natural conditions that may affect their infilling history, geologists are free to select particular end-member conditions (*eg* Syvitski 1986). The particular spectrum of fjords that we wish to highlight are those that experience high sedimentation rates and are prone to subaqueous slope failures. We are thus selecting the antithesis of the natural laboratory studied by Nelson *et al.* (1986), who avoided the effects of fluvial input on slope failure by studying the totally enclosed Crater Lake, Oregon. Slides and gravity flow processes are common to fjords located along the

TABLE 1. *Fjord statistics*

Basin characteristics	
Width	1–30 km
Length	20–300 km
Length/width	4–40
Fjord area	$<30\,000$ km^2
Water depth	<1500 m
Submarine volume	<1000 km^3
Typical slopes	
(i) Along axis	0.1°–5°
(ii) Side-wall	10°–30°
Fluvial input	
Drainage basin area	<100000 km^2
Hinterland denudation rate	≈ 10 mm a^{-1}
Freshwater input	<100 km^3 a^{-1}
Sediment input	$<0.4 \times 10^8$ t a^{-1}
Delta progradation	0.1–1 km a^{-1}
Sedimentation rate	
(i) Range	0.1 to >1000 mm a^{-1}
(ii) Average	10 mm a^{-1}
Sedimentation events	Pulsed
Rhythmic bedding	Moderately common
Sediment infill thickness	100 m to >800 m
Water characteristics	
Stratification	Relatively stable
Bottom water (b.w.) temperature	$-1.7°–15°$
Annual b.w. temperature range	1°–6°
B.w. salinity	28‰–33‰

After Syvitski *et al.* 1987.

tectonically active coastlines of Alaska, British Columbia, Quebec, Baffin Island and Norway (Table 2). The primary data supporting this paper are taken from (i) the Saguenay Fjord, Quebec, and Knight Inlet, British Columbia (Geological Survey of Canada (GSC) project SEDFLUX) and (ii) the Baffin Island fjords (GSC project SAFE).

Steady state basin models

Four system processes combine to affect the infilling of a marine basin when the main sediment source is fluvial–deltaic (Calabrese & Syvitski 1987). These include (i) hemipelagic sedimentation of particles carried seaward by river plumes, (ii) rapid deposition of bed load along delta fronts, (iii) processes that deposit sediment in deeper basins after bypassing a significant portion of the proximal prodelta and (iv) diffusive processes such as gravitational reworking and current action that combine to smear sediment into deeper water.

Four parameters affect these system processes and thus the individual basin-fill sequence: (i) the

TABLE 2. *Some examples of slope failures in fjords*

System	Comments
Trondheimsfjord, Norway	In 1888, a slide due to retrogressive liquefaction of mud and sand occurred on a slope of 8°–10°; associated giant waves caused further slides
Follafjord, Norway	In 1952, a single glide block moved 10^6 m^3 of material at a velocity of 2.8 m s^{-1}
Orkdalsfjord, Norway	In 1930 and as a result of sediment loading and an earthquake, 10^8 m^3 of material moved at 6.9 m s^{-1} at 3 km decreasing to 2.8 m s^{-1} by 18 km; 10–30 m of the upper sediment cover was affected
Vaerdalen, Norway	In 1894, a liquefaction mudflow was transported at least 11 km at 1.7 m s^{-1}
Hardangerfjord, Norway	50% of the basin infill is as turbidites
Nordfjord, Norway	Since 1967, 10^7 m^3 of sediment moved down-slope and several channels up to 6 m deep were produced
Finnvika, Norway	A 1940 slide was followed by a series of secondary slides several days later and some 5 km from the original site
Port Valdez, Alaska	The 1964 Alaska earthquake moved 10^8 m^3 of sediment down slopes greater than 10°
Lituya Bay, Alaska	In 1958, an 8.3 Richter quake set off a 10^7 m^3 rock slide that generated giant waves that reached shore heights of 524 m
Cook Inlet, Alaska	Earthquake-induced fissuring and faulting of mudflats have removed a volume of 10^8 m^3
Howe Sound, BC	In 1955, a side-entry delta failed owing to oversteepening and overloading; the coarse-grained deposit on slopes of 27° developed erosional chutes and channels with the action of debris flows
Kitimat Arm, BC	In 1975, two distinct slumps coalesced to form a continuous deposit 5 km × 2 km (about 10^8 m^3); the slide developed debris flows and turbidity currents
Knight Inlet, BC	A seafloor channel system provides an effective way of transporting failed mouth-bar sands into depths of 500 m

For further details and references see Syvitski *et al.* 1987.

initial basement geometry; (ii) the type and amount of bed load Q_b, and suspended load Q_s, brought down by rivers; (iii) the oceanographic environment of the basin, *ie* fluvial discharge, tidal and wave environment, sea-level fluctuations and climate; (iv) the tectonic regime. Figure 2 identifies three types of basin infilling where the parameters Q_s and Q_b vary in extremes. A basin may have any number of sediment point sources, each with a unique Q_s/Q_b ratio (basin sediment properties are dependent on the supply level and proximity to each particular source), and the progradation dynamics of each point source may change with time. For example, a large lake in a river's hinterland may effectively filter the bed load for thousands, possibly millions, of years. When the lake is infilled, the contribution of the bed load to the ocean margin will increase substantially. Large-scale variations in the suspended load could result from climate-induced fluctuations of vegetation cover or frost–freeze cycles in hinterland mountains. The rate of fluvial cannibalism of coastal deposits during

uplift of the hinterland is another time-varying parameter.

If a basin receives more input from suspended load than from bed load, such that $Q_s \gg Q_b$ (Fig. 2, model 1), it is dominated by the flux of hemipelagic particles, *eg* Saguenay Fjord, Quebec (Smith & Walton 1980; Schafer *et al.* 1983). The accumulation rate will depend on the residence time of each type and size of suspended particle, the circulation dynamics and the suspended load discharged from river mouths (Syvitski *et al.* 1988). River water will spread over the surface of a marine basin as a turbid plume. As the plume progresses, sediment is lost continuously and rates of sedimentation decrease exponentially out into the basin (*ie* from more than 50 cm a^{-1} to less than 0.1 cm a^{-1} within the first 100 km). Sand is deposited relatively close to the river mouth. Areas with high rates of sedimentation are dominated by mobile epifauna. Below some threshold in the sedimentation rate (*eg* 20 cm a^{-1}, Moore & Scruton 1957; 6 cm a^{-1}, Farrow *et al.* 1983), infauna activity increases. Bioturbation

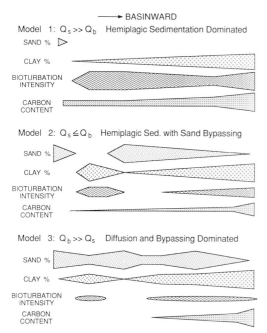

FIG. 2. A schematic representation of sediment parameters for three end-member cases of basin infilling based on both theory and field investigations (see text for details).

intensity tends to decrease basinward in response to a decrease in the flux of terrestrial carbon, only to increase eventually with contributions from the flux of marine carbon in the distal portions of the basin.

As a result of the exponential and basinward decrease in sedimentation rates, the prodelta progrades with ever-increasing slopes. Eventually the slopes will reach some critical value dependent on the ratio of the strength of the sediment and, as affected by external stimuli (earthquakes, tsunamis *etc.*), the driving force, and a prodelta mud slide will be generated. Wave or tidal current diffusion, however, may smear out the prodelta muds into deeper water, and in those cases steep gradients do not form (*eg* Turnagain Arm Delta, Cook Inlet, Alaska (Bartsch-Winkler *et al.* 1983)).

Where the bed load of a river is significant ($Q_b \geqslant Q_s$), mouth bars form and prograde onto steep foreset beds (Fig. 2, model 2). This contributes to semicontinuous failure along the delta lip where moderately sorted sand is eventually transported along submarine channels. The sand initially accelerates over some very steep gradients, easily developing into turbidity currents (*eg* Knight Inlet, British Columbia (Fig. 3)). Below some critical angle, the sand is deposited as a fan over a proportion of the basin floor. As a

consequence, sand bypasses the proximal prodelta mud deposits. In the distal portion of the fan, clay content is again high, as is the carbon content but with a lower C:N ratio. Bioturbation is normally absent in proximal areas receiving high rates of sedimentation and in areas directly affected by sediment gravity flows (Farrow *et al.* 1983).

If the marine basin receives most of its sediment input as bed load, *ie* $Q_b \gg Q_s$ (Fig. 2, model 3), the basin infill will be controlled by failure-generated diffusion of the proximal prodelta slopes with some form of sediment bypassing as an invariable consequence (*eg* Itirbilung Fiord, Baffin Island (Fig. 4)). Steep foresets will form at the delta lip if tides and waves are not effective enough to redistribute the coarse-grained bed load. Liquefaction-induced failures may occur anywhere along the steep foresets, and these may generate cohesionless debris flows and turbidity currents. As a result, sediments within the proximal prodelta are rich in sand and the down-slope bathymetry is more linear in profile (*ie* in angle of repose of the sediment). Retrogressive slide failures are common on these slopes. The run-out distance will vary with the volume of failed material and the drop height (Karlsrud & By 1981). If a failure volume is large, then even the course-grained topset deposits can travel onto the basin floor (*cf* Piper & Aksu 1987).

Bioturbation is very low for a model 3 basin, as the benthos is continually destroyed by the episodic gravity flow events (Syvitski *et al.* 1985). The carbon content within the basin sediments is highest in the distal region since there is little 'dilution' from inorganic sedimentation (*ie* low Q_s input).

In each of the above basin models, we have tacitly assumed a benthic population that is dependent on the flux of organic detritus, marine or terrestrial, and limited by particulate loading associated with high sedimentation rates or by high energy gravity flow events. However, if the basin water is occasionally isolated, eutrophication may also effectively exclude the benthos. A basin's redox discontinuity may be found within the water column, at the sediment–water interface, or within the sediment. True anoxic conditions will depend on the return interval of the deep water exchanges that these basins tend to experience (for details see Syvitski *et al.* 1987).

In terms of petroleum potential, we are most interested in models 2 and 3—scenarios that accommodate the action of prodelta failures and subsequent down-slope transport that allows the interfingering of sand (reservoir) and mud (source) units. Anoxic bottom conditions are important, for this allows for the preservation of

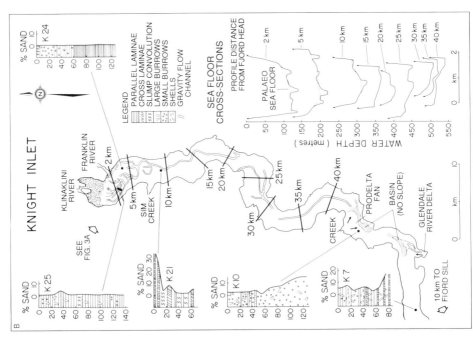

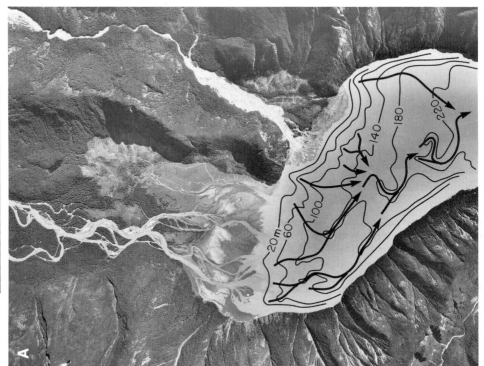

FIG. 3. (*A*) Vertical aerial photograph of the main Klinaklini Delta and the side-entry Franklin Delta, Knight Inlet, British Columbia. Indicated on the bathymetric map are the submarine channels that focus sand down the proximal prodelta slopes. (*B*) The distribution of submarine channels that meander along the seafloor in the upper half of Knight Inlet, and the core lithology in relation to these channels.

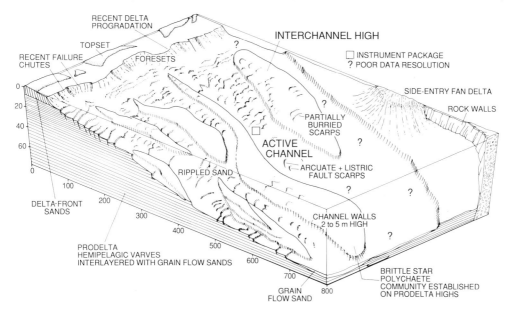

FIG. 4. Three-dimensional view of the first 0.8 km of the bed-load-dominated Itirbilung prodelta, Baffin Island, based on high resolution side-scan sonar imagery (for details see Syvitski *et al.* 1984).

good potential source rocks. Anoxia is common to both fjords and extensional half-grabens, for circulation is restricted by barriers created by glacial action or uplifted blocks respectively. Below we outline the wide variety of failure processes and products found in broadly similar depositional environments.

Scales of submarine failures and field tools

There are many possible scales associated with subaqueous slope failures. Shrimp may forage on steep basin slopes and set off microturbidity currents tens of centimetres wide and deep (Farrow *et al.* 1983). Other slides can be so large that glide blocks exposed in outcrop, or as sampled through coring, may give little indication that sediments have been translated some distance from the original slide scarp or are even part of a slide complex. For example, coherent slide blocks in McBeth Fiord, Baffin Island, are over 1 km long. Slide masses are also generally complex, and this is compounded by the fact that a once-failed mass commonly reactivates again. Often we are unable to discern whether the slide mass we are investigating is the result of the amalgamation of widely separate events or is a single contemporaneous event.

Most prodelta investigations remain biased by

the sampling tools employed. As a case in point, the upper surface of a marine slide may liquify and develop into a turbidity current that may then rework the lower reaches of a slide mass. Short cores collected from the slide surface may suggest an importance in turbidites that is not warranted. At the other extreme, multichannel seismic profiling can only resolve large failure masses. These problems have been illustrated by Field (1981).

The variety of different scalar data sets that we have used in interpreting slope failures (*cf* Fig. 6) include (i) airgun reflection seismic data (vertical resolution of 6 m) for the regional interpretation and basin geometry, (ii) deep-tow boomer profiles (*eg* HUNTEC® with a vertical resolution of 0.4 m) for detailing the upper 100 m of the section, (iii) a variety of coring techniques for surface stratigraphy (upper 10 m), (iv) various resolutions in side-scan profiling (swath width from 150 m to 3 km) and (v) photo-profiling (*eg* U-MEL sterography) using manned (PISCES IV) and unmanned (DART) submersibles.

Types of basin slope failures

Delta front and proximal prodelta failures

There are two failure styles along proximal delta slopes: (i) frequent (seasonal) failures of the sandy foreset beds as a result of oversteepening or overloading and (ii) rarer but more catastrophic failures within finer-grained prodelta sediments. In the former, numerous small-scale (10^3–10^6 m³) failures continually maintain a constant slope. In the latter, slides recur at intervals of decades or centuries and as solitary or near-contemporaneous events (a number of slides within a few years). The initial release mechanism for the more catastrophic failures is through the action of earthquakes where excess pore pressure is rapidly released and large volumes of sediment (10^6–10^9 m³) are mobilized (Syvitski *et al.* 1987).

Small-scale and frequent failures along delta fronts result in many common seafloor character-

istics: (i) channels, up to 100 m wide and 10 m deep, cover the prodelta slope; (ii) the channels originate from one or more arcuate re-entrants and chutes that have steep head-slopes cut into the delta lip; (iii) the channels, although not sinuous, may converge with or truncate one another; (iv) the channels, if active, are often floored with rippled well-sorted sand; (v) at any one time, a percentage of the channels are both inactive and being filled; (vi) interchannel areas consist of poorly sorted and weakly compacted very fine sandy muds. There is a tendency for buried channels to be reactivated.

The Itirbilung prodelta is an excellent example of seafloor morphology resulting from semicon-

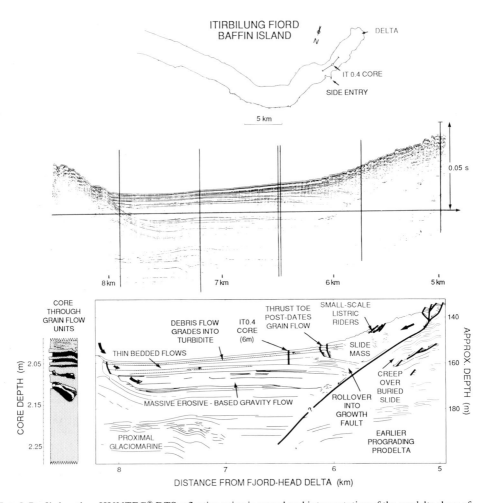

FIG. 5. Profile location, HUNTEC® DTS reflection seismic record and interpretation of the prodelta slope of Itirbilung Fiord, Baffin Island. Also shown is the lithology of a small portion of one of 25 piston and Lehigh cores, 1–10 m long, collected within the basin. The cohesionless debris flow units are massive and structureless sands.

tinuous delta-front failures (Figs 4 and 5). Bed-load input to the delta front is high (70%–90%) (Fig. 2, model 3). Along the delta front there are numerous chutes cut into the 15°–25° foresets; smaller chutes occasionally lie within larger chutes (up to 15 m wide). Chutes are commonly separated by sharp crested ridges. The foreset seafloor merges with the more gently sloping (8.5°–0.4°) bottomset beds at a water depth of 10 m. There the chutes converge into a number of distinct channels that may reach 170 m in width and 8 m in depth. The channel size eventually decreases down-slope until the channel form disappears. The channels are of low sinuosity and may converge with or truncate one another (Fig. 4). They contain dense well-sorted sand in the form of current ripples, with crests perpendicular to the channel axis. Interchannel 'islands' (Fig. 4) are stable and thus covered in mud from the hemipelagic fallout from the river plume. These submarine islands are densely populated with epifauna (*eg* brittle stars and polychaetes), as they are somewhat protected from the episodic turbidity currents that are initially contained within the channels. The entire prodelta slope is lined with retrogressive slide scarps that appear as steps, 1–3 m high, with 30° slopes and aligned perpendicular to the axis of the basin (Figs 4 and 5). These small-scale listric riders are best exposed within the channels and are more subdued over the interchannel highs where they are mantled by hemipelagic sediment.

During a 40 day period in the summer of 1985, nine significant gravity flow events were detected within one of the Itirbilung channels (see Fig. 4 for instrument position). Each event lasted between 1 and 5 h and the maximum current detected was 35 cm s^{-1} (measured 1 m above the channel floor at 10 min intervals). Sediment collected in moored traps at the current meter location contained well-sorted sand layers inter-layered with hemipelagic mud. Thus in this bed-load-dominated environment, basin infilling is controlled at present by numerous small foreset failures and subsequent sediment gravity flows. The timing of the delta-front failure may depend on strong katabatic winds that cause seiching within the basin (Syvitski 1987). The resultant internal bores and waves that impact on the oversteepened foresets could release chute-size volumes of sediment. These releases contribute to the development of cohesionless debris flows and turbidity currents.

Seismic profiles of the Itirbilung prodelta (Fig. 5) indicate the presence of a deep-seated growth fault. Failure along these slide scarps would contribute to finer-grained sediment gravity flows: submersible experiments succeeded in

generating a number of relatively low velocity turbidity currents (Syvitski *et al.* 1985). Thus the basin will receive both coarse-grained and finer-grained gravity flow sediment emplaced by cohesionless debris flows and turbidity currents.

Knight Inlet, British Columbia, is also controlled by a bed-load-dominated basin-head delta, although more representative of model 2 ($Q_b/Q_s \approx 3$) that is subject to frequent slope failures along its foreset beds (Syvitski & Farrow 1983). It has a complex submarine channel system that focuses these failed sediment masses, thus allowing sand to be moved offshore, bypassing the first 10 km of the prodelta slope (Fig. 3). Cores from the proximal prodelta environment (K25, K24) contain little or no sand, receiving sediment solely from hemipelagic fallout (Syvitski *et al.* 1988); they contain alternations of fine and coarse silt lamina, reflecting the seasonality in the river discharge cycle. The foreset sand is transported offshore by turbidity currents that have a run-out distance of about 50 km (Schafer *et al.* in press). Cores from the channel margins (K21) are not bioturbated, whereas cores from the deeper parts of the basin (K10, K7) show the highest levels of bioturbation. In both cases cores contain thin layers of both graded and non-graded sand with occasional slump deposits.

Large-scale and infrequent failures occur on prodeltas that are dominated by the vertical flux of hemipelagic mud (Fig. 2, model 1). In the case of Kitimat Arm, a 1975 slide involved more than 10^8 m^3 of material, caused a 26 m deepening of the seafloor at the slide head and resulted in the down-slope deposition of as much as 30 m of sediment (Prior *et al.* 1982). The geometry of the Kitimat slide had three main segments: (i) a short and steep head on the delta front (4°–7°); (ii) an intermediate blocky zone with 1°–2° slopes; (iii) an elongate depositional area more than 3.2 km long with slopes of less than 1° and marked by a steep toe.

Typically these slides commence with macroscopically perceptible creep, a process causing contemporaneous deformation in the form of gentle folding. Increases in the horizontal strain may lead to the break-up of feather joints in the sediment mass (Schwarz 1982). As failure progresses, antithetic precursory faults develop, followed by the development of the main shear face, often with reorientation of secondary shear faces. The main shear face or glide path usually develops as a listric fault. The main block slides down-slope, resulting in the formation of a tensional depression down-slope of the scar zone and a slide toe at the base of the slide where various fold forms and overthrust systems may result. During motion, the surfaces of slide blocks commonly undergo partial collapse forming a convolute slide or slump where internal stratification is chaotic.

In 1933 a Richter 7 earthquake hit the region of McBeth Fiord, Baffin Island. This may have initiated the major slide feature observed along its clay-rich (20–50 wt%) prodelta slope (Fig. 6). The main slide block is 1450 m long and 58 m thick and has dropped 15 m along the main slide scarp. Such tensional displacements have been accommodated through (i) internal remoulding of sensitive layers, (ii) block rotation, (iii) high angle thrusting near the base of the slide and (iv) low angle bow-wave (compressional) folding. A core taken within the main slide block shows little evidence of the slide event, except for a series of thinly bedded turbidites within the surface 50 cm, whose origin may be related to contemporaneous or subsequent failure along the various slide scarps where slopes up to 35° can be found (Fig. 6).

The prodelta slide within Tingin Fiord, Baffin Island, consists of a remarkable series of pull-aparts (caudal depressions) that have been linked by subsequent erosion from gravity flows to form a megachannel (Fig. 7). The slide has occurred within the inner of two basins, where the surface 30 m of a maximum of 100 m of infill has been affected. The megachannel has been partly infilled, initially by mass flow deposits and later through hemipelagic sedimentation (Syvitski 1985). The present channel may reach a depth of 30 m and a width of 225 m. An 8 m core collected near the base of the slide contained five units (Fig. 7): A, a 2 m thick basal unit of homogeneous mud which is of unknown origin; B, a 2 m thick burrowed unit containing different scales and intensity of bioturbation and broken with a couple of turbidite layers; C, a 1.5 m mudflow unit containing floating clay clasts; D, a 0.5 m overconsolidated and burrowed mud unit which may be the solid plug portion of the unit C mudflow; E, a 2 m thick surface unit of burrowed and very soft mud.

Basin infilling by cohesionless debris flow

Above we have described delta-front failures where portions of the failed sediment mass may liquefy and move into deeper water as turbidity currents. Another important mechanism exists for the down-slope movement of sediment— pseudo-laminar Newtonian fluids (*ie* linear viscous flows, grain flows, fluidized and liquefied flows, or some combination thereof). Such flows tend to develop when the initial slide involves large quantities of saturated loosely deposited fine sand and silt. At the point of failure there is an initial period of dilatancy, when water is expelled and the particles arrange themselves into a dense configuration (Bjerrum 1971). Once the slide mass changes to a viscous liquid and begins to flow down-slope, the face of the scar is left unsupported. This face will in turn fail and the slide will develop retrogressively slice by slice, widening the scar with each slice. Thus the dimensions of the final slide may be disproportionately large compared with the size of the initial slide (Bjerrum 1971). The sediment flow will not cease until the slide scar reaches a more dense deposit or until the rate of sediment supply to the viscous flow is greater than can be transported away.

We refer to the resulting deposits as cohesionless debris flow sequences. They tend to contain structureless clean well-sorted sand with mixed grading, *ie* both reverse and normal grading. They may contain synsedimentary clasts and there appears to be little difference between the characteristics of a pebbly grain flow and a sandy debris flow deposit, except in some arbitrary grain-size value for the matrix (Syvitski *et al.* 1987). Such grain–debris flows are non-erosive, and thus they tend to infill seafloor lows while maintaining a level surface (*cf* Figs 8 and 9). Through the action of the retrogressive development of the slide, many thinly bedded units may be deposited for each failure event. As such, amalgamated cohesionless debris flows are difficult to recognize seismo-stratigraphically since they contain poor internal stratification.

Cambridge Fiord, Baffin Island, contains a basin-fill sequence of debris flow layers between 3 and 10 m thick (about 3 m of the 9 m core in Fig. 9) that extend some 30–40 km. The base of these deposits has poor acoustic (reflection) definition, which may be related to liquefaction of the palaeo-seafloor during the rapid loading by the mass flow (Fig. 8). Sediments under the debris flow deposits have been affected by growth faults and rotational slides. (Strikingly similar growth fault and cohesionless debris flow structures occur within sediments of the Upper Ordovician at Girvan, Scotland (Fig. 8).) A core collected in the deep basin of Cambridge Fiord had two main units (Fig. 9): A, a lower 4 m unit of massive sand with floating mud clasts, attributed to a cohesionless debris flow origin; B, an upper 5 m unit of rippled turbidite sands capped with a 20 cm conglomeratic unit. These well-sorted sands were originally deposited in an ice proximal setting (Syvitski, in press).

The scale of these thick structureless sands is strikingly similar to that of the thick-bedded inverse to normally graded sandstones described from the Lower Ordovician of Quebec by Hiscott

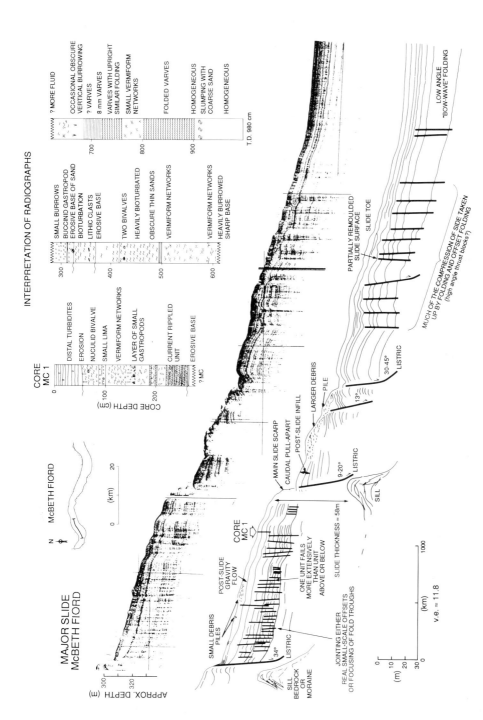

FIG. 6. Profile location, HUNTEC® DTS reflection seismic record and interpretation of the prodelta slope of McBeth Fiord, Baffin Island. Also shown is the interpretation of X-radiographs of a piston core 10 m long collected within the main slide block.

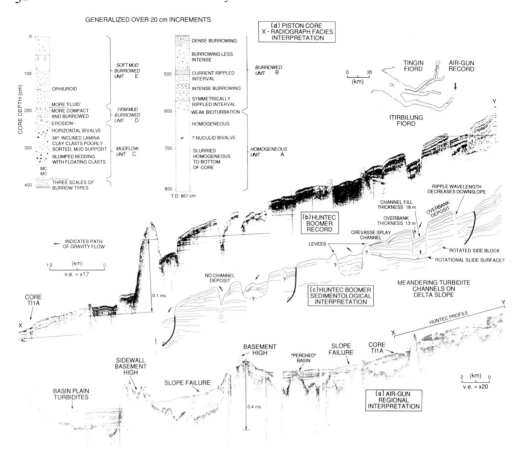

FIG. 7. Variety of scalar data sets used to describe the prodelta slide at Tingin Fiord, Baffin Island: (*a*) 0.66 L air-gun reflection seismic profile to aid in the regional interpretation including the basement geometry; (*b*) high resolution HUNTEC® DTS reflection seismic record showing details of the prodelta slide; (*c*) its sedimentological interpretation; (*d*) facies interpretation based on X-radiography of an 8 m piston core.

& Middleton (1979) and interpreted by them as accumulating in a submarine base-of-slope environment.

Basin floor slides

An important concept, often overlooked in basin analysis until recently, is the down-slope translation of cohesive but poorly consolidated sediment blocks which, in many instances, ride along very low angle glide planes. Basin infilling by slides, however, has recently been described from the tectonically active Zakynthos Straits, where subduction-related compression controls the geometry of the final deposits (Ferentinos *et al.* 1985; *cf* our Fig. 15). Furthermore, a number of significant recent oil discoveries have been made which point up the potential of this play. In the North Sea, for example, Gibbs (1984) has

described the Clyde Field, with an estimated 154 million bbl of recoverable oil, as lying above a growth fault detached from deep-seated basement faults. In the Gulf of Mexico, the Barataria Field is likewise considered to represent a block of shallow-water sand which has slid into deep water (Ventress & Smith 1985). Below we report on slides that commonly occur within basins under the influence of prograding deltas. Basin-floor slides commonly influence, or are influenced by, sediment failures along the basin margins.

Our first example is from outer Clark Fiord, Baffin Island (Fig. 10), where three slide complexes have reshaped the basin floor. The first and second complexes (Fig. 10) were generated from Quaternary sediment sliding along Pre-cambrian basement slopes of about 3°. The failed sediment masses were piled against down-slope

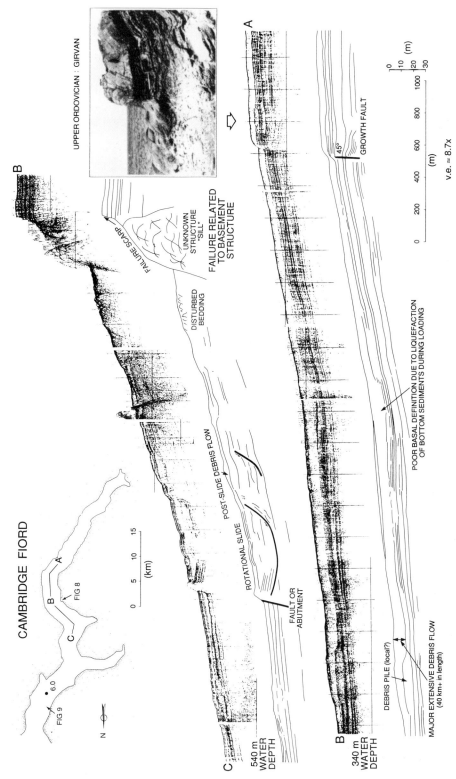

Fig. 8. A high resolution HUNTEC® DTS reflection seismic record with interpretation showing details of an extensive grain/debris flow deposit overlying an interval affected by growth faults and rotational slides, in Cambridge Fiord, Baffin Island. Note the box-fold geometry of Upper Ordovician sediment at Girvan, UK, for comparison with the Cambridge growth geometry.

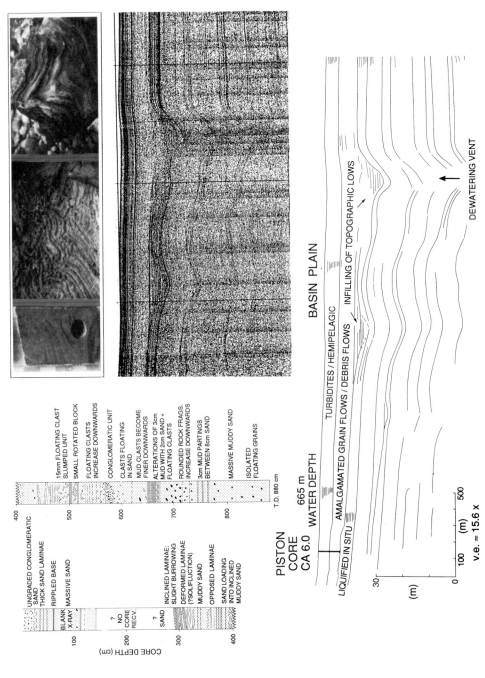

FIG. 9. A high resolution HUNTEC® DTS reflection seismic record with interpretation showing details of an extensive grain/debris flow deposit infilling lows generated by sediment dewatering (cf Fig. 8 for location within Cambridge Fiord, Baffin Island). The core interpretation is based on visual and X-radiography techniques. The photographs show dewatering structures in Kimmeridgian Allt na Cuile Sandstones, Helmsdale, Scotland.

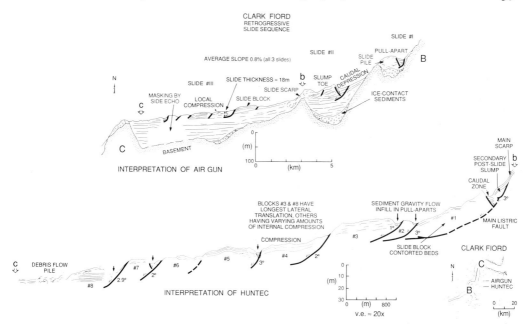

FIG. 10. Example from Clark Fiord, Baffin Island, of a slide complex with low angle glide planes based on interpretations of both 0.66 L air-gun and high resolution HUNTEC® DTS reflection seismic records.

basement 'sills' where beds were compressed into massive slump toes. Up-slope pull-apart or caudal depressions were formed. The third and most distal slide complex behaved as a retrogressive slide sequence where the translation of eight relatively coherent slide blocks, about 18 m thick, moved along a glide plane of about 0.8°. The block translation never exceeded 500 m and occurred along very low angle (1°–3°) listric faults. Blocks 3 and 8 (Fig. 10), with the longest lateral displacement, generated the largest amount of compressional folding within the next down-slope block.

In the distal part of McBeth Fiord, some 20 km from the prograding delta complex, a recent slide has translated sediment blocks 30–45 m thick along a glide plane with a 0.88° slope (Fig. 11); 10^9 m³ of sediment was affected by the slide. The tensional strain gave rise to a series of synthetic faults much steeper than the Clark slide (Fig. 10); faults have 45° angles nearest the crown of the slide but become more gentle near the toe of the slide (10°–30°). Activity along these listric faults produced two types of response: (i) block translation over distances of a few hundred metres; (ii) high angle (30°) thrust blocks. Noticeably lacking is any compressional folding so common to slides that have moved along higher angle glide planes. As in the more proximal prodelta slide of Tingin Fiord (*cf* Fig. 7) or the

Clark slide described above (Fig. 10), a mega-channel has formed by connecting a series of pull-aparts (Fig. 12), possibly through the erosive action of turbidity currents following a major longitudinal shear.

Basin megachannels share some general characteristics (Syvitski *et al.* 1987): (i) the channels are commonly found on slopes less than 2°; (ii) channel size decreases with decreasing slope; (iii) the channels are somewhat sinuous and may meander between the margins of the basin; (iv) before the channel disappears into the flat of the basin floor, levees develop; (v) up-slope, where levees are not found, the channels have near-vertical walls; (vi) if a channel is recently active, it will contain coarser sediment compared with the surrounding hemipelagic basinal muds. Many of these features are found in the McBeth megachannel (Fig. 12).

Typical of basins outside the equatorial zone is the influence of the Coriolis force on the deposition of sediment from gravity flows (see Fig. 12, profiles J and K) (the influence of the Coriolis force increases away from the equator). Similarly, river plumes will hug the Coriolis side of basins and preferentially deposit their sediment load; it is assumed that the influence of waves and tides is symmetrical across the narrow basin width. As a consequence, more marginal failures tend to be generated along the Coriolis-affected side-wall

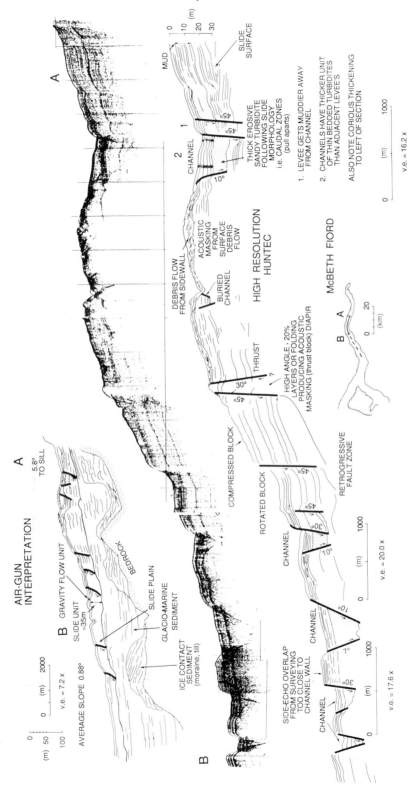

Fig. 11. Profile location and interpretation of 0.66 L air-gun reflection seismic profile over a distal slide within the basin of McBeth Fiord, Baffin Island, and a high resolution HUNTEC® DTS reflection seismic record with interpretation of a smaller portion of the slide complex.

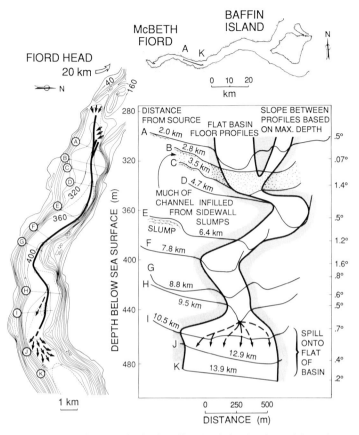

FIG. 12. Details of the megachannel in McBeth Fiord, Baffin Island, that has formed through connecting slide pull-aparts (*cf* Fig. 11).

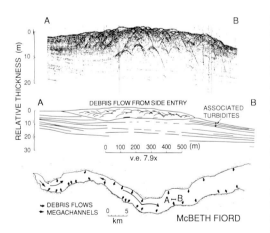

FIG. 13. The distribution of side-wall debris flows in McBeth Fiord, Baffin Island (*cf* Figs 11 and 12), based on BIO sidescan records at a 3 km swath width. There are considerably more failures along the Coriolis-influenced wall. Also shown is a high resolution HUNTEC® DTS reflection seismic record and interpretation of a typical side-entry debris flow.

(Fig. 13). Many of these mass failures generate sediment gravity flows in the later stages of their movement (Fig. 13) which become concentrated within the megachannel, thus contributing to its infill (Fig. 11). Acoustically, debris flows are characterized by a high frequency hyperbolic signature (Fig. 13) and the underlying reflections will be seen only if high band filtering is used.

Slopes on basin margins can range from 5° to more than 45°. Slide volumes in this region average between 10^3 and 10^6 m^3, although sub-aerial slides along these walls may generate enormous slides (10^8 m^3). The contribution of side-wall slumps to the basin floor increases with the rate of sedimentation and with the percentage of the seafloor consisting of basin walls. Therefore, in the early history of a basin, the number and frequency of side-wall failures may be very high. In fault-bounded basins the rate of sedimentation is directly related to the rate of uplift. In glaciated fjords the rate of sedimentation relates more to the time elapsed from the last ice advance.

The distance of lateral translation for sediment blocks from side-wall failures is much greater than for slides occurring along the axis of the basin (Fig. 10). Furthermore, marginal slide zones frequently fail more than once because of their steep slopes and the thin cover of sediment over the local basement rocks. As an example, a slide from a basin wall in the outer part of Itirbilung Fiord, Baffin Island, shows at least two stages of failure (Fig. 14). There the failed sediment mass is much more remoulded compared with slides along the basin axis.

Failures along a basin's side-wall are generally related to the development of oversteepened slopes (*ie* where the angle of internal friction is less than or equal to the seafloor slope). If the basin is truly fault bounded, however, ground accelerations from seismic activity will continually fail large volumes of sediment into the basin. In the case of the Saguenay Fjord, Quebec, the basin is situated in the highest-risk earthquake zone in eastern North America. In 1663 AD an earthquake in excess of Richter 7 hit the area with dramatic results (Schafer & Smith 1987). If precursor quakes and aftershocks are included, that event lasted about a month. In addition to moving some 10^8 m^3 of isostatically raised marine clays from land into the basin, the marine basin collapsed in on itself displacing another 10^9 m^3 of submarine sediment. There the dominant failure style was via large rotational side-wall slides (Fig. 15). In higher sedimentation areas, along the proximal prodelta slopes, the channel fill consists of a series of relatively thin onlapping slumps (Fig. 15, profile A). In the more distal environment, where the sedimentation rate has dropped by a factor of 5, the failed slide blocks became remoulded during initial failure, generating sediment gravity flows that contributed to the ponded fill sequence (Fig. 15, profile B).

Many of the acoustic structures identified from the Saguenay seismic records can be seen in Jurassic cores from the Ettrick and Brae oilfields, and at outcrop on the Helmsdale coast (Fig. 1). They include fault planes, palaeoslumps, highly folded slide masses, internal décollement, stacked debris flows and internal failure through brecciation (Fig. 15), similar in almost all respects to the structures described from synsedimentary shear zones in the deep-water Cow Head carbonates (Coniglio 1985). (It is important to stress here that, when interpreting well cores, small-scale structures are commonly the evidence for larger-scale structures. Pulham (this volume) notes that what are seen in the rock record are the lower-order structures, and therefore the higher structure of deformed sequences must be inferred.)

Significance for exploration

In the North Sea, clean well-sorted and ungraded sand beds between 1 and 4 m thick are commonly cored in Upper Jurassic–Lower Cretaceous basinal settings, giving the familiar 'box-car' log response (De'Ath & Schuyleman 1981; Turner *et al.* 1984; Stoker & Brown 1986). Well evidence shows that these sands can be correlated over distances of at least 10 km. Clearly, the mechanism by which such sands reach the basin plain is of great significance. We believe that many of these sand units, such as that which we traced for 40 km down-fjord in Baffin Island (Fig. 8) and that described by Prior *et al.* (1986), result from rapid retrogressive 'flow slides'.

Rapid deposition characterizes many fjordhead deltas (Farrow *et al.* 1983) and commonly results in loosely packed metastable deposits of well-sorted sands with high porosity. Large pore pressures may rapidly build up in these sands, reducing their strength during, for example, cyclic loading. Documented examples of the failure of such sands show that storm-wave (Prior *et al.* 1981) or tidal (Bjerrum 1971) loading is commonly responsible. In fault-bounded basins, seismicity may also act as a trigger (*eg* Helmsdale, E Greenland and Brae).

Deltas entering from a basin's margin, rather than from the head, tend to have more conglomeratic deposits that build out on steep side-wall slopes. These fanglomerates are associated with

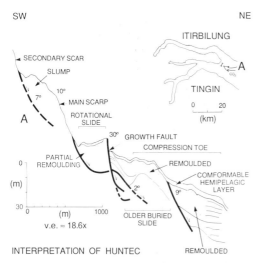

FIG. 14. Example from Itirbilung Fiord, Baffin Island, of a slide complex (average slope, 1.9%) from the margin of the basin based on interpretation of high resolution HUNTEC® DTS reflection seismic records.

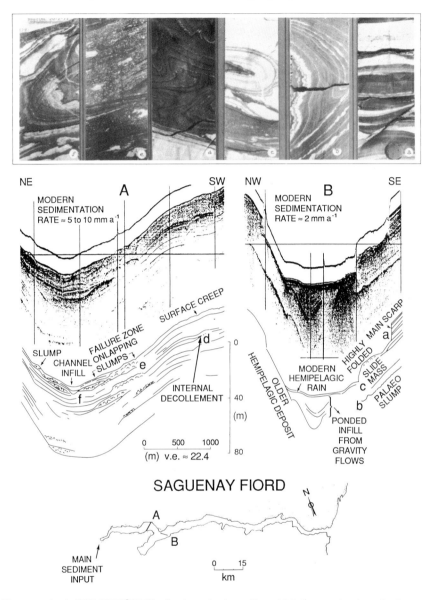

FIG. 15. Two cross-basin HUNTEC® DTS reflection seismic profiles with interpretation from the Saguenay Fiord, Quebec. Note the different sedimentation rates at each profile location. Shown at the top are core sections from the Ettrick oilfield (*cf* Fig. 1) that are the approximate lithological equivalents of structures within the acoustic profile (see text for details). Evidence for at least two phases of down-slope movement is present.

high frictional forces such that they can prograde with slopes greater than the angle of repose of the coarse material. Ultimately these deposits will fail, usually transforming into a debris flow (*cf* Prior & Bornhold 1986).

In an elongate fjord-like basin, therefore, it is probable that basin-plain sand and conglomeratic deposits do not have a common origin. The former may be far travelled by high density turbidity currents or by cohesionless debris flows following the axis of the basin; the latter may be short travelled by cohesive debris flows from the basin margins, although such debris flows can readily change direction and flow along the basin axis (Figs 11 and 12; *cf* Shelton 1966). In the light of this observation, it is astonishing how little attention has been placed on the head of the basin.

Areas like Helmsdale, on the northern flank of the inner Moray Firth Basin (Fig. 1), must have experienced regular submarine failures similar to those recorded from Canadian and Alaskan fjords (Syvitski *et al.* 1987). During the 1964 Alaskan earthquake, for example, liquefaction-type slides incorporated sand layers more than 50 m thick. Associated tsunamis and violent seiches may also initiate rock falls and submarine slides. The former would have the greatest effect on a steep shore (*eg* fault-scarp bounded) where the water height will be that of the breaking wave. (The Alaskan earthquake of 1964 caused a tsunami whose height was 7 m at the head of Alberni Inlet, 1700 km away.)

Giant waves have been produced by subaerial rock falls in fjords. In 1958, 3×10^7 m^3 of rock plunged into Gilbert Inlet, Alaska, producing an amazing bow wave that reached 524 m above sea level (Miller 1960). Both these kinds of uniformitarian catastrophe help explain the Helmsdale boulder beds and the Allt na Cuile sandstone, and may generate some of the smaller-scale cyclicity commonly described in graben-margin deposits (*eg* Turner *et al.* 1987). The Helmsdale boulder beds reach thicknesses of 650 m, and accumulated during 8 Ma in the late Jurassic. They show a remarkable cyclicity, with each boulder bed (typically 1–3 m thick) being the result of some kind of slope failure, probably triggered by seismicity along the Helmsdale fault. Many are sand-matrix debris flows, with large clasts rafted on top and projecting up to about 30 m in length. The subjacent Allt na Cuile sandstones show widespread autobrecciation which is best explained by partial liquefaction of a slightly lithified slope sand, without large-scale transport, since blocks can be 'reassembled' from neighbouring fragments.

There is a far more subtle mechanism, however, that could create cyclicity in submarine fan systems. Differential upthrow can occur between adjacent drainage basins, which may cause a major diversion of sediment from one basin to its neighbour, so that mega-prograding cycles may have little to do with down-faulting in the depocentre but more to do with differential uplift. Similarly subtle differential uplift is to be expected during (particularly oblique) extension (Hay 1978; Barr *et al.* 1985), when provenance changes should be detectable in the sedimentary record, not only amongst heavy minerals, for example, but also in the palynofacies, reflecting ecological differences between drainage basins.

In fjords, a common corollary of this uplift is the occurrence of perched alluvial terraces whose

sediment ultimately becomes cannibalized during fluvial entrenchment. The end result is that conglomeratic sediments may be surprisingly well rounded and yet have travelled only relatively short distances. Even allowing for widespread reworking of Devonian conglomerates, such an explanation is likely for some North Sea conglomerates that, on palaeogeographic grounds, can only have been derived from relatively small catchment areas similar in scale to many modern fjord drainage basins (Fig. 1). A similar space problem arises in the derivation of thick conglomerates from narrow strike-slip-related basement ridges such as the Rona Ridge W of the Shetlands (Ridd 1981, p. 422).

The stabilizing influence of terrestrial vegetation, particularly grasses, is a major factor in moderating contemporary rates of fluvial erosion. Temperate fjord-head deltas offer actualistic models of the sediment flux from land to sea when biotic effects had little importance on the architectural growth of a delta, *ie* pre-Upper Palaeozoic (Syvitski & Farrow 1983). Permafrost, however, is an important stabilizer of modern polar deltas (Syvitski 1986) and appears to perform a role similar to modern vegetation but in an environment of no vegetative cover. Thus the choice of temperate or arctic fjord delta as a modern analogue should be influenced by the particular palaeo-basin being investigated.

Stratigraphic play types

Two plays are described within this paper. The first involves the semicontinuous failure of foreset sands (Fig. 2, model 2) where, through progradation, gravity flow sands may form an areally extensive reservoir over basinal muds rich in marine organic matter. The thickness and areal extent of the reservoir depend on the distance of delta progradation and the basin dimensions, both of which will evolve through time as a function principally of tectono-thermal subsidence and eustasy. The reservoir unit will be capped by the prograding prodelta mud prism (Fig. 16a). Within this prism of mud, which contains both terrestrial and marine organic matter, are channel sands of reservoir quality (100–300 m wide, 5–20 m deep and 20–50 km long) that may or may not be intimately connected to the gravity flow sands, depending on the degree of synsedimentary faulting. The foreset sands could also become an excellent trap if rising sea level were to provide a seal of prodelta muds. Knight Inlet is a modern analogue environment (Fig. 3).

The second play is more typical of basins under

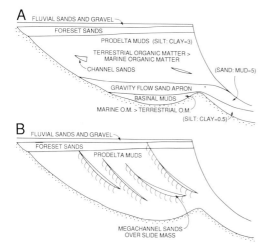

FIG. 16. Diagrams illustrating two deep basin sand plays: (*A*) a prograding delta sequence that experiences semicontinuous failure of its delta-front sands which are channelled so as to bypass the proximal prodelta muds and form a distal sand apron (*cf* Fig. 2, model 2); (*B*) a prograding basin-fill sequence that experiences periodic and catastrophic slides where sands cover the slide complex (*cf* Fig. 2, models 1 or 3).

the influence of higher energy quakes having return intervals of approximately 500 years. Here catastrophic events such as retrogressive slides may create, through pull-aparts and caudal zones, megachannels that fill with gravity flow sands during the later phase of the failure event (Fig. 16*b*). These reservoir-quality sands may come from the basin-head delta foresets or from secondary side-entry delta or beach complexes. They would appear within a prograding basin-fill sequence as relatively thick (20–50 m) units underlain by faulted and disturbed mud blocks and capped by conformable hemipelagic muds. Outer McBeth Fiord is a modern analogue environment (Figs 11 and 12). The idea of submarine failures controlling the location of the megachannel sands has already been applied to the sections within the Viking Graben (Alhilali & Damuth 1987).

Of the two plays, the second would be the more difficult to predict and to refine seismically on account of the difficulty of resolution at the typical depths encountered in North Sea deep condensate plays, *ie* 4000–7000 m.

There have recently been some encouraging discoveries which testify to the validity of these play concepts. Even well-established fields may yet contain further reserves in down-flank over-pressured stratigraphic traps. For example, additional production now comes from the Hackberry sandstones of SE Texas as a result of a proper understanding of submarine channel geometry and sand heterogeneity (Ewing & Reed 1985). Furthermore, correct recognition of the scale of deep basin sand development may on the one hand prevent costly acreage relinquishment and on the other lead to effective field development (*cf* Turner *et al.* 1987).

Conclusions

Fjords that receive high inputs of sediment experience a wide variety of slope failures. Many are appropriate modern analogues of small hydrocarbon-bearing basins that once received submarine fans. They are equally appropriate for much earlier extensional regimes such as that described recently by Arnott & Hein (1986), who interpret Late Precambrian Hector formation conglomerates from Lake Louise as filling submarine canyons cut in the seaward margin of a half-graben.

Three basin-infilling scenarios, depending on the respective inputs of bed load and suspended load, have been recognized and described (Fig. 2). Basins receiving a relatively high input of bed load provide appropriate environments for the generation and trapping of hydrocarbons. In fjords, the transport of reservoir-quality sands from the basin-head delta or from marginal environments into deep water is common.

There are two styles of apparently mutually exclusive delta failure. The first style involves small-scale and relatively frequent failure of the foreset sands. Retrogressive slides result in the transport of well-sorted sands by turbidity currents and cohesionless debris flows. The second style involves large-scale and relatively infrequent failure of the prodelta muds. Here retrogressive slides involve the down-slope translation of large sediment blocks (more than 1 km long and 50 m thick) and on glide planes less than 1°. Through the transport of sediment gravity flows within the caudal depressions or pull-aparts of the slide, a megachannel may form and remain active for a period long after the main failure event.

Portions of a basin fill are commonly affected by sediment failures along basement highs or side-wall margins.

ACKNOWLEDGMENTS: The authors wish to thank the Geological Survey of Canada (contribution 14487) and the British Geological Survey for their support. We thank our reviewers—Dennis Ardus and Stewart Brown (British Geological Survey), John Woodside and Al Grant (Geological Survey of Canada), J. M. Cohen (Imperial College) and George Postma (University of Bergen)—for their helpful comments.

References

ALHILALI, K. A. & DAMUTH, J. E. 1987. Slide block(?) of Jurassic sandstone and submarine channels in the basal Upper Cretaceous of the Viking Graben: Norwegian North Sea. *Marine and Petroleum Geology* **4**, 35–48.

ARNOTT, R. W. & HEIN, F. J. 1986. Submarine canyon files of the Hector Formation, Lake Louise, Alberta: Late PreCambrian syn-rift deposit of the Proto-Pacific miogeocline. *Bulletin of Canadian Petroleum Geology* **34**, 395–407.

BARR, D., MCQUILLIN, R. & DONATO, J. A. 1985. Footwall uplift in the Inner Moray Firth basin, offshore Scotland. *Journal of Structural Geology* **7**, 267–268.

BARTSCH-WINKLER, S., OVENSHINE, A. T. & KACHADOORIAN, R. 1983. Holocene history of the estuarine area surrounding Portage, Alaska, as recorded in a 93 m core. *Canadian Journal of Earth Sciences* **20**, 802–820.

BJERRUM, L. 1971. Subaqueous slope failures in Norwegian fjords. *Norwegian Geotechnical Institute Publication* **88**, 8 pp.

CALABRESE, E. A. & SYVITSKI, J. P. M. 1987. *Modelling the Growth of a Prograding Delta: Numerics, Sensitivity, Program Code and Users Guide.* Geological Survey of Canada Open File **1624**, 61 pp.

CONIGLIO, M. 1985. Synsedimentary submarine slope failure and tectonic deformation in deep-water carbonates, Cow Head Group, western Newfoundland. *Canadian Journal of Earth Sciences* **23**, 476–490.

DE'ATH, N. G. & SCHUYLEMAN, S. F. 1981. The geology of the Magnus Oilfield. *In:* ILLING, L. V. & HOBSON, G. D. (eds) *Petroleum Geology of the Continental Shelf of North-west Europe.* Heyden, London, 342–351.

DOWDESWELL, E. K. & ANDREWS, J. T. 1985. The fiords of Baffin Island: description and classification. *In:* ANDREWS J. T. (ed.) *Quaternary Environments: Eastern Canadian Arctic, Baffin Bay and West Greenland.* Allen & Unwin, London, 93–123.

ENGLAND, J. 1985. On the origin of high arctic fiords. *In:* VILKS, G. & SYVITSKI, J. P. M. (eds) *Arctic Land–Sea Interactions.* Geological Survey of Canada Open File **1223**, 51–53.

EWING, T. E. & REED, R. S. 1985. Depositional history gives clue to Hackberry production. *World Oil*, Feb 1, 47–52.

FARROW, G. E., SYVITSKI, J. P. M. & TUNNICLIFFE, V. 1983. Suspended particulate loading on the macrobenthos in a highly turbid fjord: Knight Inlet, British Columbia. *Canadian Journal of Fisheries and Aquatic Sciences* **40**, 273–288.

FERENTINOS, G., COLLINS, M. B., PATTIARATCHI, C. B. & TAYLOR, P. G. 1985. Mechanisms of sediment transport and dispersion in a tectonically active submarine valley/canyon system: Zakynthos Straits, NW Hellenic Trench. *Marine Geology* **65**, 243–269.

FIELD, M. E. 1981. Sediment mass-transport in basins: controls and patterns. *In:* DOUGLAS, R. G., COLBURN, I. P. & GORSLINE, D. S. (eds) *Depositional Systems of Active Continental Margin Basins.* Society of Economic Paleontologists and Mineralogists Short Course Notes, Pacific Section, San Francisco, CA, 61–83.

—— & EDWARDS, B. D. 1980. Slopes of the Southern California Borderland: a regime of mass transport. *In:* FIELD, M. E., COLBURN, I. P. & BOUMA, A. H. (eds) *Quaternary Depositional Environments of the Pacific Coast.* Society of Economic Paleontologists and Minerologists Short Course Notes, Pacific Section, Los Angeles, CA, 169–184.

GIBBS, A. D. 1984. Clyde Field growth fault secondary detachment above basement faults in North Sea. *American Association of Petroleum Geologists Bulletin* **68**, 1029–1039.

HAY, J. T. C. 1978. Structural development in the northern North Sea. *Journal of Petroleum Geology* **1**, 65–77.

HISCOTT, R. N. & MIDDLETON, G. V. 1979. *Depositional Mechanics of Thick-bedded Sandstones at the Base of a Submarine Slope, Tourelle Formation (Lower Ordovician), Quebec, Canada.* Society of Economic Paleontologists and Minerologists Special Publication **27**, 307–326.

HOLTEDAHL, H. 1965. Recent turbidites in the Hardangerfjord, Norway. *Colston Research Society Proceedings Bristol* **17**, 107–141.

KARLSRUD, K. & BY, T. 1981. Stability evaluations for submarine slopes: data on run-out distance and velocity of soil flows generated by subaqueous slides and quick-clay slides. *Norwegian Geotechnical Institute Report* **52207-7**, 67 pp.

LINDSAY, J. F., PRIOR, D. B. & COLEMAN, J. M. 1984. Distributary-mouth bar development and role of submarine landslides in delta growth, South Pass, Mississippi Delta. *American Association of Petroleum Geologists Bulletin* **68**, 1732–1743.

MILLER, D. J. 1960. Giant waves in Lituya Bay, Alaska. *U.S. Geological Survey Professional Paper 354C*, 51–86.

MOORE, P. G. & SCRUTON, P. C. 1957. Minor internal structures of some recent unconsolidated sediments. *American Association of Petroleum Geologists Bulletin* **41**, 2723–2751.

NELSON, C. H., MEYER, A. W., THOR, D. & LARSEN, M. 1986. Crater Lake, Oregon: a restricted basin with base-of-slope aprons of nonchannelized turbidites. *Geology* **14**, 238–241.

PIPER, D. J. W. & AKSU, A. E. 1987. The source and origin of the 1929 Grand Banks turbidity current inferred from sediment budgets. *Geo-Marine Letters* **7**, 177–182.

POSTMA, G. 1984. Slumps and their deposits in fan delta front and slope. *Geology* **12**, 27–30.

PRICE, L. C. 1980. Utilisation and documentation of vertical oil migration in deep basins. *Journal of Petroleum Geology* **2**, 353–387.

PRIOR, D. B. & BORNHOLD, B. D. 1986. Sediment transport on subaqueous fan delta slopes, Britannia Beach, British Columbia. *Geo-Marine Letters* **5**, 217–224.

——, —— & JOHNS, M. W. 1986. Active sand transport along a fjord-bottom channel, Bute Inlet, British Columbia. *Geology* **14**, 581–584.

——, COLEMAN, J. M. & BORNHOLD, B. D. 1982. Morphology of a submarine slide, Kitimat Arm, British Columbia. *Geology* **10**, 588–592.

——, WISEMAN, W. J. JR. & BRYANT, W. R. 1981. Submarine chutes on the slopes of fjord deltas. *Nature* **209**, 326–328.

RIDD, M. F. 1981. Petroleum geology west of the Shetlands. *In:* ILLING, L. V. & HOBSON, G. D. (eds) *Petroleum Geology of the Continental Shelf of North-west Europe.* Heyden, London, 414–425.

SCHAFER, C. T. & SMITH, J. N. 1987. Hypothesis for a submarine landslide and cohesionless sediment flows resulting from a 17th century earthquake-triggered landslide in Quebec, Canada. *Geo-Marine Letters* **7**, 31–37.

——, COLE, F. E. & SYVITSKI, J. P. M. Bio- and lithofacies in marine sediments in Knight and Bute Inlets, British Columbia. *Palaios,* in press.

——, SMITH, J. N. & SEIBERT, G. 1983. Significance of natural and anthropogenic sediment inputs to the Saguenay Fjord, Quebec. *Sedimentary Geology* **36**, 177–195.

SCHWARZ, H.-U. 1982. *Subaqueous Slope Failures—Experiments and Modern Occurrences.* Contributions to Sedimentology **11**. Elsevier, Stuttgart.

SHANMUGAM, G., DAMUTH, J. E. & MOIOLA, R. J. 1985. Is the turbidite facies association scheme valid for interpreting ancient submarine fan environments? *Geology* **13**, 234–237.

SHELTON, J. S. 1966. *Geology Illustrated.* W. H. Freeman, San Francisco, CA, 434 pp.

SMITH, J. N. & WALTON, A. 1980. Sediment accumulation rates and geochronologies measured in the Saguenay Fjord using Pb-210 dating method. *Geochimica Cosmochimica Acta* **44**, 225–240.

STOKER, S. J. & BROWN, S. S. 1986. *Coarse Clastic Sediments of the Brae Field and Adjacent Areas, North Sea: A Core Workshop* (also entiled *Lithofacies of Late Jurassic Coarse Clastic Sediments of the South Viking Graben, Northern North Sea: A Core Workshop*). British Geological Survey, 56 pp.

SURLYK, F. 1978. Submarine fan sedimentation along fault scarps on tilted fault blocks (Jurassic/Cretaceous boundary, East Greenland). *Bulletin of the Grønlands Geologiske Undersgelse* **128**, 108 pp.

SYVITSKI, J. P. M. 1985. Subaqueous slope failures within seismically active arctive fjords. *In:* VILKS, G. & SYVITSKI, J. P. M. (eds) *Arctic Land–Sea Interactions.* Geological Survey of Canada Open File **1223**, 60–63.

—— 1986. DNAG #2. Fjords, deltas and estuaries of Eastern Canada. *Geoscience Canada* **13**, 91–100.

—— 1987. Proximal prodelta investigations at two arctic deltas: Itirbilung and Cambridge Fiords, Baffin Island. *In:* SYVITSKI, J. P. M. & PRAEG, D. B. (compilers) *Sedimentology of Arctic Fjords Experiment: Data Report,* Vol. 3. Canadian Data Report of Hydrography and Ocean Sciences **54**, 468 pp.

—— On the distribution of sediment within glacier-influenced fjords: oceanographic controls. *Marine Geology,* in press.

—— & FARROW, G. E. 1983. Structures and processes in bayhead deltas: Knight and Bute Inlet, British Columbia. *Sedimentary Geology* **36**, 217–244.

——, BOUDREAU, B., SMITH, J. N. & CALABRESE, E. C. 1988. On the growth of prograding deltas. *Journal of Geophysical Research* **93**, 6895–6908.

——, BURRELL, D. C. & SKEI, J. M. 1987. *Fjords: Processes and Products.* Springer, New York, 379 pp.

——, HAY, A. E., SCHAFER, C. T. & ASPREY, K. W. 1984. SAFE: 1983 bayhead prodelta investigations. *In:* SYVITSKI, J. P. M. (compiler) *Sedimentology of Arctic Fjords Experiment: HU 83-028 Data Report,* Vol. 2. Canadian Data Report of Hydrography and Ocean Sciences **28**, 17-1–17-62.

——, SCHAFER, C. T., ASPREY, K. W., HEIN, F. J., HODGE, G. D. & GILBERT, R. 1985. *Sedimentology of Arctic Fjords Experiment: 85-062 Expedition Report.* Geological Survey of Canada Open File **1234**, 80 pp.

TURNER, C. C., COHEN, J. M., CONNELL, E. R. & COOPER, D. M. 1987. A depositional model for the South Brae oilfield. *In:* GLENNIE, K. W. & BROOKS, J. (eds) *3rd Conference on Petroleum Geology of North West Europe.* Graham & Trotman, London, 853–864.

——, RICHARDS, P. C., SWALLOW, J. L. & GRIMSHAW, S. P. 1984. Upper Jurassic stratigraphy and sedimentary facies in the Central Outer Moray Firth Basin, North Sea. *Marine and Petroleum Geology* **1**, 105–117.

VENTRESS, W. P. S. & SMITH, G. W. 1985. The 'E' sand of Barataria Field, SE Louisiana—a Miocene deep-water hydrocarbon reservoir. *Bulletin of the South Texas Geological Society* **25**, 37–43.

WALKER, R. G. 1978. Deep water sandstone facies and ancient submarine fans: models for exploration for stratigraphic traps. *American Association of Petroleum Geologists Bulletin* **62**, 923–966.

ZIEGLER, P. A. 1975. Outline of the geological history of the North Sea. *In:* WOODLAND, A. W. (ed.) *Petroleum and the Continental Shelf of Northwest Europe.* Applied Science Publishers, Barking, 165–190.

—— 1981. Evolution of sedimentary basins in north-west Europe. *In:* ILLING, L. V. & HOBSON, G. D. (eds) *Petroleum Geology of the Continental Shelf of Northwest Europe.* Heyden, London, 3–39.

JAMES P. M. SYVITSKI, Geological Survey of Canada, Bedford Institute of Oceanography, Dartmouth, Nova Scotia, Canada.
GEORGE E. FARROW, 19 Glenburn Road, Bearsden, Glasgow, UK.

Profiles of fan deltas and water depth in the receiving basin

M. Ito

SUMMARY: Profiles of prograding fan deltas are largely controlled by the water depth in the receiving basin. The initial slope of the fan-delta front with respect to that of the fan-delta plain, which is sometimes detectable in stratigraphic packages, decreases with the increase in water depth at the toe of fan-delta front deposits. The resultant relationship provides important evidence and information for restoring the depositional architecture of ancient fan deltas that may reflect the marginal geometry of sedimentary basins. Relatively steep gradients in fan-delta accretion surfaces may lead to an overestimation of the *vertical* stratigraphic thickness, particularly if it is calculated using a simple cumulative bed thickness. Furthermore, fan-delta successions which appear to be anomalously thick may not be due to progressive or incremental basin subsidence, but rather to overestimated stratigraphic thickness because the fan-delta clinoform dips were ignored. A part of the thick fan-delta deposits, therefore, can best be interpreted in terms of lateral accretion of fan-delta front deposits on the dipping clinoform surfaces, responding to the shallowing in the receiving basin.

Progradation of deltas and fan deltas generally results in the development of thick and extensive stratigraphic successions of coarse-grained deposits which may act as excellent reservoirs of fossil fuels. Lateral accretion of depositional systems on dipping clinoform surfaces is at least as important as vertical aggradation for generating thick sedimentary piles in stratigraphic records (Miall 1984*a*). Clinoforms are large-scale architectural elements of depositional systems, such as deltas and fan deltas, with depositional dips that may approach 20° or more. Recognition of depositional architecture in stratigraphic packages is important if miscalculations of stratigraphic thicknesses and tectonic dips are to be avoided. Such miscalculations may cause an overestimation of sediment accumulation rates and tectonic tilting which are important variables in understanding sedimentary basin analysis.

Fan deltas are alluvial fans that prograde into standing bodies of water from nearby highlands (Holmes 1965; McGowen 1970; Ethridge & Wescott 1984) and therefore are generally framed by subaqueous and subaerial clinoforms (Fig. 1). Fan-delta deposits, however, commonly develop along active tectonic regions where rapid temporal changes in relative elevation occur.

A method for estimating the initial slopes of fan deltas is presented in this paper, and it is applied to deformed fan-delta deposits in the late Cenozoic arc–arc collision zone of central Honshu, Japan.

Components of fan deltas

A simplified profile of a prograding fan delta comprising a subaerial and a subaqueous com-

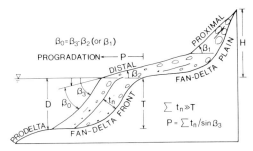

FIG. 1. Components of a prograding fan delta: H, height of the apex of an alluvial fan; T, thickness of the fan-delta front; $\sum t_n$, stratigraphic thickness of fan-delta deposits; D, depth at the toe of the fan-delta front; P, prograding distance.

ponent together with a transitional zone is illustrated in Fig. 1 (Wescott & Ethridge 1980). Alluvial fans represent the subaerial component of fan deltas (the fan-delta plain) and generally show sloping surfaces. The slope of proximal fans is commonly steeper than that of medial/distal fans ($\beta_1 > \beta_2$ in Fig. 1) and is accompanied by changes in sediment types (Blissenbach 1954; Bull 1964*a*; Boothroyd & Nummedal 1978). The major subaqueous component of fan deltas is a fan-delta front, which distally grades into a fan-delta prodelta with the subaqueous clinoform dipping at an angle β_3 (Fig. 1). Fan-delta fronts may dip steeply and represent distinct foresets of a typical Gilbert-type fan delta, although some fan deltas appear to show no marked change in slope from the subaerial to subaqueous components (Rust & Koster 1984; Nemec & Steel 1987).

The water depth D in the receiving basin is approximately equal to the vertical thickness T of the subaqueous component of fan deltas (Fig.

From WHATELEY, M. K. G. & PICKERING, K. T. (eds), 1989, *Deltas: Sites and Traps for Fossil Fuels*, Geological Society Special Publication No. 41, pp. 45–54.

1). However, a larger stratigraphic thickness for the fan-delta front deposits, exceeding the actual water depth, may easily be estimated by measuring the cumulative bed thickness perpendicular to the dipping clinoform surfaces in deformed fan-delta deposits ($\sum t_n \gg T$ in Fig. 1) (Ito & Masuda 1986b). Overestimates of stratigraphic thickness may result in a misinterpretation of the burial depth and therefore of the history of sedimentary basins.

Slope of fan deltas and water depth

A morphological variation of deltas and fan deltas, mainly in plane view, has been recognized in relation to the relative strengths of fluvial and marine processes (Bernard 1965; Wright & Coleman 1973; Coleman & Wright 1975; Galloway 1975; Hayes & Michel 1982). Prograding fan deltas are generally defined as a component of river-dominated deltas of Galloway's (1975) threefold classification of delta types (Miall 1984b). The depositional slopes of deltas and fan deltas are also controlled by the regime of delta front area as well as by the density differences between river and basin waters (Bates 1953; Wright 1977; Coleman 1981; Miall 1984b). Variations in the profiles of prograding deltas and fan deltas, however, have not yet been well recognized.

Matyas (1984) and Kenyon & Turcott (1985) have recently presented profile models of prograding deltas under a river-dominated setting. They suggested that the water depth in the receiving basin functions as one of the important factors for controlling the morphological variations in a delta front.

The principal parameters for describing the profiles of fan deltas are the gradients of the subaerial and subaqueous clinoforms (Fig. 1). However, a horizontal reference plane relative to the slopes of a fan delta (Fig. 1) cannot be determined accurately in stratigraphic packages, even for some terraced fan-delta deposits. The slope of a fan-delta front with respect to that of a fan-delta plain, here defined as the relative slope of the fan-delta front, ie $\beta_0 = \beta_3 - \beta_2$ (or $\beta_0 = \beta_3 - \beta_1$) (Fig. 1), can be measured in stratigraphic packages. The maximum water depth for the slope of the prograding fan-delta front in the receiving basin can be determined from palaeontological data. The vertical thickness of fan-delta front deposits (T in Fig. 1) can be measured in well-exposed ancient successions and can be used for estimating the palaeobathymetry.

Data on water depth D and relative slope β_0 were obtained from recent and ancient well-exposed fan-delta deposits (Fig. 2). Dips of relative slope β_0 generally tend to decrease with an increase in water depth. Thus, for sediment stability in deep waters, fan-delta fronts tend to develop a long and gentle (low gradient) profile. Such a profile may be the result of subaqueous bulk-sediment movement such as creep and landslides (Prior & Suhyda 1979; Coleman et al. 1983; Winker & Edwards 1983). Mass-flow processes on fan-delta slopes have also been recognized (Prior et al. 1981; Postma 1984). Shallow fan-delta fronts, however, may be stable with higher gradient slopes. Clearly, grain size will influence the actual slope angles, but this general relationship appears to be true. Subaerial slopes of the fan-delta plain can also affect the resultant relationship shown in Fig. 2. Where alluvial fans prograde from high-relief areas into deep water, there may be no well-developed coastal plain or distal subaerial fan-delta environment (Wescott & Ethridge 1980). The relative slope of the fan-delta front may then decrease.

The slope of alluvial fans (the fan-delta plain) generally increases with decreasing fan dimensions, such as length and width (Fig. 3), and may approach 10° or more in the proximal reaches, although the average is 3°–6° (Denny 1965; Bull 1977; Reineck & Singh 1980). In arid climatic regions the average slope of alluvial fans is generally steeper than that in humid climatic regions where stream-induced processes predominate (Bull 1964b; Ryder 1971; Kochel & Johnson 1984). If the slope of a fan-delta plain and the water depth at the toe of the fan-delta front are estimated, an approximate initial slope of the fan-delta front can be obtained (Fig. 2). The slope of the fan-delta front is described by the equation

$$\beta_3 = \tan^{-1}(aD^{b-1})$$

where a and b are coefficients determined by plotting the profile data, ie the horizontal distance and depth of fan-delta fronts from their edge, on a log–log scale following the Matyas (1984) model for prograding deltas. Furthermore, the restored slope β_3 and the apparent stratigraphic thickness $\sum t_n$ of the fan-delta front deposits are available to estimate the approximate progradation distance P (Fig. 1) as well as the average rate of fan-delta progradation in deformed stratigraphic successions.

Application of the model to Japanese examples

Fan deltas have usually developed along tectonically active regions, such as convergent, divergent

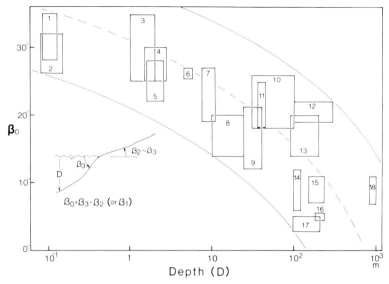

FIG. 2. Relationship between the relative slope of the fan-delta front with respect to the slope of the fan-delta plain and the water depth in the receiving basin. Each rectangle shows the range of the two variables. Principal references: 1 and 2, Ito (unpublished data); 3, Hempton *et al.* (1982); 4, Sneh (1979); 5, Rachocki (1981); 6, Casey & Scott (1979, in Ethridge & Wescott 1984); 7, Link *et al.* (1985); 8, Stanley & Surdam (1978); 9, Manspeizer (1985); 10, Yano (1986) and Ito (unpublished data); 11, Postma & Roep (1985); 12, Kondo (1985) and Ito (unpublished data); 13, Kondo (1986) and Ito (unpublished data); 14, Harms *et al.* (1982); 15, Fujii (1965) and Geographical Survey Institute (1982); 16, Sneh (1979); 17, Fujiwara *et al.* (1987); 18, Sato (1985) and Shimamura (1986).

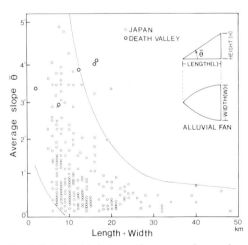

FIG. 3. Relationship between the average slope and dimension (length + width) of alluvial fans. The Japanese data are from Yazawa *et al.* (1971) and the Death Valley data are from Denny (1965).

and strike-slip plate margins, as well as along the margins of fault-bounded intracratonic seas and lakes (Ethridge & Wescott 1984; Ethridge 1985). These tectonic settings favour the development of a high-relief coastline and steeply sloping

narrow shelves, through which the distal portion of a fan-delta front tends to pass into a deep-water slope.

During the late Cenozoic, active progradation of fan deltas into deep water occurred in three major sedimentary basins, namely the Aikawa, Ashigara and Fujikawa Basins, in an arc–arc collision zone of central Honshu, Japan (Fig. 4) (Ito & Masuda 1986a, 1988). The fan-delta deposits are generally underlain by coarse-grained submarine fan and slope deposits, including volcaniclastics from submarine volcanoes above basin plain deposits (Fig. 5). Detailed descriptions and illustrations of the fan-delta deposits have been given by Ito (1985, 1987) and Ito & Masuda (1986a, 1988), to which the reader is referred for further details. Restoration of the initial slopes of the fan-delta front deposits in the deformed stratigraphic successions is discussed below.

Thick successions

The stratigraphic thickness of the fan-delta deposits studied ranges from 1300 to 3000 m (Fig. 5). The fan-delta successions generally show progressive upward shallowing with major portions represented by subaqueous fan-delta front

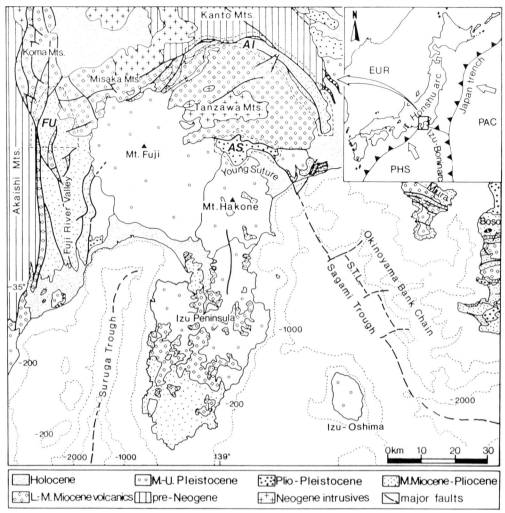

FIG. 4. Geological sketch map of the late Cenozoic arc–arc collision zone of central Honshu (simplified from Kakimi *et al.* 1982). The open arrows in the inset map show convergent directions of the Pacific (PAC) and Philippine Sea (PHS) plates to the Eurasia (EUR) plate (from Circum-Pacific Map Project 1981). AI, Aikawa Basin; AS, Ashigara Basin; FU, Fujikawa Basin.

deposits (Ito 1985, 1987; Kano *et al.* 1985). The maximum water depth D_{max} in the receiving basins can provide sufficient potential to develop the full stratigraphic superposition of sediments without subsidence of the basin floor, *ie* $T = D_{max}$, if the sediments fill the basin. However, the thickness of fan-delta front deposits, as well as that of submarine fan and slope deposits, generally exceeds the possible original thickness (Fig. 5). Lateral accretion and/or progradation of fan-delta front deposits on dipping clinoform surfaces may have played an important role in generating the thicker stratigraphic successions. The thick-

ness of the subaerial component of fan deltas (the fan-delta plain) usually exceeds about 650 m and is a few times thicker than that of modern alluvial fans (Kochel & Johnson 1984), suggesting that basin subsidence may also have been important in generating the thicker deposits. It seems impossible to explain the origin of the thick gradually shallowing-up successions of fan-delta deposits only by the interplay of progressive subsidence and infilling of sedimentary basins. An interplay between subsidence and progradation, unless critically balanced, will tend to generate stacked shallowing-up sequences.

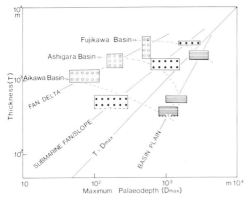

FIG. 5. Stratigraphic thickness and maximum palaeodepth of the late Cenozoic deposits in three major basins in an arc–arc collision zone of central Honshu. The deepest basin has the potential for developing thicker stratigraphic successions. The stratigraphic thickness of fan-delta deposits, as well as submarine fan and slope deposits, usually exceeds the value of the maximum palaeodepth in the receiving basin. Stratigraphic data on the Fujikawa Basin are mainly from Akiyama (1957), Matsuda (1961) and Soh (1986). Palaeobathymetric data are mainly from Sugita (1962), Matsushima (1982), Huchon & Kitazato (1984), Kano *et al.* (1985) and Kitazato (1986).

Architecture of fan deltas

Facies analysis and basin mapping are the principal methods for the description of the stratigraphic architecture of depositional systems (Miall 1984*a*). Data on the large-scale architectural components of a depositional system, such as dipping clinoforms, can provide useful constraints for modelling the depositional environments and basin geometry.

Reconstruction of the depositional architecture of the late Cenozoic fan-delta successions in the Aikawa and Ashigara Basins is discussed here. By using the available data on palaeobathymetry at the toe of the fan-delta front deposits (Fig. 5) and on the slope of modern alluvial fans (Fig. 3), the initial slope of the fan-delta front can be restored using the graph in Fig. 2.

Figure 6 shows an outcrop of the lower portion of the fan-delta front deposits (about 480 m thick) in the Aikawa Basin. They overlie slope and/or high gradient shelf deposits (about 63 m thick), which indicate a water depth of 50–120 m (Sugita 1962), and pass upwards into distal fan-delta plain deposits (about 310 m thick) overlain in turn by proximal fan-delta plain deposits (about 570 m thick) (Fig. 7). To restore the palaeocurrent data, the tectonic tilt should be corrected until the beds are horizontal or, in this example (Fig. 6), dipping southwards by up to a few degrees in

FIG. 6. Deformed fan-delta front deposits in the Aikawa Basin. The arrow shows the palaeocurrent direction determined from clast imbrication.

the probable original attitude of the fan-delta front deposits. The original slope β_3 of the fan-delta front in the Aikawa Basin can be estimated at 15°–20°, based on the graph in Fig. 2, if it is assumed that the slope of the distal fan-delta plain is about 1° ± in Fig. 3. The restored attitude of the fan-delta front deposits shows that the alternating mudstone and sandstone beds thicken basinwards and drape the steeply inclined conglomerate units (Fig. 6). The original attitude of the dipping clinoforms of the fan delta in the Ashigara Basin can also be restored using the same methods. In the Ashigara Basin the fan-delta front deposits (about 950 m thick) overlie slope deposits 580–1250 m thick, which indicate a water depth of 150–200 m (Matsushima 1982; Huchon & Kitazato 1984; Kitazato 1986), and are overlain abruptly by proximal fan-delta plain deposits (about 650 m thick) (Fig. 7). The slope

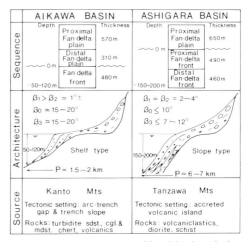

	AIKAWA BASIN		ASHIGARA BASIN	
Sequence	Depth	Thickness	Depth	Thickness
	Proximal Fan-delta plain	570 m	Proximal Fan-delta plain	650 m
	Distal Fan-delta plain	310 m	Proximal Fan-delta front	490 m
	Fan-delta front	480 m	Distal Fan-delta front	460 m
Architecture	$\beta_1 > \beta_2 \simeq 1°\pm$ $\beta_0 \simeq 15\sim20°$ $\beta_3 \simeq 15\sim20°$ Shelf type P ≃ 1.5~2 km		$\beta_1 = \beta_2 \simeq 2\sim4°$ $\beta_0 \leqslant 10°$ $\beta_3 \leqslant 7\sim12°$ Slope type P ≃ 6~7 km	
Source	Kanto Mts Tectonic setting: arc-trench gap & trench slope Rocks: turbidite sdst., cgl. & mdst. chert, volcanics		Tanzawa Mts Tectonic setting: accreted volcanic island Rocks: volcaniclastics, diorite, schist	

FIG. 7. Principal components of fan-delta deposits in the Aikawa and Ashigara Basins. See Fig. 1 for abbreviations.

of the subaerial component is estimated to be about 2°–4° and the resultant slope of the fan-delta front to be 7°–12° (Fig. 7). Even the basinward advance of the fan-delta front, by as much as 6–7 km, may have given rise to the deposits (950 m thick) without invoking any tectonic subsidence.

The difference in water depth in the Aikawa and Ashigara Basins, which are within the same tectonic setting of an arc–arc collision boundary, can be explained largely by variation in shelf width and slope at the basin margins. The bottom slope of sedimentary basins has also been interpreted as one of the important controls for fluvial and deltaic processes (Coleman & Wright 1975; Miall 1981). The difference in basin margin geometry was probably governed by the geological composition of the source terranes (Fig. 7). The Kanto Mountains are a major source for the Aikawa Basin fill and comprise mainly pre-Neogene arc–trench gap and trench slope sediments (Ogawa & Horiuchi 1978). The Ashigara Basin, however, developed on the southern margin of the Tanzawa Mountains, an accreted volcanic island (Niitsuma & Matsuda 1984; Ito & Masuda 1986a) composed mainly of Lower to Middle Miocene subaqueous volcaniclastic rocks.

This terrane may have been flanked by a narrow and steep shelf, similar to the present-day volcanic island of Izu Oshima (Fig. 3).

Reading the slope change of a fan-delta front

The deformation style of the fan-delta deposits in the Ashigara Basin (Fig. 8A) has resulted in a generally westward tilt of 45°–85° with a strike of N20°–E30° (Fig. 8B). The degree of deformation seems to be almost equivalent over the fan-delta front deposits, except around the basin margin fault (Kannawa Fault) (Fig. 8A) where the mainly proximal fan-delta plain deposits have been highly deformed and locally overturned. Palaeo-currents of the fan-delta front deposits are mainly from NE to SW (Fig. 8C) and indicate that the fan delta prograded southwestwards across the present NNE tectonic strike. The WNW component of the original dips of the fan-delta front deposits may therefore be reflected in the tectonic dips (Fig. 8D).

The fan-delta front deposits in the Ashigara Basin show an overall shallowing upwards (Ito 1985), and the WNW-dipping component of their initial slope may have increased in response to the shallowing, as predicted from Fig. 2. Figure

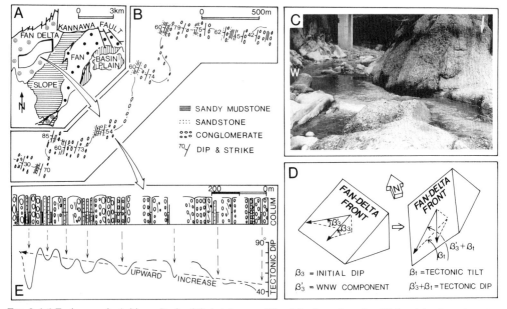

FIG. 8. (A) Facies map in Ashigara Basin; (B) sketch map of fan-delta front deposits; (C) fan-delta front deposits tilted WNW (the arrow indicates the palaeocurrent direction determined from cross-bedding at the base of conglomerate units); (D) an interpretation of the superposition of the initial dip and tectonic tilt which results in the present tectonic dip; (E) upward-increasing tectonic dip of fan-delta front deposits (several stratigraphic intervals, where sandy mudstones and sandstones are developed, generally show gentler dip than the general trend).

8*E* shows an overall trend of upward increasing dips of the fan-delta front deposits. This pattern may reflect the changes in gradient of the prograded fan-delta front. Tectonic dips of several stratigraphic intervals, where sandy mudstones and sandstones are well developed, are generally lower than those of conglomeratic deposits (Fig. 8*E*). The variation in tectonic dips may represent the gentler slope of sandy mudstones and sandstones compared with that of the conglomerates on the fan-delta front. Development of sandy mudstone and sandstone intervals may represent quiescence of active progradation of the gravelly fan delta in response to ephemeral retreat of the b⸴sin margin scarps, erosion of the uplifted source (Heward 1978; Rust & Koster 1984; Whateley & Jordan, this volume) and/or lateral switching of fan lobes, to generate fining-upward sequences 60–110 m thick within an overall coarsening-upward sequence. While the fan-delta topography returned to equilibrium, alternating sandy mudstones and sandstones may have draped the more steeply dipping conglomeratic deposits, as seen in the restored sections from the Aikawa Basin (Fig. 6).

Conclusions

The morphology of fan deltas is controlled by the interaction of river and marine processes. The geometry of the receiving basin, which is governed by the tectonic setting, largely affects these dynamic processes (Coleman & Wright 1975; Miall 1981). Water depth is a factor in determining the bottom slope (one of the geometric variables) of sedimentary basins. It may also govern the progradation rate of fan deltas. The relationship between water depth in the receiving basin and the relative slope of the fan-delta front with respect to the slope of the fan-delta plain has been mainly examined in this paper. As a response to deepening in the receiving basin, the relative slope of the fan-delta front generally decreases, and this relationship allows the restoration of the profiles of fan deltas in deformed stratigraphic packages. The restoration of large-scale architectural components of fan deltas, together with facies analysis and basin mapping, can provide a clear interpretation of the morphological setting for the receiving basin. The approximate distance and rate of forward growth or progradation can be estimated using the restored slope and measured stratigraphic thickness of fan-delta front deposits.

The dip of the clinoforms can partly explain the anomalously large stratigraphic thickness of many fan-delta deposits, *ie* lateral accretion of fan-delta front deposits without substantial progressive subsidence. However, subsidence may be important, as in juvenile rift systems. Recognition of the original slopes of fan deltas also offers some refinement for tilt correction of palaeomagnetic and palaeocurrent directions. Morphological analysis of fan-delta deposits in stratigraphic successions can be used as a primary key for interpreting the large-scale depositional architecture which governs the distribution of fossil fuels.

ACKNOWLEDGMENTS: I would like to thank Professor Hiroshi Noda and Dr Fujio Masuda, University of Tsukuba, for their critical discussion. I am grateful to Dr Kevin T. Pickering, University of Leicester, for his interest in my research and for inviting me to contribute to this volume. An early version of the manuscript was critically reviewed by Dr Pickering.

References

AKIYAMA, M. 1957. The Neogene strata along the upper course of Fujikawa River, Yamanashi Prefecture, Japan. *Journal of the Geological Society of Japan*, **63**, 669–683 (in Japanese with English abstract).
BATES, C. C. 1953. Rotational theory of delta formation. *American Association of Petroleum Geologists Bulletin* **37**, 2119–2162.
BERNARD, H. A. 1965. A résumé of river delta types (abstract). *American Association of Petroleum Geologists Bullein* **49**, 334–335.
BLISSENBACH, E. 1954. Geology of alluvial fans in semiarid regions. *American Association of Petroleum Geologists Bullein* **65**, 175–190.
BOOTHROYD, J. C. & NUMMEDAL, D. 1978. Proglacial braided outwash: a model for humid alluvial-fan deposits. *In*: MIALL, A. D. (ed.) *Fluvial Sedimen-tology*. Canadian Society of Petroleum Geologists Memoir **5**, 64–668.
BULL, W. B. 1964a. History and causes of channel trenching in western Fresco County, California. *American Journal of Science* **262**, 249–258.
—— 1964b. Alluvial fans and near-surface subsidence in Western Fresco County, California. *U.S. Geological Survey Professional Paper* **437-A**, 71 pp.
—— 1977. The alluvial fan environment. *Progress in Physical Geography* **1**, 222–270.
CASEY, J. M. & SCOTT, A. J. 1979. Pennsylvanian coarse-grained fan deltas associated with Uncompahgre uplift, Talpa, New Mexico. *New Mexico Geological Society Guidebook, 30th Field Conference, Santa Fe County*, 211–218.
CIRCUM-PACIFIC MAP PROJECT. 1981. *Plate Tectonic*

Map of the Circum-Pacific Region, Northwest Quadrant. American Association of Petroleum Geologists.

COLEMAN, J. M. 1981. *Deltas, Processes of Deposition and Models for Exploration* (2nd edn). Burgess, Minneapolis, MN, 124 pp.

—— & WRIGHT, L. D. 1975. Modern river deltas: variability of processes and sand bodies. *In*: BROUSSARD, M. L. (ed.) *Deltas. Models for Exploration.* Houston Geological Society, 99–150.

——, PRIOR, D. B. & LINDSAY, J. F. 1983. Deltaic influences of shelf edge instability processes. *In*: STANLEY, D. F. & MOORE, J. T. (eds) *The Shelfbreak: Critical Interface on Continental Margins.* Society of Economic Paleontologists and Mineralogists Special Publication **33**, 121–137.

DENNY, C. S. 1965. Alluvial fans in the Death Valley region, California and Nevada. *U.S. Geological Survey Professional Paper* **466**, 62 pp.

ETHRIDGE, F. G. 1985. Modern alluvial fans and fan deltas. *In*: *Recognition of Fluvial Depositional Systems and their Resource Potential.* Society of Economic Paleontologists and Mineralogists Short Course **19**, 101–126.

—— & WESCOTT, W. A. 1984. Tectonic setting, recognition and hydrocarbon reservoir potential of fan-delta deposits. *In*: KOSTER, E. H. & STEEL, R. J. (eds) *Sedimentology of Gravels and Conglomerates.* Canadian Society of Petroleum Geologists Memoir **10**, 217–235.

FUJII, S. 1965. The development of the Kurobe fan and the submarine forests around the Toyama Bay. *Earth Science* **78**, 11–20 (in Japanese with English abstract).

FUJIWARA, Y., SATO, T., ITO, M. & AONO, H. 1987. Sedimentary environment of the Uonumo Group in Niigata Prefecture. *Annual Reports of the Institute of Geoscience, University of Tsukuba* **13**, 80–82.

GALLOWAY, W. E. 1975. Process framework for describing the morphologic and stratigraphic evolution of deltaic depositional systems. *In*: BROUSSARD, M. L. (ed.) *Deltas, Models for Exploration.* Houston Geological Society, 87–98.

GEOGRAPHICAL SURVEY INSTITUTE, 1982. *Report of Basic Survey on Coastal Area (Eastern Part of The Toyama Bay).* Geographical Survey Institute Data D-3 **40**, 224 pp (in Japanese).

HARMS, J. C., SOUTHARD, J. B. & WALKER, R. G. 1982. *Structures and Sequences in Clastic Rocks.* Society of Economic Paleontologists and Mineralogists Short Course **9**, 3-44–45.

HAYES, M. O. & MICHEL, J. 1982. Shoreline sedimentation within a forearc embayment, Lower Cook Inlet, Alaska. *Journal of Sedimentary Petrology* **52**, 251–263.

HEMPTON, M. R., DUNNE, L. A. & DEWEY, J. E. 1982. Sedimentation in an active strike-slip basin, southeastern Turkey. *Journal of Geology* **91**, 401–412.

HEWARD, A. P. 1978. Alluvial fan sequence and megasequence models: with example from West-

phalian D–Stephanian B coalfields, northern Spain. *In*: MIALL, A. D. (ed.) *Fluvial Sedimentology.* Canadian Society of Petroleum Geologists Memoir **5**, 669–702.

HOLMES, A. 1965. *Principles of Physical Geology* (2nd edn). Roland Press, New York, 1288 pp.

HUCHON, P. & KITAZATO, H. 1984. Collision of the Izu Block with central Japan during the Quaternary and geological evolution of the Ashigara area. *Tectonophysics* **110**, 201–210.

ITO, M. 1985. The Ashigara Group: a regressive submarine fan–fan delta sequence in a Quaternary collision boundary, north of Izu Peninsula, central Honshu, Japan. *Sedimentary Geology* **45**, 261–292.

—— 1987. Middle to late Miocene foredeep basin successions in an arc–arc collision zone, northern Tanzawa Mountains, central Honshu, Japan. *Sedimentary Geology* **54**, 61–91.

—— & MASUDA, F. 1986a. Evolution of clastic piles in an arc–arc collision zone: Late Cenozoic depositional history around the Tanzawa Mountains, central Honshu, Japan. *Sedimentary Geology* **49**, 223–259.

—— & —— 1986b. Progradation and syndepositional deformation of the Late Cenozoic basin fills around the Tanzawa Mountains, central Honshu, Japan. *Earth Monthly* **8**, 616–620 (in Japanese).

—— & —— 1988. Late Cenozoic deep-sea to fan-delta sedimentation in an arc–arc collision zone, central Honshu, Japan: sedimentary response to varying plate-tectonic regime. *In*: NEMEC, W. & STEEL, R. J. (eds) *Fan-deltas: Sedimentology and Tectonic Settings.* Blackie, Glasgow, 425–443.

KAKIMI, T., YAMAZAKI, H., SANGAWA, A., SUGIYAMA, Y., SHIMOKAWA, K. & OKA, S. 1982. *Neotectonic Map of Tokyo, 1:500 000.* Geological Survey of Japan.

KANO, K., SUZUKI, I. & KITAZATO, H. 1985. Paleogeography of the Shizukawa Group in the Nakatomi area in the upper reaches of Fuji River, South Fossa Magna region, central Japan. *Geoscience Report of Shizuoka University* **11**, 135–153 (in Japanese with English abstract).

KENYON, P. M. & TURCOTT, D. L. 1985. Morphology of a delta prograding by bulk sediment transport. *Geological Society of America Bulletin* **96**, 1457–1465.

KITAZATO, H. 1986. Palaeogeographic changes in the South Fossa Magna. *Earth Monthly* **10**, 605–611 (in Japanese).

KOCHEL, R. G. & JOHNSON, R. A. 1984. Geomorphology and sedimentology of humid-temperate alluvial fans, central Virginia. *In*: KOSTER, E. H. & STEEL, R. J. (eds) *Sedimentology of Gravels and Conglomerates.* Canadian Society of Petroleum Geologists Memoir **10**, 109–122.

KONDO, Y. 1985. Stratigraphy of the Upper Pleistocene in the Udo Hills, Shizuoka Prefecture, Japan. *Journal of the Geological Society of Japan* **91**, 121–140 (in Japanese with English abstract).

—— 1986. Shallow marine gravelly deltas and associated faunas from the Upper Pleistocene Negoya

Formation, Shizuoka, Japan. *Journal of the Faculty of Science, University of Tokyo* **21**, 169–190.

LINK, M. H., ROBERTS, M. T. & NEWTON, M. S. 1985. Walker Lake Basin, Nevada: an example of Late Tertiary (?) to recent sedimentation in a basin adjacent to an active strike-slip fault. *In*: BIDDLE, K. T. & CHRISTIE-BLICK, N. (eds) *Strike-slip Deformation, Basin Formation, and Sedimentation.* Society of Economic Paleontologists and Mineralogists Special Publication **37**, 105–125.

MANSPEIZER, W. 1985. The Dead Sea rift: impact of climate and tectonism on Pleistocene and Holocene. *In*: BIDDLE, K. T. & CHRISTIE-BLICK, N. (eds) *Strike-slip Deformation, Basin Formation, and Sedimentation.* Society of Economic Paleontologists and Mineralogists Special Publication **37**, 143–158.

MATSUDA, T. 1961. The Miocene stratigraphy of the Fuji River Valley, central Japan. *Journal of the Geological Society of Japan* **67**, 79–96 (in Japanese with English abstract).

MATSUSHIMA, Y. 1982. Molluscan fauna from the middle and upper parts of the Ashigara Group. *Memoirs of the Natural Science Museum, Tokyo* **15**, 53–62 (in Japanese with English abstract).

MATYAS, E. L. 1984. Profiles of modern prograding deltas. *Canadian Journal of Earth Science* **21**, 1156–1160.

McGOWEN, J. H. 1970. Gum Hollow fan delta, Nueces Bay, Texas. *Bureau of Economic Geology, University of Texas at Austin, Report of Investigation* **69**, 91 pp.

MIALL, A. D. 1981. Alluvial sedimentary basins: tectonic setting and basin architecture. *In*: MIALL, A. D. (ed.) *Sedimentation and Tectonics in Alluvial Basins.* Geological Association of Canada Special Paper **23**, 1–33.

—— 1984a. *Principles of Sedimentary Basin Analysis.* Springer, New York, 490 pp.

—— 1984b. Deltas. *In*: WALKER, R. G. (ed.) *Facies Models.* Geoscience of Canada Reprint Series **1**, 105–118.

NEMEC, W. & STEEL, R. J. 1987. Convenors' address: What is a fan delta and how do we recognize it? *In*: NEMEC, W. (ed.) *International Symposium, Fan Deltas—Sedimentology and Tectonic Settings, Bergen, Norway.* Blackie, Glasgow, 11-7 (abstract).

NIITSUMA, N. & MATSUDA, T. 1984. Collision in the South Fossa Magna area, central Japan. *Recent Progress of Natural Sciences in Japan* **9**, 41–50.

OGAWA, Y. & HORIUCHI, K. 1978. Two types of accretionary fold belts in central Japan. *Journal of Physics of the Earth, Supplement* **26**, 321–336.

POSTMA, G. 1984. Mass-flow conglomerates in a submarine canyon: Abrioja fan-delta, Pliocene, southeast Spain. *In*: KOSTER, E. H. & STEEL, R. J. (eds) *Sedimentology of Gravels and Conglomerates.* Canadian Society of Petroleum Geologists Memoir **10**, 237–258.

—— & ROEP, T. B. 1985. Resedimented conglomerates in the bottomsets of Gilbert-type gravel deltas. *Journal of Sedimentary Petrology* **55**, 874–885.

PRIOR, D. B. & SUHAYDA, J. M. 1979. Submarine landslides—geometry and nomenclature. *Zeitschrift für Geomorphologie* **23**, 415–426.

——, WISEMAN, W. J. JR. & GILBERT, R. 1981. Submarine slope processes on a fan delta, Howe Sound, British Columbia. *Geo-Marine Letters* **1**, 85–90.

RACHOCKI, A. 1981. *Alluvial Fans.* Wiley, New York, 161 pp.

REINECK, H.-E. & SINGH, I. B. 1980. *Depositional Sedimentary Environments* (2nd edn). Springer, New York, 549 pp.

RUST, B. N. & KOSTER, E. H. 1984. Coarse alluvial deposits. *In*: WALKER, R. G. (ed.) *Facies Models.* Geoscience of Canada Reprint Series **1**, 53–69.

RYDER, J. M. 1971. The stratigraphy and morphology of paraglacial alluvial fans in south-central British Columbia. *Canadian Journal of Earth Science* **8**, 279–298.

SATO, T. 1985. Geology of Suruga Bay. *In*: COASTAL OCEANOGRAPHY RESEARCH COMMITTEE (ed.) *Coastal Oceanography of Japanese Islands.* Tokai University Press, 429–437 (in Japanese).

SHIMAMURA, K. 1986. Topography and geological structure in the bottom of the Suruga trough: a geological consideration of the subduction zone near the collision plate boundary. *Journal of Geography* **95**, 317–338 (in Japanese with English abstract).

SNEH, A. 1979. Late Pleistocene fan-deltas along the Dead Sea rift. *Journal of Sedimentary Petrology* **49**, 547–552.

SOH, W. 1986. Reconstruction of Fujikawa trough in Mio-Pliocene age and its geotectonic implication. *Memoirs of the Faculty of Science, Kyoto University, Geology and Mineralogy Series* **52**, 1–68.

STANLEY, K. O. & SURDAM, R. C. 1978. Sedimentation on the front of Eocene Gilbert-type deltas, Washakie Basin, Wyoming. *Journal of Sedimentary Petrology* **48**, 557–573.

SUGITA, M. 1962. The fossils and their habitats in the Neogene Katsuragawa Group in the south Kanto region. *Transactions of the Palaeontology Society of Japan, New Series* **47**, 281–290.

WESCOTT, W. A. & ETHRIDGE, F. G. 1980. Fan-delta sedimentology and tectonic setting—Yallahs fan delta, southeast Jamaica. *American Association of Petroleum Geology Bulletin* **64**, 374–399.

WINKER, C. D. & EDWARDS, M. B. 1983. Unstable progradational clastic shelf margins. *In*: STANLEY, D. J. & MOORE, G. T. (eds) *The Shelfbreak: Critical Interface on Continental Margins.* Society of Economic Paleontologists and Mineralogists Special Publication **33**, 139–159.

WRIGHT, L. D. 1977. Sediment transport and deposition at river mouths. *Geological Society of America Bulletin* **88**, 857–868.

—— & COLEMAN, J. M. 1973. Variations in morphology of major river deltas as functions of ocean wave and river discharge regimes. *American Association of Petroleum Geologists Bulletin* **57**, 370–398.

YANO, S. 1986. Stratigraphy, depositional environment and geologic ages of the southern part of the Oiso Hills, Kanagawa Prefecture. *Geoscience Reports of Shizuoka University* **12**, 191–208 (in Japanese with English abstract).

YAZAWA, D., TOYA, H. & KAIZUKA, S. (eds) 1971. *Alluvial Fans*. Kokon Shoin, Tokyo, 318 pp (in Japanese).

MAKOTO ITO, Geological Institute, College of Arts and Sciences, Chiba University, Chiba 260, Japan.

Subsurface and Geophysical
Techniques

Diagenesis of Upper Carboniferous sandstones: southern North Sea Basin

G. Cowan

SUMMARY: Upper Carboniferous sediments from the central part of the southern North Sea Basin were deposited on a broad fluvio-deltaic plain. Potential reservoir rocks occur within distributary channel facies sandstones, but porosity and permeability have been reduced by authigenic mineral growth and improved by at least two separate phases of porosity enhancement: firstly by the infiltration of meteoric waters during post-Carboniferous uplift, and secondly by the action of acidic fluids generated during burial. The depth of invasion of meteoric fluids is highly variable, ranging from a few metres to hundreds of metres, and the best porosity is preserved in sediments which have undergone both phases of porosity enhancement. Virtually all the observed porosity in Carboniferous sandstones can be interpreted as having a secondary origin. Integrating the deduced diagenetic history with a burial and temperature history for specific wells allows the timing of diagenetic events to be estimated.

An outline of the stratigraphy of the Carboniferous is shown in Fig. 1.

It has long been recognized that the source of gas discovered in Rotliegendes and Bunter sandstones in the southern North Sea Basin is in deeply buried organic-rich mudrocks and coals of Carboniferous age, and gas shows have frequently been recorded from Carboniferous sandstones. Although oil has been produced from the Namurian and Westphalian fluvio-deltaic sediments of the E Midlands since the end of the First World War, rocks of similar age from the southern North Sea Basin have poor reservoir qualities in comparison with the overlying Rotliegendes sandstones (Glennie 1986). Consequently, the Carboniferous has been considered to be economic basement in the basin, and most exploration wells halt drilling when rocks of Carboniferous age are encountered. Rotliegendes sandstones are absent from the central and northern part of the basin (Marie 1975). This fact, combined with the recognition of widespread secondary porosity development (Schmidt & McDonald 1977) and the advances in understanding of basin evolution and thermal maturation modelling (MacKenzie 1978; Waples 1980), has led to a re-evaluation of the reservoir potential of the Upper Carboniferous.

The work presented here is based upon 'in-house' work carried out to determine the controls on reservoir potential of Upper Carboniferous fluvio-deltaic sediments in an attempt to produce a working diagenetic model for central southern North Sea hydrocarbon exploration.

Major Subdivisions of the Carboniferous

SUB SYSTEM	SERIES	STAGES
SILESIAN	STEPHANIAN	STEPHANIAN C B A
		CANTABRIAN
	WESTPHALIAN	WESTPHALIAN D
		WESTPHALIAN C
		WESTPHALIAN B
		WESTPHALIAN A
	NAMURIAN	YEADONIAN
		MARSDENIAN
		KINDERSCOUTIAN
		ALPORTIAN
		CHOKIERIAN
		ARNSBERGIAN
		PENDLEIAN
DINANTIAN	VISEAN	BRIGANTIAN
		ASBRIAN
		HOLKERIAN
		ARUNDIAN
		CHADIAN
	TOURNASIAN	COURCEYAN

FIG. 1. Carboniferous stratigraphy.

Data base

A total of 580 m of core from six wells was logged at a 1:50 scale to determine the depositional facies, and 142 samples were taken for thin-section analysis. Each thin section was stained for carbonates with alizarin red S and potassium ferricyanide, and for K-feldspar with sodium

From WHATELEY, M. K. G. & PICKERING, K. T. (eds), 1989, *Deltas: Sites and Traps for Fossil Fuels,*
Geological Society Special Publication No. 41, pp. 57–73.

57

cobaltinitrite, prior to point counting (200 points) to determine the modal percentages of the detrital and authigenic phases present as well as the visible porosity. Selected samples were studied using a scanning electron microscope equipped with a semiquantitative energy dispersive analyser in both the secondary electron and back-scattered electron imaging modes to allow determination of the clay and authigenic mineral morphologies. Whole-rock and clay mineral X-ray diffraction (XRD) was carried out on selected sandstone samples and five mudstone samples to determine the clay mineralogy.

Fifteen thin sections were studied using a cold-cathode luminescence system.

Depositional environments

During the Westphalian, sedimentation occurred on a broad upper delta plain between the Grampian–Fenno-Scandian high to the N and the newly uplifted Rheno-Hercynian zone to the S. The bulk of the sediment was derived from the N, although sediment was supplied locally from the London–Brabant and Welsh Massifs (Fig. 2).

A summary log showing the aggregate depositional facies of Westphalian rocks encountered in the study area is shown in Fig. 3. In general, the sediments are similar to those described in the Durham Coalfield by Fielding (1984) and on the Northumberland coast by Haszeldine & Anderton (1980). Fielding's depositional model can be applied to the bulk of the Westphalian sediments encountered in the central part of the southern North Sea Basin with the addition of a pebbly braided distributary channel facies.

During the Westphalian, deposition was predominantly by overbank flooding and crevassing into interdistributary bays and lakes, producing argillaceous sequences which often coarsen upwards. Thin sands interbedded with mudstones were deposited in distal crevasse-splay and overbank settings, and proximal crevasse splays (minor mouth bars) produce upward-coarsening fine- to medium-grained sand bodies, often cut into and overlain by distributary channel sandstones. The channel sandstones have distinctive log motifs, forming blocky or upward-fining sequences, and were deposited in shoestring and sheet-like geometries. The distributary channel sand bodies have the best reservoir potential. These sands show variable thickness and internal structures but can form stacked multistorey channel bodies over 30 m thick. Grain size is highly variable, ranging from fine grained to microconglomeratic (Fig. 3).

Present-day porosities are highly variable, ranging from zero to 15%, with correspondingly low permeabilities, ranging from zero to a few hundred millidarcies.

Detrital mineralogy

Representative point-count data for Carboniferous distributary channel sandstones from the study area are presented in Table 1. Table 2 shows quantitative whole-rock XRD data from five representative shale samples.

Quartz

Detrital quartz grains are predominantly mono-crystalline and show weak to strong strain extinction, with Boehm lamellae in some samples. Polycrystalline grains account for 10%–30% of the detrital quartz population. Where it is possible to observe the shape of the detrital grains, they are usually well rounded, but the shape is masked by later overgrowths and replacive cementation. Inclusions of rutile and zircon(?) are common. Cathodoluminescence (CL) photomicrographs show violet, dull grey and grey–brown emission colours, suggestive of derivation from a mixed igneous–metamorphic terrain.

Feldspar

Untwinned albite was the dominant feldspar type observed in the samples with minor amounts (10% of the feldspar population) of twinned plagioclase. K-feldspar was recorded in trace amounts in a few samples only. Feldspar shows replacement by siderite and dolomite and its abundance is severely reduced in rocks which have undergone flushing by meteoric waters during the pre-Permian uplift phase, regardless of the stratigraphic age of the sediments. It is therefore probable that detrital feldspar suffered dissolution by meteoric fluids introduced at this time. Twinned plagioclase grains often show fracturing due to compaction but appear relatively fresh. Owing to its limited abundance, no attempt was made to estimate twinned plagioclase composition. Untwinned feldspars are often inclusion rich and show incipient (hydrothermal?) alteration to illite. The limited composition of the feldspar grains is typical of an authigenic origin. Hawkins (1978) and Huggett (1984) have shown that the dominant feldspar type in the less deeply buried Carboniferous sandstones of the E Midlands is alkali feldspar. Walker (1984) has recorded the diagenetic alteration of alkali feld-

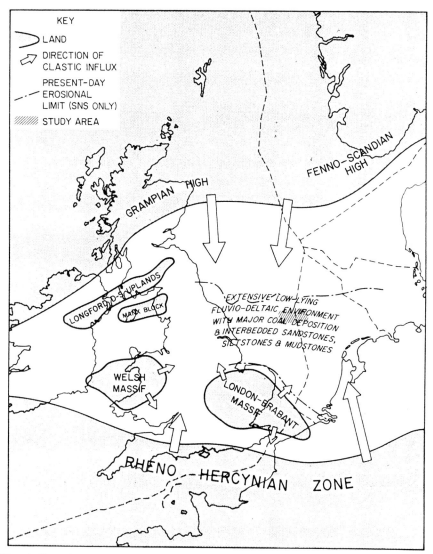

FIG. 2. Westphalian palaeogeography and location map. (From Johnson 1982; Ziegler 1982.)

spar to albite in Tertiary sandstones from the Gulf Coast of the USA. Such alteration is thought to have occurred through an intermediate phase of replacement by carbonate or evaporite cements, and the result of alteration can produce grain textures similar to those found in microcline grains. Walker (1984) noted a progressive increase in albite modal percentages at temperatures in excess of 100 °C. It is probable that the dominance of K-feldspar in the Carboniferous rocks of the E Midlands is a reflection of the shallower burial of the E Midlands sediments.

Mica

Fresh muscovite is the dominant mica type with traces of highly altered biotite. Muscovite exhibits a variety of stages of alteration and deformation, ranging from fresh straight blades to buckled grains showing alteration to kaolinite. In some samples very large (30 μm) kaolinite crystals, pseudomorphing muscovite, can be seen. Splitting and fanning of grains also occurs, mainly in sediments which have undergone deep Carboniferous burial. In argillaceous sandstones, where

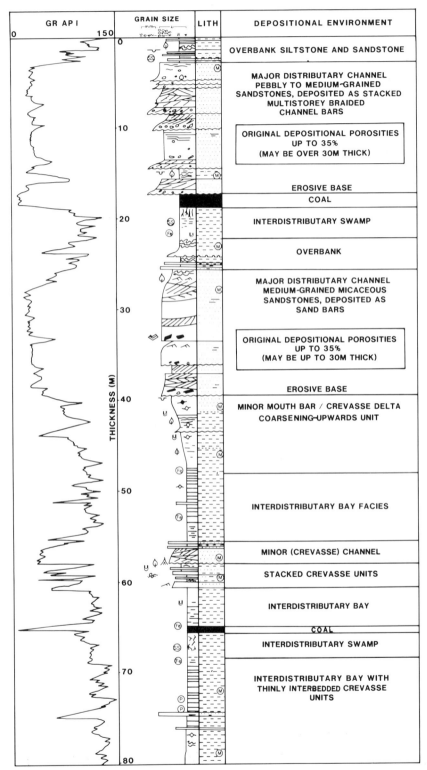

FIG. 3. Typical Upper Carboniferous facies associations encountered in the southern North Sea Basin, based on Westphalian A to C cored sections.

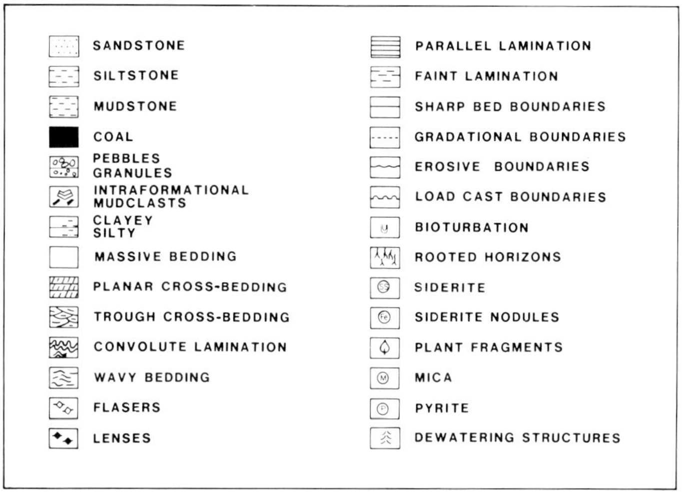

SANDSTONE		PARALLEL LAMINATION	
SILTSTONE		FAINT LAMINATION	
MUDSTONE		SHARP BED BOUNDARIES	
COAL		GRADATIONAL BOUNDARIES	
PEBBLES GRANULES		EROSIVE BOUNDARIES	
INTRAFORMATIONAL MUDCLASTS		LOAD CAST BOUNDARIES	
CLAYEY SILTY		BIOTURBATION	
MASSIVE BEDDING		ROOTED HORIZONS	
PLANAR CROSS-BEDDING		SIDERITE	
TROUGH CROSS-BEDDING		SIDERITE NODULES	
CONVOLUTE LAMINATION		PLANT FRAGMENTS	
WAVY BEDDING		MICA	
FLASERS		PYRITE	
LENSES		DEWATERING STRUCTURES	

KEY TO FIG. 3

TABLE 1. *Point-count data*

	Sample								
	A	B	C	D	E	F	G	H	I
Quartz	67.0	52.0	71.5	67.0	59.0	63.5	75.5	43.5	68.0
Feldspar*		0.5			5.0	3.0	3.5	4.0	3.5
Mica	1.5	10.5	2.0	0.5	6.0	1.5	2.5	5.7	2.5
Lithics	2.0	4.0	2.0	3.0	15.5	3.5	2.0	1.0	2.0
Opaques					0.5		0.5		
Organic matter						2.0		1.0	
Matrix clays		15.0							
Authigenic quartz	9.5	N/C	N/C	N/C	N/C	N/C	N/C	N/C	N/C
Haematite	1.0	16.5	1.5	7.0					
Anhydrite– barite				2.0					
Ferroan dolomite			3.5	0.5	1.0	11.0	6.0	18.0	1.5
Dolomite	5.0	2.0		1.5	0.5		1.5		
Siderite	2.0				2.0	2.0	1.0	20.0	4.0
Kaolinite	7.0		9.5	5.5	6	8.0	4.0	5.0	17.0
Illite	1.0		2.5	1.0	3	2.0	0.5	3.0	2.0
Others				0.5†					
Primary porosity						1.0			
Secondary porosity	9.5		7.0	12.0	1.0	?	3.0	—	0.5
Grain size	Coarse	Fine	Medium	Medium	Medium	Medium	Coarse	Very fine	Medium
Sorting	Moderate	Very poor	Moderate	Moderate	Poor	Moderate	Poor	Good	Moderate

Samples A, B, C and D come from the zone which has been affected by meteoric flushing during post-Carboniferous uplift.
N/C, not counted.
*Dominantly albite.
†Zircon clay.

TABLE 2. *Semiquantitative whole-rock X-ray diffraction of five representative shale samples*

Sample	Mica–illite–MLC	Kaolin	Anhydrite	Albite	Siderite	Quartz	Total	Illite (Kubler)	Crystallinity (Weaver)
A	28	37	1	3	1	35	(104)	1.17	1.26
B	37	14	4	4	5	30	(94)	0.83	2.02
C	28	30	2	11	5	20	(96)	1.52	1.03
D	43	5	4	9	9	30	(100)	0.87	1.62
E	50	4	2	6	7	30	(99)	0.91	1.32

In addition all samples contained 1%–2% calcite and (except for C) 1%–2% orthoclase. C has 3% chlorite.

compaction has been allowed to take place by ductile deformation of detrital clays, micas are usually more highly deformed and altered.

Rock fragments

Metamorphic rock fragments consist of quartz schist grains, quartz–mica phyllites and mica phyllites. The quartz schist grains are stable and show little deformation or alteration, whereas the mica schist and phyllite grains show varying degrees of deformation, alteration (mainly to kaolinite and illite) and in some cases dissolution. Ductile deformation of these lithics is common.

Intraformational argillaceous rip-up clasts were the dominant sedimentary rock fragments observed in the sandstones studied, but minor amounts of siltstone were also observed. The claystone rip-up clasts are concentrated at the channel bases and have undergone early diagenetic alteration to siderite.

Other detrital grains

Apatite occurs as large framework-sized grains, not obvious in thin section but clearly observed in the backscatter mode of the scanning electron microscope. Apatite is derived from acid igneous rocks and occurs as an accessory mineral in limestones. Others include zircon, rutile and green pleochroic tourmaline as well as scattered opaque minerals.

Organic matter

Organic matter occurs associated with mica in argillaceous laminae within the sandstones, and is more abundant in crevasse and overbank sandstones. Organic matter is not present in rocks which have undergone secondary reddening, having been oxidized during uplift. It occurs as brown–black lenses which have been compacted between framework grains and also disseminated throughout the crevasse and overbank facies. Discrete plant fragments are often observed.

Provenance

The presence of the accessory minerals noted above, together with the quartz CL data, strongly suggest a plutonic igneous source, whereas the presence of metamorphic rock fragments indicates a metamorphic source. It is therefore more likely that the sediments were derived from the Fenno-Scandian shield to the N than from the Variscan forelands to the S.

Authigenic mineralogy

Authigenic quartz

Authigenic quartz occurs to varying degrees in all the samples studied. Since dust rims were not preserved in most samples, it was not generally possible to quantify the amount of authigenic quartz using the polarizing microscope. Where dust rims are preserved, up to 21% of the rock volume can be seen to consist of authigenic quartz, and CL photomicrographs show an authigenic quartz content of up to 20% (Plate 1 D, F). Euhedral quartz overgrowths are best developed in coarse-grained porous channel sands immediately below the Permian erosion surface. They are least well developed in fine-grained argillaceous crevasse sands, where the compaction of detrital clays has inhibited the formation of quartz overgrowths, and in rare cases where haematite or clay rims have inhibited overgrowths.

CL work shows that a three-stage zonation of quartz overgrowths is most common, but, owing to distinctive luminescence, up to six zones can be picked out, with up to five being recorded in any one sample. The zonation can be caused by either variations in rate of growth or pore fluid chemistry or by discrete phases of authigenesis separated by periods of non-growth.

Quartz overgrowths show euhedral faces in empty pores (Fig. 4C) but have interlocking relationships with carbonates and kaolinite. Quartz overgrowths have clearly had a long history in the samples studied. They occur as one of the earlier recorded cements and as a late-stage cement. It is arguable that quartz precipitation can be considered as an essentially continuous process from early diagenesis to deepest burial, only interrupted by the precipitation of other cements.

Early siderite

Early siderite cement occurs as pore-filling microcrystalline masses associated with argillaceous laminae in the channel sandstones but as more continuous laminae and rare spherulitic forms in the overbank and crevasse sandstones (Plate 2A).

Mudstone clasts are also extensively siderite cemented, and CL work has shown that some discrete calcite is also present in these clasts;

however, whole-rock XRD shows that calcite is extremely rare. Early siderite is conspicuous in hand specimens, often picking out cross-lamination, and can be seen as oversized uncompacted masses in thin section. Its abundance is severely reduced in samples which have undergone secondary reddening, where haematite is present. Where present in such rocks, it occurs as skeletal partially dissolved crystals (Plate 3D).

Pyrite

Small amounts of apparently detrital opaque minerals were observed in most samples. Subsequent analysis showed that the bulk of it was pyrite, and therefore it was probably authigenic in origin. It occurs in masses as an early cement

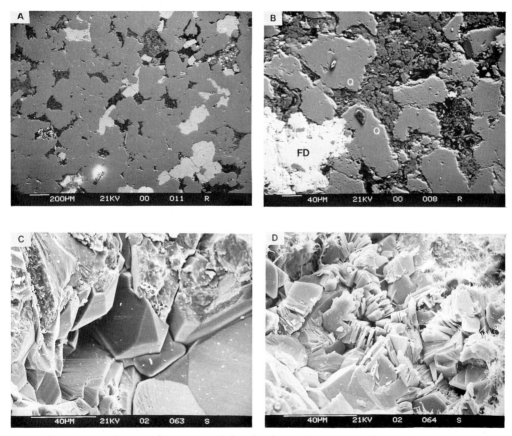

FIG. 4. (A) Backscattered electron image: general view showing patchy unconnected porosity distribution (dark grey) and carbonate cements (light grey). (B) Backscattered electron image: etched ferroan dolomite (FD) and quartz (Q) on the edge of a secondary pore system which has been partially infilled by kaolinite. (C) Secondary electron image: interlocking quartz overgrowths in an open pore. Note the stepped crystal morphology caused by etching (bottom centre). (D) Secondary electron image: well-crystallized pore-blocking kaolinite books, intergrown with quartz. Such intergrowths can be caused by either later or syngenetic precipitation of quartz, or by aggressive reactions between the kaolinite and quartz. In some cases kaolinite can be seen to impinge upon the detrital quartz core.

and is also seen rimmed by haematite and siderite. Whole-rock XRD shows that pyrite is at the limit of detection and therefore is not an abundant cement. The availability of sulphate is low in non-marine sediments. Under reducing conditions iron precipitates as siderite after the sulphate is consumed by pyrite precipitation.

Haematite

Haematite occurs as grain-rimming crystals and as polycrystalline masses associated with clay minerals and rare calcite masses (identified by CL, but not in thin section). It is found in a few samples as an early diagenetic cement which precipitated directly onto the detrital grain surfaces. The more common occurrence of haematite is as patchily distributed masses and partial replacement of an earlier siderite cement (Plate 3A).

These replacive haematite cements formed as a consequence of meteoric flushing during the late Carboniferous–early Permian uplift by replacement of siderite and pyrite, and post-date the earliest phase of quartz overgrowth and early siderite precipitation. Haematite is absent from sediments which have not undergone this flushing, except in pebbly braided channel sandstones from one well which contains an early diagenetic grain-rimming form which precipitated directly on the grain surfaces with no earlier cement precursor.

Non-ferroan dolomite

Non-ferroan dolomite precipitated in poikilotopic and rhombic pore-filling forms and often overgrows haematite. In some cases the haematite is seen rimming an earlier non-ferroan dolomite cement. It therefore appears that some dolomite was precipitated prior to uplift and the precipitation of haematite, but the bulk of it is demonstrably a post-haematite cement. This dolomite now has a patchy distribution but was probably more abundant prior to the formation of burial-related (mesogenetic) secondary porosity. It is overgrown by ferroan dolomite in some pores.

Non-ferroan dolomite is found both in sediments which have and in sediments which have not been exposed to meteoric fluids during the late Carboniferous uplift and therefore probably precipitated after this event.

Ferroan dolomite

At least two phases of ferroan dolomite precipitation occurred, one during early burial diagenesis and the other during late burial diagenesis.

Ferroan dolomite precipitation post-dates the formation of non-ferroan dolomite cements in most wells, but in others the relationship between ferroan and non-ferroan dolomite and the formation of secondary porosity is unclear. In most of the wells studied secondary porosity formation was achieved by the dissolution of ferroan dolomite (Plate 1A), but in other samples fresh ferroan dolomite rhombs can be seen occupying secondary pores (Plate 1B). Some samples show ferroan dolomite overgrowing haematite with the total reduction of porosity. It appears that ferroan and non-ferroan dolomite are syngenetic in other samples. High modal percentages of ferroan dolomite correspond to poikilotopic cements, which are more abundant in sediments which have not undergone secondary reddening and are therefore generally more common in rocks which have suffered deep Carboniferous burial.

Barite–anhydrite

Barite and anhydrite occur as pore-filling cements. Never abundant (less than 1%) and with patchy distribution, they can be seen to post-date the earliest phase of quartz precipitation. It is usually assumed that barite and anhydrite cements in the southern North Sea are precipitated from sulphate derived from Zechstein evaporites, but their stratigraphic position (hundreds of metres below the pre-Permian erosion surface) together with their patchy distribution suggest that the sulphate was derived from a local source, possibly by the oxidation of sulphides (pyrite) in the surrounding mudstones.

Authigenic feldspar

Authigenic albite was observed as overgrowths on detrital feldspars. It occurs in trace amounts, however, since it coats detrital grains; it appears to be common from scanning electron microscopy (SEM) examination.

Clay minerals

Clay minerals were identified by XRD of the fraction below 2 μm, and the crystal morphologies were identified using the secondary electron imaging mode of the scanning electron microscope. Typical diffractograms of the clay mineral fraction of the rocks studied are shown in Fig. 5.

Kaolinite

Kaolinite is conspicuous in almost all the samples. It occurs as patchily distributed, loosely packed

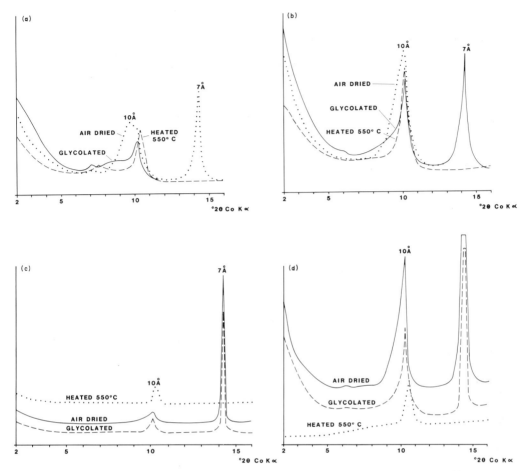

FIG. 5. X-ray diffractograms of the < 2 μm fraction of typical sandstone and shale samples. (*a*) Shale sample: the clay fraction contains kaolinite and illite/illite–smectite in comparable amounts. Traces of chlorite are also present. The presence of an expandable interstatified illite–smectite as well as discrete illite is indicated by the glycolated XRD trace. (*b*) Shale sample: kaolinite is indicated by the 7 Å peak which collapses on heating to 550 °C. Illite (10 Å) is also indicated. The 10 Å illite peak sharpens and shows some broadening on glycol solvation. This behaviour is caused by the presence of some interstatified smectite layers within the illite. (*c*) Sandstone sample: kaolinite is the dominant clay mineral (7 Å), and the small amount of illite is unaffected by glycolation or heating and is therefore pure 10 Å material. (*d*) Sandstone sample: the sharp 10 Å illite peak is unaffected by auxiliary treatments, and the 7 Å peak represents kaolinite.

pore-filling books, replacing micas, lithics and feldspar. SEM examination revealed that two morphologies are present: small poorly crystalline books (Fig. 6*A*), and a coarser-grained well-crystallized form having a higher Al:Si ratio and a morphology similar to that of dickite (Fig. 4*D*). This morphology can be seen to occupy secondary carbonate dissolution pores. As yet no detailed work has been done on the kaolinite distribution, but a conjectural hypothesis is that the poorly crystallized type formed from the alteration of feldspar during the uplift phase or during early

diagenesis. The 'dickite' form clearly precipitated after the formation of secondary porosity, since it occurs in secondary pores. In some samples the later form of kaolinite effectively infills macro-porosity, leaving only microporosity between the individual books.

Illite

Illite occurs as a detrital mineral in rocks with abundant detrital matrix clays and as an authigenic phase via the alteration of kaolinite, micas

FIG. 6. (*A*) Secondary electron image: small poorly crystalline kaolinite books. This form is quite distinct from the coarsely crystalline books which can be seen to occupy secondary dissolution pores. This morphology is less abundant than the latter type and possibly formed during post-Carboniferous uplift from ions released by dissolution of feldspar grains. (*B*) Secondary electron image: fibrous illite in the pore bridging mode. Illite is not as abundant as kaolinite and its morphology is highly variable. It appears to have precipitated late in association with rhombic dolomite. (*C*) Secondary electron image: platy illite on the pore walls giving way to hairy illite in the pore centre. Such hairy forms are highly susceptible to damage during sample preparation. It is probable that the hairy morphology has collapsed onto the pore walls and was more extensively developed *in situ*. (*D*) Secondary electron image: detail of illite morphology. It appears that the thickness of the illite fibres and laths decreases towards the pore centres, and therefore that the hairy forms precipitated last.

and feldspar, as well as a direct precipitate. Its morphology is highly variable, with bladed and platy forms present as well as box work and rare hairy forms (Fig. 6). As a replacement of kaolinite minerals, it occurs as a late cement, but the age of the mica illitization is less certain. Clay-fraction XRD shows this illite to have only a minor mixed-layer component, which is too small to be quantified. Weaver illite crystallinity indices (Weaver 1960) are consistently high (1.5–2.3), commensurate with the deep burial (over 4000 m) to which the rocks have been subjected.

Associated sediments

Five representative shale samples were studied using XRD to determine the stage of clay mineral diagenesis to which the rocks had been subjected. Quantitative whole-rock XRD (Table 2 and Fig. 5) showed that the mineral assemblage of the clay fraction is dominated by illite and kaolinite, with traces of illite–smectite. The illite in the shales has a high Weaver crystallinity index (varying from 1.26 to 2.02), which is lower, however, than those recorded in the sandstones. The lower values probably reflect the fact that the illite in the shales was mainly derived by transformation of a pre-existing smectite or mixed-layer clay mineral, whereas a significant proportion of the illite in the sandstones formed by direct precipitation during the deepest period of burial and is likely to have a composition closer to that of muscovite (Woodward & Curtis 1988) and to show a higher Weaver crystallinity index.

Maturity modelling and diagenesis

Curtis (1978, 1983) and Surdam *et al.* (1984) have described the connections between the maturation of organic matter in mudrocks and sandstone diagenesis. In particular, it has generally been accepted that the maturation of organic matter and the transformation of clay minerals have a profound effect on sandstone diagenesis by producing fluids and ions which migrate into the pore systems in the sandstones. The study of the maturity of associated organic-rich rocks therefore has significant implications for any diagenetic study.

Curtis (1983) has shown that carbonate supersaturation can occur by balanced combinations of sulphate and iron reduction and of fermentation in mudrocks at temperatures of less than 80 °C; the latter mechanism favours iron-rich carbonate precipitation. Carothers & Kharaka (1980) and Surdam *et al.* (1984) have shown that bacterial activity is halted at temperatures in excess of 80 °C and there is a consequent increase in the abundance of organic acid complexes as well as acids produced by thermal decarboxylation of organic matter.

The 80 °C temperature coincides with the onset of smectite-to-illite transformations as well as the very beginning of hydrocarbon generation (Boles & Franks 1979; Hunt 1979). Organic acid complexes (Surdam *et al.* 1984) significantly increase the solubilities of silicate minerals and allow the relatively insoluble alumina to be carried in solution, whilst acids produced by thermal decarboxylation have long been postulated as an agent of carbonate dissolution (Schmidt & McDonald 1979).

The 80 °C isotherm marks the temperature at which the pore fluids in Carboniferous sediments changed from being carbonate saturated to being acidic and carbonate and silicate aggressive, since it defines the temperature at which the abundance of acid-consuming bacteria decreases. The timing of this change can be used to estimate the beginning of secondary porosity development, and this can be related to hydrocarbon generation by using maturity modelling techniques. For this reason, it is vital to relate the diagenetic history deduced from the study of the sediments to the maturity of the surrounding rocks. Maturity modelling gives an indication of the temperature history of the well and has the advantages that it can be carried out quickly and gives a result which can be converted to allow comparison with vitrinite reflectance data.

The maturity modelling used in this study was carried out using the Lopatin time–temperature index (TTI) technique described by Waples (1980), and the dimensionless TTI values are converted to equivalent vitrinite reflectance values R_0(est) for comparison with measured values. For the purposes of this study, a TTI of 75 was taken to be equivalent to an R_0 of 1%. Most published studies of maturity take the present-day geothermal gradient and extrapolate this back through time, but extensional models of basin evolution (MacKenzie 1978) have shown that, during periods of rapid sediment accumulation associated with crustal thinning, palaeo-geothermal gradients are likely to have been considerably higher than at the present day.

Leeder (1982) has postulated that Carboniferous basin formation was the result of crustal thinning, and Fisher (1986) has shown that the Triassic was a period of rifting and rapid sediment accumulation. The present-day geothermal gradients of the southern part of the southern North Sea vary from 20 to 30 °C km^{-1}, somewhat lower than the present-day global average of around 30 °C km^{-1}. The geothermal gradient of the study area is 25 °C km^{-1} (Harper 1974). In rapidly subsiding rift basins the geothermal gradients are higher than the global average, varying from 40 to 50 °C km^{-1} (Perodon & Masse 1984). A palaeogeothermal gradient of 45 °C km^{-1} was taken to be representative of the period of Triassic rifting, decaying exponentially to the present-day geothermal gradient of 25 °C km^{-1}. A slightly lower figure of 35 °C km^{-1} was taken to be representative of the Carboniferous, since sediment accumulation rates were lower than those of the Triassic.

The thermal model presented here is very simplistic, using only linear geothermal gradients chosen for three key periods, but it does offer a significant improvement over extrapolating the present-day gradients into the past. More sophisticated models can be constructed, but the above model gives good agreement between estimated and measured vitrinite reflectance values for specific wells. Figure 7 shows the burial history plot of a well in the study area with isotherms superimposed (broken curves). The burial history of the top Carboniferous strata is shown as a thick open line. The amount of post-Permian

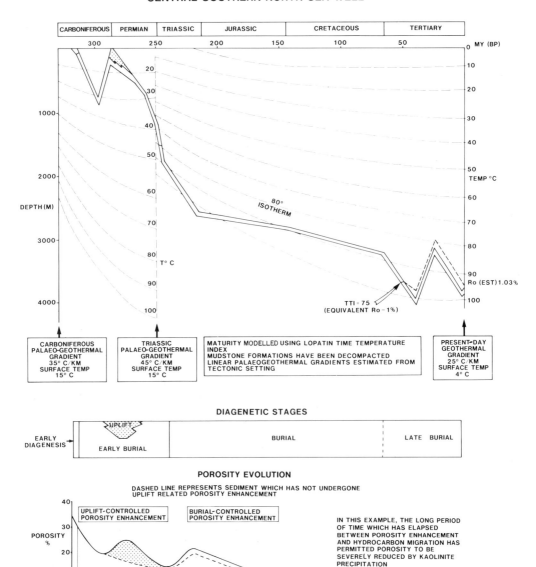

**BURIAL / MATURITY MODEL FOR TOP CARBONIFEROUS STRATA :
CENTRAL SOUTHERN NORTH SEA WELL**

FIG. 7. Burial history and thermal maturation plot.

uplift was calculated using the Bunter shale velocity method described by Marie (1975) and updated by Cope (1987).

The amount of post-Carboniferous uplift was estimated by subtracting the thickness of Westphalian section present in any well from an estimated total Westphalian thickness of 1000 m over the area. The iso-TTI value of 75, corresponding to a vitrinite reflectance value of 1% (gas generation), is plotted as a bold broken line in Fig. 7. The stippled area in Fig. 7 represents sediments which have been exposed to meteoric fluids during uplift and have undergone two periods of porosity enhancement. This plot allows the depth and temperature to which sediments have been subjected to be estimated from the time of deposition to the present day.

An estimate of the porosity evolution and the diagenetic stages are shown below the burial plot.

Diagenetic history

Five diagenetic zones can be recognized on the basis of the relative timing of cement precipitation and the mineral assemblage present (Fig. 7). A schematic representation of the major diagenetic effects is shown in Fig. 8.

Early diagenesis

By analogy with modern tropical fluvial waters, the original pore fluids would have been oxidizing and weakly acidic upon deposition of the sandstones. The constant recharge due to fluid flow would have kept the distributary channel sandstones charged with oxidized acidic fluids for longer than the overbank and crevasse sands, which would only be charged during crevassing and flood events. The minor quartz overgrowths which can be seen to pre-date the development of early microcrystalline siderite were probably produced by the action of these pore waters on unstable detrital grains.

To allow the precipitation of siderite a neutral to alkali pH is required, and it is probable that quartz was precipitated during the increase in pH prior to the precipitation of siderite. The reduction of iron releases hydroxyl ions into the pore fluid which have the effect of increasing the pH (Curtis & Spears 1968). The presence of unoxidized organic matter in both the channel and overbank sandstones (Plate 2C) indicates that an inadequate supply of oxygenated water was fed to the sandstones to halt the establishment of reducing conditions, which were initiated by the oxidation of the abundant organic matter in the interdistributary facies mudstones and within the sandstones themselves.

Early microcrystalline siderite is much more common in the overbank and crevasse facies where it occurs in laminae associated with clays and organic matter and as replacement of intraformational mudclasts (Plate 2B, C). The early nature of these cements is highlighted by the fact that they have prevented the compaction of these intraformational mudclasts.

Early burial diagenesis

The boundary between early diagenesis and the early burial diagenetic phase is difficult to define.

The early siderite cements clearly pre-date the effects of compaction (which has bent and fractured mica and plagioclase and squeezed the organic matter and soft lithics between the grains), whilst the quartz and carbonate cements which characterize the early burial stage appear to have precipitated prior to and after compaction. CL photomicrographs show that there has been remarkably little mechanical or physiochemical compaction of the framework grains (Plate 2A). Normal thin-section analysis gives the impression of a highly compacted rock with interpenetrant grain contacts, but under CL it can be seen that this is caused by interpenetrant quartz overgrowths. The junctions between the overgrowths and the detrital cores are invisible in plane polarized light (Plate 1D, F). Up to six different zones of quartz overgrowth can be recorded in some samples.

It is probable that the quartz cementation was essentially a continuous process throughout diagenesis and the zones indicated by differing CL response represent variations in trace-element chemistry or pressure and temperature conditions rather than discrete phases of overgrowth. These quartz overgrowths have halted the compaction of the sediments after only minor compaction of the micas, soft lithics and organic matter.

Carbonate cements precipitated during this phase of diagenesis consist of non-ferroan and ferroan dolomite, although rarer siderite and calcite are present. These carbonates can account for up to 30% of the modal composition of the rock and may show replacive boundaries with the framework grains. They were probably produced by bacterial fermentation reactions in surrounding mudrocks.

Uplift diagenesis

Variscan deformation resulted in the uplift and erosion of Carboniferous strata prior to the deposition of the overlying Rotliegendes. The uppermost Carboniferous rocks were exposed to meteorically derived oxidizing fluids which caused the visible reddening of the top Carboniferous. This event oxidized the organic matter in the sediments and altered the early siderite to haematite. This uppermost 'red' Carboniferous section is often erroneously referred to as 'Barren Red Measures' on oil company logs since the oxidizing event also removed the palynomorphs used in biostratigraphic dating. The depth of reddening is variable, ranging from a few metres to over 200 m, and there seems to be a relationship between the amount of erosion and the depth of invasion of the oxidizing fluids. The most deeply eroded sections, with the oldest pre-Permian

70

G. Cowan

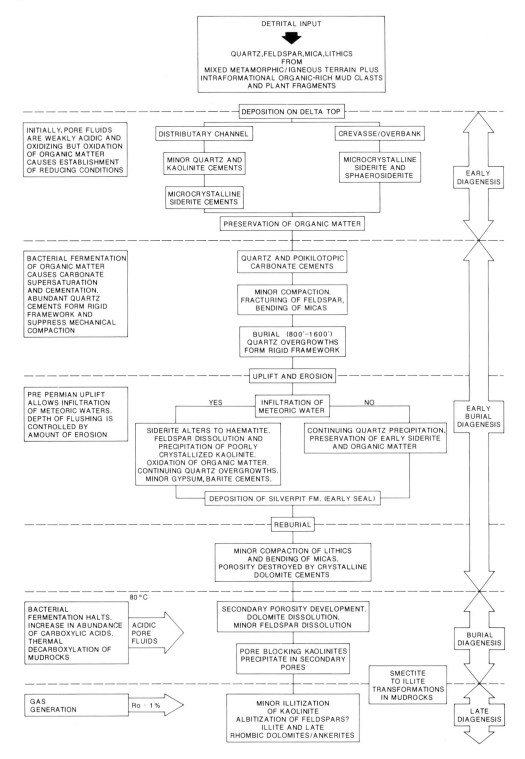

Fig. 8. Schematic diagram of the principal diagenetic reactions for the Upper Carboniferous sandstones of the central southern North Sea.

subcrop, show less reddening than younger sediments, possibly because the more deeply eroded sections were exposed to the fluids for a shorter time or because their permeabilities were reduced by Carboniferous burial, impeding the through-flow of meteoric waters. The lithology and orientation of the strata which form the pre-Permian subcrop also exert a control. Permeable sands show reddening to greater depths than mudstones, and steeply dipping strata allow a greater depth of invasion than flat-lying strata.

As well as the oxidation of organic matter and the alteration of early siderite to haematite, the oxidizing fluids also caused the dissolution of feldspars. It is not clear whether this dissolution was accompanied by the precipitation of kaolinite. Two distinct morphologies of kaolinite exist: a blocky late type, up to 20 μm in diameter, and a poorly crystalline form around 5 μm in diameter which may have been precipitated during the uplift event (Fig. 6A). The textural evidence for this origin, however, is unclear.

Carbonate dissolution also occurred at this time, and it is probable that porosity was increased by over 10% in some cases. During subsequent burial, diagenesis continued in a manner similar to that which had occurred prior to uplift, except that the grains most highly susceptible to alteration—feldspars and ferromagnesian grains—had been removed along with the organic matter.

Burial diagenesis

The burial diagenetic phase is characterized by the dissolution of feldspars and carbonate and the precipitation of pore-filling kaolinite in secondary pores. It is therefore possible to fix the boundary between the burial diagenetic phase and early burial diagenesis at the 80 °C isotherm. Surdam *et al.* (1984) have shown that silicate dissolution occurs at temperatures in excess of 80 °C as a result of the action of organic acids, the abundance of which is greatly increased above this temperature (Carothers & Kharaka 1980). The 80 °C isotherm was crossed by the uppermost Carboniferous strata at a depth of less than 1500 m during the early Triassic. The rapid burial during the Triassic (Fig. 7), combined with the high estimated palaeogeothermal gradient, caused rapid heating of the Carboniferous strata. Because of this rapid burial, adding or subtracting 10 °C from the palaeogeothermal gradient does not significantly alter the timing of this event.

It is always difficult to estimate the amounts of cement removed during dissolution, but in the majority of the Carboniferous rocks studied the distribution of the secondary dissolution pores is rather patchy. Where dissolution has not occurred, carbonate and quartz cements reduce the porosity to almost zero, and around 20% of the volume of the rock can be accounted for by carbonate cements. It appears that virtually all the observed porosity can be interpreted as having a secondary origin.

The amount of quartz cementation is difficult to estimate because of the lack of visible dust rims around the detrital cores, but where it is possible to estimate it, up to 21% of the rock volume can be accounted for by quartz cementation. Cement precipitation is the major agent of porosity destruction in the sandstones studied; CL photomicrographs show that little physiochemical compaction has taken place. Pore-filling kaolinite cements occupy up to 18.5 vol. % of the sandstones, and kaolinite is the dominant clay mineral present although it is under-represented in the X-ray diffractograms for the fraction below 2 μm because of its coarse crystal size. The pore fluids would have been acidic during the burial phase of diagenesis since acids are produced by thermal carboxylation reactions. When acidic fluids carrying alumina in solution react with soluble minerals (carbonates or feldspars) in the sandstones, the pH may be increased sufficiently to cause the alumina to precipitate, forming kaolinite (Curtis 1983).

Kaolinite can also form by the *in situ* alteration of feldspar, but this mechanism appears to have been less common than direct precipitation in secondary pores. Kaolinite crystals usually have interpenetrant relationships with quartz. CL photomicrographs show the kaolinite impinging upon the detrital cores of quartz grains. The interpenetrant contacts between kaolinite and quartz therefore appear to be caused by aggressive mineral reactions rather than by late-stage quartz overgrowth.

Late burial diagenesis

The last minerals to precipitate in the pores were illite, ferroan rhombic dolomite and trace amounts of albite overgrowths, noted using the scanning electron microscope but not in thin section. The dolomite can be seen as fresh rhombs in secondary dissolution pores. Precise dating of the timing of precipitation of these late cements is difficult, and the burial plot (Fig. 7) shows that temperatures in excess of 100 °C were reached during the Jurassic as well as during maximum burial during the Tertiary for the example well. In other wells maximum temperatures coincided with maximum burial and a similar mineral assemblage is recorded. It is probable that the last phase of cementation coincided with deep burial during the Tertiary. The onset of the burial diagenetic phase was probably approximately contemporaneous with the onset of gas migration.

Illite is stable under alkaline conditions and at high temperatures. For a given K^+/H^+ ratio, illite becomes stable in preference to kaolinite with increasing temperature. It is not possible to predict accurately the temperature at which illite precipitation will occur in preference to kaolinite since the pH of the sediments as well as the K^+/H^+ ratio also affects stability. The fact that illite is present in both hydrocarbon- and water-saturated sandstones implies that at least some precipitation occurred prior to gas migration.

Conclusions

The oil industry has only recently revived its interest in the hydrocarbon potential of the Carboniferous, but it is already apparent that the reservoir potential of the Carboniferous is adversely affected by diagenetic factors. The work presented here is an attempt to produce a diagenetic model for Carboniferous fluvio-deltaic sediments using limited data. To summarize briefly, the following conclusions can be drawn.

(1) Virtually all the porosity preserved in Carboniferous sandstones is secondary in origin.
(2) Two phases of porosity enhancement have occurred, the first in response to pre-Permian uplift, and the second in response to organic acid generation and thermal decarboxylation of organic matter at temperatures in excess of 80 °C during burial.

(3) Porosity reduction has been achieved by precipitation of quartz and carbonate cements without major physiochemical compaction.
(4) Integrating the deduced diagenetic history with source-rock maturity modelling allows the absolute timing of diagenetic events to be postulated.
(5) Work on Carboniferous diagenesis in the southern North Sea Basin is at an early stage but is of vital importance for understanding the complex porosity distributions encountered. An integrated sedimentological, geochemical and basin analysis approach is required to understand these porosity distributions and to optimize exploration and, eventually, development efforts in Carboniferous sediments. Future work of direct application includes K–Ar dating of illite, stable isotope studies of carbonate phases, fission track analysis and fluid inclusion studies on quartz overgrowths.

ACKNOWLEDGMENTS: I would like to thank the Directors of British Gas plc for permission to publish this work. My colleagues in the Geology Department of British Gas (Karen Woodward, Mark Goodchild, Malcolm Brown and Alan Levison) helped in discussion of various aspects of this project. I would also like to thank Ramendra Sinha, Christine Shepperd and John McQuilken from the British Gas London Research Station for carrying out the SEM and XRD analyses. Thanks are due to the British Gas drawing office at Marble Arch for drafting the diagrams. Cathodoluminescence work was carried out by Geochem Ltd, Chester. This work was greatly improved by the comments of the referees and the editors.

References

BOLES, J. R. & FRANKS, S. G. 1979. Clay diagenesis of Wilcox Sandstones in South West Texas: implications of smectite diagenesis on sandstone cementation. *Journal of Sedimentary Petrology* **49**, 55–70.

CAROTHERS, W. W. & KHARAKA, Y. K. 1980. Stable carbon isotopes of HCO_3 in oil field waters—implication for the origin of CO_2. *Geochimica et Cosmochimica Acta*, 323–332.

COPE, M. J. 1987. An interpretation of vitrinite reflectance data from the southern North Sea Basin. *In:* BROOKS, J., GOFF, J. & VAN HOORNE, B. (eds) *Habitat of Palaeozoic Gas in N.W. Europe.* Geological Society of London Special Publication **23**, 85–100.

CURTIS, C. D. 1978. Possible links between sandstone diagenesis and depth related geochemical reactions occurring in enclosing mudstones. *Journal of the Geological Society of London* **135**, 107–117.

—— 1983. Geochemistry of porosity enhancement and reduction in clastic sediments. *In:* BROOKS, J. (ed.)

Petroleum Geochemistry and Exploration of Europe. Geological Society of London Special Publication **12**, 113–125.

—— & SPEARS, D. A. 1968. The formation of sedimentary iron minerals. *Economic Geology* **63**, 357–370.

FIELDING, C. 1984. Upper delta plain lacustrine and fluviolacustrine facies from the Westphalian of the Durham Coalfield. *Sedimentology* **31**, 547–564.

FISHER, M. J. 1986. Triassic. *In:* GLENNIE, K. W. (ed.) *Introduction to the Petroleum Geology of the North Sea* (2nd edn). Blackwell Scientific Publications, Oxford, 113–132.

GLENNIE, K. W. (ed.) 1986. *Introduction to the Petroleum Geology of the North Sea* (2nd edn). Blackwell Scientific Publications, Oxford, 278 pp.

HARPER, M. L. 1974. Approximate geothermal gradients in the North Sea Basin. *Nature* **230**, 235–236.

HASZELDINE, R. S. & ANDERTON, R. 1980. A braidplain

facies model for the Westphalian B Coal Measures, North East England. *Nature* **243**, 51–53.

HAWKINS, P. J. 1978. Relationship between diagenesis, porosity reduction and oil emplacement in late Carboniferous sandstone reservoirs, Bothamsall Oilfield, East Midlands. *Journal of the Geological Society, London* **135**, 7–24.

HUGGETT, J. 1984. Controls on mineral authigenesis in Coal Measures sandstones of the East Midlands, U.K. *Clay Minerals* **19**, 343–358.

HUNT, J. M. 1979. *Petroleum Geochemistry and Geology.* W. H. Freeman, San Francisco, CA, 716 pp.

JOHNSON, G. A. L. 1982. Geographical changes in Britain during the Carboniferous. *Proceedings of the Yorkshire Geological Society* **44**, 181–203.

LEEDER, M. R. 1982. Upper Palaeozoic basins of the British Isles—Caledonide inheritance versus Hercynian plate margin processes. *Journal of the Geological Society, London* **139**, 481–494.

MACKENZIE, D. 1978. Some remarks on the development of sedimentary basins. *Earth and Planetary Science Letters* **40**, 25–32.

MARIE, J. P. 1975. Rotliegendes stratigraphy and diagenesis. *In:* WOODLAND, A. (ed.) *Petroleum and the Continental Shelf of Europe.* Institute of Petroleum, London, 205–210.

PERODON, A. & MASSE, P. 1984. Subsidence, sedimentation and petroleum systems. *Journal of Petroleum Geology* **7**, 5–26.

SCHMIDT, V. & MCDONALD, D. A. 1979. Texture and recognition of secondary porosity in sandstones. *In:* SCHOLLE, P. A. & SCHLUGER, P. R. (eds) *Aspects of Diagenesis.* Society of Economic Paleontologists and Mineralogists Special Publication **26**, 175–207.

SURDAM, R. C., BOESE, S. W. & CROSSEY, L. J. 1984. The chemistry of secondary porosity. *In:* MCDONALD, D. A. & SURDAM, R. C. (eds) *Clastic Diagenesis.* American Association of Petroleum Geologists Memoir **37**, 127–150.

WALKER, T. R. 1984. Diagenetic albitisation of potassium feldspar in arkosic sandstones. *Journal of Sedimentary Petrology* **54**, 3–16.

WAPLES, D. W. 1980. Time and temperature in petroleum formation: an application of Lopatin's method to petroleum exploration. *American Association of Petroleum Geologists Bulletin* **64**, 916–926.

WEAVER, C. E. 1960. Possible uses of clay minerals in the search for oil. *American Association of Petroleum Geologists Bulletin* **44**, 1505–1508.

WOODWARD, K. & CURTIS, C. D. 1988. Predictive modelling for the distribution of production constraining illite, Morecambe Bay gas field, Irish Sea. *In:* GLENNIE, K. W. (ed.) *Proceedings of the 3rd World Conference on Petroleum Geology,* in press.

ZIEGLER, P. A. 1982. *Geological Atlas of Western and Central Europe.* Shell International Petroleum, Maatschappij.

G. COWAN, British Gas plc, 59 Bryanston Street, London W1A 2AZ, UK.

Detailed sedimentology and gamma-ray log characteristics of a Namurian deltaic succession
I: Sedimentology and facies analysis

C. S. Bristow and K. J. Myers

SUMMARY: The aim of this study was to investigate the relationship between sedimentary facies and the gamma-ray log chacteristics of the Upper Namurian Rough Rock exposed in a road section at Elland, Yorkshire. The overall succession coarsens upwards and is divided into five facies which record a prograding deltaic system. Coarse sands of a major braided fluvial distributary (Facies 1) rest upon channelled coarse sands of minor distributaries or crevasse channels (Facies 2) which overlie mouth-bar deposits (Facies 3) and more distal deltaic deposits (Facies 4). Abandonment deposits (Facies 5) occur above and between the channellized sands and the mouth-bar sands. Thus the delta front featured a series of small distributaries characterized by periodic switching in discharge and sediment transport. The shoal-water delta may have been similar to the Lafourche Delta on the Mississippi.

The aim of this two-part study was to investigate the relationship between sedimentary facies and the gamma-ray log characteristics of the Upper Namurian Rough Rock Group. The section studied is a road cutting at Elland, Yorkshire, which exposes a prograding deltaic succession of the Rough Rock and Rough Rock Flags (Fig. 1).

The Rough Rock Group is the youngest group in the Millstone Grit Series and is equivalent in age to the Yeadonian Stage of the Namurian (Ramsbottom *et al.* 1978) (Fig. 2). The Rough Rock Group as a whole is considered to represent a sheet delta (Elliot 1978). The Rough Rock occurs at the top of the group and is interpreted as a braided river sheet sandstone deposit (Shackleton 1962; Heath 1975; Bristow 1987). It is underlain at Elland by coarse- to fine-grained sandstones and siltstones of the Rough Rock Flags. To the W of the studied section, the Rough Rock is underlain by the Lower and Upper Haslingden Flags (Fig. 2) which are interpreted

as bar-finger sand deposits (Collinson & Banks 1975).

This study is presented in two parts. In Part I the sedimentology of the rocks is described and interpreted with a discussion of the controls on deltaic sand-body geometries in the Namurian of northern England. The results of outcrop gamma-ray spectrometry are discussed in Part II. The distributions of potassium (K), uranium (U) and thorium (Th) are described and interpreted with reference to the five facies defined in Part I.

Outcrop description

The outcrop in the road cutting for the A629 Elland bypass (SE102215) (Fig. 1) exposes a section 500 m long and between 15 and 30 m high through the Rough Rock and Rough Rock Flags. The cutting was made around 1970 and the exposure is fresh and relatively unweathered. Exposure is 100% and, because the road cutting is made in a series of steps, almost all the rock face is accessible. The outcrop is oriented NNE–SSW (Fig. 1) and provides a unique flow-parallel section through the prograding Rough Rock and top Rough Rock Flags succession. A true-scale two-dimensional outcrop diagram of the Elland road cutting has been constructed from measured sections spaced at 10 m intervals (Figs 3 and 4). Five facies have been recognized and they are described and interpreted below.

Facies 1

Facies 1 consists of thick-bedded coarse to very coarse grained sandstones with some granule-

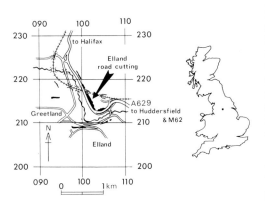

FIG. 1. Location of the Elland road cutting in England.

From WHATELEY, M. K. G. & PICKERING, K. T. (eds), 1989, *Deltas: Sites and Traps for Fossil Fuels,* Geological Society Special Publication No. 41, pp. 75–80.

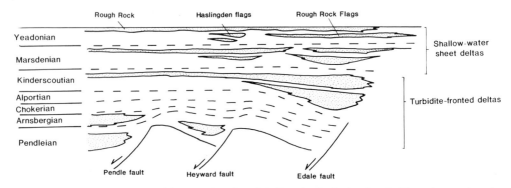

FIG. 2. Generalized cross-section of the central section of the Pennine Basin showing the Namurian stratigraphy and location of major sand bodies within the basin. There appears to be an overall southerly migration of sedimentation with subsidiary changes in facies across tilt blocks within the basin. The Rough Rock is the most widespread of the Namurian sandstones.

grade quartz and wood fragments. The sharp erosional base of Facies 1 is overlain by an intraformational conglomerate of mudstone rip-up clasts, wood fragments and granule-grade quartz clasts. The top of Facies 1 is not exposed but probably occurs just above the top of the present outcrop.

Internal sedimentary structures are dominated by large-scale trough cross-stratification with sets up to 2 m thick which are laterally continuous for at least 200 m in a flow-parallel direction. Some planar cross-stratification also occurs and deformed foresets are present in two beds near the top of the outcrop. Several concave-up internal erosion surfaces are overlain in places by minor intraformational conglomerates. Palaeocurrent indicators in this facies have a vector mean of 163° (from 136 measurements) (Fig. 3). The vector mean swings along the outcrop from 170° between logs 1 and 15, to 160° between logs 16 and 23, and 156° between logs 24 and 42.

Facies 1 correlates with the Rough Rock which extends as a sheet sandstone over approximately

10 000 km^2 in northern and central England. It has been previously interpreted as a braided river deposit from other outcrops (Shackleton 1962; Heath 1975; Bristow 1987). The sets of cross-stratification are interpreted as the product of subaqueous dune bedforms migrating in response to unidirectional flow. Sets of trough cross-stratification were deposited by sinuous crested dunes while sets of planar cross-stratification were deposited by straight crested dunes.

The cross-stratification style, unidirectional palaeocurrents and sheet-like geometry of Facies 1 are all compatible with deposition in a braided river. There are no vertical or downstream changes in grain size or sedimentary structures within the Elland road cutting which would indicate the presence of macroforms. The swing in palaeocurrents along the outcrop may indicate a broad curvature of the channel but there is no good evidence of fluvial style. The interpretation of Facies 1 as a braided river deposit is dependent on comparison with other outcrops. Lateral-accretion surfaces exposed in Greetland Quarry

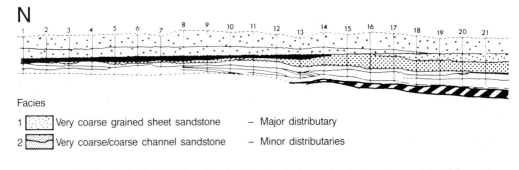

Facies

1 [ ] Very coarse grained sheet sandstone   – Major distributary

2 [ ] Very coarse/coarse channel sandstone   – Minor distributaries

FIG. 4. Facies distribution in the Elland road cutting showing the internal and external geometries of the sand bodies based on interpretation of the graphic logs in Fig. 3.

(SE095216) 2 km across the valley indicate that individual channels in the Rough Rock river were at least 6.5 m deep and the width of the river is estimated to be about 1 km (Bristow 1987).

Facies 2

Facies 2 underlies Facies 1 and is dominantly coarse-grained cross-stratified sandstone occurring as two channel sand bodies (Fig. 4). These are both about 300 m wide and up to 7 m deep. There is no evidence of macroforms within these channels and the internal geometry is of very simple vertical cut and fill. Sedimentary structures are dominated by trough cross-stratification. Sets up to 1.5 m thick can be traced for 150 m in a downcurrent direction. Occasional beds appear to be massive and one of these contains a tree trunk 5 m long (Fig. 3, log 11). Wood fragments are also very common in the basal channel lags. There are several minor channels at the N end of the section which have been included in Facies 2. These are simple scours up to 2 m deep with concentrations of granules around log jams.

The base of both of the larger channel sandstones is a concave-up erosion surface. The top is either truncated by the overlying Facies 1 or overlain conformably by fine-grained sandstones and siltstones of Facies 5. At the 17 m level of logs 33–39 (Fig. 3) fine-grained sandstones and siltstones overlie an erosive lag deposit and partially fill the channel. The palaeocurrent vector mean for the upper channel sandstone is 145° ($n = 44$) and that for the lower channel sandstone is 138° ($n = 48$). This is a divergence of up to 30° from the sets in the underlying and overlying sandstones.

Facies 2 is also interpreted as a fluvial channel deposit, but the channel sandstone geometry is different from the overlying Facies 1 sheet sandstone. The channels in which Facies 2 accumulated were probably much smaller but they contain similar lithologies and sedimentary structures to Facies 1. They are interpreted as the deposits of minor distributary channels. This interpretation is supported by the small palaeocurrent divergence, entrenched basal erosion surfaces and simple channel fill. There is no evidence of macroforms or lateral-accretion surfaces within either channel even though the whole channel form is exposed. The fine-grained fill in the upper channel sandstone in logs 33–39 (Fig. 3) indicates temporary abandonment. This suggests that discharge in these channels was irregular because of either seasonal changes in discharge or, more probably, channel switching or avulsion between minor distributaries.

Facies 3

Facies 3 contains medium-grained trough cross-stratified and occasionally plane-bedded sandstones. This facies is not generally channellized but several small channels up to 3 m deep and 50 m wide do occur within it at the northern (upstream) end of the outcrop (Fig. 3, logs 1–6). The sand body is dominated by downcurrent dipping 'bedsets' which are generally non-erosive and dip at angles between 1° and 10°. There are some minor internal erosion surfaces of limited lateral extent which occur on the scale of individual dune-like bedforms. Minor erosion surfaces occur at the base of Facies 3 but there is no evidence of channellization.

Sedimentary structures are dominated by trough cross-stratification. Foresets are generally very low angle (about 5°) with swept-out toesets. The maximum thickness of individual sets is 1 m, but most are around 0.2 m and die out downcurrent in less than 20 m. The palaeocurrent vector

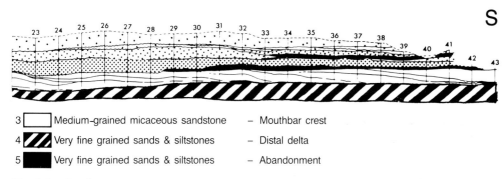

3 ☐ Medium-grained micaceous sandstone   – Mouthbar crest

4 ▨ Very fine grained sands & siltstones   – Distal delta

5 ■ Very fine grained sands & siltstones   – Abandonment

Fig. 4. (*continued*)

mean for Facies 3 is 171.5° from 155 measurements.

Facies 3 is finer grained than Facies 1 and 2, which suggests a lower energy or more distal environment of deposition. The dominance of internal downcurrent-dipping bedding surfaces indicates extensive progradation. Combining this with the position of Facies 3 between the coarse-grained fluvial deposits of Facies 1 and 2 and the fine-grained deposits of Facies 4 supports the interpretation of this facies as the proximal part of a mouth bar. The channels at the upstream end of the section (Fig. 3) may mark a transition from the minor distributaries into the mouth-bar facies. The swept-out foresets are attributed to a transition from dune to upper-phase plane bed and the presence of a high suspended sediment load which would damp turbulence in the lee of dunes. This interpretation is supported by the high mica content of these sandstones.

Facies 4

Facies 4 occurs at the base of the outcrop and contains micaceous fine-grained sandstones and siltstones. The dominant sedimentary structures are planar and ripple lamination. The ripples are mostly wave-modified current ripples with some simple current ripples and very occasional wave ripples. Trough cross-stratification is rare. Facies 4 occurs as 1–2 m scale alternations of laterally continuous beds of fine-grained sandstones and shales. Individual beds can be traced laterally for over 300 m with almost no change in thickness. The palaeocurrent vector mean is 189° from 17 measurements. One bed contains small-scale dewatering structures and load casts. Bioturbation is rare and no body fossils have been discovered although there are some comminuted plant fragments.

The rippled fine-grained sands and occasional beds of trough cross-stratification in Facies 4 indicate some current activity with palaeoflows towards the SSE. Most of the current ripples have been modified by wave activity, although true wave ripples only occur in one horizon. The large amounts of mica and comminuted plant fragments within the sandstones suggest high rates of deposition from suspension. The siltstone beds are also interpreted as suspension deposits. The alternation of siltstones and sandstones may be due to seasonal fluctuations in discharge. There are no channels in Facies 4 and there is no evidence of deposition on a slope within the scale of the outcrop. Facies 4 is interpreted as more distal deltaic deposits. The lack of bioturbation within these beds is probably due to the high rates of suspended sediment deposition. The

absence of evidence for reversing tidal flow regimes or extensive wave reworking of these deposits indicates that the delta was fluvially dominated.

Facies 5

Facies 5 contains fine-grained sandstones, siltstones and clays. The siltstones are generally light grey and less micaceous than the dark grey siltstones of Facies 4. Facies 5 forms thin sheets or laterally confined lenses (50–100 m wide) within or between the sandstone-dominated Facies 1, 2 and 3. The dominant sedimentary structures are plane lamination and current–ripple cross-lamination. Some impressions of plants are preserved in the siltstones but there are no rootlet horizons within the outcrop. In logs 36–39 (Fig. 3), 17 m above the base, Facies 5 occurs as a channel fill overlying a basal erosion surface and lag deposit. Facies 5 overlies Facies 3 between logs 29 and 41 and occurs between Facies 1 and 2 in logs 1–13.

Facies 5 is interpreted as abandonment deposits. These occur at three different horizons within the outcrop (Figs 3 and 4). Two of these are within Facies 1 and 2, and the other overlies the mouth-bar crest deposits (Facies 3). The presence of abandonment deposits both between the mouth bar and the minor distributaries and within the minor distributaries indicates that sediment supply switched between several different distributaries on the delta front. Each lens of fine-grained sediment marks the abandonment of one distributary in favour of another.

Style of deltaic deposition

The outcrop in the Elland road cutting is an overall coarsening-upward succession which is interpreted as the product of a prograding delta. The contacts between the different facies described above are not gradational but are locally erosive or covered by abandonment deposits so that, while the facies are superimposed in a prograding sequence, they were not originally contiguous. The contact between the distal deltaics (Facies 4) and the more proximal mouth-bar crest (Facies 3) is conformable but locally erosive on the scale of individual bedforms. The top of the mouth bar is channellized in the northern (upstream) portion but overlain by abandonment deposits in the S which indicates abandonment of the mouth-bar lobe before the incision of the minor distributary channels (Fig. 5). There are also abandonment deposits (Facies 5) between the minor distributary channel sandstones (Facies 2) and beneath the major distributary sandstone (Facies 1). This implies that the delta front consisted of a series of smaller distributaries rather than a single large distributary (Fig. 5). Discharge was periodically switched from one distributary to another (Fig. 5).

The internal structure and lateral continuity of the facies indicates deposition by a lobate shoal-water delta like the Laforche Delta of the Mississippi River (Fisk 1955; Gould 1970). This style of deltaic deposition contrasts strongly with that described by Collinson & Banks (1975) in the laterally equivalent Upper Haslingden Flags in the Rossendale area. They describe elongate sand bodies which Collinson (1988) compares with the modern bird's foot delta of the Mississippi. The exposure in the Elland road cutting may also be analogous to the modern Atchafalaya Delta which has a complex bifurcating delta front (Roberts *et al.* 1980; Van Heerden & Roberts 1980). The change in delta style is believed to be related to water depth which may indicate differential subsidence across tilt blocks with the Pennine Basin in the Yeadonian (Bristow 1988).

Comparison with other Namurian deltas

The lobate and elongate deltaic sand bodies of the Yeadonian in the Pennine Basin show some similarities with the underlying Marsdenian deltaic deposits described by Benfield (1969). However, these contrast strongly with the earlier Kinderscoutian turbidite-fronted deltas described by Collinson (1968, 1969), McCabe (1976) and Elliot (1978). The overall change in deltaic style in the Namurian of the Pennine Basin is attributed to a progressive reduction in water depth and depositional slopes within the basin. This was achieved through a combination of a gradual decrease in tectonic subsidence and change to thermal subsidence (Leeder & Mac-Mahon 1988) and progressive filling of half-grabens within the Pennine Basin (Jones 1980; Collinson 1988; Steele 1988) (Fig. 2). As a result the Rough Rock delta prograded into a broad shallow-water basin and deposited a laterally continuous sheet of sandstone across most of the Pennine Basin.

Controls on sand-body geometry

The Rough Rock delta prograded into a basin with very little wave or tidal activity and was apparently fluvially dominated. Depositional models for fluvial-dominated deltas emphasize

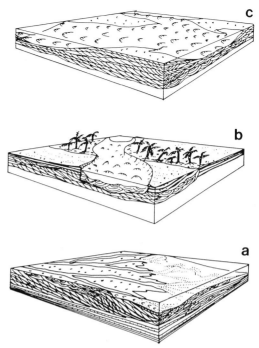

FIG. 5. Reconstruction of environments of deposition in the Rough Rock Group exposed in the Elland road cutting: (*a*) mouth bar of a shoal-water lobate delta prograding across more distal deltaic deposits (Facies 3 and 4); (*b*) minor distributary channels advance across the top of the mouth bar, and frequent switching of distributaries leads to the preservation of abandonment deposits (facies 5) within and between the minor distributary sand bodies; (*c*) cross-stratified sheet sand deposited by the major distributary channel (Facies 1) which may have had a braided channel system.

the importance of a bird's foot distributary pattern and the deposition of elongate sand bodies (Elliot 1978). Elongate deltaic sand bodies have been described in the Yeadonian by Collinson & Banks (1975), and these are compared with the elongate bar-finger sands of the modern Mississippi by Collinson (1988). However, Bristow (1988) has suggested that the sand-body geometry in the Upper Haslingden Flags is controlled by subsidence and that the predominant delta style in the Rough Rock Group is lobate. Analogues for lobate fluvially dominated deltas include the Lafourche Delta of the Mississippi River (Fisk 1955; Gould 1970) and the modern delta of the Atchafalaya River (Roberts *et al.* 1980; Van Heerden & Roberts 1980). In both cases the deltas prograde into relatively shallow water. Fast rates of sediment deposition at the river mouth result in the development of a mouth bar. This leads to channel bifurcation and construction of a lobate delta with a complex pattern of distributary channels (Roberts *et al.* 1980). A similar pattern of mouth-bar deposition, channel bifurcation and switching of distributaries across the delta is believed to have controlled sand-body geometries in the Rough Rock Group exposed in the Elland road cutting.

Conclusions

The Elland road cutting exposes an overall coarsening-upward succession with five facies which are interpreted as the deposits of a prograding delta front. Very coarse grained sands of a major braided fluvial distributary (Facies 1) prograded across several minor distributary or crevasse channel sands (Facies 2). These in turn overlie a mouth-bar deposit (Facies 3) and more distal deltaic deposits (Facies 4). Fine-grained abandonment deposits (Facies 5) occur above and between the channellized sandstones of Facies 1 and 2 and the mouth-bar sandstones. This suggests that the delta front consisted of a series of small distributaries characterized by periodic switching of discharge and sediment transport. The interpreted shoal-water Rough Rock delta may be analogous to the Lafourche Delta on the Mississippi.

References

BENFIELD, A. C. 1969. The Huddersfield White Rock Cyclotherm in the Central Pennines (field report). *Proceedings of the Yorkshire Geological Society* 37, 181–187.

BRISTOW, C. S. 1987. *Sedimentology of Large Braided Rivers Ancient and Modern.* Unpublished PhD thesis, University of Leeds, 274 pp.

—— 1988. Controls on sedimentation in the Rough Rock Group. *In:* BESLY, B. M. & KELLING, G. (eds) *Sedimentation in a Synorogenic Basin Complex.* Blackie, Glasgow, 114–131.

COLLINSON, J. D. 1968. Deltaic sedimentation units in the Upper Carboniferous of northern England. *Sedimentology* 10, 233–254.

—— 1969. The sedimentology of the Grindslow Shales and the Kinderscout Grit: a deltaic complex in the Namurian of northern England. *Journal of Sedimentary Petrology* 39, 194–221.

—— 1988. Controls on Namurian sedimentation in the Central Province basins of northern England. *In:* BESLY, B. M. & KELLING, G. (eds) *Sedimentation in a Synorogenic Basin Complex.* Blackie, Glasgow, 85–101.

—— & BANKS, N. L. 1975. The Haslingden Flags (Namurian G₁) of South East Lancashire: barfinger sands in the Pennine Basin. *Proceedings of the Yorkshire Geological Society* 40, 431–458.

ELLIOT, T. 1978. Deltas. *In:* READING, H. G. (ed) *Sedimentary Environments and Facies.* Blackwell Scientific Publications, Oxford, 97–142.

FISK, H. N. 1955. Sand facies in Recent Mississippi Delta deposits. *4th World Petroleum Congress, Rome, 1955, Proceedings, Section 1.* Carlo Colombo, Rome, 377–398.

GOULD, H. R. 1970. *The Mississippi Delta Complex.* Society of Economic Paleontologists and Mineralogists Special Publication 15, 3–31.

HEATH, C. W. 1975. *A Sedimentological and Palaeogeographical Study of the Namurian Rough Rock in the Southern Pennines.* Unpublished PhD thesis, Keele University.

JONES, C. M. 1980. Deltaic sedimentation in the Roaches Grit and associated sediments (Namurian R₂b) in the south-west Pennines. *Proceedings of the Yorkshire Geological Society* 43, 39–67.

LEEDER, M. R. & MCMAHON, A. 1988. Tests and consequences of the extensional theory for Carboniferous basin evolution in Northern England. *In:* BESLY, B. M. & KELLING, G. (eds) *Sedimentation in a Synorogenic Basin Complex.* Blackie, Glasgow, 43–51.

MCCABE, P. J. 1976. Deep distributary channels and giant bedforms in the Upper Carboniferous of the Central Pennines, Northern England. *Sedimentology* 24, 271–290.

RAMSBOTTOM, W. H. C., CALVER, M. A., EAGAR, R. M. C., HODSON, F., HOLLIDAY, D. W., STUBBLEFIELD, C. J. & WILSON, R. B. 1978. *A Correlation of Silesian Rocks in the British Isles.* Geological Society of London Special Report 10, 81 pp.

ROBERTS, H. H., ADAMS, R. D. & CUNNINGHAM, R. H. W. 1980. Evolution of sand-dominant subaerial phase, Atchafalaya Delta, Louisiana. *American Association of Petroleum Geologists Bulletin* 64, 264–279.

SHACKLETON, J. S. 1962. Cross-strata of the Rough Rock (Millstone Grit Series) in the Pennines. *Liverpool and Manchester Geological Journal* 3, 109–118.

STEELE, R. P. 1988. The Namurian sedimentary history of the Gainsborough Trough. *In:* BESLY, B. M. & KELLING, G. (eds) *Sedimentation in a Synorogenic Basin Complex.* Blackie, Glasgow, 102–113.

VAN HEERDEN, I. L. & ROBERTS, H. H. 1980. The Atchafalaya Delta: rapid progradation along a traditionally retreating coast (South Central Louisiana). *Zeitschrift für Geomorphologie* 34, 188–201.

C. S. BRISTOW, BP Exploration and Production Department, Britannic House, London EC2Y 9BU, UK.

K. J. MYERS, BP Research Centre, Sunbury-on-Thames, Middlesex TW16 7LN, UK.

Detailed sedimentology and gamma-ray log characteristics of a Namurian deltaic succession II: Gamma-ray logging

K. J. Myers and C. S. Bristow

SUMMARY: A portable gamma-ray spectrometer was used to measure the total radioactivity, potassium content (% K), uranium content (ppm U) and thorium content (ppm Th) in the Upper Namurian Rough Rock at an outcrop in a road section at Elland, Yorkshire. The total gamma radiation generally decreases upwards in the succession, reflecting an upward increase in grain size. Gamma-ray peaks within the succession are due partly to facies/grain-size changes and partly to concentrations of thorium-bearing monazite in mouth-bar sandstones and basal channel lags. Failure to identify the true nature of these peaks could lead to erroneous correlations in the subsurface. The Th/K ratio decreases upwards in association with the increase in grain size and a decrease in the quartz/feldspar ratio. The Th/K ratio and Th–K cross-plots are effective in distinguishing some of the deltaic facies.

The gamma-ray log measures the natural radioactivity of rocks and is one of the fundamental tools of the petroleum geologist. It is most commonly used for interwell correlation and lithology determination.

The gamma-ray log has also been used to interpret grain-size profiles in sedimentary successions (eg Cant 1983; Selley 1985; Selley, this volume). The gamma-ray reading is observed to increase with decreasing grain size in proportion to the amount of clay minerals present in the formation. However, the gamma-ray reading is *not* a function of grain size but of mineralogy, eg the gamma-ray reading of a clean (clay-free) well-sorted fine sandstone will be similar to that of a coarse sandstone of the same mineralogy. A clean sandstone can, in fact, give a high gamma-ray response if it contains potassium feldspars, micas, glauconite, monazite or uranium minerals.

The natural gamma-ray spectrometry (NGS) log is a considerable improvement on conventional total-count gamma-ray logs as it measures the elemental concentrations of the three main contributors to natural radioactivity, ie potassium (^{40}K), uranium (^{238}U) and thorium (^{232}Th). Despite its now widespread use, its potential has not been fully realized because of the limited understanding of the distribution of these elements in sedimentary rocks.

In this study a portable gamma-ray spectrometer was used to measure the potassium, uranium and thorium contents of some fluvio-deltaic sediments at outcrop. This somewhat novel technique has been described in detail by Myers (1987) and has been applied recently to the study of organic-rich mudrocks (Myers & Wignall 1987) and chamositic oolitic ironstones (Myers,

in press). The aim of this study is to observe the distribution of potassium, uranium and thorium in a well-exposed deltaic succession and to investigate the potential uses of gamma-ray spectrometry in correlation and facies discrimination. The section chosen for study is a road cut in the Upper Namurian Rough Rock in South Yorkshire, which exposes a prograding deltaic succession. The sedimentology and facies analysis of the section have been described in Part I of this work.

Field procedure

Profiles of gamma radiation have been measured every 20 m along the exposure using a Geometrics GR-410A gamma-ray spectrometer. The field procedure is described fully by Myers (1987) and Myers & Wignall (1987), and so only a brief account is included here.

The spectrometer is calibrated for a flat surface about 3 m in diameter (cf Myers & Wignall 1987, Fig. 1), so, where possible, the detector was placed perpendicular to the rock face with enough rock surface exposed to approximate a 2π (semi-infinite) measuring geometry. Measurements within 0.5 m of the base (or the edge) of a cliff (Fig. 1) were avoided as this affects the amount of rock the detector 'sees' and causes an increase (or decrease) in gamma-ray counts which is unrelated to lithology. The circle of influence of the detector is theoretically a radius of 40 cm from the detector but beds thinner than 40 cm can still be resolved if there is sufficient contrast in radioactivity. Measurements were taken about every 50 cm, although less frequently where

From WHATELEY, M. K. G. & PICKERING, K. T. (eds), 1989, *Deltas: Sites and Traps for Fossil Fuels,*
Geological Society Special Publication No. 41, pp. 81–88.

81

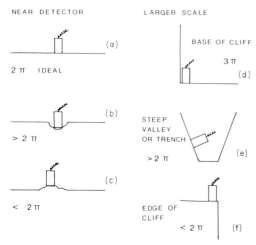

FIG. 1. (a) The ideal source–detector geometry together with examples of deviations from 2π geometry which can result in (b, d, e) increases or (c, f) decreases in gamma-ray counts which are unrelated to lithology. Continuous vertical (or horizontal) sections give the most representative readings.

FIG. 2. This photograph shows part of log 32 (Plate 1). The measurement is being taken on the fine-grained abandonment facies, which is underlain by medium-grained mouth-bar facies sandstones. The erosive base of a minor distributary channel sandstone which down-cuts into the abandonment facies can be seen near the top of the ladder.

exposure is poor. Discretion was necessary in choosing measuring sites as ideal source–detector geometries were not always available. Ladders and abseiling were used to reach the higher cliff faces (Fig. 2). A counting time of 2 min was used which resulted in errors of approximately $\pm 5\%$ for potassium, $\pm 25\%$ for uranium and $\pm 10\%$–15% for thorium.

Gamma-ray log characteristics

Total gamma-ray profiles (Plate 1)

The gamma-ray profiles for logs 2–42 are plotted against the graphic logs in Plate 1. The first point to note is that there is a broad upward decrease in gamma radiation through the entire outcrop punctuated by peaks which usually correlate with changes in facies and grain size. In general, the gamma radiation increases with decreasing grain size. However, there are also two other groups of gamma-ray peaks which are not related to changes in grain size: (a) above the basal erosion surface of the major distributary channel sand in logs 16, 18, 20, 22, 24, 32, 34 and 36; (b) within the mouth-bar sands in logs 22, 24, 28 and 36.

The highest gamma radiation of the entire outcrop was recorded in a medium-grained mouth-bar sand (marked in log 28). The contact between the Facies 4 distal delta and the overlying Facies 3 mouth bar is always clearly marked by a

sharp decrease in gamma radiation. The upper boundary of Facies 3 is marked only where the fine-grained abandonment Facies 5 is developed and a sharp increase in gamma radiation occurs. Where the abandonment facies is missing, there is little contrast between Facies 3 and 2. The contact between Facies 2 and 1 is marked in some parts of the outcrop by a gamma-ray peak coincident with the basal erosion surface.

Some important implications for interwell correlation by gamma-ray logs are apparent when logs 26 and 32, spaced 60 m apart, are compared. In log 32 there are three correlatable peaks above the mouth-bar facies sandstones. The lower two peaks correspond to fine-grained delta abandonment deposits, whilst the upper peak represents the basal erosion surface. In contrast, log 26 has no correlatable peaks as the abandonment facies is missing and the basal erosion surface peak is only poorly developed. The hazards of down-hole gamma-ray correlation in deltaic sediments are clear. Table 1 summarizes the gamma-ray properties of the different facies.

Gamma-ray spectrometry

General trends

The trends in distribution of potassium, uranium and thorium are similar across the whole outcrop and are summarized in Table 1 and Fig. 3. The general decrease in total gamma radiation upwards reflects a gradual decrease in the abundance of all three radioelements. However, there are also some important changes in the *relative* concentrations of potassium, uranium and thorium. The decrease in thorium (and uranium) content upwards is more pronounced than that of potassium. This is reflected in a decreasing upward trend in the Th/K ratio (Table 1 and Fig. 3). For example, the potassium content does not change significantly between Facies 3 and Facies 1, whereas both thorium and uranium decrease by some 40%. The net result is that the mouth-bar and distal delta sands have Th/K ratios greater than 5, whilst the distributary channels have Th/K ratios less than 5.

The upward decrease in the Th/K ratio can be correlated with an increase in grain size and, more importantly, with a decrease in the quartz/feldspar ratio (Fig. 3). The change in mineralogy is a consequence of a change in grain size, and it is the presence of large fresh K-feldspars in the coarse-grained sands which produces the low Th/K ratio. The difference in mean Th/K ratios between Facies 1, 2 and 3 is significant at the 95% confidence level, but Facies 3, 4 and 5 cannot be discriminated using the Th/K ratio alone. The most coarse-grained sandstones of Facies 1 with the greatest reservoir potential have the lowest Th/K ratios.

The relative contributions of potassium, uranium and thorium to the total gamma-ray flux have been estimated using the unit of radioelement concentration (ur) (IAEA 1976) (Table 2). The results show that potassium contributes about half of the total gamma radiation in Facies 1 but declines to only 35% in Facies 4. Conversely, thorium's contribution increases from 35% in Facies 1 to 42% in Facies 3, 4 and 5 where it is the single largest contributor to the total gamma-ray flux.

Facies 1—major distributary channels

The major distributary sands have the lowest thorium contents and Th/K ratios and form a tight compositional field on Th–K cross-plots across the whole outcrop (Fig. 4a). The overall high potassium content of Facies 1 is associated with a high K-feldspar content and these sediments are classified as subarkoses. The basal erosion surface of Facies 1 is characterized by a high total gamma count and a high Th/K ratio. This is particularly strongly developed in logs 16, 18, 20, 22, 24, 32, 34 and 36 (Plate 1). The increase in thorium content is explained most easily by a concentration of thorium-bearing heavy minerals in a basal lag. The upward decrease in total gamma radiation in Facies 1 could be falsely interpreted as a coarsening-upward trend within the sandstone. This does not fit the field data and could lead to erroneous correlations and interpretations in the subsurface.

Facies 2—minor distributary channels

The composition field of Facies 2 straddles those of Facies 1 and 3 (Fig. 4b), and so the composition of the minor distributary channel sands is intermediate between the mouth-bar and major distributary channel sandstones. Despite the overlap, there is still statistical significance in the different mean Th/K ratios of Facies 1, 2 and 3. The wide spread of data within Facies 2 is explained by the range of grain size from very coarse grained sand, similar to Facies 1, to medium-grained sand, similar to Facies 3.

Facies 3—mouth bar

The thorium content of the mouth-bar facies is higher and more variable than in the overlying distributary channel sands, whereas the potassium contents are similar (Table 1 and Fig. 4c). Several points on Fig. 4c are highly enriched in thorium and plot well outside the general composition field. They appear to correlate with mica-rich foreset laminae and are best explained by concentrations of fine-grained heavy minerals. The outcrop's highest single gamma-ray reading of 100 cps was recorded in this facies (log 28), with a corresponding thorium content of 31.6 ppm. Thorium accounts for over 70% of the radiation at this point. The compositional field of the Facies 3 mouth bar is quite distinct from Facies 1 and 2 and is transitional to the distal deltaic Facies 4 sediments.

Facies 4—distal delta

Average thorium and potassium contents both increase by about 40% from the Facies 3 mouth-bar sands (Table 2). Taken as a whole, this facies shows a positive correlation between potassium

K. J. Myers & C. S. Bristow

TABLE 1. *Mean total radioactivity, potassium (% K), uranium (ppm U) and thorium (ppm Th) contents, and Th/K ratio in the five facies of the Rough Rock*

	n	Total counts per second		% K		ppm U		ppm Th		Th/K ratio	
		Mean ± s.d.	Range	Mean ± s.d.	Range	Mean ± s.d.	Range	Mean ± s.d.	Range	Mean ± s.d.	Range
Facies 1	81	39 ± 6.1 (16%)	27–48	1.59 ± 0.24 (15%)	1.17–2.33	1.1 ± 0.4 (36%)	0.3–2.3	6.0 ± 1.3 (22%)	3.6–9.4	3.8 ± 0.8 (21%)	2.3–6.9
Facies 2	78	40 ± 6.6 (17%)	27–55	1.55 ± 0.16 (10%)	1.17–2.03	1.2 ± 0.4 (33%)	0.5–2.1	7.0 ± 1.7 (24%)	3.5–14.1	4.5 ± 1.1 (24%)	2.2–9.4
Facies 3	102	49 ± 12 (24%)	32–100	1.53 ± 0.20 (13%)	1.17–2.08	2.0 ± 0.7 (35%)	0.7–5.1	9.8 ± 3.7 (38%)	5.1–31.6	6.5 ± 2.7 (42%)	3.8–25
Facies 4 (sandstones)	28	60 ± 4 (7%)	49–66	1.60 ± 0.28 (18%)	1.07–2.19	2.9 ± 0.6 (21%)	1.9–3.7	11.9 ± 1.3 (11%)	9.2–14.7	7.7 ± 1.7 (22%)	5.6–9.7
Facies 4 (siltstones)	40	74 ± 6 (8%)	59–89	2.19 ± 0.26 (12%)	1.49–2.73	3.6 ± 0.6 (17%)	2.7–5.0	13.5 ± 1.7 (13%)	10.2–17.4	6.2 ± 0.9 (15%)	4.5–7.8
Facies 5	33	74 ± 12 (16%)	51–95	2.42 ± 0.48 (20%)	1.28–3.32	3.3 ± 1.0 (33%)	1.5–6.1	14.6 ± 2.7 (18%)	9.4–20.8	6.2 ± 1.4 (22%)	4.5–9.9

Facies 1, major distributary channel; Facies 2, minor distributary channel; Facies 3, mouth bar; Facies 4, distal delta; Facies 5, abandonment. *n*, number of samples; s.d., standard deviation.

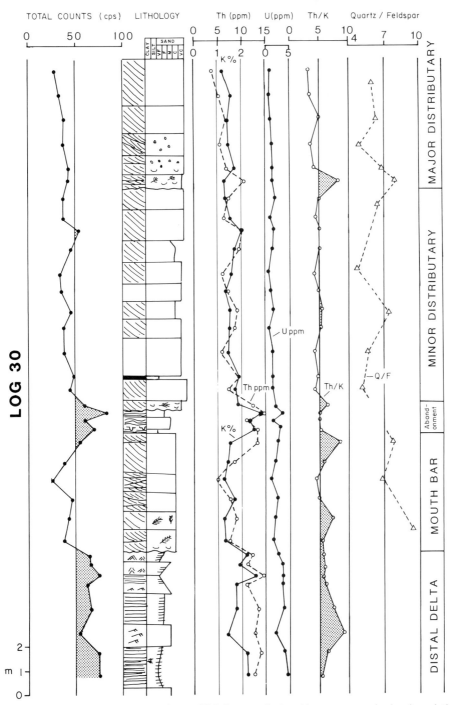

FIG. 3. Trends in gamma-ray spectrometry data and lithology are displayed in a representative log through the exposure (log 30 in Plate 1). Facies 1 (the Rough Rock) generally has the lowest radiation with an increase in the basal conglomerate which gives a false 'fining-upward' log motif. Facies 3 has the most variable gamma radiation owing to a variable heavy mineral component. The quartz/feldspar ratio was measured by point-count analysis of thin sections (300 points per slide).

TABLE 2. *Percentage contributions of potassium, uranium and thorium to the total gamma-ray flux*

	K (%)	U (%)	Th (%)
Facies 1	51	14	35
Facies 2	47	14	39
Facies 3	38	19	43
Facies 4	35	23	42
Facies 5	38	20	42

Calculated from average values in Table 1 using the definition of the ur unit of radioelement concentration (IAEA 1976), where 1 ppm U = 1 ur, 1 ppm Th = 0.47 ur and 1% K = 2.6 ur.

and thorium ($r = 0.58$) (Fig. 4d) similar to, but weaker than, those reported for other mudstones (Myers & Wignall 1987). This is because Facies 4 consists of two distinct subgroups: (a) ripple cross-laminated fine-grained sandstones, and (b) homogeneous mudstones and siltstones (Table 2). The sandstones are laterally continuous with similar potassium contents to the coarser-grained sandstones but with 30%–40% more uranium and thorium than the medium-grained mouth-bar sands. As a result the average Th/K ratio of 7.7 is the highest of any facies.

Plate 1 demonstrates the lateral continuity of the sandstone beds within Facies 4 which can be traced for 180 m along the section. The relative standard deviations of potassium, uranium and thorium for the series of assays taken along strike are similar to those predicted by counting statistics. This shows both that the bed is laterally very homogeneous in composition and that the field procedure had adequately maintained a constant source–detector geometry.

The siltstones and mudstones of Facies 4 show a 37% increase in average potassium content over the sandstones but only a 10%–20% increase in thorium and uranium. This results in a decrease in the Th/K ratio to 6.2. The Th/K ratio is lower than that for medium- and fine-grained sandstone facies and similar to that for the abandonment mudstones.

Facies 5—abandonment

The 'abandonment facies' is, on average, the most radioactive, with the highest average potassium, uranium and thorium contents. It also shows the greatest amount of scatter on the Th–K cross-plots and cannot be said to form a distinct composition field. Many of the points fall within the Facies 4 composition field which contains similar lithologies deposited in a different environment. Although the average potassium con-

tent is higher and the Th/K ratio is lower than those of the distal delta facies (Table 1), the differences between the two facies are not statistically significant.

Interpretation and discussion

Thin-section analyses show the coarse-grained Facies 1 and 2 sandstones to be subarkosic in composition with 12%–14% K-feldspar; Na-feldspar and micas are rare. All the potassium measured in the Facies 1 and 2 sandstones can be accounted for by K-feldspar containing 12% K. K-feldspar and quartz contain about 2 ppm Th on average (Myers & Wignall 1987, Fig. 7), so that only a maximum of 2 ppm Th can be accounted for by these minerals. Since in all cases the thorium content of Facies 1 and 2 sands exceeds 2 ppm, another thorium-bearing phase must be present, most probably a thorium-bearing heavy mineral.

Thorium is transported in the sedimentary cycle as a component of detrital resistate minerals, such as monazite, zircon and thorite, or adsorbed onto natural colloidal-sized materials, particularly clays (Langmuir & Herman 1980). Potassium and thorium carried in clay minerals will tend to be concentrated in suspension-deposited sediments, whereas potassium in larger minerals like K-feldspar will tend to be concentrated in bed-load current deposits. Thorium-bearing heavy minerals can be deposited by either process depending on grain size.

The thorium-rich heavy mineral monazite is a prominent component of the heavy mineral suite in the Rough Rock at Elland (Drewery et al. 1987) and elsewhere in the Pennines, where it has sometimes also been found to be highly concentrated in extremely thin (6–12 mm) layers (Gilligan 1919). Monazite contains, on average, 25 000 ppm Th, and so very small amounts of monazite can make a significant contribution to the whole-rock thorium content. Heavy minerals are often concentrated in channel-lag deposits by hydraulic sorting, and so the recorded thorium anomalies in Facies 1 can be attributed to concentrations of monazite. The thorium anomalies in the mouth bar are also attributed to monazite. However, the mechanism of monazite concentration is unclear from field evidence and requires further investigation.

The Facies 3 mouth-bar sands contain slightly less K-feldspar and have higher quartz/feldspar ratios than the Facies 1 and 2 sandstones. They are medium grained and micaceous with a higher silt content than the coarser-grained sands. The higher thorium contents of these sands can be

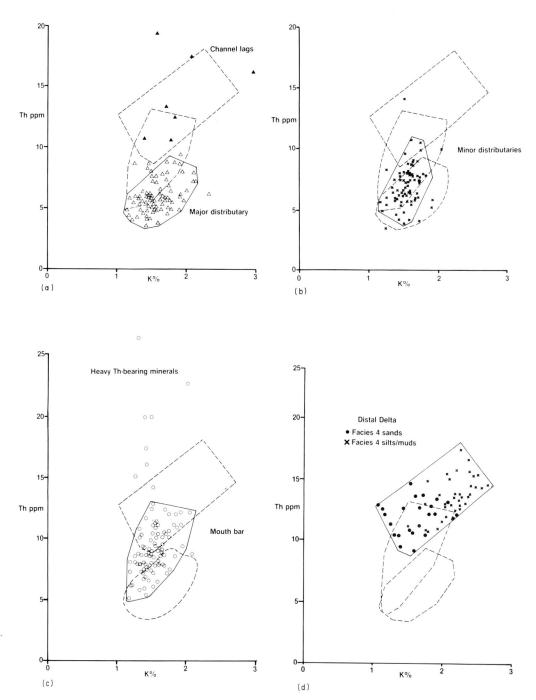

FIG. 4. K–Th cross-plots can be used to distinguish the different facies identified from sedimentary analysis: (*a*)

explained by deposition of thorium-bearing accessory minerals in the silt fraction.

The quartz/feldspar ratio, which is often cited as a measure of mineralogical maturity (Pettijohn 1975), increases with decreasing grain size owing to a progressive loss of feldspar. However, the average potassium content does not change significantly from the very coarse (Facies 1) to fine-grained sand (Table 1), indicating that the loss of potassium in K-feldspar is balanced by a gain from micas and other clay minerals. It is only in the silt/clay grain size of the distal delta and abandonment facies that the gain in potassium from micas and clays outweighs the loss from decreased feldspar, resulting in a net increase in potassium content.

The thorium and potassium contents and the Th/K ratios appear to be strongly influenced by grain size. The grain size is, in turn, related to sedimentary facies and processes. The changes in grain size are accompanied by changes in mineralogy, and it is these compositional changes which can be used to discriminate between different depositional environments. It is important to note that, while different depositional environments have been successfully defined using gamma radiation in this study, the same gamma-radiation facies cannot be used to interpret depositional environments in other basins where the detrital mineralogy (largely controlled by provenance) is different.

More outcrop studies of this type on a wide range of sedimentary environments are required before the full potential of gamma-ray spectrometry in subsurface facies analysis is realized.

Conclusions

Gamma radioactivity in the Rough Rock is controlled by the concentration of K-feldspar, clay minerals and monazite. There is a general upward decrease in gamma radiation through the entire outcrop which correlates with an upward increase in grain size. The change in gamma radiation is a punctuated trend which can be correlated with changes in facies, enabling lateral correlation to be made between units. Several total gamma-radiation peaks correlate with the abandonment deposits (Facies 5). Three of the five different facies can be discriminated using the Th/K ratio which reflects changes in the original mineralogy. The Th/K ratio decreases upwards (roughly paralleling the quartz/feldspar ratio). There are also other laterally discontinuous gamma-radiation peaks which do not correlate with facies or grain-size changes. These are caused by concentrations of the thorium-bearing mineral monazite in mouth-bar sandstones and basal channel lags. The presence of these peaks could lead to erroneous correlations in the subsurface. Gamma-ray spectrometry allows these non-correlatable peaks to be distinguished.

ACKNOWLEDGMENTS: This work was undertaken as a part of PhD research projects at the University of Leeds and the Imperial College of Science and Technology. C.S.B. would like to thank Shell and Texaco for funding this study. K.J.M. would like to thank Fina Exploration Ltd for funding his research. Both C.S.B. and K.J.M. thank British Petroleum for permission to publish the results.

References

CANT, D. J. 1983. Subsurface sedimentology. *Geoscience Canada* **10**, 115–121.

DREWERY, S., CLIFF, R. A. & LEEDER, M. R. 1987. Provenance of Carboniferous sandstones from U–Pb dating of zircons, *Nature* **325**, 50–53.

GILLIGAN, A. 1919. The petrography of the Millstone Grit of Yorkshire. *Quarterly Journal of the Geology Society of London* **75**, 251–294.

LANGMUIR, Y. & HERMAN, J. S. 1980. The mobility of thorium in natural waters at low temperatures. *Geochimica et Cosmochimica Acta* **44**, 1753–1766.

MYERS, K. J. 1987. *Onshore Outcrop Gamma-ray Spectrometry as a Tool in Sedimentological Studies.* Unpublished PhD thesis, University of London.

——. The origin of the Cleveland Ironstone formation of North East Yorkshire—new evidence from portable gamma ray spectrometry. In: *Phanerozoic Ironstones.* Geological Society of London Special Publication, in press.

—— & WIGNALL, P. 1987. Understanding Jurassic organic-rich mudrocks: new evidence using gamma ray spectrometry and palaeo-ecology: examples from the Kimmeridge clay of Dorset and the Jet Rock of Yorkshire. In: LEGGETT, J. K. (ed.) *Marine Clastic Sedimentology: Developments and Case Studies.* Graham & Trotman, London.

PETTIJOHN, F. J. 1975. *Sedimentary Rocks* (3rd edn). Harper & Row, New York, 628 pp.

SELLEY, R. C. 1985. *Ancient Sedimentary Environments* (3rd edn). Chapman & Hall, London.

K. J. MYERS, BP Research Centre, Sunbury-on-Thames, Middlesex TW16 7LN, UK.

C. S. BRISTOW, British Petroleum, Exploration and Production Department, Britannic House, London EC2Y 9BU, UK.

Deltaic reservoir prediction from rotational dipmeter patterns

R. C. Selley

SUMMARY: The dipmeter is a useful logging tool for predicting the distribution of petroleum reservoirs in deltaic sediments, particularly in the Tertiary deltas of the Gulf Coast petroleum province of the USA. Conventionally upward-coarsening genetic increments are defined on gamma or sonic potential logs. Then progradational dip motifs are back-plotted to define the apex of the delta lobe and hence the highest sand-to-shale ratio of the reservoir. This method can also be applied to lobate deposits of alluvial and submarine fan origin.

The dipmeter technique can be extended by noting how depositional dips rotate as a fan lobe progrades. When viewed down-fan the depositional dip slope rotates clockwise on the right flank and anticlockwise on the left flank. These rotational dips may be detectable on the dipmeter log. Rotating dips are poorly displayed on the conventional tadpole plot but are readily apparent on graphic vector plots. Thus it may be possible to determine which side of a fan lobe is well penetrated. Dip data from two wells may define the origin and termination of the fan axis and hence locate the parent distributary channel that may contain the best reservoir sand development.

Because deltas are such important reservoirs much work has been done in applying geophysical logs to their subsurface facies analysis (*eg* Weber 1971; Selley 1975; Rider & Laurier 1979). Particular attention has been paid to the interpretation of the dipmeter log, particularly in the Tertiary sediments of the Gulf Coast of the USA (Gilreath 1960; Gilreath & Maricelli 1964; Gilreath *et al.* 1969; Gilreath & Stephens 1971). The basic gamma, sonic potential (SP) and dipmeter log motifs of deltaic mouth bars, distributary channels and crevasse splays are reviewed in textbooks by Klein (1980), Galloway & Hobday (1983) and Selley (1985).

Conventionally the first step in the subsurface facies analysis of any petroleum reservoir is to prepare a detailed sedimentological log of all available cores. These can then be used to identify depositional environments and to calibrate the geophysical well logs. Gamma and/or SP logs can then be used to define the genetic increments. Major sand bodies can be recognized, integrated with seismic data and correlated from well to well. The dipmeter log can then be used to attempt to interpret palaeocurrent patterns so as to predict reservoir geometry and trend.

The dipmeter tool normally consists of four spring-loaded arms bearing pairs of focused electrodes of various numbers and spacings. These measure the resistivity of the formation as the sonde is drawn up the borehole. Variations in resistivity often reflect lithological changes and, if correlated from pad to pad, may define the amount and dip of bedding planes. Early dipmeter logs only gave a bed resolution of about 1 m. Modern devices can compute dips down to a resolution of 3 cm. There are many different computer programs to make resistivity correlations and dip computations. New programs are constantly evolving in response to the development of new dipmeter sondes. Programs and processing parameters are selected according to the tool type, reservoir facies and purpose of the survey. Dipmeter computations may be presented in a tabular format. It is more helpful to display them in what is conventionally termed a tadpole plot (*eg* Fig. 6). In this log format the head of a tadpole plotted at a given depth shows the amount of dip according to a horizontal scale, while the tadpole tail points in the dip direction. For further details of the dipmeter log see Selley (1985, pp. 27–35) for a simple résumé, or the manuals of the wireline logging service companies for exhaustive accounts.

The dipmeter log is as prone to misinterpretation as seismic data, with its own particular pitfalls of data acquisition, processing and interpretation. It cannot be too strongly emphasized that the dipmeter log must never be used in isolation. As with any geophysical log it can only be used when integrated with all other available sedimentological, palaeontological, seismic and regional data. Attempts can then be made not only to identify structural tilt but, after this has been removed, to see cross-bedding dips and thus to interpret channel orientation and trend. This practice is often extremely difficult for several reasons. Most channels are infilled with homogeneous sands that lack good resistivity contrast for the focused micrologs of the dipmeter to measure. Furthermore, cross-bedded channel sands commonly have polyplanar fabrics that are difficult to interpret. Too few good-quality dips to be statistically significant may be obtained in

From WHATELEY, M. K. G. & PICKERING, K. T. (eds), 1989, *Deltas: Sites and Traps for Fossil Fuels,* Geological Society Special Publication No. 41, pp. 89–95.

any one channel. However, successful interpret-ations of channel dipmeter patterns have been reported from the US Gulf Coast, Nigeria, eastern Venezuela and Indonesia (J. A. Gilreath, pers. comm.).

It may often be easier and safer to identify the upward-increasing 'blue' motifs associated with the progradational phase of delta outbuilding. These dips are commonly well defined, being picked out by interlaminated sand and shale that give good resistivity contrast. Furthermore, the dips are normally consistent in amount and azimuth within any one increment. It is often remarked that the amount of dip may be of the order of 10°–15° or even more, and that these amounts are in excess of the angle of slope observed on recent delta slopes and submarine fans. The progradational dips in one carefully correlated increment can then be back-plotted from two or more wells. Thus it may be possible to locate the apex of the delta lobe, and hence drill wells where the sand-to-shale ratio is at a maximum.

The purpose of this paper is to describe an extension of this technique. It was specially developed during a consulting assignment on the subsurface facies analysis of the Brae field in the British sector of the North Sea. The Brae field and associated discoveries occur in a complex series of Upper Jurassic submarine fans along the western side of the Viking Graben (Stow *et al.* 1982; Turner *et al.* 1987). The technique of dipmeter interpretation described in this paper is equally applicable, however, to fluvial and deltaic fan systems.

Rotating palaeocurrents and depositional dips introduced

The concept of rotational palaeocurrents is an old one, although perhaps unfamiliar to many geologists. It was first recognized by Chamberlin (1888), who noted how radial flow directions rotate in a prograding ice lobe. He pointed out that, when viewed down the lobe, the palaeo-currents rotate anticlockwise on the left flank and clockwise on the right flank.

Rotating palaeocurrents have been recorded in the braided alluvial Applecross Group of the Torridonian (Precambrian) of NW Scotland. In a regional study Williams (1969) noted that palaeocurrents determined from cross-bedding rotated in consistent directions when plotted on vertical sections. He related these patterns to the retreat of a scarp as the fans regressed across a pediment. In this case, unlike a prograding fan,

palaeocurrents rotate clockwise on the left-hand side of the fan axis and anticlockwise on the right-hand side. The concept that palaeocurrents rotate as fans prograde or regress is applicable not only to glacial and alluvial sediments but also to other environments that generate lobate sedi-ments, notably deltas and submarine fans.

Figure 1 illustrates the single genetic increment of a prograding sediment lobe. The scale is immaterial and the setting could be a submarine fan, a delta front, a crevasse splay or even a washover fan on the lagoonal side of a barrier island. Note how, as the lobe progrades from W to E, the palaeoslope remains unaltered along the axis of the lobe but rotates on its flanks. On the northern (left-hand) side the rotation is anticlock-wise. On the southern (right-hand) side the rotation is clockwise. Consider the logs of the three wells A, B and C shown in Fig. 1. The gamma logs show the typical upward-coarsening sequence of a progradational increment. The dipmeter shows the upward-increasing 'blue' pattern expected in such a situation. The dipmeter picks out the gently inclined sand and clay laminae that were successively deposited on the surface of the fan, with the amount of dip increasing up the increment. Notice how the tadpoles in well A rotate anticlockwise up the increment. In well B, which is on the fan axis, the dips all point in the same easterly direction, and in well C, which is drilled on the right flank of the fan, they rotate clockwise upward. Figure 2 shows how the highest and lowest dips in wells A and C can be used to locate the origin and termination of the fan axis. The approximate extent of the distributary channel can thus be defined. Wells drilled along this trend may be expected to locate good-quality reservoir sands. The theory must now be tested in recent and ancient sediments.

Rotating dips in recent deltas

Figure 3 shows cross-sections through two recent deltas. They were constructed using successive bathymetric surveys to show how the depositional slopes have evolved with time. Both cross-sections show upward-increasing angles of dip at any one location, albeit with considerable local fluctuations. The Cubits Gap lobe is a crevasse splay on the eastern side of the Head of Passes of the Mississippi Delta. Figure 4 illustrates succes-sive bathymetric charts of the Bay Rondo, into which sediment splayed from Cubits Gap. The first map shows bathymetry in 1838, 22 years before the crevasse opened. The map of 1870, 10 years after the gap opened, shows little change.

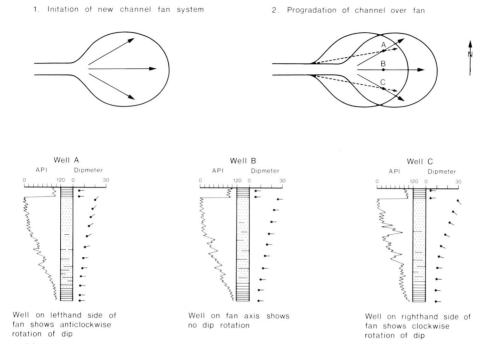

FIG. 1. Diagrams to illustrate the rotating slopes of a prograding fan lobe. Note the anticlockwise and clockwise rotation of the slopes on the northern and southern flanks. The lower part of the figure shows gamma and dipmeter logs for wells A, B and C.

However, by 1877, 17 years after the event, a major lobe of sediment had prograded northeastwards into Bay Rondo. The growth rate had slowed by 1903 and had terminated by 1953, when the crevasse splay had built up to sea level and had aquired a swamp veneer.

Figure 5 shows how the depositional dip varied through time as the crevasse splay prograded. A number of 'well' locations have been arbitrarily selected. The 1838 map reflects dips prior to

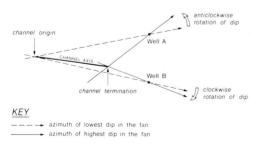

KEY

– – – – ➤ azimuth of lowest dip in the fan
————➤ azimuth of highest dip in the fan

Channel axis, origin and termination found by plotting highest and lowest dips in each well.

FIG. 2. Diagram showing how two wells A and B drilled on the left and right flanks of a fan can define the origin and termination of the fan axis, and hence can locate distributary channel sand reservoirs.

progradation and the 1877 map shows dips at the culmination of progradation. The third map in Fig. 5 shows the amount of rotation of the depositional dips at the various well locations. Rotation is absent away from the crevasse splay and along its axis. It is anticlockwise on its northern (left-hand) flank and clockwise on its southern (right-hand) flank. The amount of rotation observed is as much as 90° at one well location. These rotational changes support the thesis previously advanced and clearly demonstrate that rotational depositional dips occur on prograding sediment lobes.

Rotational dips in ancient sediments

The dipmeter has been extensively used in the petroliferous Tertiary sediments of the US Gulf of Mexico. An example from Montgomery County, Texas, shows how rotational dips are consistent with the orientation of a prograding mouth bar. Figure 6 shows SP and dipmeter logs through the sand. There is no clearly defined motif on the SP curve. However, the dipmeter log shows an upward-increasing 'blue' motif through the lower part of the sand and an upward-decreasing 'red' motif towards the top. A 90° dip

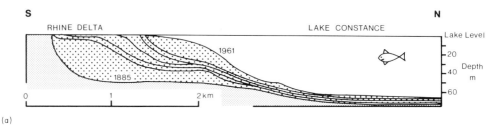

(a)

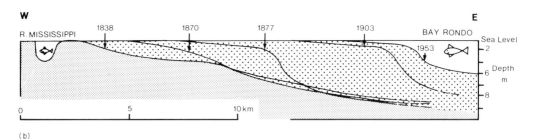

(b)

FIG. 3. (a) Cross-section through the delta produced where the River Rhine flows into Lake Constance (from Waibel 1962); (b) cross-section through Bay Rondo on the Mississippi Delta plain showing the progradation of the Cubits Gap crevasse splay (for location see Fig. 4).

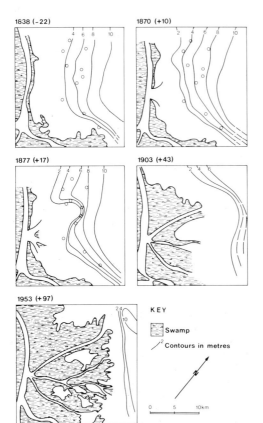

switch occurs at the 'blue–red' transition. This is typical of the transition from down-slope prograding fan to thalweg inclined channel fill in both deltaic and submarine fans (*eg* Selley 1985, pp. 141, 278). Figure 7 shows the dips through the sand plotted in vertical sequence using the graphic vector mean method (Raup & Miesch 1957). When looking for rotational dips this is a far superior way of displaying dipmeter data than the conventional tadpole plot. The tadpole plot is suitable for illustrating vertical changes in the amount of dip but is ineffective for displaying vertical dip switches and rotations. In the graphic display, dips are plotted in sequence up (or down) the well. The azimuth of each dip is plotted for a consistent unit length (*eg* 1 cm). The line drawn for each dip originates at the termination of the previous dip line in the sequence. No account is taken of the amount of dip. Graphic vector plots can be used to identify mean dip directions by drawing a line from the origin of the first reading to the termination of the final one. Computer graphic software for plotting dipmeter data using

FIG. 4. Bathymetric maps of Bay Rondo on the Mississippi Delta plain showing the progradation of the Cubits Gap crevasse splay: ○, location of arbitrarily selected 'wells' whose significance is shown in Fig. 5. (Modified from Welder 1959.)

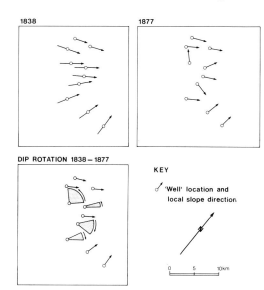

DIP ROTATION 1838 – 1877

KEY

↗ 'Well' location and local slope direction

0   5   10km

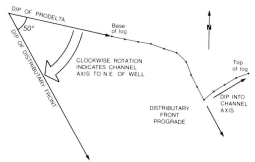

FIG. 7. Graphic vector plot of dipmeter data through the Montgomery County sand body illustrated in Fig. 6. Note the clockwise rotation up the prograde and the 90° dip switch at the prograde–channel contact. These indicate deposition on the southern flank of a southeasterly prograding mouth bar. The overlying parent channel trend is coincident with this trend.

FIG. 5. Maps of Bay Rondo showing the changing direction of the depositional slope at the various 'well' locations during the progradation of the Cubits Gap crevasse splay. Note the clockwise rotation on the southern (right) flank, the anticlockwise rotation on the northern (left) flank and the absence of rotation down the fan axis.

the vector mean method has long been available. The main use of this display is that it highlights vertical dip rotations and dramatic vertical dip switches. This is well illustrated in Fig. 7. Moving up from the base of the sand the dips show a gradual clockwise rotation from N105°E at the base to N150°E. At this depth there is a 90° dip switch to the NE. Reference back to Fig. 6 shows that this occurs at the 'blue–red' transition. According to the model the clockwise dip rotation in the lower part of the sand reflects progradation on the right flank of a southeasterly prograding mouth bar. The northeasterly dips in the upper 'red' section would be conventionally interpreted as indicating that the well penetrated the SW flank of a channel with a NW–SE axis. Figure 8 is an isopach map of the sand body that confirms that this interpretation is indeed correct.

Dipmeter logs are very useful in the US Gulf Coast petroleum province, as the Montgomery County example illustrates. Such good results are not always obtained in the North Sea, partly because of poor hole conditions, related to overpressuring, and the use of oil-based drilling muds. Figure 9 shows how the concept of rotational dips was used in the development drilling of the hypothetical Laverock oilfield in Block 211/9. A seismically defined tilted fault block was tested by the 211/9–1 well. A thick oil column in a saw-toothed upward-coarsening sandy increment with a well-defined 'blue' dip motif was found. An alert facies analyst noted that the dips rotated anticlockwise from N72°E at the base of the increment to N60°E at the top. In an attempt to locate a better-quality reservoir the 211/9–2 well was then drilled on a crestal position further S. This well also found an oil-

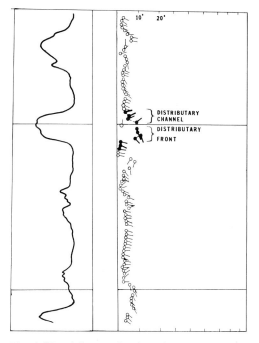

DISTRIBUTARY CHANNEL

DISTRIBUTARY FRONT

FIG. 6. SP and dipmeter log through a Tertiary mouth bar sand, Montgomery County, Texas. (© J. A. Gilreath; reproduced with permission.)

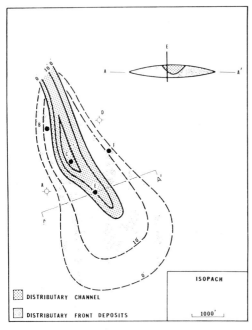

FIG. 8. Isopach map of the Montgomery County mouth bar sand. The log of well E is shown in Fig. 6. Note that the geometry and trend of the sand are consistent with the interpretation of the dipmeter log, *ie* well E penetrated the right-hand flank of a southeasterly prograding fan lobe with a NW–SE trending distributary channel. (© J. A, Gilreath; reproduced with permission.)

bearing, but poorly developed, reservoir. Again a 'blue' motif occurred on the dipmeter log over the reservoir interval, but this time the dips rotated clockwise from N92°E at the base to N102°E at the top. The 211/9–3 well was then drilled on the western flank of the structure. The rationale for this location was that it would penetrate the proximal part of the fan, where a thick reservoir should be developed. This indeed was the case. Not only was the prograde very sandy, but the southern margin of a feeder channel was also encountered. Sadly, however, the well was dry as it hit the top of the reservoir below the oil–water contact. There are no further plans for development drilling on the Laverock field.

Conclusions

Integrated dipmeter analysis has long been used to develop deltaic petroleum reservoirs. Cross-bedding is seldom easy to identify and interpret from dipmeter logs because of the homogeneous nature of many sands and because of their

polyplanar fabrics. The progradational dip motifs of lobate sands, such as mouth bars and crevasse splays, are more amenable to dipmeter interpretation. Thinly interlaminated sand and shale give good resistivity contrast, and the dips are generally uniplanar and vertically consistent. Conventionally progradational dips have been back-plotted from two or more wells within a single prograding sand body. Thus the apex of the lobe can be identified and wells can be drilled in the area with the highest sand-to-shale ratio. This technique can be refined by taking cognizance of the fact that the slope of a prograding fan rotates anticlockwise on its left flank and clockwise on its right flank. These rotating palaeoslopes can be detected on the dipmeter log. It may thus be possible to determine not only that a well has penetrated a fan but also whether it has penetrated the left or right flank. Two wells can be used to pinpoint the origin and termination of the fan axis and hence the location of the parent feeder channel. This technique is applicable to lobate sand bodies in deltaic and other depositional environments.

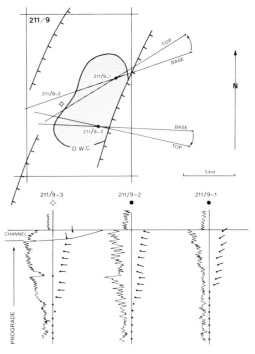

FIG. 9. Maps and logs of the hypothetical Laverock oil field. The 211/9–1 and 211/9–2 wells penetrated poor-quality sands on the left and right flanks of a prograding fan. Dips rotate anticlockwise and clockwise on the left and right flanks of the lobe respectively. The 211/9–3 well found thick sand at the fan apex but was below the oil–water contact.

ACKNOWLEDGMENTS: I am grateful to Marathon International Petroleum (GB) Ltd and their partners for the opportunity of studying the Brae field and thus initiating the geospasm presented in this paper. I am grateful to J. A. Gilreath of Schlumberger for critically reviewing the manuscript and for information on the interpretation of dipmeters in the US Gulf Coast province in general and the Montgomery County example in particular.

References

CHAMBERLIN, T. C. 1888. Rock scourings of the great ice invasion. *7th Annual Report of the U.S. Geological Survey*, 218–223.

GALLOWAY, W. E. & HOBDAY, D. K. 1983. *Terrigenous Clastic Depositional Systems: Applications to Petroleum, Coal and Uranium Exploration*. Springer-Verlag, Berlin, 423 pp.

GILREATH, J.A. 1960. Interpretation of dipmeter surveys in Mississippi. *Transactions of the Gulf Coast Association of Petroleum Geologists* **10**, 267–275.

—— & MARICELLI, J. M. 1964. Detailed stratigraphic control through dip computations. *American Association of Petroleum Geologists Bulletin* **48**, 1902–1910.

—— & STEPHENS, R. W. 1971. Distributary front deposits interpreted from dipmeter patterns. *Transactions of the Gulf Coast Association of Petroleum Geologists* **21**, 2–8.

——, HEALY, J. S. & YELVERTON, J. N. 1969. Depositional environments defined by dipmeter interpretation. *Transactions of the Gulf Coast Association of Petroleum Geologists* **19**, 101–111.

KLEIN, G. DE V. 1980. *Sandstone Depositional Models for Exploration for Fossil Fuels* (2nd edn). Burgess, Minneapolis, MN, 149 pp.

RAUP, O. B. & MIESCH, A. T. 1957. A new method for obtaining significant average directional measurements in cross-stratification studies. *Journal of Sedimentary Petrology* **27**, 313–321.

RIDER, M. H. & LAURIER, D. 1979. Sedimentology using a computer treatment of well logs. *Transactions of the Society of Professional Well Log Analysts, 6th European Symposium, London*, Paper J.

SELLEY, R. C. 1975. Subsurface diagnosis of deltaic deposits with reference to the northern North Sea. *In*: FINSTAD, K. & SELLEY, R. C. (eds) *Proceedings of the Jurassic Northern North Sea Conference*. Norwegian Petroleum Society, Oslo.

—— 1985. *Ancient Sedimentary Environments and their Subsurface Diagnosis* (3rd edn). Associated Book Publishers, London, 317 pp.

STOW, D. A. V., BISHOP, C. D. & MILLS, S. J. 1982. Sedimentology of the Brae oilfield, North Sea. Fan models and controls. *Journal of Petroleum Geology* **5**, 129–148.

TURNER, C. C., COHEN, J. M., CONNELL, E. R. & COOPER, D. M. 1987. A depositional model for the South Brae oilfield. *In*: BROOKS, J. & GLENNIE, K. (eds) *Petroleum Geology of North West Europe*. Graham & Trotman, London, 853–864.

WAIBEL, F. 1962. Das Rheindelta im Bodensee. *Bei richt des Osterreichischen Rheinbauleiters, Seegrundaufrahme, 1961.*

WEBER, K. J. 1971. Sedimentological aspects of the Niger delta. *Geologie Mijnbouw* **50**, 559–576.

WELDER, F. A. 1959. Processes of deltaic sedimentation in the lower Mississippi River. *Technical Report* **12**, Coastal Studies Institute, Louisiana State University, Baton Rouge, LA, 90 pp.

WILLIAMS, G. E. 1969. Characteristics and origin of a Precambrian pediment. *Journal of Geology* **77**, 183–207.

R. C. SELLEY, Department of Geology, Imperial College, University of London, London, UK.

Selected Delta Case Studies

Nile Delta: a review of depositional environments and geological history

G. Sestini

SUMMARY: The distribution of the Holocene–Recent sedimentary environments of the Nile Delta is related to secular variations of the semi-arid climate and of the Nile River sediment input and flux, and to coastal subsidence and high wave energy, together with an eastward littoral drift. The delta started to form in the Late Pliocene, with its main development taking place in the Pleistocene, associated with progradation of large volumes of coarse coastal delta sands and the offshore accumulation of turbidites (Nile Cone). In the Oligocene–Early Miocene, an ancestral Nile discharged to the NW. Fluvio-deltaic deposits appeared in the present delta area in the Late Miocene, where they built a narrow prism close to an active faulted flexure, which in the subsurface now separates the South Delta Block from the deep North Delta Basin. The stratigraphy, structure and depocentre development of the region are functions of the interaction between the E–W- to WSW–ENE-trending Mediterranean margin structures, active since the end of the Cretaceous, and the NW-, NNW- and NNE-trending Red Sea–Gulf of Suez structural fabric associated with arching and rifting from the Oligocene to Pleistocene. The main phases of tectonic activity and of clastic deposition, which also correlate with lowered sea levels, were in Late Oligocene–Early Miocene, late Middle Miocene, Tortonian–Messinian, Middle Pliocene and Early–Middle Pleistocene. Petroleum exploration in the Nile Delta has produced modest results to date with a few medium-sized gas fields and minor oil discoveries.

No delta could be more classical in shape than that of the Nile River which was compared by Herodotus in the fifth century BC to the shape of the Greek letter 'delta'. Despite this classical history, the geographical and geological study of the Nile Delta has been limited until recently. For instance, the Nile Delta has not been subjected to systematic surface and subsurface studies of the type that were undertaken in the deltas of the Mississippi, Niger and Rhone in the 1950s and 1960s when delta studies became relevant to models for petroleum exploration (Kruit 1955; Scruton 1960; Allen 1970; Oomkens 1970).

Geologically the Nile Delta also includes the continental shelf from 80 km W of Alexandria to N Sinai, the continental slope and the Nile Submarine Fan (Nile Cone) (Fig. 1). Other areas in the SE Mediterranean genetically related to the Neogene–Recent Nile sediments include the continental shelf of S Israel, the Herodotus Abyssal Plain and the southern part of the Levant Platform. Turbidites derived from the Nile Delta also occur in the Mediterranean Ridge (Bartolini et al. 1975).

Tectonics has played a dominant role in the location and in the structural and depositional history of the Nile Delta. The region occupies a key position in the context of the plate-tectonic evolution of the eastern Mediterranean and the Red Sea. It lies on the moderately deformed external margin of the African Plate (the 'Unstable Shelf' of Said 1962), which extends well N

of the Mediterranean coast (Sancho et al. 1973; Woodside 1977; Sestini 1984). Its structural development is largely contemporaneous with the break-up of the plate margin consequent on the opening of the Red Sea and the northward translation of the Arabian Peninsula. In response to these movements there has been much post-Messinian deformation (including halokinetic phenomena) in the SE Mediterranean (Neev et al. 1976; Ross et al. 1978). The African Plate still appears to be actively subducting under the Aegean and Turkish Plates (cf the Crete and Cyprus arches), with the Mediterranean Ridge being the product of the ensuing compression (Mulder et al. 1975; Finetti 1976).

Geographical and geomorphological studies of the Nile Delta have been few and generally superficial (Said 1958; Sestini 1976; Abdel Kader 1982; Frihy et al. 1988). There are no comprehensive reviews, except for information on the location of former Nile channels (Daressy 1929; Tousson 1934; Ball 1942; Bietak 1966; Sneh & Weissbrod 1973; Butzer 1976; Coutellier & Stanley 1987) and coastal changes (UNESCO/UNDP 1976; UNDP/UNESCO 1978; Frihy 1987). Information on the subsurface remains limited. Early knowledge was derived from boreholes drilled for water (Attia 1954; Shata & El Fayoumi 1970). Later petroleum exploration has produced over 90 wells and hundreds of kilometres of reflection seismic records, but regrettably most of the resulting information remains confidential, particularly with regard to

G. Sestini

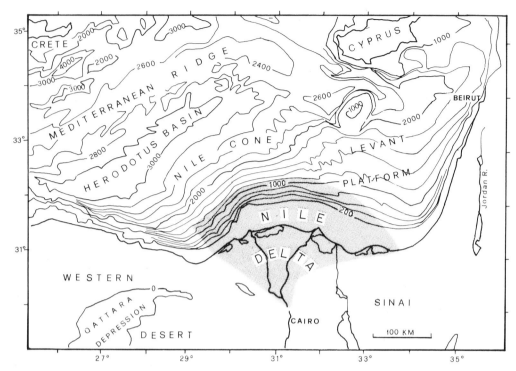

FIG. 1. Principal geographical features of the Nile Delta region and of the Levant Basin of the eastern Mediterranean Sea.

structure. Nevertheless, a certain amount has become available (Viotti & Mansour 1969; Omara & Ouda 1972; Abu Sakr 1973; Khairy 1974; Zaghloul 1976; Bentz & Gutman 1977; Korrat 1977; El Nahass 1978; Rizzini *et al.* 1978; Kora 1980; Barber 1981; Deibis 1982; Elf Aquitaine 1982; Taylor & Jones 1982; Effat & Gezeiry 1986) and has generated stratigraphic and regional interpretations (Hantar 1975; Salem 1976; Zaghloul *et al.* 1977*b*, 1979*a, b*; Barber 1980; Said 1981; El Heiny 1982; Deibis *et al.* 1986).

Information has been gathered on the nature and distribution of sediments and on the structure of the submarine parts of the Nile Delta, particularly the Nile Cone and the Levant Platform (Maldonado & Stanley 1976; Neev *et al.* 1976; Summerhayes *et al.* 1977; Woodside 1977; Nur & Ben Avraham 1978; Ross *et al.* 1978; Moreau-Martin 1979; Ross & Uchupi 1979), the continental shelf of Egypt (El Wakeel *et al.* 1974; Summerhayes *et al.* 1978; El Sayed 1979; Coleman *et al.* 1980), and Israel–N Sinai (Derin & Reiss 1973; Neev *et al.* 1976; Gvirtzman & Buchbinder 1978; Nur & Ben Avraham 1978; Ben Avraham & Mart 1981; Mart & Ben Gai 1982; Garfunkel 1984; Mart 1984, 1988).

The Nile Delta lies in a subtropical desert environment, is microtidal and 'wave dominated' (Wright & Coleman 1973) and at present is in an abandonment phase. With the control of Nile floods, the advent of permanent irrigation and particularly after the closure of the Aswan High Dam in 1964, summer deposition of sediment no longer compensates for delta plain subsidence and the large-scale shoreline erosion.

To what extent can this modern delta be used as a model for the interpretation of ancient deltaic successions? The evidence is that the present and Holocene Nile Delta environments are quite superficial. In the subsurface of the Nile Delta the Pleistocene reveals other delta types related to a larger river input of coarser clastics and to a different climatic setting in both the hinterland and the receiving basin.

Exploration for hydrocarbons has been conducted in the Nile Delta during the last 25 years on the assumption of a depositional and structural history (*eg* continuous progradation, subsidence and subsidence-related structures) comparable with that of the oil-rich deltas of the Mississippi and Niger. The results have been only moderately encouraging, with a few medium-sized gas discoveries and some minor oil discoveries (Ansary

& Deibis 1977; Ayouti 1980; Shawky Abdine 1981; Deibis 1982).

The aim of this paper is to present a summary, based on available data, of what is known about the Nile Delta today regarding modern environments and processes, and subsurface stratigraphy and structure. The geological history of the region and its bearing on petroleum habitats are also reviewed.

The factors of deltaic development

The Nile

The Nile is 6690 km long, drains a basin of area 2 880 000 km^2 and derives its waters almost entirely from the E African highlands, supplied by summer monsoon rains, especially over the Ethiopian Plateau (the Kenya Plateau contribution, through the White Nile is largely lost in the Sudd swamps of southern Sudan). After the Atbara, N of Khartoum, it receives no effluents, yet still flows for more than 2000 km to the sea.

The Nile crosses no less than five vast regions of different structural and geological history, namely the Lakes Plateau, the Sudd, Central Sudan, the Ethiopian Highlands, the cataracts region of Sudan–upper Egypt and Egypt itself. The present river course through these regions consists of different segments which became connected rather recently in geological history (Butzer & Hansen 1968; Said 1981). The Ethiopian source does not appear to have become relevant for the Nile in Egypt until the Pleistocene (Hassan, cited by Wendorf & Schild 1978).

The sediment discharged by the Nile at the Mediterranean coast is strongly affected by the hydrodynamic regime generated by the winter storm and wind circulation systems of the northern hemisphere. The Nile has a relatively low discharge rate—eight to ten times smaller than rivers of comparable drainage basin area (*eg* the Mississippi, Congo and Parana) and the same mean discharge as the Po and Rhone, which drain basins 40 times smaller (Coleman 1982).

From 1902 to 1964, discharge at Aswan averaged 84 billion m^3 of water (Hurst 1931–1966), mostly (93%) between July and November with a peak in the flood months of August and September. The average sediment load was 160×10^6 tonnes a^{-1} (varying between 50 and 300×10^6 tonnes a^{-1} (Quelennec & Kruk 1976)), largely carried in suspension, in the ratio of 25% fine to very fine grained to medium silt (0.125–0.008 mm) and 75% fine silt and clay. The amount of bed load, although never accurately measured,

has been estimated as possibly 1%–10% (Inman & Jenkins 1984). Between Aswan and the sea, as much as 60% of water discharge is lost through irrigation, evaporation and seepage (Sharaf El Din 1977). The loss of silt–clay over agricultural land in the Nile Valley and the Delta may have been up to 30%–35% of the Aswan load, with possibly 5–10 tonnes (or more?) of sediment deposited annually over the delta surface. Thus, before 1964, actual water discharge into the sea averaged 50×10^9 m^3 a^{-1} (65% through the Rosetta Nile branch) and sediment discharge averaged $(100–115) \times 10^6$ m^3 a^{-1} of which only a quarter was sand to coarse-silt grade.

Before 1902, however, the Aswan discharge was in general 25% greater (average 120×10^9 m^3 a^{-1}), with correspondingly greater sediment loads and more silt–clay deposition over the delta, where flood basin irrigation was still practised (controlled annual irrigation began with the construction of the Delta Barrage in 1846). The higher discharge was due to heavier rainfall over Ethiopia (Rossignol-Strick 1983).

The most significant aspect of the Nile River discharge for delta construction was its variability. Records of flood heights are available from the Roda (Cairo) Nilometer from 622 AD, and from the Delta Barrage gauge from 1846 (Tousson 1934). These records permitted the calculation of flood volumes and suspended load variability (Quelennec & Kruk 1976). Despite large amplitude variations (30% or more of annual flood discharge), there is a recognizable and statistically valid small-scale cyclicity between high and low peaks of 8–12 years and 18–20 years (Bell 1975; Hassan 1981; Hamid 1984).

Larger cycles include 60–90 year periods which can be verified back to the tenth century AD. Years of large Nile floods are associated with considerable tropical monsoon rains and also with high pressure and more stormy weather over northern Europe and the eastern Atlantic. In addition, there are longer-term variations spanning several centuries; for instance, higher floods occurred in the period 500 BC–400 AD, and from about 1400 AD to the late 1800s, while low floods occurred from the fifth to the ninth centuries AD (Butzer 1976; Said 1981).

Over the last 1000 years there has been a substantial change in the position of the distributary systems. Before the second to fifth centuries AD, there were several Nile branches with at least six channel mouths (Canopic, Saitic, Sebennytic, Tanitic, Mendesian and Pelusiac). According to Herodotus and Strabo (fifth to third centuries BC), the main river flow was through the western and central branches, while the later Rosetta and Damietta branches were still mere canals. Information on Nile discharge in Dynastic times (3050–330 BC) is less reliable; however, historical and archaeological records indicate the recurrence of the 60–100 year periodicity and secularly declining floods (Hassan 1981; Said 1981).

Much larger variations have been inferred from the Late Pleistocene sediments of the Nile Valley (Butzer & Hansen 1968; Wendorf & Schild 1978; Said 1981). The following are the most notable.

(1) Periods of high discharge, *eg* 11 500–10 800 BP (Adamson *et al.* 1980) which was an exceptionally wet period over E Africa and caused a rise in the water level of Lake Victoria so that it spilled into the White Nile, a threefold increase in rainfall in the Sudan and enormous aberrant floods in the Nile. The older cross-bedded pebbly Qena sands of upper Egypt also suggest a large bed load, most of which was transported by traction (Said 1981).

(2) Periods of low discharge, such as the arid interval 19 000–13 000 BP (during the Late Würmian), with a sluggish Nile which was perhaps dry for most of the year. Long periods of very reduced or zero sediment contribution from the Nile, associated with a very arid climate, also occurred in the early Late Pleistocene (0.7–1.85 Ma) (Said 1981).

Coastal and marine processes

Before 1964 sediment discharge at the Mediterranean coast was concentrated in the summer months, with reworking and dispersal processes mainly occurring in the winter (Fig. 2). The controlling factors of sedimentation on the delta surface were, instead, local climate (temperature, rainfall and winds) and the variability of the Nile floods.

Atmospheric circulation over the Mediterranean in summer produces constant NW surface winds that generate swells with periods of 9–10 s, heights $H^{1/3}$ of 0.40–0.75 m and lengths of 100–200 m. Wave fetch can be as much as 1000–1500 km, from the Ionian and Aegean Seas. During the winter (November–April), however, depressions moving WNW–ESE cause cyclonic storms that last 4–7 days, and up to 10 or 12 of these almost regularly strike the coast of NE Egypt each winter. Storm waves have periods of 7–8 s, wave heights of 1.5–2 m, and sometimes up to 3 m, and generally approach the coast from the NW, secondarily from the N, NNE and NE. The energy level at the coast is moderate, with a mean shoreline wave power calculated at 1.16×10^6 erg s^{-1} per metre of coast (Wright & Coleman 1973).

The Nile Delta is affected by NW to N offshore winds most of the year, with greater velocities in summer, and by S to SW onshore winds from

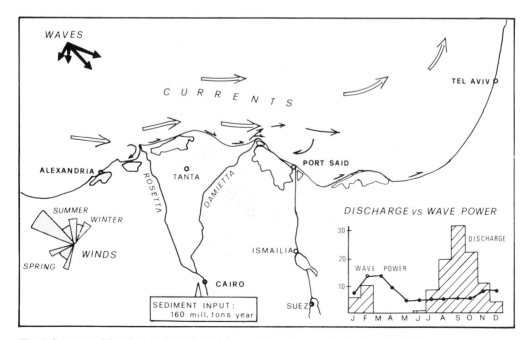

FIG. 2. Summary of the climatic, dynamic and river discharge factors affecting the Nile Delta.

December to March (including the hot sand-storm-bearing *Khamessin* in early spring).

Water circulation in the Mediterranean displays a counterclockwise gyre with surface currents off the Nile Delta commonly flowing at 0.25 cm s^{-1} (Sharaf El Din 1973). These currents have been responsible for transporting much of the fine silt–clay grade of the Nile flood to the Levantine Basin (turbid flood water could be traced as far as the Lebanon a few weeks after discharge).

Near the delta a weak clockwise eddy in Abuqir Bay and a stronger eddy (60 cm s^{-1}, Murray *et al.* 1980, 1981) off Damietta have developed. The oblique wave approach generates an active longshore current system with an eastward flow and mean velocities of 20–50 cm s^{-1}; maxima of 80–140 cm s^{-1} have been recorded off the breaker zone (ASRT/UNESCO/UNDP 1977; Sharaf El Din 1977; Manohar 1981; Fanos 1986).

These currents and littoral drift drive beach and near-shore sand along the entire SE shore of the Mediterranean as far as Israel (Emery & Neev 1962). The erosive power and magnitude of sand transport are considerable, but coastwise rates are uneven because wave refraction produces a complex pattern with pronounced zones of convergence and divergence, owing to both the variability of wave approach and intensity, and local shoals (Inman 1976; ASRT/UNESCO/UNDP 1977). Estimates vary from 10^6 m^3 of sand to 60 000 m^3 of sand moved in the shadow zones and embayments E of promontories.

Additional large amounts of sand have been and are being removed from the beaches by the offshore winds, with the build-up of extensive dunes, particularly in the central-western part of the coast (Kholief *et al.* 1977; El Fishawi & El Askary 1981) and in NW Sinai (Tsoar 1974). Estimates of sand losses are of the order of 200 000 m^3 a^{-1} W of Rosetta mouth and 400 000 m^3 a^{-1} for the Burullus–Ras El Barr coast.

The high proportion of fine-grained sediment carried in suspension by the River Nile and the high salinity of the Mediterranean (38‰, reduced to 30‰ at flood time) resulted in the formation of large eastward-deflected surface plumes off the Rosetta and Damietta distributary mouths. Turbid plumes are formed today, as shown on satellite images (ASRT/UNESCO/UNDP 1977; Klemas & Abdel Kader 1982); most sediment in suspension is in a near-shore belt 2.5–7 km wide where coastal processes are most active, with onshore, offshore and along-shore sand movement through a system of submerged bars (Manohar 1981). During winter storms, deeper fine-grained bottom sediment is stirred into suspension and moved seaward, mainly to the E (Summerhayes *et al.* 1978).

Information about bottom currents and the extent and pattern of sediment movement on the continental shelf is still limited and apparently contradictory. Manohar (1976) has concluded that summer swells and winter waves should be capable of lifting sediment to depths of about 60 m. NE of the Damietta promontory strong bottom currents have apparently created a field of actively migrating sand ridges at depths of 25–60 m (Coleman *et al.* 1980; Murray *et al.* 1981). Elsewhere (*eg* Abuqir Bay, N of Burullus and N of Bardawil), however, long-period measurements have indicated that near-bottom current velocities are greater than 10 cm s^{-1} (maximum of 20–30 cm s^{-1}) only 10%–15% of the time (ASRT/UNESCO/UNDP 1977; Gerges 1981; Manohar 1981). Suggestions for sediment dispersal patterns have also been advanced on the basis of bottom-sediment distribution (Misdorp & Sestini 1976) and changes in sea-bottom profiles (Toma & Salama 1980). Inman & Jenkins (1984) have proposed the existence of two major sand sinks N of Port Said and the Bardawil Lagoon respectively.

The role of sea level in the contemporary coastal dynamics of the Nile Delta appears to be moderate (Manohar 1981). The tidal range is small, with average spring tides of 30–40 cm. The tide may be relevant only in controlling water exchange at lagoon outlets. Winter storm surges and the wave set-up are more significant because, in combination with high tides and low barometric pressure, they can raise the sea level by 1–1.5 m, causing the regular flooding of several low-lying stretches of the coast.

The modern delta

The Nile Delta (area 12 500 km^2) is very flat (18 m above sea level at Cairo, 150 km S of the coast) and the delta plain has been cultivated for several millennia. By the mid-1950s only the coastal barrier complex, the lagoons and a zone 5–15 km wide further S were still in a natural or barely modified condition (today, much of the latter zone has been reclaimed). Study of these areas in the field, on air photographs and on topographical maps dating back to the early nineteenth century (*cf* Fig. 3 which shows the scale 1:100 000 map surveyed in the wake of Napoleon's invasion of Egypt (Jacotin 1826)) have provided substantial information on the nature and extent of depositional environments (Sestini 1976; Frihy *et al.* 1988).

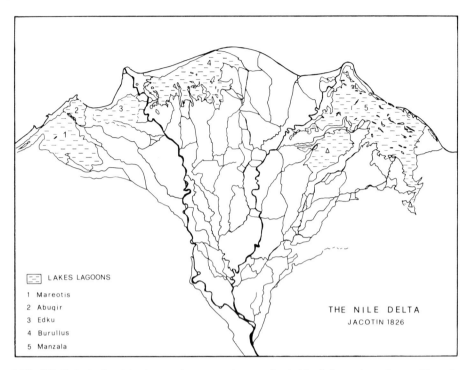

FIG. 3. The Nile Delta in the early nineteenth century: △, areas flooded for 8–9 months each year. (From Jacotin 1826.)

The Nile Delta includes the following physio-graphical–depositional provinces (Fig. 4): (i) the upper ('abandoned') delta plain with channel–interchannel fluviatile deposition; (ii) the lower ('active') delta plain characterized by a lagoon belt with its transitional environments; (iii) the delta front and beach–dune complex shaped by coastal longshore drift; (iv) the inner continental shelf (to depth −50 m) characterized by the muddy Rosetta and Damietta lobes; (v) the middle to outer continental shelf (depth 50–100 m), dominated by Late Pleistocene–Holocene relict sediments and erosional surfaces; (vi) the muddy prodelta of the continental slope and rise (Nile Cone).

The boundary between the upper and lower delta plain (Fig. 4) follows the maximum extent of lakes and lagoons in the early nineteenth century, prior to artificial irrigation, when the distribution of delta environments probably represented a state little changed during the previous 1000 years.

The upper delta plain

The delta plain has a broad fan-like shape (Fig. 5) abruptly delineated at the sides by scarps 10–

20 m high of Late Pleistocene (locally Pliocene) raised sediments (Said 1981). The delta plain is dominated by a system of distributary channels and natural levees, with a braided pattern about the main Nile branches. The Nile Delta appears to have developed mainly by channel extension and secondarily by channel switching. Flooding from the numerous distributaries, canals and artificial drains produced lower flood crests and lower levees than those upstream of Cairo. During floods, the delta was a vast expanse of water, reticulated by an endless network of levees.

The contours of the delta surface reflect the direction of former alluvial ridges, possibly meander belts of old distributary channels (Abdel Kader 1982). Apart from the Nile channels, the present water courses are generally unrelated to contours. The two main Nile branches have a well-developed meander belt (3–4 km wide) with a high sinuosity index (1.25–1.9), undoubtedly related to the high percentage of suspended load and the low gradient of the plain. Only local sections of minor streams, now canalized, still show meandering.

Sedimentation in overbank areas has been a function of the present semi-arid environment (the mean annual rainfall is 170 mm a^{-1} at the

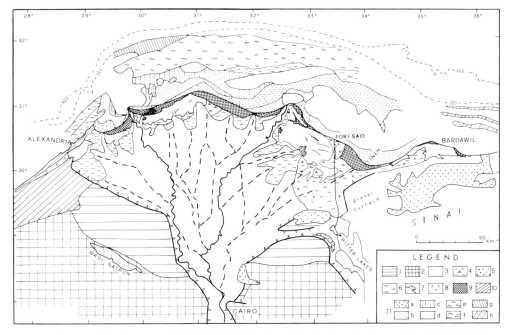

FIG. 4. Sediments and environments of the Nile Delta (onshore, situation in early nineteenth century) including data from Sestini (1976), Summerhayes *et al.* (1978) and Coleman *et al.* (1980): 1, raised Pleistocene uplands (20–50 m elevation) with marginal escarpment; 2, uplands (>50 m), Pliocene and older terrains; 3, upper delta plain with main distributary channels; 4, lower delta wetlands (lagoons, marshes, lakes and basins flooded for most of the year); 5, inland dunes; 6, dunes–lakes–salt plains; 7, lagoonal deltas; 8, salinas; 9, coastal beach–dune complex; 10, raised Pleistocene beach–dune ridges; 11, offshore sediments ((a) sands; (b) silty sands; (c) sand–silt–clay; (d) mud; (e) bioclastic sands; (f) muddy bioclastic sands; (g) calcareous muds; (h) *kurkar* ridges).

coast, rapidly decreasing to 19 mm a^{-1} at Cairo, with average summer temperatures of 27 °C at Alexandria and 32 °C at Cairo). Under natural (pre-1950) conditions the flood basins S of the Burullus Lagoon (Fig. 6) were silt plains with streams at various stages of summer desiccation, lacustrine to swampy areas and some salt flats. Silt-grade wind-formed dunes were a common feature at the SE margin of various plains (Sestini 1976; Frihy *et al.* 1988). The interchannel areas closer to the lagoons, particularly the Manzala Lagoon, were prone to form seasonal or perennial lakes and swamps, certainly more extensive in former periods of greater winter rainfall and/or greater Nile flooding, even distant from the lakes. Between these basins were elevated (0–2 m) N–S island-like areas of silts and fine sand covered with permanent vegetation. Several of these, S of Burullus, show traces of former streams attributed to the old Saitic and Sebennytic branches (Sestini 1976), with well-developed meanders, levees and point bars, and higher mounds containing remains of late Dynastic–Ptolemaic settlements.

The SE region between Ismailia and the Manzala Lagoon is characterized by hypersaline flats and salt marshes bordered by aeolian dunes. Remnants of the Pelusiac branch were mapped by Sneh & Weissbrod (1973) and Sneh *et al.* (1986) W of Tineh.

The lower delta plain

The coastal lakes and lagoons of the Nile Delta were larger in the last century (Jacotin 1826); their reduction in size is due to extensive siltation and, more recently, to reclamation. Salt pans have formed in cut-off portions of Lake Mariut and the Manzala Lagoon. The Idku, Burullus and Manzala Lagoons are brackish because they are fed by the Nile drains (Kerambrun 1986), but the Bardawill Lagoon is entirely outside the Nile influence and has salinities of 40‰–70‰, exceptionally rising to 100‰ when the outlets are blocked (Levy 1974).

The lagoons are mostly shallow (less than 2 m) and large parts near the islands and inner shores

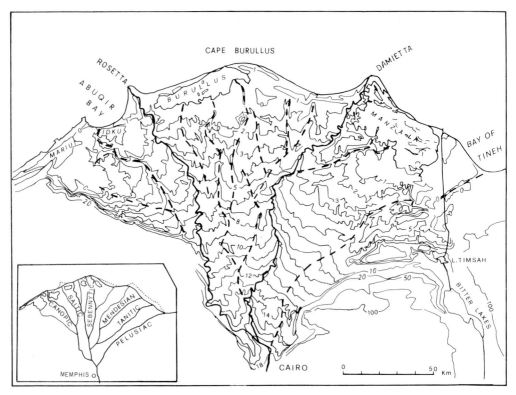

FIG. 5. Topography of the Nile Delta with probable former distributary channels. Inset: ancient Nile branches according to Tousson (1934).

are covered with reeds. Bottom sediments are predominantly structureless silty clays, with a high organic content in parts, but large areas of shelly to silty–muddy sand also occur in the Manzala (El Wakeel & Wahby 1970), Burullus and Idku Lagoons, possibly in relation to older sands (Saad 1981). In Burullus and Manzala some of the more active drains had built small (3–5 km wide) digitate prograding deltas.

The many islands in the lagoons, and in the former Lake Mariut, have various origins; remnants of former strandlines (Manzala) (Daressy 1931; El Askary & Lofty 1980; Coutellier & Stanley 1987), riverbanks and dunes (Idku and Burullus). The original morphology of many islands has been modified by wind and lake waves, in general with well-defined sandy beaches in the WNW, while the down-wind extremities have long curved spits that enclose marshy shores.

Delta-front morphology

The 400 km long coastal barrier complex (Abuqir to Bardawil spit) includes the beach zone to the lower shoreface (−7 m), the backshore plains with large dunes and the two projecting subdeltas of the Rosetta and Damietta distributary mouths. The typical morphology of these distributaries is well displayed in pre-1964 air photographs and maps of the 1810–1910 period of active progradation (Fig. 7). In the late eighteenth to early nineteenth centuries the fast-growing Rosetta promontory was characterized by a bifurcated channel and a middle-ground mouth bar, which in 1957 had become a large E–W lunate bar (4 km long by 1 km wide). River-dominated morphology is evidenced by lagoons and marshy areas and, on the E side, by the rapid addition of spits and bars with the active formation of beach ridges and marshy lagoons. In this century the promontory has gradually receded, with a sharp acceleration of retreat to over 200 m a^{-1} since 1980. The Rosetta mouth tends to be blocked by a spit-submerged bar extending southwestwards from the eroding foreland (Frihy 1987).

The shape of the NE-oriented Damietta mouth was more strongly influenced by the northwestward wave approach (*eg* submerged spit and

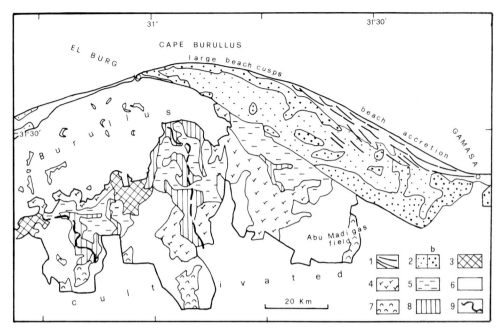

FIG. 6. Geomorphology S of Cape Burullus (based on 1956 air photographs, Sestini 1976): 1, beach ridges and backshore plains; 2, wind-blown sands (b, high dunes); 3, lakes; 4, areas flooded for several months; 5, marshes; 6, silt plains; 7, silt dunes; 8, sand–silt 'islands'; 9, abandoned river channels.

linear bars on the W side). Active growth occurred on the E side with beach ridges enclosing lagoons, which remained partly connected to both the sea and the river. The Damietta headland has retreated by 140–160 m a^{-1} since 1964, and the Ras El Barr outlet is now dammed by a spit. In the extensive beach-ridge system E of Damietta, discrete groups of former strandlines are related to stages of advance and retreat (Fig. 7b). Beach ridges are added where the shoreline turns SE; a conspicuous new set of spits has grown more than 7.5 km since the late 1950s, with a coastal advance of more than 2.5 km (Frihy 1987). The ridge complex is 5–8 m thick (Anwar *et al*. 1984).

In the rest of the arcuate coast beaches may be straight and smooth or, in wind-parallel stretches, strongly undulating owing to the migration of large (300–1500 m long) asymmetric cusps (*eg* between Cape Burullus and Gamasa) and larger (5–6 × 2 km) swells (*eg* W of Port Said).

The backshore plains are vast expanses of beach sand, which are particularly developed on the accreting stretches of the coast by the addition of beach ridges or spits and the subsequent filling of the narrow intervening lagoons by washovers and wind-blown sand (traces of former shorelines are evidenced by narrow flat ridges of shelly sand (Frihy *et al*. 1988)). The backshore plains of the

W and N parts of the delta are usually flooded by winter storm surges and their inner depressed areas, which in many places are below sea level, tend to have salt deposition (but they are not real sabkhas, as stated by Said (1981), Inman & Jenkins (1984) and El Buseili & Frihy (1984)).

The larger dune belts are generally located at the southern margins of these plains, bordering the lakes or (former) marshy areas. They are continuous belts of complex dunes and deformed barkhans 10–15 m (up to 25 m) long.

The beach sands at the Nile Delta coast are mainly fine to very fine grained (average median grain size, 0.15 mm), but the coast-wise distribution of grain size is abnormal (Sestini *et al*. 1976) because coarser beach sands (average 0.25 mm, coarse fraction 0.5 mm) occur down-drift of the Rosetta mouth, in the western part of Abuqir Bay, in the stretch 35 km W to 20 km E of Cape Burullus and immediately E of Gamasa. They have been interpreted as relict sands.

Heavy minerals, especially opaques (mainly magnetite and ilmenite), are particularly abundant at the Rosetta mouth (where exploitation feasibility studies have been made for ilmenite, zircon, rutile and monazite (Nahla 1958)) and at Cape Burullus and Ras El Barr (El Fishawi & Molnar 1985). Other typical 'Nile minerals' are

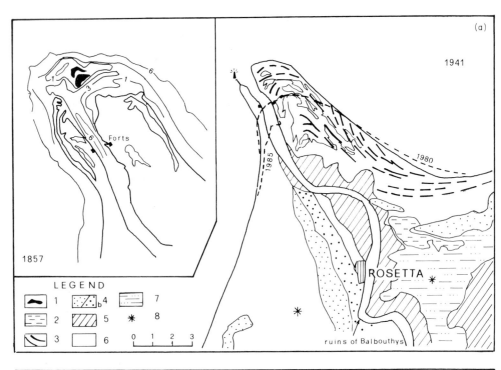

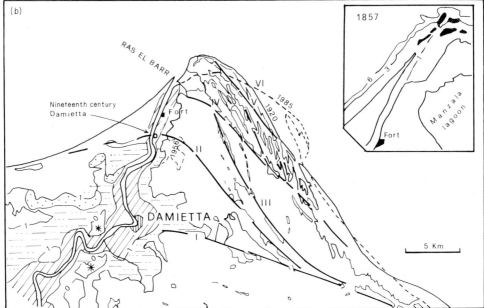

FIG. 7. Morphology of the Rosetta and Damietta subdeltas: 1, submerged sand bars; 2, lakes; 3, beach ridges; 4, dunes (b, >10 m); 5, channel–levee complex; 6, backshore plains; 7, flood basins; 8, areas <0 m; I–VI, stages of shoreline migration (depths in metres).

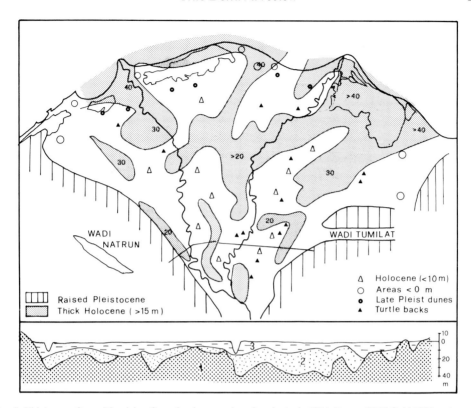

FIG. 8. Thickness of post-Flandrian Complex (summarizes data in Attia 1954, Butzer 1976, Said 1981 and Coutellier & Stanley 1987). The cross-section is modified after Butzer (1976). 1, Pleistocene sands and gravel; 2, Late Pleistocene sand; 3, Holocene Nile mud.

hornblende, augite, diopside, epidote and garnet (Hilmy 1951).

The position of the coast about 2000 years BP can be estimated with reasonable approximation from historic records (Ball 1942; Goby 1954). In the W, the Canopic delta lobe seems to have disappeared between the first and third centuries AD. The former location of the Canopic mouth is indicated by historic information, with supporting archaeological evidence (Tousson 1934), and by various marine morphological and sedimentological features (Sestini 1976; El Buseili & Frihy 1984). The coast off Rosetta has prograded no less than 14 km away from the Roman Balbouthys, which was located near the sea in the first century AD (Fig. 7b).

The tracing of dated shorelines near Tineh northwestwards through the relict beach ridges of the Manzala Lagoon (Sneh & Weissbrod 1973) suggests a total advance since 0 AD of about 15 km. The coastal bulge at Port Said appears to result from the advancement of the ancient Mendesian and Tanitic branches from 3000 to

1400 BP (Coutellier & Stanley 1987). The progradation of the cuspate subdeltas is supported by borehole evidence, where beach sand overlies marine clays with a coarsening-upward sequence. In the centre of the delta, the Burullus coast has been retreating such that beach deposits rest on lagoonal clays with peats (see Fig. 9). Total retreat since Roman times may amount to at least a few kilometres.

The submarine delta

The continental shelf of the Nile Delta is much steeper than the subaerial plain (26.5 cm km^{-1} versus 8.8 cm km^{-1}). Unlike the coast it is broadly asymmetric and wider at the eastern end. It is characterized by a series of flat surfaces, separated by gentle slopes and local scarps, which mainly trend E–W, in contrast with the shape of the coast. The main surface levels are at depths of 15–27 m (upper terrace), 40–55 m (intermediate terrace) and 65–139 m (lower terrace complex) (Misdorp & Sestini 1976). The latter is very

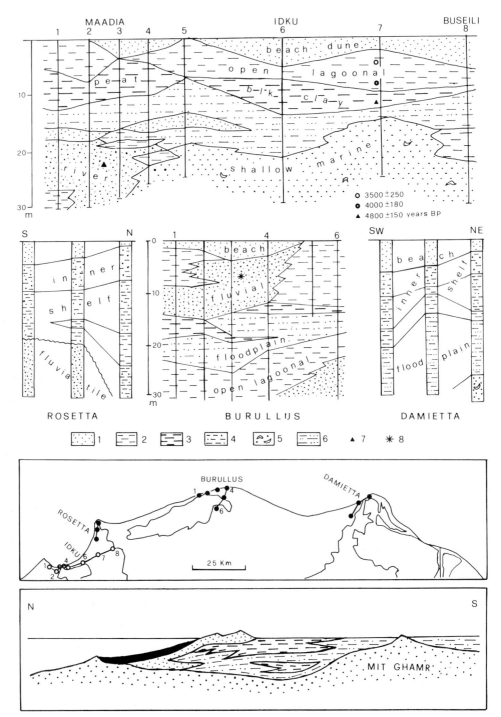

FIG. 9. The subsurface Holocene Complex near the coast (based on Sestini 1976; UNESCO/UNDP 1976; Anwar *et al.* 1984): 1, beach–dune complex; 2, open lagoonal deposits; 3, marshy-restricted lagoonal deposits; 4, marine clays and silts; 5, marine sands (medium to fine grained); 6, fluviatile sand–silt–clay; 7, sands of the Canopic distributary channel; 8, sands of the Sebennytic distributary channel. The schematic model (bottom) shows the regressive inner neritic (mud)–beach–dune–lagoon complex near the Burullus Lagoon.

irregular: in addition to several probable ero-sional surfaces, there are slump blocks, angular algal mounds and possible mud volcanoes (Summerhayes *et al*. 1978; Coleman *et al*. 1980). The edge of the continental shelf lies at 80–100 m in the W (where it is incised by the Rosetta canyon) and at 180–700 m in the centre and E. The western (off Rosetta) and the eastern (off Port Said) margins are fault bounded (Bardawil Escarpment, Neev *et al*. 1976; Almagor & Hall 1978).

The upper terrace is mostly covered with terrigenous sands (delta platform), except where transgressed by the muds of the Rosetta and Damietta delta lobes (which are 25–30 m thick) and by mud blankets that can be related to older Nile mouths. In contrast, the lower terraces are covered with bioclastic sediments ranging from pure algal coralline debris, shelly sands and gravel to muds with bioclastic debris and a high $CaCO_3$ content (Summerhayes *et al*. 1978). This remarkable bioclastic facies occurs extensively at the same depth range as the shelf break in many parts of the western Mediterranean and elsewhere (Milliman 1974). The outer shelf and the continental slope are covered with calcareous muds, particularly on the western part of the slope.

Other genetically interesting features of the submarine delta include the following.

(1) W of Abuqir, the submarine morphology is characterized by several flat terraces and ridges, which may be in the same system of Pleistocene submarine bars that are now elevated along the coast W of Alexandria (Butzer 1960). The bottom sediments are entirely bioclastic (El Sayed 1979) and oolitic W of Alexandria (Hilmy 1951).

(2) Small asymmetric sand ridges that rise 5–20 m above the margin of the upper terrace (NE of Rosetta and N of Gamasa) have been interpreted as submerged dunes (*cf* the *Kurkar* cemented-sand ridges of the Sinai–Israel shelf (Almagor 1979)).

(3) In contrast with the fine-grained angular heavy-mineral-rich terrigenous sands near the coast, the sand patches on the upper terrace from Rosetta to N of Damietta and on the intermediate terrace N of Burullus (a former delta lobe?) are medium to coarse grained, well sorted and rounded, quartzose and often iron stained.

(4) Sub-bottom reflections have revealed not only the occurrence of relict sands under the Rosetta and Damietta mud blankets, but also the presence of buried intermediate and lower terraces W of longitude 31°E (Summerhayes *et al*. 1978; Ross & Uchupi 1979). However,

they have not demonstrated any buried Nile distributaries.

It seems likely (Misdorp & Sestini 1976) that the 50–90 m, 40–50 m and 15–27 m terraces can be related to still-stands and minor reversals of the Flandrian transgression that occurred at 14 000–15 000 BP, 12 500–10 000 BP and 9000–8500 BP (Mörner 1976). The distribution of relict sediments can be attributed to climatically conditioned Nile discharge. In the period 18 000–12 000 BP no terrigenous clastics were deposited on the outer shelf, whereas between 12 000 and 6000 BP there was a considerable sand influx, related to a large Nile discharge (Adamson *et al*. 1980). The period from 6000 BP to the present has been associated with the formation of mud blankets on the upper shelf, with sand being confined to the beaches and the delta front.

Holocene subsurface stratigraphy

The development of the modern delta complex began with the Flandrian transgression following the Würmian glacial stage and represents deposition during the last 6000 years. In the delta subsurface as many as 350 boreholes have clearly delineated a post-Flandrian 10–40 m thick complex of interbedded medium- to fine-grained sands, silts, clays and peats (Bilqas Formation, Rizzini *et al*. 1978; Neo-Nile 'delta', Said 1981), that lies unconformably upon coarser-grained Late Pleistocene clastics (Fig. 8). In the S and SE of the delta plain, the latter crop out as low sandy hills ('turtle backs', Kholeif *et al*. 1969) or lie below a thin cover of silts (Butzer 1976). Offshore, the coarser relict sands N of Burullus and in western Abuqir Bay are also likely to be exposed pre-Holocene sands rather than sands delivered from the Nile mouths in historic times, as previously suggested by Sestini *et al*. (1976).

The thickness distribution of the post-Flandrian delta complex (Fig. 8) emphasizes the role of the western, central and eastern Nile branches in historical times and the occurrence of several sandy areas at shallow depth S of latitude 31°N, with deposits that are generally thicker N of this latitude. Knowledge of the subsurface Holocene deposits from sedimentological studies of cores, with the support of radiocarbon dating, is so far limited to the eastern delta (Sneh *et al*. 1976; Stanley & Liyanage 1986; Coutellier & Stanley 1987; Frihy & Stanley 1988) and to parts of the coast between Abuqir and Damietta (UNDP/UNESCO 1978; El Askary *et al*. 1984; El Fishawy 1985; El Askary & Frihy 1986).

East of the Damietta branch a generalized

sequence, starting at depths of 20–40 m, displays delta-plain channel and interchannel deposits (sands, peats) that pass upwards and laterally towards the NE to delta-front sand and silt–clays, the whole prograding over marine prodelta muds (Coutellier & Stanley 1987, Figs 2, 5 and 7). This deltaic regression represents the development of several delta lobes in the period 8000–5000 BP, when the rate of sea-level rise began to decline and the Flandrian transgression had reached its maximum extent, as far as 10–20 km S of the present lagoons. In detail the stratigraphy is complex owing to the lateral overlap and the different size of each delta lobe; for instance the later Tanitic lobe overlapped the earlier more southeasterly Mendesian and Pelusian lobes, while the recent Damietta lobe has transgressed over the Mendesian and Tanitic lobes. These lobes were probably cuspate subdeltas, typically deflected eastwards, with a notable asymmetric growth of beach ridges.

The sequence at the Abuqir Bay and Burullus margin is also complex (Fig. 9), with lagoonal, swamp and beach–marine facies interbedded with Nile flood-plain deposits formed by the Rosetta and Sebennytic distributaries. About 5000 BP western Abuqir Bay had a lower deltaic environment, at the beginning of lagoonal development. The Late Pleistocene beach-related dunes that lie S of the Idku Lagoon (Sestini 1976; Frihy *et al.* 1988) and the isolated sand hills that are aligned immediately S of the Burullus Lagoon (Fig. 8) could be indicative of the shoreline at the maximum extent of the Flandrian transgression.

The subsurface facies record offshore (*cf* oil exploration wells (Khairy 1974; Elf Aquitaine 1982; Poumot & Bourullec 1986)) also suggests that transitional marine environments shifted N to S relative to the present coast, possibly with minor sea-level oscillations between 5000 and 3000 BP (Fairbridge 1968).

The Holocene thickness map points to considerable subsidence in the coastal–lower delta plain in the last 6000 years. Dated sequences suggest rates of subsidence of 15 mm per 10 years. Further evidence for continued subsidence during the last 2000 years includes (i) Roman ruins (Alexandria to Abuqir) which are now 5–8 m below sea level, deeper than can be accounted for by a sea-level rise of 1.5 cm per 100 years, (ii) the occurrence in the lagoons of drowned (or subsided) river banks and beach ridges and of submerged ruins and (iii) many areas that are at present 1–3 m below sea level (Fig. 8). N of Ismailia, precise measurements have shown a subsidence of 0.50 cm in 80 years (Goby 1952), a value close to that suggested by Stanley (1988).

Although in late Dynastic-Ptolemaic times much of the northern delta was lakes and swamps, as shown by archaeological records, the Burullus and Manzala Lagoons were not as extensive as today. Smaller lagoons or lakes may have represented interdistributary bays closed in by the coastwise extension of spits issued from the various cuspate subdeltas (*cf* upper Adriatic Sea lagoons and the Po Delta (Fabbri 1985)). The expansion of the lagoons during the last 2000 years has been due to continuous, although irregular, subsidence behind a stable beach barrier. There are records of the rapid expansion of swamps and lagoons since the fifth to tenth centuries, and also following major earthquakes (Daressy 1928; Tousson 1934) (see Ben Menachem (1979) for a review of earthquake activity in the Middle East).

In conclusion, two broad sedimentary models can be envisaged for the Nile Delta: (i) Holocene deltaic environments characterized by ribbons of distributary sands that pass laterally to floodplain laminated silts (or to peats or evaporites as the deposits of wetter or drier periods respectively) and seaward to lagoonal–marshy deposits with marine clay intervals and transversal bodies of beach–dune sands; (ii) Late Pleistocene environments, with predominant sheet sands formed during periods of alternating large sand input with braided stream channels and very weak Nile supply associated with intermittent erosion and much aeolian reworking.

The structural setting of delta growth

Geologically, the greater Nile Delta region covers the following structural–sedimentary provinces (Fig. 10): (i) the South Delta Block, (ii) the North Delta Basin; (iii) the Nile Cone; (iv) the Levant Platform. Exterior to the actual delta, but relevant to its history, are the Sinai and the Suez Rift.

The boundary between the South Delta Block and the North Delta Basin is a major zone of flexure at about latitude 31°N, along which faults down-throw to the N by 5000–6000 m the Cretaceous–Middle Eocene carbonates of N Egypt (Salem 1976; Korrat 1977; Barber 1981; Said 1981). Tertiary sediments below the northern Nile Delta are at least 5000–7000 m thick, whereas the correlated succession in the S is only 500–1000 m thick. S of the Flexure Zone the principal pre-Oligocene structures are tilted fault blocks and horsts oriented E–W to NNE, which show considerable stratigraphic gaps owing to repeated uplift from the Senonian to early Eocene (Said 1962; El Shazly 1977; Sestini 1984, Fig. 4). In the N, the Sinai–Negev equivalent structures and surface folds, onshore and offshore, are

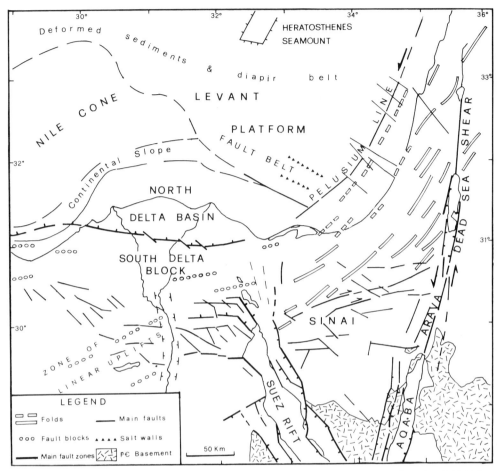

FIG. 10. Principal structures and tectonic regions of NE Egypt and the SE Mediterranean (based on Said 1962; Neev 1975; Neev *et al.* 1976; El Shazly 1977; Garfunkel & Bartov 1977; Ross & Uchupi 1979; Barber 1980; Ben Avraham & Mart 1981; Mart & Ben Gai 1982; Sestini 1984). Broken curves indicate subsurface structures.

oriented NE (Said 1962; Ginzburg *et al.* 1975*a*; Neev 1975; Neev *et al.* 1976). Superimposed on these structures in the Suez Canal area is a possible northward continuation of the Gulf of Suez Rift (Garfunkel & Bartov 1977) and a general eastward rise to the Sinai Block along NNW–SSE normal faults (El Shazly & Abdel Hady 1975).

The Faulted Flexure may be not only a tectonic feature but also a major facies boundary between platform and slope carbonates, analogous with the Jurassic–Cretaceous hinge line in Israel (Bein & Gvirtzman 1977). A possible extension of this feature through N Sinai is suggested by the pattern of basement structure in NE Egypt (El Shazly 1977), but immediately W of the Nile Delta wells show no indication of a major facies change (Sestini 1984).

The structure and stratigraphy of the Nile Cone and the Levant Platform are well defined, mainly above Reflector M at the top of the Messinian evaporites (Mulder *et al.* 1975; Neev *et al.* 1976; Ross & Uchupi 1979). Post-Messinian sediments, averaging 2000 m thick (but up to 3000 m) with an estimated volume of 387 000 km^3 (Ross *et al.* 1978), overlie a channelled erosion surface (Gvirtzman & Buchbinder 1978; Ryan 1978; Ben Avraham & Mart 1981). Numerous piston cores, two DSDP holes (131 and 132) and seismic studies (Bartolini *et al.* 1975; Maldonado & Stanley 1978) have indicated that the Nile Submarine Fan and the Levant Platform were constructed by Nile-derived clastics, particularly during the Pleistocene eustatic sea-level drops and low-stands.

In the W, the smooth slope of the Nile Cone

corresponds to an undisturbed structure of the Plio-Pleistocene sediments (Ross & Uchupi 1979), but in the Herodotus Basin the smooth bottom masks a considerable subsurface irregularity caused by salt diapirism (Smith 1975). Salt diapirs also characterize the toe of the Nile Submarine Fan near the Mediterranean Ridge, where they are associated with increasing intensity of sediment deformation, and the Levant Platform, where the submarine topography (hummocky relief, fault blocks, linear valleys and ridges) is also strongly controlled by salt tectonics, *ie* salt walls, plugs and superficial collapse structures (Kenyon *et al.* 1975; Ross & Uchupi 1979). The overall elevated position of the Levant Platform has been attributed to the rise and lateral spread of the Messinian evaporites. The continental slope off the NE delta around N Sinai is characteristically affected by numerous slope-parallel young normal faults and by sliding (Neev *et al.* 1976; Almagor & Hall 1978; Almagor & Garfunkel 1979; Ross & Uchupi 1979; Ben Avraham & Mart 1981).

The relations between the main tectonic elements of the region, particularly E of the Nile Delta, are complex and ill defined. The structural history has involved several stages and types of deformation (Ginzburg *et al.* 1975a; Neev 1975; Stanley 1977; Ben Avraham 1978; Nur & Ben Avraham 1978; Ross & Uchupi 1979; Bartov *et al.* 1980; Mart 1984, 1988; Scott & Govean 1985) which appear to have been as follows.

(1) *Late Cretaceous–Eocene:* arching of the Nubian Shield ('Red Sea Swell').
(2) *Oligocene:* initiation of Suez rifting and of Aqaba–Jordan shearing; strong differential movements along the Levant Platform margin.
(3) *Early Miocene:* first stage of Gulf of Suez rifting.
(4) *Middle Miocene:* intense tensional deformation involving the Levant–North African margin, with uplift inland (Western Desert, Sinai–Cairo area *etc.*), erosion of fault blocks and development of widespread unconformities (*eg* End Burdigalian, Late Serravallian to Early Tortonian); second stage of Gulf of Suez rifting.
(5) *Late Miocene:* major Alpine compression to the N (Cyprus, S Turkey *etc.*) shaping the present eastern Mediterranean basins, coinciding with tensional events in the foreland, the beginning (to Middle Pliocene) of the main sea-floor spreading in the Red Sea, the uplift of rift margins, Aqaba shearing and the northward movement with clockwise rotation of Africa.

(6) *Middle Pliocene:* intense structural event with considerable down-faulting and subsidence affecting the eastern Mediterranean basins and the Gulf of Suez.
(7) *Late Pliocene–Early Pleistocene:* further northward movement of the African Plate and continued opening of the Red Sea.

The tectonic deformation is complex because of the interaction of differently oriented tectonic fabrics, such as the following.

(1) The differential vertical movement of E–W fault blocks and grabens along the passive margin of the African Plate, as well as the overall subsidence of the Tethyan basin versus the uplift of the Afro-Arabian margin.
(2) The uplift of the Red Sea region, followed by NW–SE to NNW–SSE rifting and sea-floor spreading with the northeastward movement of Arabia relative to Africa. The ensuing shear system (dextral E–W and NW–SE, and sinistral NNE, NE and NNW strike-slip) created a series of en-echelon depressions, or depocentres, in the Gulf of Suez and the delta region (Garfunkel & Bartov 1977) (*cf* the NW- and NE-bounded basement depressions in the gravity-interpreted maps of Riad (1977) and Meshref (1982)). Extension to produce rifts has also been controlled by earlier basement fracture systems, particularly those with a NNE (Garson & Krs 1976) and N20°E to N20°W orientation (Crane & Bonatti 1987).
(3) The effect of sediment loading to drive subsidence at the Plio-Pleistocene delta front, with the formation of growth faults, shelf-margin sliding, normal faulting and halo-kinesis (Ben Avraham 1978; Ross & Uchupi 1979; Mart & Ben Gai 1982). It is likely, however, that this subsidence did not result exclusively from isostatic adjustment, but also from regional tectonic tensile stresses during the break-up of the rigid Sinai margin (Ben Avraham & Mart 1981; Mart 1984, 1988), from transtension between the sinistral and dextral strike-slip faults (Neev 1975; Neev *et al.* 1980) and, in the W, because of subduction and the associated intraplate extension during plate flexure.

Subsurface stratigraphy of the Nile Delta

The subsurface stratigraphy of the Nile Delta is still incompletely known, despite many drilled wells. The stratigraphic scheme for the North

Delta Basin proposed by Rizzini *et al.* (1978) has been substantiated by later wells and by improvements in biostratigraphy, with palynomorph and nannoplankton zonation in addition to foraminifera zonation. A summary of present knowledge is given in Figs 11 and 12. The time interval represented by some formations has been disputed in a few central-western wells (Ansary & Deibis 1977; El Ayad 1980; Kora 1980). However, difficulties with stratigraphic subdivision

INNER NILE CONE
NORTH DELTA BASIN
SOUTH DELTA BLOCK
PLEISTOCENE
Mit Ghamr
El Wastani
PLIOCENE
Kafr el Sheikh
MESSINIAN
Abu Madi
TORTONIAN
Qawasim
LANGH - SERRAV.
Sidi Selim
Mohra
SHELF MARGIN STRUCTURES
FAULTED FLEXURE
M-U Eo-Olig
L. Eoc
UK
LK
J

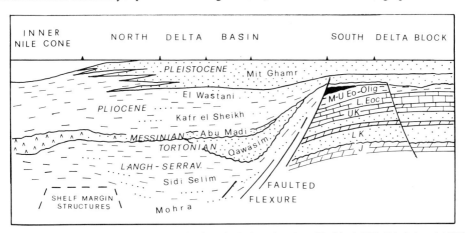

FIG. 11. Scheme of the stratigraphic relations in the Nile Delta subsurface (Zaghloul 1976; Rizzini *et al.* 1978).

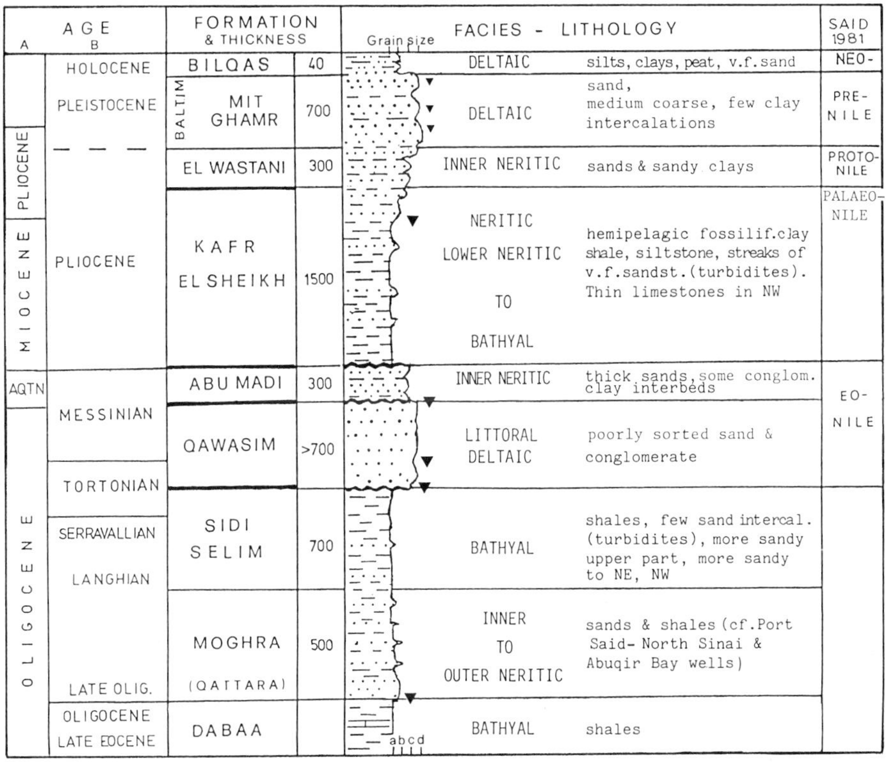

AGE
A
B
FORMATION & THICKNESS
Grain size
FACIES - LITHOLOGY
SAID 1981
HOLOCENE
BILQAS
40
DELTAIC
silts, clays, peat, v.f.sand
NEO-
PLEISTOCENE
BALTIM
MIT GHAMR
700
DELTAIC
sand, medium coarse, few clay intercalations
PRE-NILE
PLIOCENE
EL WASTANI
300
INNER NERITIC
sands & sandy clays
PROTO-NILE
PALAEO-NILE
MIOCENE
PLIOCENE
KAFR EL SHEIKH
1500
NERITIC
LOWER NERITIC
TO
BATHYAL
hemipelagic fossilif.clay shale, siltstone, streaks of v.f.sandst.(turbidites). Thin limestones in NW
AQTN
MESSINIAN
ABU MADI
300
INNER NERITIC
thick sands, some conglom. clay interbeds
EO-NILE
QAWASIM
>700
LITTORAL DELTAIC
poorly sorted sand & conglomerate
TORTONIAN
OLIGOCENE
SERRAVALLIAN
LANGHIAN
SIDI SELIM
700
BATHYAL
shales, few sand intercal. (turbidites), more sandy upper part, more sandy to NE, NW
LATE OLIG.
MOGHRA (QATTARA)
500
INNER TO OUTER NERITIC
sands & shales (cf.Port Said- North Sinai & Abuqir Bay wells)
OLIGOCENE
LATE EOCENE
DABAA
BATHYAL
shales
abcd

FIG. 12. Summary of the stratigraphy of the Nile Delta (based on Rizzini *et al.* 1978; Zaghloul *et al.* 1979*b*; Elf Aquitaine 1982; El Heiny 1982; Poumot & Bourullec 1984; Deibis *et al.* 1986): ▼, main episodes of sea-level fall; age A, B, contrasting interpretations (see text); bold lines, main seismic markers; wavy lines, main unconformities. Bulk grain size: a, clay; b, silty clay–very fine sand; c, fine to medium sand; d, coarse-grained sand.

and correlations have arisen where seismic interpretation has involved indistinct clastic facies with few sharp seismic reflectors.

In the North Delta Basin the outstanding seismic and stratigraphic discontinuities occur at the top of the Sidi Salem Shales, the unconformity beneath the Abu Madi Formation, the erosional discontinuity under the Early Pliocene *Sphaero-idinellopsis* Zone and the base of the El Wastani sands (Deibis 1982). The stratigraphic succession represents three main sedimentary cycles, which appear to be related to external events of tectonic deformation, rates of clastic input and changes of global sea level. The regressive episodes correspond to increases in grain size and feldspar/quartz ratios and to heavy mineral peaks (Zaghloul *et al.* 1979*a*). The three cycles are as follows.

Cycle 1: the Late Oligocene–Early Miocene Qattara–Moghra regression and ensuing Sidi Salem transgression. The Qatrani–Moghra del-taic–inner neritic facies occurs W and SW of the Nile Delta (Said 1962; Omara & Ouda 1972; Omara & Sanad 1975). Facies in the North Delta Basin are outer neritic–bathyal with little sand, whereas there are middle neritic and more sandy sediments in the E (Derin & Reiss 1973).

Cycle 2: the Late Miocene Qawasim–Abu Madi regression followed an Early Tortonian fall of sea level and the 'Serravallian–Tortonian tectonic event'. The Kafr El Sheikh transgression corresponds to a global rise of sea level. The Qawasim is a clastic unit of variable thickness, in a belt approximately 30 km wide N of the Faulted Flexure, comprising a progradational succession of coalescing fan deltas to submarine slope (mass flow) deposits, suggesting rapid deposition to the N and NE (Zaghloul *et al.* 1977*b*; Rizzini *et al.* 1978; Deibis *et al.* 1986). The Abu Madi sands are deltaic onshore and lagoonal, marginal, marine and evaporitic off-shore to the NW (Rosetta Formation), but to the N and NE the clastics are entirely marine without evaporites (Khairy 1974; Rizzini *et al.* 1978) and there are limestones in NW Sinai (Ziglaq Formation, Derin & Reiss 1973). In some wells the top of the Abu Madi may be Early Pliocene (transgressive reworked deposits?). The Qawasim and Abu Madi formations thin northwards. A precise regional correlation of the non-evapor-itic, partly evaporitic and entirely evaporitic (Mavqim Formation, Gvirtzman & Buchbinder 1978; Mart & Ben Gai 1982) Messinian deposits and of their seismic and stratigraphic boundaries has not been published.

Cycle 3: shallowing and an increased clastic input appears to correlate with a lowered sea level in the Late Pliocene. The main delta advance

in the Pleistocene occurred during the low-stands of sea level associated with glacial periods. The El Wastani Formation is a transitional unit characterized by large delta-foreset progradation. The Mit Ghamr Formation is a sheet-like sand body with considerable lateral continuity. Abundant pyroxene suggests an early Ethiopian contribution (Said 1981). The facies indicate lagoons, beach, near-shore in the lower part and generally more continental environments at the top (coastal plain to littoral). In offshore wells (*eg* Baltim and Rosetta) thick intervals of marine sands and clays alternate in the lower half (Baltim Formation), whereas the upper half (especially in the N and NE) is quite sandy and conglomeratic and of fluviatile–littoral origin.

Said (1981) has correlated the subsurface Plio-Pleistocene Nile Delta sequence with the exposed Nile Valley sections, recognizing four main fluvial intervals (Palaeo-, Proto-, Pre- and Neo-Nile) separated by unconformities due to long periods of erosion associated with very reduced or no fluvial activity. The correlation between the Nile Delta and upper Egypt is rather conjectural. There is no absolute or relative age dating of the Pleistocene in the delta subsurface, and the stratigraphic record of the exploration wells is insufficiently detailed. The studies by Zaghloul *et al.* (1975, 1976, 1977*a, b*) have nevertheless emphasized four transgressions and regressions with brief intervening periods of subaerial exposure. Pollen studies (El Beialy 1980) have suggested six humid periods alternating with dry periods, possibly related to glaciations.

The Plio-Pleistocene sediments under the continental slope and Nile Submarine Fan are predominantly clays with subordinate turbiditic sands (Ryan & Hsu 1973; Bartolini *et al.* 1975). Detailed studies of the upper post-50 000 BP section (Maldonado & Stanley 1976; Stanley & Maldonado 1977; Summerhayes *et al.* 1977) have demonstrated a characteristic cyclic sequence of turbiditic sands and muds, hemipelagic muds and calcareous oozes, and sapropel layers. The latter have been related to periods of pronounced water stratification and bottom stagnation during stages of falling or rising sea level, whereas the calcareous oozes were formed in the cooler periods at low-stands of sea level. Four main cycles have been correlated with major and minor stages of cooling–warming in the last glacial stages (Stanley & Maldonado 1979).

The stratigraphy of the South Delta Block is different. The Mesozoic sequence is akin to that of the Western Desert and N Sinai (Said 1962; Salem 1976; Zaghloul 1976) with Late Jurassic, Aptian and Late Cretaceous (Turonian–Santonian) limestones and dolomites, Senonian–Early Eocene open marine lime–mudstones and chalks, early Middle Eocene Nummulitic limestones in the S, grading northwards to calcilutites, and Middle Jurassic, Early Cretaceous and Albian–Cenomanian sands. The pre-Pliocene Cenozoic section is 50–450 m thick. It is characterized not only by a northward transition from shallow to deeper neritic clastics, but also by several unconformities, namely between Cretaceous and Late Eocene, between Oligocene and Early Miocene, and in middle Late Miocene times. Late Oligocene, Aquitanian and Burdigalian discontinuities due to uplift and lowered sea level resulted in the dissection of a basalt-covered tableland (*cf* Abu Zabal Basalt, Taylor & Jones 1982). The amalgamation of these unconformities has produced the partial to total absence, in parts, of Palaeocene–Eocene, lower Middle Miocene and Tortonian–Messinian deposits (Zaghloul 1976; Barber 1981; El Heiny 1982; Taylor & Jones 1982).

Structure of the Nile Delta

The subsurface structure of the Nile Delta region can only be outlined in general terms (Fig. 13), with information taken from El Shazly *et al.* (1975), Neev *et al.* (1976), Bentz & Gutman (1977), El Nahass (1978), Ross & Uchupi (1979), Barber (1981), Ben Avraham & Mart (1981), Deibis (1982), Elf Aquitaine (1982), Taylor & Jones (1982), Mart (1984, 1988) and El Kholy & Ramadan (1986).

The South Delta Block is characterized by a gradual northward dip of the top of the Middle Eocene carbonates which is gently folded and dissected by approximately E–W normal faults. In the corridor between the Rosetta and Damietta branches of the Nile the carbonate surface is depressed owing to Plio-Pleistocene subsidence along NNW, NE and N–S faults (not shown in Fig. 13, other than the Cairo Graben). Some of the main faults of the Faulted Flexure extend through the Pliocene sediments. Just N of the

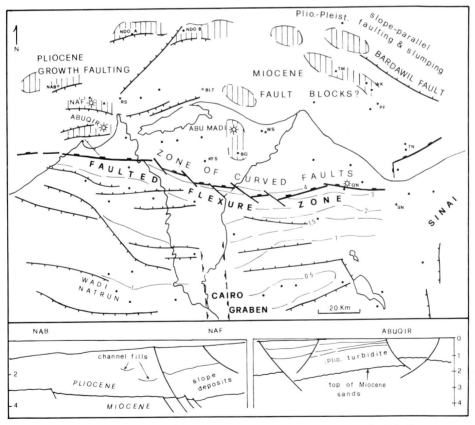

FIG. 13. Summary of the subsurface structure of the Nile Delta region. Contours: top of middle Eocene carbonates (seconds, two-way time). Ruled areas: Oligocene ?–Miocene fault blocks. Exploration wells: BQ, Bilqas; BLT, Baltim; KFS, Kafr El Sheikh; PF, Port Fouad; QN, El Qantara; RS, Rosetta; SN, Sneh; TM, Temsah; TN, Tineh; WK, El Wakara; WS, El Wastani. (Cross-section simplified from Elf Aquitaine 1982.)

flexure, the Miocene deposits are dislocated by faults that bound rotated fault blocks and extend to the Serravallian–Tortonian unconformity. They are probably growth faults due to the greater thickness of Miocene sediments in that belt.

The structure of the 'carbonate top' under the North Delta Basin remains unknown. It is possible that highs exist under some of the Tertiary positive structures that have been recognized on seismic sections (Ross & Uchupi 1979; Deibis 1982; Elf Aquitaine 1982) and by well drilling (*eg* the structure of Abuqir, Naf-1, NDO-A and NDO-B; Temsah, El Wakara, Port Fouad; Abu Madi and Bilqas) (*cf* Khairy 1974; Ansary & Deibis 1977; Deibis *et al.* 1986). No details of these structures or their age are available, although condensed sections suggest a pre-Tortonian development. Their common orientation is E–W, but the structures NE of Damietta probably trend NW, in agreement with the Sinai fault belt (Neev *et al.* 1976; Mart 1982, 1984). The Abu Madi and Bilqas structures trend N–S, as suggested by the thickness distribution of the Sidi Salem, Qawasim and Abu Madi Formations (Zaghloul *et al.* 1977b).

An indirect picture of Miocene structural trends and post-Miocene structural control is afforded by the Mio-Pliocene unconformity surface (Fig. 14). This has a pronounced erosional profile with a northward directed 'drainage' from a major E–W-oriented fragmented escarpment immediately S of the Faulted Flexure. An analysis of morpho-structural trends suggests the influence of NW- and ENE-oriented faults that pre-date the Plio-Pleistocene deformation. The latter mainly consisted of growth and antithetic normal faults (*eg* to the N and NW offshore) with uplift in the NE, a part that was generally subsiding in the Miocene with an eastward palaeoslope (Zaghloul *et al.* 1977b), and of subsidence in the axial part of the delta.

The Abuqir and W Burullus grabens are Pliocene features with associated growth faulting (Ansary & Deibis 1977; Deibis 1982). Closely spaced normal faults and possible growth faults also affect the continental slope N of Baltim and Damietta (Summerhayes *et al.* 1977; Ross & Uchupi 1979), whereas E of longitude 31°E the slope shows a step-like drop (Bardawil Escarpment) with several steep NNE-facing escarp-

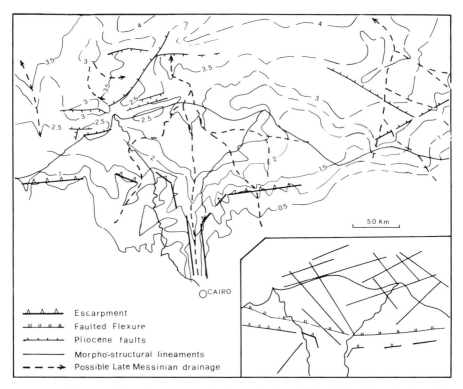

FIG. 14. Features of the Messinian–Pliocene unconformity surface (compiled with simplifications from Neev *et al.* 1976; Ansary & Deibis 1977; El Nahass 1978; Ryan 1978; Barber 1981; Ben Avraham & Mart 1981; Garfunkel 1984).

ments and SSW tilting of sediments. Salt walls parallel these faults (Mart & Ben Gai 1982; Mart 1984).

A preliminary conclusion of this structural review is that the North Delta Basin is characterized by two structural motifs: (i) a deep pre-Tortonian (possibly to Eocene or Late Cretaceous) series of mainly E–W fault blocks, prominent among which are the shelf-margin structures, which to the N appear to have bounded a belt of Miocene subsidence and sedimentation while in the S the active Faulted Flexure induced the formation of synsedimentary listric (growth?) faults; (ii) a more shallow post-Messinian fabric genetically related to sedimentary loading at the unstable delta margin, which caused growth faulting, slumping and normal faults as well as diapirism of uncompacted Pliocene clays and Messinian evaporites.

Both sets of structures, however, appear to have been modified and governed by NW- and NE-trending fractures, undoubtedly the result of the tensional opening of pull-apart basins and grabens normal to the regional ENE to NNE sinistral movements associated with the rifting between Africa and Sinai-Arabia. In the central part of the delta (South Delta Block) such grabens are clearly suggested by the tectonic map of Riad (1977), in agreement with Said's (1981) maps of sediment thickness.

The continued influence of tensional faulting on sedimentation is evidenced by the uplifted versus subsided Late Pleistocene deposits at the margins of the present delta plain, and by the down-faulting of the Alexandria Pleistocene carbonate ridges between Abuqir and the Edku Lagoon (Butzer 1960).

Geological history

Figure 15 summarizes the geological history of the Nile Delta from Late Eocene times. A synopsis of this history is given below.

Late Eocene–Oligocene

Clastic sedimentation in N Egypt resumed in the Late Eocene (Said 1962) with mainly mud and silt deposition (Dabaa Formation). Reconstructed shorelines (Said 1962; Sestini 1984) emphasize the emergence of the eastern versus the western regions, but there are no data to indicate whether the North Delta Basin was already more subsident, by analogy with subsidence at the Israel margin. The Oligocene deltaic facies (Qatrani Formation) occurs W of the present Nile Delta, whereas the Cairo–Suez region was occupied by shallow marine sands (Said 1962). This clastic phase is related to the uplift of the Red Sea–Sinai region, accompanied by basaltic extrusion, with uplift and/or active erosion of fault blocks in Sinai and the Eastern Desert.

Early Miocene

The clastic discharge of an ancestral Nile River continued to be towards the eastern part of the Western Desert. The Moghra Formation of the Qattara Depression eastern margin, and its subsurface equivalents (Marzouk 1970), may be the final regressive stage of the Qatrani Delta. Considerable subsidence is recorded, however, only near the Arab Gulf coast (700–900 m of outer neritic prodeltaic sandy shales). In the Nile Delta region the Flexure Zone was probably already active, as seen in the facies and subsidence contrasts between the North Basin (*eg* deep marine with few sandy fine-grained turbidites, although more sandy in the eastern and western depocentres) and the South Delta Block.

In the Early Miocene a marine transgression due to both a global sea-level rise and rifting in the Gulf of Suez caused a marine connection between the Gulf of Suez and the Mediterranean (Said 1962). Stratigraphic evidence suggests a shallow sill, rather than a continuation of the northern Gulf of Suez depocentre (Reynolds 1979).

Middle Miocene (Langhian–Serravallian)

In the North Delta Basin down-faulting accentuated the Flexure Zone with the development of listric faults and half-grabens in the central and eastern areas. The environment continued to be outer neritic to bathyal, except that facies in wells drilled over the shelf-margin structures suggest much shallower conditions. At this time Sinai, the area SE of Cairo and the Gulf of Suez margin, and all the Western Desert were uplifted (after the Marmarica Limestone deposition). As a consequence of this uplift, and of the probable development of NW or NNW grabens in the present S delta region, deltaic deposition shifted to the E.

Late Miocene (Tortonian–Messinian)

The Tortonian Qawasim clastics suggest marked syndepositional movements along the Flexure Zone, with the development of at least three depocentres. The prograding Qawasim clastics mark the first fluvio-deltaic influx in the North Delta Basin (the Neo-Nile of Said 1981)—an

120 *G. Sestini*

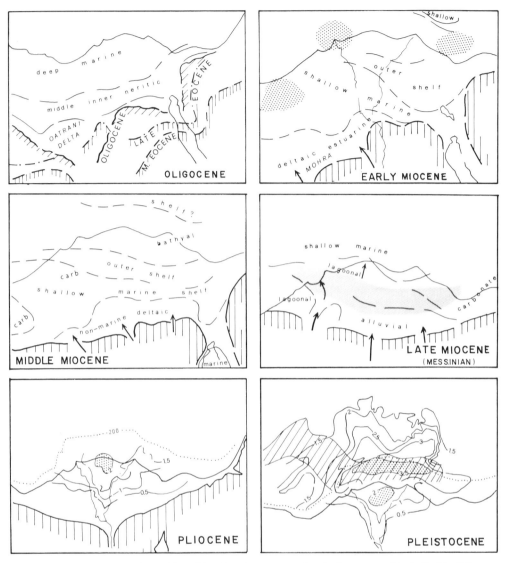

FIG. 15. Schematic reconstructions of depositional environments and main depocentres (shaded) of the Nile Delta region from Late Oligocene to Pleistocene. Late Miocene shading refers to the Qawasim Formation. Pliocene and Pleistocene isopachs (onshore) are in kilometres, and Pleistocene offshore contours are in seconds (Ross & Uchupi 1979). Ruling (Pleistocene) refers to areas with rates of deposition greater than 15 cm per 1000 years (Stanley & Maldonado 1977).

unusually large supply of coarse sands undoubt-edly reflecting an intensified uplift of the SE hinterland with the active erosion of Cretaceous Nubian sandstones and basement rocks. The unconformable deposition of the Abu Madi sands could be associated with a renewed lowering of sea level in the Messinian and marginal tectonic uplift. Distribution of facies (Zaghloul *et al.* 1977*b*; Said 1979; Deibis 1982; Poumot & Bourullec 1984) suggests considerable variability of marginal deltaic to lagoonal and evaporitic environments. In the N, evaporites intermittently accumulated in the shallow Herodotus, Levantine and Antalia Basins (Gvirtzman & Buchbinder 1978).

The hypothesis of a single large-scale fall of sea level in the eastern Mediterranean, by analogy with that suggested for the western Mediterranean (Hsu *et al.* 1973; Ryan & Cita 1978), followed by pronounced incision of the margins of topographically deep basins (Ryan 1978; Barber 1981) appears to be inconsistent with both Messinian stratigraphy and the early Middle Pliocene tectonic deepening of the Mediterranean basins (Ginzburg *et al.* 1975*b*; Bein & Gvirtzman 1977; Stanley 1977; Woodside 1977; Gvirtzman & Buchbinder 1978; Mart & Ben Gai 1982; Rouchy 1982). In the southern Nile Delta stream erosion certainly deepened a Palaeo-Nile valley (*cf* the 'Thebes Canyon', Ryan 1978), but a precursor to this had probably already been carved in response to the repeated Middle and Late Miocene differential uplifts and sea-level changes. Furthermore, the depression of the Cairo Graben, as evidenced by the Messinian unconformity surface, is associated with Plio-Pleistocene deformation and subsidence.

Pliocene–Pleistocene

The Early Pliocene marine transgression extended inland up a down-faulted Nile Valley, although not necessarily as far as Aswan, as claimed by Ryan (1978) and Said (1981), and to the Dead Sea Rift (Gvirtzman & Buchbinder 1978). In the Middle Pliocene the regional subsidence of the eastern Mediterranean floor amplified the former evaporite sinks.

Subsidence patterns changed. Whereas Miocene subsidence had been concentrated in the North Delta Basin, mainly with a latitudinal trend, this spread in the Pliocene to the South Delta Block with a N–S depocentre (Zaghloul *et al.* 1977*b*; Said 1981). In the Pleistocene, this axial deepening continued, but with the rapid advancement and associated spread of clastic deposition from the Late Pliocene when a main E–W depocentre developed under the present outer continental shelf and S of the E–W Miocene structures (Summerhayes *et al.* 1977; Ross & Uchupi 1979). Significant sand accumulation also affected areas to the E (N Sinai–Israel offshore), where previous Early Pleistocene deposition had been essentially calcareous–pelitic (Yafo Formation, Gvirtzman & Buchbinder 1978).

The Nile Submarine Fan was constructed principally in the Pleistocene. Isopachs suggest two main dispersal patterns, by N- and NE-trending distributaries, during the Pleistocene eustatic low-stands. The Rosetta Canyon may have been incised during the periods of falling sea level, while high-stands should have been responsible for the elevated Pleistocene deposits of the present delta margins (*eg* W of Alexandria,

Wadi Natrun and Wadi Tumilat). Late Pleistocene sand dispersal (channelled turbidites) shifted to the NW, possibly in response both to plate subduction versus the halokinetic uplift of the Levant Platform and to greater sediment discharge to the NW versus restriction of sand to small shelf delta lobes in the E.

The petroleum habitats of the Nile Delta

The Nile Delta has been superficially likened to the Niger Delta. However, in addition to important climatic, hydrodynamic and geotectonic differences, Niger Delta deltaic regression was continuous from the Eocene to the Pleistocene, with off-lapping deltaic, prodeltaic and marine to continental slope sediments (Evamy *et al.* 1978). Hydrocarbon pools are located principally in channel and barrier sands in the thick upper fluvio-marine facies, mainly in rollover anticlines. In contrast, the petroleum habitats of the Nile Delta, in so far as known (Ansary & Deibis 1977; Shawky Abdine 1981; Deibis 1982), are quite different. The hydrocarbon discoveries have been mainly gas and condensate (Abu Madi, with estimated reserves of 1 TCF and 20 Mbbls of condensate, Abuqir and NAF-1), principally in the Messinian Abu Madi sands that are sealed by the Pliocene clays. Pool control is structural and stratigraphic. Minor gas shows have been found in many thin lenticular sands throughout the Mio-Pliocene succession. The probable source rocks are deeply buried Oligocene and Miocene shales (Deibis *et al.* 1986), and the location of the main gas occurrences near the delta coast (onshore and offshore) is probably related to a belt of optimal depth of burial and maturation located in the E–W Miocene basin. The offshore of the Nile Delta has not been sufficiently tested. The recent oil discoveries in Oligocene and early Miocene sands in the E (El Wakara, Port Fouad Marine and Tineh) suggest scope for testing zones deeper than Middle Miocene, and this is also the case in the central and western parts of the northern Nile Delta. However, the non-deltaic possible sand reservoirs in the pre-Tortonian Moghra and Sidi Salem Formations are thin and discontinuous.

Exploration along the Flexure Zone belt and in the South Delta Block has been negative, at least with regard to Cretaceous and Tertiary sediments, despite stratigraphic and facies similarities to the Western Desert oil province.

References

ABDEL KADER, A. 1982. *Landsat Analysis of the Nile Delta, Egypt.* MSc thesis, University of Delaware, 260 pp.

ABU SAKR, M. M. 1973. *Studies on the Subsurface Geology of the Nile Delta.* MSc thesis, Faculty of Sciences, Mansoura University.

ADAMSON, D. A., GASSE, F., STREET, F. A. & WILLIAMS, M. A. J. 1980. Late Quaternary history of the Nile. *Nature* **288**, 50–55.

ALLEN, J. R. L. 1970. Sediments of the modern Niger Delta: a summary and review. *In:* MORGAN, J. P. (ed.) *Deltaic Sedimentation, Modern and Ancient.* Society of Economic Paleontologists and Mineralogists Special Publication **15**, 138–151.

ALMAGOR, G. 1979. Relict sandstone of Pleistocene age on the continental shelf of northern Sinai and Israel. *Israel Journal of Earth Sciences* **28**, 70–76.

—— & GARFUNKEL, Z. 1979. Submarine slumping in continental margin of Israel. *Bulletin of the American Association of Petroleum Geologists* **63**, 324–340.

—— & HALL, S. K. 1978. Morphology of the continental margin off NE Sinai and southern Israel. *Israel Journal of Earth Sciences* **27**, 138–142.

ANSARY, S. & DEIBIS, S. 1977. The geology and hydrocarbon potentialities of the Abu Qir and Abu Madi gas fields. *Proceedings of the 12th Arab Petroleum Congress, Cairo,* Paper 133 (B-3).

ANWAR, Y. M., EL ASKARY, M. A. & FRIHY, O. E. 1984. Reconstruction of sedimentary environments of Rosetta and Damietta promontories in Egypt, based on textural analysis. *Journal of African Earth Sciences* **2**, 17–29.

ASRT/UNESCO/UNDP. 1977. *Proceedings of Seminar on Nile Delta Coastal Processes (with Special Emphasis on Hydrodynamical Aspects).* Academy of Scientific Research and Technology, Cairo, 150 pp.

ATTIA, M. I. 1954. *Deposits of the Nile Valley and Delta.* Geological Survey, Cairo, Publication **65**, 356 pp.

AYOUTI, M. K. 1980. Oil and gas prospects in Egypt. *In:* BUROLLET, P. F. & ZIEGLER, V. (eds) *Energy Resources, 35th International Geological Congress, Colloquium C2.* BRGM, Paris, 131–141.

BALL, J. 1942. *Egypt in the Classical Geographers.* Survey of Egypt, Government Press, Cairo, 203 pp.

BARBER, P. M. 1980. Paleogeographic evolution of the Proto-Nile Delta, during the Messinian salinity crisis. *Géologie Méditerranée* **7**, 13–18.

—— 1981. Messinian subaerial erosion of the Proto-Nile Delta. *Marine Geology* **44**, 353–372.

BARTOLINI, C., MALESANI, P. G., MANETTI, P. & WEZEL, F. C. 1975. Sedimentation and petrology of Quaternary sediments from the Hellenic Trench, Mediterranean Ridge and Nile Cone, DSDP Leg 13 cores. *Sedimentology* **22**, 205–236.

BARTOV, Y., STEINITZ, G., EYAL, H. & EYAL, Y. 1980. Sinistral movement along the Gulf of Aqaba, its age and relation to the opening of the Red Sea. *Nature* **285**, 220–221.

BEIN, A. & GVIRTZMAN, G. 1977. A Mesozoic fossil

edge of the Arabian Plate along the Levant coastline and its bearing on the evolution of the Eastern Mediterranean. *In:* BIJOU-DUVAL, B. & MONTADERT, L. (eds) *Structural History of the Mediterranean Basins.* Technip, Paris, 95–109.

BELL, B. 1975. The oldest records of Nile floods. *Geographical Journal* **136**, 569–573.

BEN AVRAHAM, Z. 1978. The structure and tectonic setting of the Levant continental margin, Eastern Mediterranean. *Tectonophysics* **44**, 313–331.

—— & MART, Y. 1981. Late Tertiary structure and stratigraphy of North Sinai continental margin. *Bulletin of the American Association of Petroleum Geologists* **65**, 1135–1145.

BEN MENACHEM, A. 1979. Earthquake catalogue for the Middle East (92 BC–1980 AD). *Bollettino di Geofisica Teorica Applicata* **21**, 245–313.

BENTZ, F. P. & GUTMAN, S. J. 1977. *Landsat* data contributions to hydrocarbon exploration in foreign regions. *US Geological Survey Professional Papers* **1015**, 83–92.

BIETAK, M. 1966. Zur Palaeogeographie des oestlichen Nildeltas. *Oesterreiches Akademie Wissenschaft, Denkschrifte Gesamtakademie* **4**, 47–112.

BUTZER, K. W. 1960. On the Pleistocene shorelines of the Arab's Gulf, Egypt. *Journal of Geology* **68**, 626–637.

—— 1975. Delta. *In:* HELK, W. & OTTO, E. (eds) *Lexicon der Aegyptologie,* Vol. 1. Otto Harrasowitz, Wiesbaden, 1043–1052.

—— 1976. *Early Hydraulic Civilization in Egypt. A Study in Cultural Ecology.* University of Chicago Press, Chicago, IL, 134 pp.

—— & HANSEN, C. L. 1968. *Desert and River in Nubia.* University of Wisconsin Press, Madison, WI, 562 pp.

COLEMAN, J. M. 1982. *Deltas: Processes of Deposition and Models of Exploration.* Human Resources Development Corporation, Boston, MA, 124 pp.

——, ROBERTS, H. H., MURRAY, S. P. & SALAMA, M. 1980. Morphology and dynamic sedimentology of the eastern Nile Delta shelf. *Marine Geology* **41**, 325–339.

COUTELLIER, V. & STANLEY, D. J. 1987. Late Quaternary stratigraphy and paleo-geography of the eastern Nile Delta, Egypt. *Marine Geology* **27**, 257–275.

CRANE, K. & BONATTI, E. 1987. The role of fracture zones during early Red Sea rifting. *Journal of the Geological Society of London* **144**, 407–420.

DARESSY, M. G. 1929. Les branches du Nil sous la XVIII Dynastie. *Bulletin de la Société Royale de Géographie d'Egypte* **16**, 293–329.

—— 1931. Les branches du Nil sous la XVIII Dynastie. La région du Lac Manzala. *Bulletin de la Société Royale de Géographie d'Egypte* **17**, 189–223.

DEIBIS, S. 1982. Abu Qir Bay, a potential gas province area, offshore Mediterranean Egypt. *Proceedings of the 6th Exploration Seminar.* EGPC, Cairo.

——, FUTYEN, A. R. I., INCE, D. M., MORLEY, R. J., SEYMOUR, W. P. & THOMPSON, S. 1986. The stratigraphic framework of the Nile Delta and its implications with respect to the region's hydrocarbon potential. *Proceedings of the 8th Exploration Conference.* EGPC, Cairo.

DERIN, B. & REISS, Z. 1973. Revision of marine Neogene stratigraphy in Israel. *Israel Journal of Earth Sciences* **28**, 199–210.

EFFAT, A. & GEZEIRY, M. 1986. New aspects of Abu Qir stratigraphic and sedimentologic model. *Proceedings of the 8th Exploration Conference.* EGPC, Cairo.

EL ASKARY, M. A. & FRIHY, O. E. 1986. Depositional phases of Rosetta and Damietta promontories on the Nile Delta coast. *Journal of African Earth Sciences* **5**, 627–633.

—— & LOTFY, M. F. 1980. Depositional environment of the islands in Manzala Lake, Egypt. *Bulletin of the Faculty of Sciences, King Abdul Aziz University, Jeddah* **4**, 241–254.

——, ABDEL AAL, A. A. & FRIHY, O. E. 1984. Study of subrecent foraminifera and some macrofossils in the subsurface sediments of the Rosetta and Damietta promontories. *Neues Jahrbuch Geologie Palaeontologie, Abhandlungen* **167**, 347–374.

EL AYAD, F. M. 1980. *Biostratigraphic Studies on the Tertiary Sediments of Kafr El Dawar Well, NW Area, Nile Delta, Egypt.* MSc thesis, Faculty of Science, Mansoura University.

EL BEIALY, S. Y. M. 1980. *Palynological Studies on the Subsurface Plio-Pleistocene Sediments of Abu Qir Well No. 2X, Northern Coast of Nile Delta.* MSc thesis, Faculty of Science, Mansoura University.

EL BUSEILI, A. & FRIHY, O. 1984. Textural and mineralogical evidence denoting the position of the mouth of the old Canopic Nile branch on the Mediterranean coast of Egypt. *Journal of African Earth Sciences* **2**, 103–107.

ELF AQUITAINE. 1982. North Alexandria area: geological and geophysical study. *Proceedings of the Sixth Exploration Seminar.* EGPC, Cairo.

EL FISHAWI, N. M. 1985. Textural characteristics of the Nile Delta coastal sands: an application in reconstructing the depositional environments. *Acta Mineralogica-Petrographica, Szeged* **27**, 71–88.

—— & EL ASKARY, M. A. 1981. Characteristic features of the coastal sand dunes along the Burullus–Gamasa stretch, Egypt. *Acta Mineralogica-Petrographica, Szeged* **5** (25/1), 63–73.

—— & MOLNAR, B. 1985. Mineralogical relationships between the Nile Delta coastal sands. *Acta Mineralogica-Petrographica, Szeged* **27**, 89–100.

EL HEINY, I. 1982. Neogene stratigraphy of Egypt. *Newsletters on Stratigraphy* **2** (2), 41–54.

EL KHOLY, H. & RAMADAN, R. I. 1986. Seismic complexities in Abu Quir Bay area. *Proceedings of the 8th Exploration Conference.* EGPC, Cairo.

EL NAHASS, H. A. M. 1978. *Geophysical and Structural Studies on the Nile Delta Area, Egypt.* MSc thesis, Faculty of Sciences, Mansoura University.

EL SAYED, M. K. 1979. The inner shelf of the Nile Delta: sediment types and depositional environments. *Oceanologica Acta* **2**, 249–252.

EL SHAZLY, E. M. 1975. *Geologic Interpretation of Landsat Satellite Images of West Nile Delta Area, Egypt.* Remote Sensing Research Project, Academy of Scientific Research Technology, Cairo, 34 pp.

—— 1977. Geology of the Egyptian region. *In:* NAIRN, A. E. M., STEHLI, P. F. & KANES, W. H. (eds) *The Ocean Basins and Margins, IV, The Eastern Mediterranean.* Plenum, New York, 479–544.

—— & ABDEL HADY, M. A. (eds) 1975. *Geological and Geophysical Investigations of the Suez Canal Zone.* Remote Sensing Project, Academy of Scientific Research Technology, Cairo, 42 pp.

EL WAKEEL, S. K. & WAHBY, S. D. 1970. Bottom sediments of Lake Manzala. *Journal of Sedimentary Petrology* **40**, 605–611.

——, ABDOU, H. F. & MOHAMED, M. A. 1974. Texture and distribution of recent marine sediments of the continental shelf of the Nile Delta. *Bulletin of the Geological Society of Iraq* **7**, 15–34.

EMERY, K. O. & NEEV, D. 1962. Mediterranean beaches of Israel. *Geological Survey of Israel Bulletin* **26**, 1–24.

EVAMY, B. D., HARAMBOURE, J., KAMERLING, P., KNAAP, W. A., MOLLOY, F. A. & ROWLANDS, P. H. 1978. Hydrocarbon habitats of Tertiary Niger Delta. *Bulletin of the American Association of Petroleum Geologists* **62**, 1–39.

FABBRI, P. C. 1985. Coastline variations in the Po Delta since 2500 BP. *Zeitschrift für Geomorphologie (Geomorphology of Changing Coastlines),* Supplement 57, 121–138.

FAIRBRIDGE, R. W. 1968. Holocene. *In:* FAIRBRIDGE, R. W. (ed.) *The Encyclopedia of Geomorphology,* Vol. 3. Reinhold, New York, 525–536.

FANOS, A. M. 1986. Statistical analysis of longshore current data along the Nile Delta coast. *Water Science Journal* **1**, 45–55.

FINETTI, I. 1976. The Mediterranean Ridge: a young submerged chain associated with the Hellenic Arc. *Bollettino di Geofisica Teorica e Applicata* **19** (69), 31–65.

FRIHY, O. E. 1987. Nile Delta shoreline changes: aerial photographic study of a 28-year period. *Journal of Coastal Research,* in press.

—— & STANLEY, D. J. 1988. Texture and coarse fraction composition of Nile Delta deposits: facies analysis and stratigraphic correlation. *Journal of African Earth Sciences,* 237–255.

——, EL FISHAWI, N. M. & EL ASKARY, M. A. 1988. Geomorphological features of the Nile Delta coastal plain: a review. *Journal of African Earth Sciences,* in press.

GARFUNKEL, Z. 1984. Large scale submarine rotational slumps and growth faults in the Eastern Mediterranean. *Marine Geology* **55**, 305–324.

—— & BARTOV, Y. 1977. The tectonics of the Suez rift. *Geological Survey of Israel Bulletin* **71**, 44 pp.

GARSON, M. S. & KRS, M. 1976. Geophysical and geological evidence of the relationship of the Red Sea transverse tectonics to ancient fractures. *Bulletin of the Geological Society of America* **87**, 169–181.

GERGES, M. A. 1981. Recent observations of currents from moorings in the Egyptian Mediterranean waters off the Sinai coast. *Ocean Management* **6**, 159–171.

GINZBURG, A., COHEN, S. S., HAY-ROE, H. & ROSSEN-ZWEIG, A. 1975a. Geology of the Mediterranean shelf of Israel. *Bulletin of the American Association of Petroleum Geologists* **59**, 2142–2160.

——, GVIRTZMAN, G. & BUCHBINDER, B. 1975b. The desiccation events in the Eastern Mediterranean during the Messinian times, as compared with other desiccation events in basins around the Mediterranean. *In:* BIJOU-DUVAL, B. & MONTAD-ERT, L. (eds) *Structural History of Mediterranean Basins.* Technip, Paris, 411–420.

GOBY, J. E. 1952. Histoire des nivellements de l'Isthme de Suez. *Bulletin de la Société d'Etudes Historiques et Géographiques de l'Isthme de Suez* **4**, 99–190.

—— 1954. Modifications des rivages de la Mer Rouge et de la Méditerranée à l'époque historique. *Bulletin de la Société d'Etudes de l'Isthme de Suez* **5**, 23–43.

GVIRTZMAN, G. & BUCHBINDER, B. 1978. The Late Tertiary of the coastal plain and continental shelf of Israel and its bearing on the history of the Eastern Mediterranean. *In: Initial Reports. DSDP Project* **42** (2), 1195–1222.

HAMID, S. 1984. Fourier analysis of Nile flood levels. *Geophysical Research Letters* **11**, 843–858.

HANTAR, G. 1975. Contribution to the origin of the Nile delta. *Ninth Arab Petroleum Congress, Dubai,* Paper 117 (B-3).

HASSAN, F. A. 1981. Historical Nile floods and their implications for climatic change. *Science* **212**, 1142–1145.

HILMY, M. E. 1951. Beach sands of the Mediterranean coast of Egypt. *Journal of Sedimentary Petrology* **21**, 109–120.

HSU, K. J., RYAN, W. B. F. & CITA, M. B. 1973. Late Miocene desiccation of the Mediterranean. *Nature* **242**, 239–243.

HURST, E. H. 1931–1966. *The Nile Basin,* Vols 1–9. Physical and Nile Control Department, Ministry of Public Works, Cairo.

INMAN, D. L. 1976. Application of nearshore processes to the Nile Delta. *Proceedings of the Seminar on Nile Delta Sedimentology.* UNESCO/ASRT/UNDP, Alexandria, 205–255.

—— & JENKINS, S. A. 1984. The Nile littoral cell and Man's impact on the coastal littoral zone in the SE Mediterranean. *Proceedings of the 17th International Coastal Engineering Conference.* ASCE, Sidney, 1600–1617.

JACOTIN, E. 1826. *Atlas Géographique: Description de l'Egypte.* Panokouke, Paris, 48 plates.

KENYON, N. H., STRIDE, A. H. & BELDERSON, R. H. 1975. Plan view of active faults and other features on the lower Nile Cone. *Geological Society of America Bulletin* **86**, 1733–1739.

KERAMBRUN, P. (ed.) 1986. *Coastal Lagoons of the Southern Mediterranean Coast (Algeria, Egypt, Libya, Morocco, Tunisia): Description and Bibliography.* UNESCO Reports in Marine Science **34**, 182 pp.

KHAIRY, M. A. 1974. *Geological and Mineralogical Studies on the Sedimentary Succession of the Continental Shelf of the Nile Delta.* PhD thesis, Faculty of Sciences, Mansoura University.

KHOLEIF, M. M., FRIHY, O. & NAWAR, A. 1977. Sedimentological features and primary structure of the Nile Delta coastal dunes, Egypt. *Proceedings of the 6th Colloquium on Geology, Aegean Region,* Athens, 843–857.

——, HILMY, E. & SHAHAT, A. 1969. Geological and mineralogical studies of some sand deposits in the Nile Delta. *Journal of Sedimentary Petrology* **39**, 1520–1529.

KLEMAS, V. & ABDEL KADER, A. M. 1982. Remote sensing of coastal processes, with emphasis on the Nile Delta. *Proceedings of the International Symposium on Remote Sensing of Environments, Cairo, Egypt,* 27 pp.

KORA, M. A. N. 1980. *Geology of the Tertiary–Quaternary Subsurface Sedimentary Succession, West of the Nile Delta Area.* MSc thesis, Faculty of Sciences, Mansoura University.

KORRAT, I. M. I. 1977. *Geophysical Studies on Some Subsurface Formations in the Nile Delta Region.* MSc thesis, Faculty of Sciences, Mansoura University.

KRUIT, C. 1955. Sediments of the Rhone Delta. *Geologie en Mijnbouw* **15**, 304–357.

LEVY, Y. 1974. Sedimentary reflections of depositional environment in the Bardawil Lagoon. *Journal of Sedimentary Petrology* **44**, 219–227.

MALDONADO, A. & STANLEY, D. J. 1976. The Nile Cone submarine fan development by cyclic sedimentation. *Marine Geology* **20**, 27–40.

—— & —— 1978. Nile Cone depositional processes and patterns in Late Quaternary. *In:* STANLEY, D. J. & KELLING, G. (eds) *Sedimentation in Submarine Canyons, Fans and Trenches.* Dowden, Hutchinson & Ross, Stroudsburg, PA, 239–260.

MANOHAR, M. 1976. Dynamic factors affecting the Nile Delta coast. *In: Proceedings of the Seminar on Nile Delta Sedimentology.* UNESCO/ASRT/UNDP, Alexandria, 104–129.

—— 1981. Coastal processes at the Nile delta coast. *Shore and Beach* **49**, 8–15.

MART, Y. 1982. Quaternary tectonic patterns along the continental margin of the southeastern Mediterranean. *Marine Geology* **49**, 327–344.

—— 1984. The tectonic regime of the southeastern Mediterranean. *Marine Geology* **55**, 365–386.

—— 1988. Superimposed tectonic patterns along the continental margin of the southeastern Mediterranean. *Tectonophysics,* in press.

—— & BEN-GAI, Y. 1982. Some depositional patterns of the continental margin of the southeastern Mediterranean Sea. *Bulletin of the American Association of Petroleum Geologists* **66**, 460–470.

MARZOUK, I. 1970. Rock stratigraphy and oil potentialities of the Oligocene and Miocene in the Western Desert, U.A.R. *7th Arab Petroleum Congress, Kuwait,* Paper 54 (B-3).

MESHREF, W. M. 1982. Regional structural setting of

northern Egypt. *Proceedings of the 6th Exploration Seminar.* EGPC, Cairo.

MILLIMAN, J. D. 1974. *Marine Carbonates.* Springer, Berlin, 475 pp.

MISDORP, R. & SESTINI, G. 1976. Topography of the continental shelf of the Nile Delta. *Proceedings of the Seminar on Nile Delta Sedimentology.* UNESCO/ASRT/UNDP, Alexandria, 145–161.

MOREAU-MARTIN, C. 1979. *Sédimentation sur le Cone Sous-marin du Delta du Nil.* Thèse 3me cycle, Université Paris Sud.

MÖRNER, N. A. 1976. Eustatic changes during the last 8000 years. *Palaeogeography, Palaeoclimatology, Palaeoecology* 19, 63–85.

MULDER, C. J., LEHNER, P. & ALLEN, D. C. K. 1975. Structural evolution of the Miocene salt basins of the Eastern Mediterranean and Red Sea. *Geologie en Mijnbouw* 54, 208–221.

MURRAY, S. P., COLEMAN, J. M., ROBERTS, H. H. & SALAMA, M. 1980. Eddy currents and sediment transport off the Damietta Nile. *Proceedings of the 17th International Coastal Engineering Conference.* ASCE, Sidney, 1680–1700.

——, ——, —— & —— 1981. Accelerated currents and sediment transport off Damietta Nile promontory. *Nature* 293, 51–54.

NAHLA, F. A. 1958. Mineralogy of Egyptian black sands and its application. *Egyptian Journal of Geology* 2, 12–21.

NEEV, D. 1975. Tectonic evolution of the Middle East and the Levantine basin (easternmost Mediterranean). *Geology* 3, 683–686.

——, ALMAGOR, G., ARAD, A., GINZBURG, A. & HALL, S. K. 1976. The geology of the southeastern Mediterranean. *Geological Survey of Israel Bulletin* 68, 55 pp.

——, GREENFIELD, L. & HALL, J. 1980. Slice tectonics in the Eastern Mediterranean Sea. *In:* STANLEY, D. J. & WEZEL, C. F. (eds) *Geological Evolution of the Mediterranean Basin.* Springer, Berlin, 249–272.

NUR, A. & BEN AVRAHAM, Z. 1978. The Eastern Mediterranean and the Levant: tectonics of continental collision. *Tectonophysics* 46, 279–311.

OMARA, S. M. & OUDA, K. 1972. A lithostratigraphic revision of the Oligocene Miocene succession in the northern Western Desert, Egypt. *8th Arab Petroleum Congress, Algiers,* Paper 94 (B-3).

—— & SANAD, M. I. 1975. Rock stratigraphy and structural features of the area between Wadi El Natrun and the Moghra depression (Western Desert, Egypt). *Geologische Jahrbuch* 16, 45–73.

OOMKENS, E. 1970. Depositional sequences and sand distribution in the post glacial Rhone delta complex. *In:* MORGAN, J. P. (ed.) *Deltaic Sedimentation, Modern and Ancient.* Society of Economic Paleontologists and Mineralogists Special Paper 15, 198–214.

POUMOT, C. & BOURULLEC, J. 1984. Palynoplanktological contribution to the Miocene–Oliocene environmental and palaeoclimatological conditions of the Nile Delta area. *Proceedings of the 7th Exploration Seminar.* EGPC, Cairo, 325–335.

QUELLENEC, R. E. & KRUK, C. B. 1976. Nile suspended

load and its importance for Nile Delta morphology. *Proceedings of the Seminar on Nile Delta Sedimentology.* UNESCO/ASRT/UNDP, Alexandria, 130–145.

REYNOLDS, M. L. 1979. Geology of the northern Gulf of Suez. *Annals of the Geological Survey of Egypt* 9, 322–343.

RIAD, S. 1977. Shear zones in north Egypt interpreted from gravity data. *Geophysics* 42, 1207–1214.

RIZZINI, A., VEZZANI, F., COCOCCETTA, V. & MILAD, G. 1978. Stratigraphy and sedimentation of the Neogene–Quaternary section in the Nile delta area. *Marine Geology* 27, 327–348.

ROSS, D. A. & UCHUPI, E. 1979. The structure and sedimentation history of the SE Mediterranean Sea. *American Association of Petroleum Geologists Bulletin* 61, 872–890.

——, ——, SUMMERHAYES, C., KOELSCH, D. F. & EL SHAZLY, M. 1978. Sedimentation and structure of the Nile Cone and the Levant Platform area. *In:* STANLEY, D. J. & KELLING, G. (eds) *Sedimentation in Submarine Canyons, Fans and Trenches.* Dowden, Hutchinson & Ross, Stroudsburg, PA, 261–275.

ROSSIGNOL-STRICK, M. 1983. African monsoons, an immediate climatic response to orbital insulation. *Nature* 303, 46–49.

ROUCHY, J. M. 1982. Paleogeographic interpretation of the Mediterranean area during the deposition of the Messinian evaporites, based on the study of the peri-Messinian erosional surfaces. *Bulletin de la Société Géologique de France Séries 7,* 24, 653–677.

RYAN, W. F. B. 1978. Messinian badlands on the SE margin of the Mediterranean Sea. *Marine Geology* 27, 389–363.

—— & CITA, M. B. 1978. The nature and distribution of Messinian erosional surfaces—indicators of several km deep Mediterranean in the Miocene. *Marine Geology* 27, 193–230.

—— & HSU, K. J. 1973. *Initial Report of Deep Sea Drilling Project* 13. U.S. Government Printing Office, Washington, DC, 514 pp.

SAAD, M. A. H. 1981. Core sediments from four Egyptian lakes. *Rapports de la Commission Internationale de la Mer Méditerranée* 25/26, 113–119.

SAID, R. 1958. Remarks on the geomorphology of the deltaic coastal plain between Rosetta and Port Said. *Bulletin de la Société Géographique d'Egypte* 31, 58–67.

—— 1962. *The Geology of Egypt.* Elsevier, Amsterdam, 377 pp.

—— 1979. The Messinian in Egypt. *Annales Géologiques des Pays Helléniques* 1, 1083–1090.

—— 1981. *The River Nile.* Springer, Berlin, 151 pp.

SALEM, A. 1976. Evolution of Eocene–Miocene sedimentation patterns in Northern Egypt. *Bulletin of the American Association of Petroleum Geologists* 60, 34–64.

SANCHO, J., BIJOU-DUVAL, B., COURRIER, I. P., LETOUZEY, J., MONTADERT, L. & WINNOCK, E. 1973. New data on the structure of the Eastern Mediterranean basin from seismic reflection. *Earth and Planetary Science Letters* 18, 189–204.

SCOTT, R. W. & GOVEAN, F. M. 1985. Early depositional history of a rift basin: Miocene in Western Sinai. *Palaeogeography, Palaeoclimatology, Palaeoecology* **52**, 143–158.

SCRUTON, P. C. 1960. Delta building and deltaic sequence. *In:* SHEPARD, F. P., PHLEGER, F. B. & VAN ANDEL, T. H. (eds) *Recent Sediments of the Northwest Gulf of Mexico.* American Association of Petroleum Geologists, Tulsa, OK, 89–102.

SESTINI, G. 1976. Geomorphology of the Nile Delta. *Proceedings of the Seminar on Nile Delta Sedimentology.* UNESCO/ASRT/UNDP, Alexandria, 12–24.

—— 1984. Tectonics and sedimentary history of the NE African margin (Egypt–Libya). *In:* DIXON, J. D. & ROBERTSON, A. H. F. (eds) *The Geological Evolution of the Eastern Mediterranean.* Blackwell, Oxford, 151–126.

——, FISHAWY, N. M., FAHMI, M. & SHAWKY, A. 1976. Grain size of Nile Delta beach sands. *Proceedings of the Seminar on Nile Delta Sedimentology.* UNESCO/ASRT/UNDP, Alexandria, 79–94.

SHARAF EL DIN, S. H. 1973. Geostrophic currents in the southeastern sector of the Mediterranean Sea. *Acta Adriatica (Symposium on Eastern Mediterranean Sea)* **18**, 221–235.

——1977. Effect of Aswan High Dam on Nile flood and on the estuarine and coastal circulation pattern along the Mediterranean Egyptian coast. *Limnology and Oceanography* **22**, 194–207.

SHATA, A. & EL FAYOUMI, I. 1970. Remarks on the regional geological structure of the Nile delta. *In:* Hydrology of Deltas, IASH/UNESCO Bucharest Symposium, Vol. 1. UNESCO, Paris, 189–197.

SHAWKY ABDINE, A. 1981. Egypt's petroleum geology: good grounds for optimism. *World Oil,* December, 99–112.

SMITH, S. G. 1975. Diapiric structures in the Eastern Mediterranean Herodotus Basin. *Earth and Planetary Science Letters* **32**, 62–68.

SNEH, A. & WEISSBROD, T. 1973. Nile Delta: the defunct Pelusian branch identified. *Science* **180**, 59–61.

——, —— & EHRLICH, A. 1976. Holocene evolution of the northeastern corner of the Nile Delta. *Quaternary Research* **26**, 194–206.

STANLEY, D. J. 1977. Post-Miocene depositional patterns and structural displacements in the Mediterranean. *In:* NAIRN, A. E. N., KANES, W. H. & STEHLI, F. G. (eds) *The Ocean Basins and Margins,* Vol. 4a. Plenum, New York, 77–150.

—— 1988. Subsidence in the northeastern Nile Delta: rapid rates, possible causes, and consequences. *Science* **240**, 497–500.

—— & LIYANAGE, A. N. 1986. Clay-mineral variations in the northeastern Nile Delta, as influenced by depositional processes. *Marine Geology* **73**, 263–283.

—— & MALDONADO, A. 1977. Nile Cone: Late Quaternary stratigraphy and sediment dispersal. *Nature* **266**, 129–135.

—— & —— 1979. Levantine Sea–Nile Cone lithostratigraphic evolution: quantitative analysis and correlation with paleoclimatic and eustatic oscil-

lations in the Late Quaternary. *Sedimentary Geology* **23**, 37–65.

SUMMERHAYES, C. P., ROSS, D. A. & STOFFERS, P. 1977. Nile submarine fan: sedimentation, deformation and oil potential. *Proceedings of the 9th Offshore Technology Conference,* Vol. 1. OTC 2731, Houston, TX, 35–40.

——, SESTINI, G., MISDORP, R. & MARKS, N. 1978. Nile Delta: nature and evolution of continental shelf sediment system, a preliminary report. *Marine Geology* **27**, 43–65.

TAYLOR, P. W. & JONES, S. L. 1982. Tertiary studies of the Beheira area. *Proceedings of the 6th Exploration Seminar.* EGPC, Cairo, 18 pp.

TOMA, S. A. & SALAMA, M. S. 1980. Changes in bottom topography of the western shelf of the Nile delta since 1922. *Marine Geology* **36**, 325–339.

TOUSSON, O. 1934. Mémoire sur les anciennes branches du Nil. *Mémoires de l'Institut d'Egypte* **4**, 144 pp.

TSOAR, H. 1974. Desert dune morphology and dynamics, El Arish, Northern Sinai. *Zeitschrift für Geomorphologie, Neue Folge, Supplement* **20**, 41–61.

UNDP/UNESCO 1978. *Arab Republic of Egypt: Coastal Protection Studies. Final Technical Report,* Vol. 2. UNESCO, Paris, 206–483.

UNESCO/UNDP 1976. *First Technical Report (Geology).* UNESCO/UNDP, Alexandria.

VIOTTI, C. & MANSOUR, A. 1969. Tertiary planktonic foraminiferal zonation from the Nile delta, Egypt. Part I, Miocene zonation. *Proceedings of the 3rd African Micropaleontological Colloquium, Cairo.* Geological Survey, Cairo, 425–460.

WENDORF, F. & SCHILD, R. 1978. *Prehistory of the Nile Valley.* Academic Press, New York, 414 pp.

WOODSIDE, J. M. 1977. Tectonic elements and crust of the Eastern Mediterranean Sea. *Marine Geophysical Research* **3**, 317–354.

WRIGHT, L. D. & COLEMAN, J. M. 1973. Variations in morphology of major river deltas as functions of ocean waves and river discharge. *Bulletin of the American Association of Petroleum Geologists* **57**, 370–398.

ZAGHLOUL, Z. M. 1976. Stratigraphy of the Nile Delta. *Proceedings of the Seminar on Nile Delta Sedimentology.* UNESCO/ASRT/UNDP, Alexandria, 40–49.

——, ANDRAWIS, S. F. & AYAD, S. N. 1979a. New contribution to the stratigraphy of Tertiary sediments of Kafr El Dawar well no. 1, NW Nile Delta, Egypt. *Annals of the Geological Survey of Egypt* **9**, 292–307.

——, ESSAWY, M. I. & EL SHIRBINI, S. I. 1976. Grain size and texture analysis of the subsurface sediments of Sidi Salim well no. 1, Nile Delta, Egypt. *Bulletin of the Faculty of Sciences, Mansoura University* **4**, 79–99.

——, HINNAWI, E. & AB SAKR, M. 1975. The grain size analysis and sedimentological significance of some subsurface sediments from the Nile Delta, Egypt. *Bulletin of the Faculty of Sciences, Mansoura University* **3**, 125–148.

——, TAHA, A. A. & GHEITH, A. M. 1977a. Microlithofacies studies and paleoenvironment trends of the subsurface sediments of Kafr El Dawar well no. 1. *Bulletin of the Faculty of Sciences, Mansoura University* **5**, 113–138.

——, —— & —— 1979a. On the Quaternary–Neogene cyclic sedimentation in the Nile Delta area. *Annals of the Geological Survey of Egypt* **9**, 779–803.

——, ——, HEGAB, O. & EL FAWAL, F. 1977b. The Neogene–Quaternary sedimentary basins of the Nile Delta. *Egyptian Journal of Geology* **21**, 143–161.

——, ——, —— & —— 1979b. The Plio-Pleistocene Nile Delta subenvironments, stratigraphic section and genetic class. *Annals of the Geological Survey of Egypt* **9**, 282–291.

G. SESTINI, Via Della Robbia 28, 50132 Florence, Italy.

Arcuate and bird's foot deltas in the late Pleistocene Palaeo-Tokyo Bay

H. Okazaki & F. Masuda

SUMMARY: Two types of prograding deltaic sequences are recognized in the upper part of the late Pleistocene Shimosa Group, which crops out in southern Ibaraki Prefecture and northern Chiba Prefecture, eastern Kanto, Japan. The older sequence is interpreted as a wave- and tide-dominated arcuate delta and the younger sequence as a fluvial-dominated bird's foot delta. The older sequence is characterized by thick molluscan shell beds (c. 10 m) in the middle part. In contrast, the younger sequence is characterized by the rhythmic alternation of fine-grained sand and mud with ripples in the lower part and coarse-grained sandy deposits with trough cross-stratification in the upper part. Both sequences show distinct coarsening-upward sequences. The difference between these two types of delta was caused by the water depth in the delta front during the lowering of the sea level from the interglacial to the glacial period during the last stage of Palaeo-Tokyo Bay.

Kanto Plain, the largest plain in Japan, is situated at the centre of the Pacific side of the Japanese Islands. The plain geomorphologically comprises hills, uplands and lowlands. Joso Upland is 20–40 m high and occupies the widest area in the eastern part of the Kanto Plain (Fig. 1). The upland mainly comprises the late Pleistocene Simosa Group as thick units of sands and thin mud and gravel beds, deposited in Palaeo-Tokyo Bay (Yabe 1931; Narita Research Group 1962) and now widely distributed over the modern Kanto Plain.

Aoki & Baba (1973, 1980) have divided the Shimosa Group into the following formations, in decreasing age: Kongochi Formation (30 m thick), Naruto Formation (20 m thick), Jizodo Formation (30–70 m thick), Semata Formation (20–30 m thick), Kamiizumi Formation (25 m thick), Narita Formation (20–40 m thick; Kiyokawa, Kamiiwahashi and Kioroshi Members), Ryugasaki (or Anegasaki) Formation (1–10 m thick) and Itabashi Formation (0.5–3 m thick) (Fig. 2). Recently, an unconformity was discovered at the base of the Kioroshi Member, and therefore the member was redefined as the Narita Formation (Kikuchi 1981) or Kioroshi Formation (Shimosa-daichi Research Group 1984; Tokuhashi & Endo 1984). In this paper we tentatively use the division proposed by Aoki & Baba (1973). The Simosa Group is unconformably overlain by the Musashino volcanic ash bed; Sugihara (1979) described the marker tephra in the Simosa Group. The fission-track ages of some marker tephra were reported or cited by Machida & Suzuki (1971) and by Sugihara et al. (1978). All boundaries between the formations and members of the Shimosa Group correspond to lower sea-level stages or glacial periods (Masuda 1985). For instance, the unconformity between the Kami-

iwahashi and Kioroshi Members correlates with stage 6 of Emiliani's (1978) oxygen isotope curve shown in Fig. 2. The interglacial period or higher sea level of stage 6 is the depositional period of the Kioroshi Member, the Ryugasaki and Anegasaki Formations, and the Itabashi Formation. The rapid sea-level rise between stage 6 and stage 5, which is called the Shimosueyoshi transgression, resulted in the last stage of marine deposition in Palaeo-Tokyo Bay. The sea-level fall between stage 5 and stage 2 was slow, and included several small fluctuations. This early regressive stage was the time of active progradation of the delta systems described here (Masuda & Okazaki 1983b). In this paper we propose a new interpretation for two different types of prograding deltas developed in the final marine depositional phase of Palaeo-Tokyo Bay during the Shimosueyoshi high-stand in sea level.

Wave- and tide-dominated arcuate delta

A coarsening-upward sequence, 10–20 m thick, as shown in Fig. 3(a) is well developed around Kioroshi (type locality of the Kioroshi Member) in the northern part of Chiba Prefecture (Fig. 4). Three main units (A, B and C) are distinguished from base to top in the sequence.

Unit A consists of muds with thin sheet-sand layers and is more than 3 m thick. The sand layer shows current-ripple lamination and forms lenticular bedding. Intense bioturbation is present (Fig. 5). The in situ shells of Theola lubrica and Raeta yokohamensis occur occasionally in the muds. The unit also includes the marine diatoms Melosira sulcata and Triceratium favus. Around

From WHATELEY, M. K. G. & PICKERING, K. T. (eds), 1989, Deltas: Sites and Traps for Fossil Fuels, Geological Society Special Publication No. 41, pp. 129–138.

H. Okazaki & F. Masuda

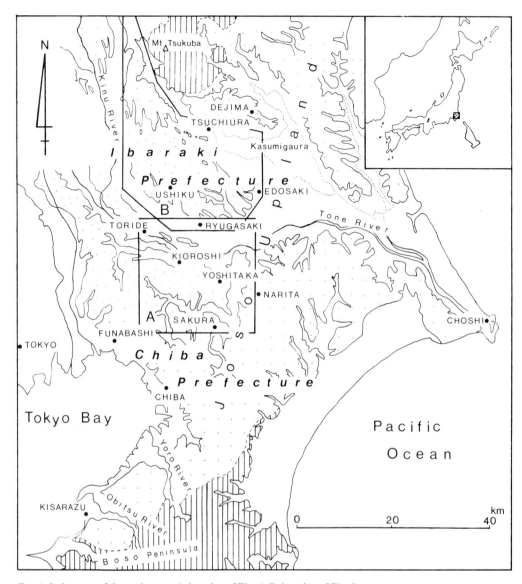

FIG. 1. Index map of the study area: A, location of Fig. 4; B, location of Fig. 9.

Yoshitaka (Fig. 4, locations 8 and 9) poorly sorted sandy muds, about 5 m thick, are developed below this unit. The muds are burrowed and include mud clasts, gravels, drift pumice grains, plant debris and shells.

Unit B is divided into two subunits B_a and B_b. Subunit B_a consists of well-sorted fine to very fine grained sands with molluscan shells and is 3–6 m thick. Parallel and trough cross-stratification are present. The *in situ* molluscan shells of *Fluvia mutica, Raeta yokohamensis, Fabulina nitidula,*

Siliqua pulchella and *Solen krusensterni* and transported shells of *Mactra sulcataria* are dominant. Bioturbation is common.

Subunit B_b consists of medium-grained sands with very abundant molluscan shells and is 3–6 m thick. The shell beds are the so-called 'Kioroshi shell bed' (Kojima 1958). The subunit shows parallel or large-scale low angle tabular cross-stratification (Fig. 6). Molluscan shells of *Mactra sulcataria, Tapes variegata, Gomphina neastartoides* and *Glycymeris vestita* are dominant.

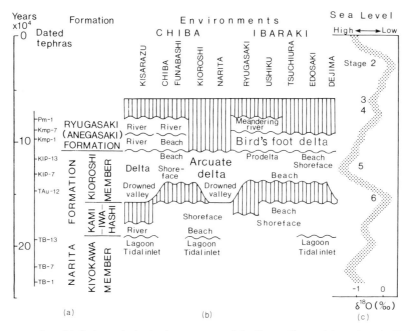

FIG. 2. Chronostratigraphic facies evolution in the upper part of the Simosa Group: (*a*) stratigraphy (Aoki & Baba 1973) and the marker tephra dated by the fission-track method (Machida & Suzuki 1971; Sugihara *et al.* 1978); (*b*) facies distributions in the Chiba and Ibaraki Prefectures; (*c*) oxygen isotope curve (Emiliani 1978) showing the global sea-level change.

Vertical burrows of the *Ophiomorpha* type are abundant in the shell beds.

Unit C is divided into three subunits C_a, C_b and C_c. Subunit C_a consists of fine- to medium-grained sand and is 2–3 m thick. Trough cross-stratification with angular foresets is predominant and commonly shows 'herring-bone cross-bedding' (Fig. 7). Transported shells, either intact or as fragments, are contained in the basal part above erosive channel-like scours. The upper part contains some bioturbation and commonly consists of alternations of sand and mud (0.5–1.5 m thick) (Fig. 4, locations 1 and 6) with tabular beds of broken shells of *Echinarachnius mirabilis*.

Subunit C_b consists of fine- to medium-grained sands with silt layers and is 1–2 m thick. Wave ripples and parallel lamination are the dominant structures (Fig. 8). Convolute bedding and dish structures are also present. Bioturbation is common and is dominated by *Ophiomorpha*-type burrows; casts of *in situ* shells are also included.

Subunit C_c is 1–4 m thick and consists of muds with peaty layers and rootlets. Thin beds of very fine-grained sand and laminated mud with 'inverse grading' (Iseya 1982) are also present. Poorly developed bioturbation is present in the lower part.

Stratigraphically, units A and B are correlated with the Kioroshi Member of Aoki & Baba (1980), with the Kioroshi Formation of Sugihara (1979) and the Shimosa-daichi Research Group (1984), and with the Narita Formation of Kikuchi (1977). Unit C is correlated with the Itabashi and Ryugasaki Formations of Aoki & Baba (1973,

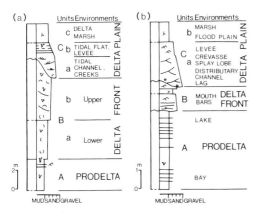

FIG. 3. Two types of prograding deltaic sequence showing coarsening upwards: (*a*) wave- and tide-dominated arcuate delta; (*b*) fluvial-dominated bird's foot delta. The structural symbols used in the columnar sections are defined in Fig. 4*c*.

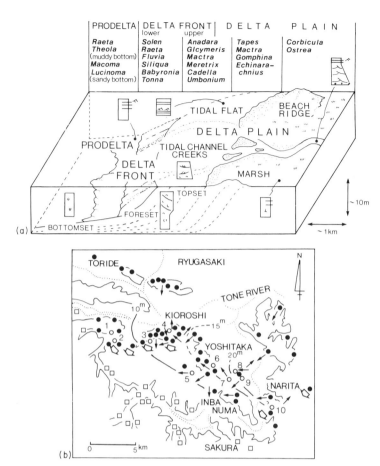

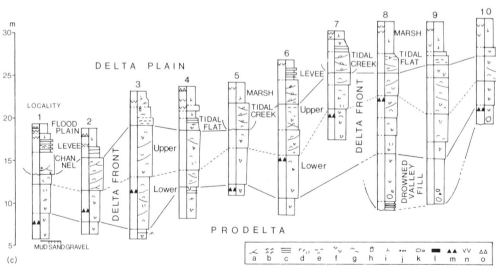

FIG. 4. Arcuate delta system: (*a*) facies model; (*b*) distribution of the deltaic succession (○, localities for the columnar sections; ○, ●, localities observed in the delta-front facies; □, localities which the prograding delta did not reach and which remained in deeper water; contour lines, position of the scoria horizon; →, palaeocurrent direction on the delta plain; ⇒ foreset direction of the delta front); (*c*) columnar sections (1, Kizaki; 2, Funato; 3, Izumi; 4, Kioroshi; 5, Tsukuriya; 6, Sunaoshi; 7, Otake; 8, Yoshitaka; 9, Kabuwada; 10, Sogo; a, cross-beds; b, ripples; c, parallel beds; d, molluscan shells; e, *Echinarachnius*; f, bioturbation; g, *Excirolana*; h, *Ophiomorpha*; i, roots; j, gravel; k, mud clast; l, peat; m, scoria (marker tephra); n, K1P tephra; o, Pm-1 tephra).

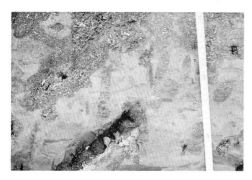

FIG. 5. Intensely burrowed muds in the prodelta facies of unit A in the arcuate delta (Yoshitaka, location 8). Scale, 37 cm long.

FIG. 8. Wave ripples and parallel lamination with bioturbation in the tidal flat facies (Toride): subunit C_b in the arcuate delta. The trowel is 30 cm long.

FIG. 6. Large-scale low angle tabular cross-stratification in the middle part of the cliff that dips gently to the left (Funato, location 2): delta-front facies of subunits B_b in the arcuate delta. The height of the cliff is about 17 m. The base of the cross-stratified units is arrowed.

FIG. 7. Trough cross-stratification with tabular foreset in the tidal channel facies (Sogo, location 10): subunit C_a in the arcuate delta. Intense bioturbation can be seen in the lower part of the facies. The trowel is 30 cm long.

1980), with the lower part of the Joso Clay of Kikuchi (1981) and with the lower clay member of the Joso Formation of Kodama *et al.* (1981).

Environmental interpretation

The depositional environments of the units are best interpreted by an arcuate delta model as proposed here (Fig. 4*a*). This model also explains the distribution of the dominant molluscs, and is based on our facies analysis in this area and the studies of the modern delta in Tokyo Bay (Kaizuka *et al.* 1979). On the coast of the Boso Peninsula in Tokyo Bay, deltaic sand bodies from the modern Yoro and Obitsu rivers are formed with thicknesses of about 13 m and 20 m respectively. The delta sands above the prodelta muds consist of fine- to medium-grained sand.

The muds of unit A indicate a low energy prodelta environment where thin sand layers were introduced by storm-generated currents that included density currents or underflows. The poorly sorted sandy muds at Yoshitaka (locations 8 and 9) were deposited as a drowned valley fill in the early stage of a rapid transgression. Unit B corresponds to the delta-front deposits. The lower subunit B_a represents deeper water conditions than the upper subunit B_b, on the basis of the molluscan ecology. The large-scale low angle foreset tabular cross-beds indicate the original dip of the deltaic slope. Unit B was developed by the progradation of a delta sand body. Unit C was formed on the delta plain. The sandy deposits of subunits C_a and C_b are characterized by wave ripples and herring-bone cross-bedding with distinct bimodal dispersal patterns formed as a result of the interaction of waves and tides. The wave ripples show the same wave conditions as the ripples formed on the modern tidal flats of

Tokyo Bay (Masuda & Makino 1987). Subunits C_a and C_b are the deposits of tidal channels or creeks and tidal flats respectively. The alternations of sand and mud in subunit C_a represent the levee deposits of tidal channels or creeks. The high energy coarse-grained layer with a unidirectional pattern, intercalated in subunit C_b, corresponds to the washover deposits on tidal flats and beaches. The depositional environment of subunit C_c corresponds to a delta marsh characterized by the muddy layers deposited from suspension with roots showing plant colonization. Thin sand layers with 'inverse grading' show a characteristic structure of flood deposits (Masuda & Iseya 1985).

The palaeocurrents, from NE to SW in the delta front and from E to W on the delta plain, are the dominant directions (Fig. 4b).

The original height of the delta front of unit B, which varies between 5 and 12 m, can be estimated from the thickness of the delta-front facies based on the delta model in Fig. 4a. Masuda & Okazaki (1983b) discovered that the marker tephra layer of scoria described by Sugihara (1979) occurs in the various units. In the western and southern areas, the tephra layer is intercalated in the prodelta facies of unit A (Fig. 4, locations 1 and 5) or the lower delta-front facies of subunit B_a (location 2). In the northern Kioroshi area it is in the upper part of the lower delta-front facies of subunit B_a (location 3), and in the eastern area it occurs in the uppermost part of the lower delta-front facies of subunit B_a (locations 6–8). Differences in the location of the datum tephra at various localities are shown in Figs 4b and 4c. These observations suggest that there was a difference in water depth of about 10–15 m between the northwestern and southeastern areas. An initial inclination of the delta slope of approximately 1/100 to 3/1000 is calculated from these observations.

Three geomorphic surfaces of lower (19–20 m at locations 1 and 2), middle (23–24 m, locations 3–5) and higher (30–32 m, locations 7–10) altitude are distinguishable in this area (Fig. 4c). The palaeo-depth, estimated from the thickness of the delta-front facies, decreases stepwise. Two rapid falls of sea level, of several metres, occurred during the progradation of this delta. These sea-level changes can be correlated with the small fluctuations in the oxygen isotope curve during the interglacial stage 5 (Fig. 2).

The river that constructed the arcuate delta from the SE is unknown, but it can be assumed to have flowed from the S of Boso Peninsula. A river source in the N cannot be acceptable because the palaeogeography of the northern district suggests widespread rather deep water environments (Watanabe et al. 1987). The river

was similar to the modern Yoro River in distributional area and thickness of the delta-front sands.

The age of the progradation of the arcuate delta is estimated at $(13–11) \times 10^4$ years BP on the basis of fission-track ages of the K1P tephra (Sugihara et al. 1978) intercalated in the deposits (Fig. 2).

Fluvial-dominated bird's foot delta

A coarsening-upward sequence 5–10 m thick (Fig. 3b) is well developed from Tsuchiura to Ryugasaki in Tsukuba Upland, located in the northwestern part of Joso Upland (Fig. 9). Three main units (A, B and C) are distinguished from base to top in the succession.

Unit A consists of alternations of fine to very fine grained sands and muds and is 1–5 m thick. Small-scale wave ripples and parallel lamination are the dominant structures in the sand layers (Fig. 10). Poorly developed bioturbation, including small burrows, is observed. Faecal pellets and plant debris are common. Marine diatoms, such as *Melosira sulcata*, *Cyclotella stylorum* and *Coscinodiscus* spp, and fresh water diatoms, such as *Stephanodiscus niagarae* and *Cyclotella comta*, are dominant in the mud layers. The thickness of the sand layers, or the sand-to-mud ratio in the beds, varies from a sand-laminated mud bed to a thick sand bed with thin mud layers. This variation corresponds to the change in storm-generated sand layers from distal to proximal facies in the downcurrent direction (Fig. 9, locations 15–18). This unit is widely distributed from Ushiku to Dejima in southern Ibaraki (Okazaki et al. 1984; Masuda & Makino 1987).

Unit B consists of fine-grained sands and mud layers and is 1–4 m thick. Wave ripples, parallel lamination, trough and tabular cross-stratification and swaley to hummocky cross-stratification are the dominant structures in the sand layers (Fig. 11). Bidirectional palaeocurrents with opposing flow directions are present (Fig. 9a). The mud layers show bioturbation.

Unit C is divided into two subunits C_a and C_b. It shows a typical fining-upward sequence. Subunit C_a mainly consists of fine- to medium-grained sands and is 0.5–4 m thick. The basal part of the subunit is composed of granule-bearing coarse-grained sands. Trough cross-stratification, epsilon cross-stratification and channel-shaped erosional bases are the dominant structures. This subunit is typified by a unidirectional palaeocurrent pattern (Fig. 12). The upper part frequently contains sand and mud beds with 'inverse grading'. Subunit C_b consists of muddy

(a)

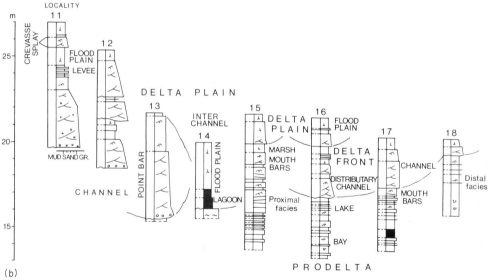

(b)

Fig. 9. (*a*) Bird's foot delta system (→, palaeocurrent direction; a, distributary channels; b, delta plain; c, prodelta; d, land); (*b*) columnar sections (11, Takegaki; 12, Ozone; 13, Hirooka; 14, Yatabe; 15, Okami; 16, Besho; 17, Koya; 18, Umamiyama; the structural symbols are defined in Fig. 4*c*).

layers with roots and plant debris and is 0.5–1.5 m thick. Thin beds of very fine grained sand and sand-streaked laminated mud are common (Fig. 13).

Stratigraphically, subunit C_a is correlated with the Ryugasaki Formation of Aoki & Baba (1977) and with the middle sand member of the Joso Formation of Kodama *et al.* (1981). Subunit C_b is correlated with the Itabashi Formation of Aoki & Baba (1980), with the upper part of the Joso

FIG. 10. Alternation of fine to very fine grained sands and muds in the prodelta facies: subunit A in the bird's foot delta. Small wave ripples and parallel lamination in the sands. The locality is 3 km from Koya (location 17) in the N of Kasumigaura. The pencil is 15 cm long.

FIG. 11. Mouth-bar deposits of swaley to hummocky cross-stratification (Okami, location 15): unit B in the bird's foot delta. The stratification is slightly disturbed by dehydration (arrowed). The trowel is 15 cm long.

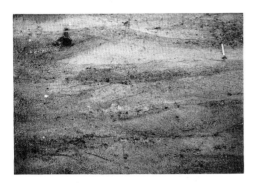

FIG. 12. Trough cross-stratification in the fluvial distributary channel facies (Ozone, location 12): subunit C_a in the bird's foot delta. The trowel is 30 cm long.

FIG. 13. Muds with roots and plant debris in the flood-plain facies (Yatabe, location 14): subunit C_b in the bird's foot delta. The scale is 15 cm long.

Clay of Kikuchi (1981) and with the upper clay member of the Joso Formation of Kodama *et al.* (1981). Units A and B are also included in the Ryugasaki Formation of Aoki & Baba (1977).

Environmental interpretation

Ikeda *et al.* (1982), using evidence obtained from a detailed topographical analysis, identified a bird's foot delta plain landform on the surface of Tsukuba Upland, located in the northwestern part of Joso Upland. Masuda & Okazaki (1983*a*) clarified the palaeocurrent systems of this bird's foot delta from measurements of cross-stratifications (Fig. 9*a*).

The facies of the upper reaches in the delta comprise a fining-upward sequence from gravelly sands to muds (Fig. 9, locations 11 and 12). In contrast, the facies of the lower reaches form a fining-upward sequence of sands (Fig. 9, location 13) and thick muds (Fig. 9, location 14). These thick muds include plant roots, peat layers, driftwood, and freshwater and brackish-water diatoms such as *Melosira granulata, Eunotia* spp,

Pinnularia spp, *Cymbella* spp, *Gomphonema* spp, *Diploneis interrupta* and *Rhopaiodia gibberula*.

The depositional environment of the sandy facies corresponds to a distributary channel and the muddy facies corresponds to an interdistributary channel area, such as a flood plain and/or a distributary bay. Unit A shows relatively low energy facies. Coarse clastics of this unit were deposited under the interaction of sand-laden tractional river currents, wave reworking and storm-generated currents, with the fine-grained clastics settling from suspension (Katsura *et al.* 1985). The depositional environment of this unit is interpreted as a bay or lake of a delta plain.

Unit B is characterized by wave-formed structures, such as wave ripples, swaley to hummocky cross-stratification and trough cross-stratification. Unit B comprises mouth bars at the front of distributary channels. Trough cross-stratification, with bidirectional current patterns, is formed by the migration of ripple and dune fields by river currents and oscillatory flow at a delta front. Unit C is delta-plain deposits. Sandy deposits of subunit C_a were deposited under unidirectional currents on point bars of fluvial distributary channels. Fine-grained deposits of subunit C_b settled from suspension in a marsh or flood-plain environment. The height of the delta front is 1–4 m, estimated by the thickness of delta-front facies of unit B.

The prodelta deposits of unit A were developed in an area that remained deep with a slow rate of deposition which continued during the regressive stage. During the regressive stage, a bird's foot delta of the 'Palaeo-Kinu River' prograded from NW to SE (Fig. 9a). This succession was built by a prograding fluvial-dominated bird's foot delta $(10-8) \times 10^4$ years BP. The age is estimated from the fission-track ages of intercalated tephra of the Km-P and overlying Fm-1 tephra (Sugihara *et al.* 1978).

The difference between the two types of delta, the older wave- and tide-dominated arcuate delta and the younger fluvial-dominated bird's foot delta, can be attributed to a difference of water depth at the delta front rather than the amount of sediment supply from rivers. In other words, deeper and more extensive marine influences constructed an arcuate type delta, whereas shallower water favoured a bird's foot type delta.

The sea area was widespread in the Pleistocene Kanto district or what has been called Palaeo-Tokyo Bay. A deep and wide 'valley' was formed during the low-sea-level stage of a glacial period and infilled during the transgression associated with the following interglacial period. With the initial post-interglacial lowering of sea level, a wave- and tide-dominated arcuate delta prograded in the drowned valley. As the water depth continued to shallow, a fluvial-dominated bird's foot delta was developed. Both types of delta were developed in Palaeo-Tokyo Bay at different times during the regressive period of the post-Shimosueyoshi high-stand in sea level.

References

AOKI, N. & BABA, K. 1973. Compilation of the stratigraphy and molluscan fossil assemblages of the Pleistocene Shimosa Group, eastern part of Kanto Plain, central Japan. *Journal of the Geological Society of Japan* **79**, 453–464.

—— & —— 1977. Ryugasaki Formation of the southwestern part of Ibaraki Prefecture. *Tsukuba Environmental Studies* **2**, 114–120.

—— & —— 1980. Pleistocene molluscan assemblage of the Boso Peninsula, Central Japan. *Science Reports of the Institute of Geosciences, University of Tsukuba, Section B* **1**, 107–148.

EMILIANI, C. 1978. The case of the ice ages. *Earth and Planetary Science Letters* **37**, 349–352.

IKEDA, H., MIZUTANI, K., SONODA, Y. & ISEYA, F. 1982. Geomorphic development of the Tsukuba Upland, Ibaraki Prefecture. *Tsukuba Environmental Studies* **6**, 150–156.

ISEYA, F. 1982. A depositional process of reverse graded bedding in flood deposits of the Sakura River. *Geographical Review, Japan* **55**, 567–613.

KAIZUKA, S., AKUTSU, J., SUGIHARA, S. & MORIWAKI, H. 1979. Geomorphic development of alluvial plains and coasts during the Holocene in Chiba

Prefecture, Central Japan, with a note on diatom assemblages in Holocene deposits near the junction of rivers Miyako and Furuyama. *Quaternary Research, Japan* **17**, 189–205.

KATSURA, Y., MASUDA, F., OKAZAKI, H. & MAKINO, Y. 1985. Storm deposits developed in the Pleistocene strata around the Tsukuba Upland. *Tsukuba Environmental Studies* **9**, 54–60.

KIKUCHI, T. 1977. Pleistocene sea level changes and tectonic movements in the Boso Peninsula, Central Japan. *Geographical Report of Tokyo Metropolitan University* **12**, 77–103.

—— 1981. Sedimentary environment of the Joso Clay overlying the marine Pleistocene deposits in the Kanto district, central Japan. *Memoirs of the Geological Society of Japan* **20**, 129–145.

KODAMA, K., HORIGUCHI, M., SUZUKI, Y. & MITSUNASI, T. 1981. Late Pleistocene crustal movement of the Kanto plain, central Japan. *Journal of the Geological Society of Japan* **20**, 113–128.

KOJIMA, N. 1958. Geological study of the Kioroshi district, Chiba Prefecture, Japan. The studies on the Narita Group (1). *Journal of the Geological Society of Japan* **64**, 165–171.

MACHIDA, H. & SUZUKI, M. 1971. Chronology of the late Quaternary Period as established by fission track dating. *Kagaku (Science)* **41**, 263–270.

MASUDA, F. 1985. Evolution of depositional system in Paleo-Tokyo Bay. *Report of Grant-in-Aid, Science Research (C), 5854502*, 47 pp.

—— & ISEYA, F. 1985. 'Inverse grading': a facies structure of flood deposits in meandering rivers. *Journal of the Sedimentology Society of Japan* **22/23**, 108–116.

—— & OKAZAKI, H. 1983*a*. Directional structures observed in the Pleistocene strata of Tsukuba Upland. *Tsukuba Environmental Studies* **7C**, 99–110.

—— & —— 1983*b*. Two types of prograding deltaic sequence developed in the late Pleistocene Paleo-Tokyo Bay. *Annual Report of the Institute of Geosciences, University of Tsukuba* **9**, 56–60.

—— & MAKINO, Y. 1987. Paleo-wave condition reconstructed from ripples in the Pleistocene Paleo-Tokyo Bay deposits. *Journal of the Geographical Society of Japan* **96**, 23–45.

NARITA RESEARCH GROUP 1962. The Shimosueyoshi transgression and the Paleo-Tokyo Bay. *Chikyu-kagaku (Earth Science)* **60/61**, 8–15.

OKAZAKI, H., ISHIKAWA, T. & SHINDO, S. 1984. Study on groundwater flow system in the Dejima area, the north Kasumigaura Upland (1). *Journal of the Japan Association of Groundwater and Hydrology* **26**, 97–110.

SHIMOSA-DAICHI RESEARCH GROUP 1984. The relation between the shape of the base of the Kioroshi Formation and its facies, around Tega-numa, Chiba Prefecture, Central Japan. *Chikyu-Kagaku (Earth Science)* **38**, 226–234.

SUGIHARA, S. 1979. Stratigraphy and bedrock topography on Narita Formation at Shimosa Upland, Kanto Plain. *Institute of Cultural Science, Meiji University, Memoir* **18**, 1–41.

——, ARAI, F. & MACHIDA, H. 1978. Tephrochronology of the Middle to Late Pleistocene sediments in the northern part of the Boso Peninsula, central Japan. *Journal of the Geological Society of Japan* **84**, 583–600.

TOKUHASHI, S. & ENDO, H. 1984. *Geology of the Anegasaki District. Quadrangle Series, Scale 1:50 000*. Geological Survey of Japan, 136 pp.

WATANABE, K., MASUDA, F., KATSURA, Y. & OKAZAKI, H. 1987. Paleoenvironmental development in the Kanto region (1), (2). *Japan Society of Earth Science Education* **4C**, 1–12, 79–90.

YABE, H. 1931. Geological growth of the Tokyo Bay. *Earthquake Research Institute Bulletin* **9**, 333–339.

H. OKAZAKI, Central Museum of Chiba Prefecture, 955-2 Aobacho, Chiba City, Chiba 280, Japan.

F. MASUDA, Institute of Geoscience, University of Tsukuba, 1-1-1 Tennodai, Tsukuba, Ibaraki 305, Japan.

A fluvial-dominated lacustrine delta in a volcanic province, W Greenland

G. K. Pedersen

SUMMARY: A 50 m thick sequence of dark grey to black non-marine shales are exposed in a 12 km section along the NE coast of Disko in W Greenland. The shales are sandwiched between sandy fluvial deposits, and the sedimentological logs demonstrate a marked lateral continuity of facies within the shale unit. The shale is fissile and micaceous, and contains locally comminuted plant debris. The average total organic carbon (TOC) content is 6%, the organic matter corresponds to type III kerogen and the dominant clay mineral is kaolinite. The shale unit constitutes one coarsening-upward sequence interpreted as lacustrine prodelta mud, with thin sheets of sand deposited from density underflows which pass transitionally up into cross-laminated mouth-bar sand finally overlain by cross-bedded sand deposited in fluvial channels. This sequence represents progradation of a fluvial-dominated delta into a lake partially bounded by subaqueous hyaloclastic breccias.

In relation to oil exploration off the W Greenland coast in the mid-1970s it was demonstrated by Schiener & Leythaeuser (1978) that organic-rich, albeit immature, lacustrine source rocks existed in the Nûgssuaq Embayment, particularly in the Naujât Member of the Upper Atanikerdluk Formation. This report provided an incentive to undertake the present study with the aim of elucidating the sedimentology of these lacustrine sediments.

Non-marine black shales are reported from various depositional settings. Deep stratified lakes contain laminated shales with excellent preservation of organic matter, often rich in type I and type II kerogen (Demaison & Moore 1980; Kelts 1985; Talbot 1985). Shales with high total organic carbon (TOC) contents may also accumulate in interdistributary bays and flood plains (Fielding 1984; Guion 1984; Haszeldine 1984; McCabe 1984). These settings are characterized by terrestrial plants, which eventually are a source of type III kerogen, and by wave-reworked sand streaks, crevasse-splay deposits, rootlets, association with coal seams and signs of occasional subaerial exposure such as desiccation cracks (Elliott 1974; Fielding 1984).

The current sedimentological study indicates that the dark grey to black shales represent non-marine basinal and prodelta facies, whereas the intercalated sand layers formed from density underflows. Their upward-increasing frequency is interpreted as delta progradation and is accompanied by increasing silt content and decreasing TOC values in the shale.

Geological setting

The Cretaceous–Tertiary clastic sequence of W Greenland is more than 1 km thick (Henderson et al. 1976). The sediments were deposited in the Nûgssuaq Embayment which constitutes a smaller fault-bounded basin linked northwards to the W Greenland Basin and the Baffin Bay Rift (Henderson et al. 1981; Rolle 1985). The Nûgssuaq Embayment is bounded in the E by a major fault towards the Precambrian basement terrains of Greenland and to the W by the Disko Gneiss Ridge, a possible basement horst (Fig. 1). The sedimentary sequence increases in thickness northwards along the axis of the Embayment (Henderson et al. 1981, Fig. 4) and is overlain by Tertiary plateau basalts (Fig. 1).

During the Cretaceous the Nûgssuaq Embayment contained a large delta characterized by vertical accretion rather than progradational growth and northward sediment transport (Schiener 1975; Henderson et al. 1976; Hansen 1980; Shekhar et al. 1982; Surlyk 1982).

During the Tertiary, fluvial sediments were supplied from the S and E, while marine environments prevailed in central and northern Nûgssuaq. Deposition took place on a topography determined by volcanic eruptions and late Danian extensional block faulting which tilted the blocks slightly (Rosenkrantz & Pulvertaft 1969; Schiener & Leythaeuser 1978; Hansen 1980). Subaqueous volcanic breccias overlain by subaerial plateau basalts prograded eastwards during the early Tertiary (Clarke & Pedersen 1976). The contemporaneous non-marine sediments constitute the Upper Atanikerdluk Formation which was described in SE Nûgssuaq and divided into five members by Koch (1959). Dark grey to black fissile shales are dominant in the Naujât Member and the Aussivik Member which both grade up into fluvial sand (Koch 1959). Sedimentological data are presented on a shale sequence in a well-exposed coastal section 12 km long on NE Disko

From WHATELEY, M. K. G. & PICKERING, K. T. (eds), 1989, *Deltas: Sites and Traps for Fossil Fuels,* Geological Society Special Publication No. 41, pp. 139–146.

FIG. 1. Geological location map showing the distribution of sediments deposited in the Nûgssuaq Embayment. Section AB is shown in Fig. 3.

(Figs 1–3). The lithostratigraphic correlation to the shale members on Nûgssuaq awaits further data on the correlation of the volcanic rocks.

Sedimentary facies

The shales studied range from black fissile mudshale to grey silty mudstone. Sand streaks and thin continuous layers of sand occur intercalated in the shale (Fig. 4).

Shale

The dark grey to black, weakly laminated mudshale has a well-developed fissility. Continuous or patchy cementation by siderite occurs in graded locally cross-laminated layers 1–3 cm thick of very fine grained sand. Neither pyrite nor jarosite have been observed in the shale. Well-preserved plant fossils are locally common, but no invertebrate remains or trace fossils have been found. There is a gradational transition to silty slightly fissile grey mudstone, with much comminuted plant debris locally. The lack of lamination may be the result of bioturbation, although no distinct trace fossils or rootlet horizons have been observed. Thin tuff layers (0.5–2.5 cm) can be traced 100–500 m through the exposures (Fig. 5), irrespective of shale facies, and indicate that bioturbation was restricted to the uppermost few millimetres of the sediment.

The shales formed through settling of mud

FIG. 2. The shale sequence at Pingo. The lowermost 12 m of shale are locally covered with scree. The vertical cliff reflects the sand-streaked shale (outlined in white). The coarsening upward sequence from shale to horizontally bedded sand can be seen.

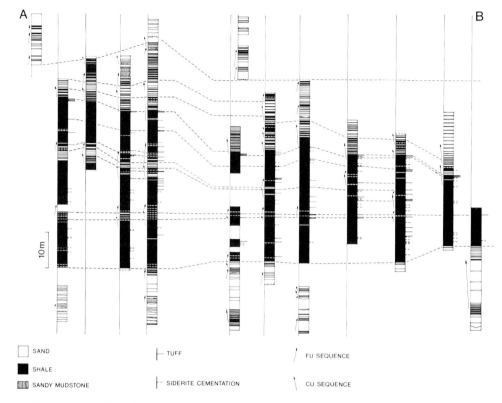

FIG. 3. Sedimentological logs demonstrating the sharp base and the coarsening upward trend of the shale sequence. The middle pulse of underflow sand sheets decreases in thickness westwards. Tuff layers are correlated by the stippled lines. The arrows designating fining-upward (FU) and Coarsening-upward (CU) sequences are placed to the left of the logs. The location of section AB is shown in Fig. 1.

from suspension at depths greater than the wave base. The intercalated thin graded layers of very fine grained sand may represent distal underflows similar to those discussed by Haszeldine (1984, Fig. 2c).

The clay mineral assemblage of the shales is dominated by kaolinite, with illite occurring in very small amounts in all samples. This suggests provenance in deeply weathered soils developed in a temperate to warm humid climate, and this is also indicated by the flora (Koch 1963). Identification of the clay minerals is based on X-ray diffraction analysis (parallel-oriented mounts) of samples subjected to chemical pretreatment following a procedure based on the work of Anderson (1962), Follett *et al.* (1965), Alexiades & Jackson (1966) and Roth *et al.* (1969).

Combustion and pyrolysis analyses show average TOC values of 6% (Fig. 6) in outcrop samples of shales mainly from section A–B (Figs 1 and 3). Cross-plots of the oxygen and hydrogen indices show a dominance of type III kerogen (Fig. 7), which indicates a superabundance of allochthonous organic matter. Katz (1985) stressed that pyrolysis data give the bulk composition and that smaller amounts of autochthonous organic matter may not be detected owing to matrix effects or dilution by allochthonous plant debris. Supplementary analyses of the pyrolysis data (Fig. 7) are not available.

Sand-streaked shale

Streaks of coarse to very coarse-grained sand are separated by 5–20 mm of silty mudstone. The streaks are often only one grain thick but are laterally continuous over several metres. Locally they are associated with cross-laminated erosively based layers 2–3 cm thick of coarse-grained sand. Tuff-layer correlation demonstrates that the sand-streaked shale facies may be laterally continuous for several kilometres, particularly in a level

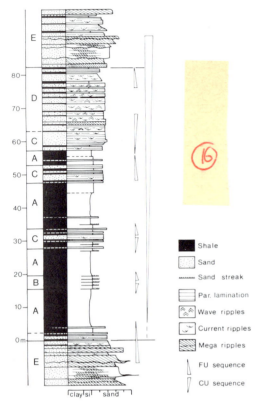

FIG. 4. Sedimentological log measured at Pingo. The lacustrine sequence (0–82 m) is sandwiched between fluvial channel deposits. Note the overall coarsening-upward trend. Minor CU and FU sequences are indicated. A, shale; B, sand-streaked shale; C, shale with sand layers; D, cross-laminated sand; E, large-scale cross-bedded sand.

deposition of sand, probably from density under-flows. Towards the top of the shale unit the sand beds increase in thickness and frequency and pass transitionally up into cross-laminated sand,

FIG. 5. Shale with laterally continuous beds of fine-grained sand interpreted as two pulses of density underflow deposits. Detail of the sequence shown in Fig. 2. Thin layers of tuff can be seen (T).

about 10 m above the base of the shale (Figs 2–4). The sand streaks may possibly have been generated by discontinuous sediment influx into a stratified lake as discussed by Sturm (1979).

Shale with sand layers

Thin layers of fine-grained sand are well exposed at two levels in the shale at Pingo (Figs 2 and 5). The sand contains mica and comminuted plant debris and occurs as beds 1–10 cm thick with sharp but not erosive bases. Thin beds are structureless and locally graded, while thicker beds are parallel or cross-laminated (Fig. 8). Individual sand layers can be traced 100–500 m along the outcrop and retain constant thicknesses. The distinct sheet geometry of the sand layers and the alternation with shale suggest an episodic

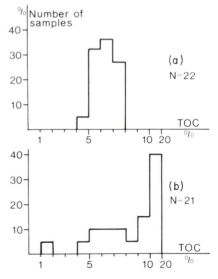

FIG. 6. Measured TOC values in shales from Tertiary non-marine shales: (*a*) average TOC values of 6% characterize the prodelta lacustrine shales from section AB; (*b*) TOC values from distal lacustrine shales are redrawn from Schiener & Leythaeuser (1978, Fig. 3).

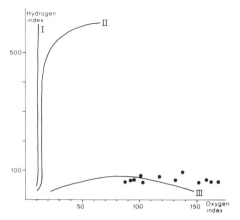

FIG. 7. A cross-plot of oxygen and hydrogen indices shows the bulk composition of the kerogen. Type III is dominant, indicating a high proportion of comminuted plant debris.

while the interbedded shale is lighter grey, more silty and more micaceous than in the lower parts of the member.

Cross-laminated sand

Fine-grained sand was deposited in beds 10–70 cm thick often separated by 1–2 cm of silty shale (Fig. 9). The sand is small-scale trough cross-bedded with subordinate wave-ripple lamination. It is difficult to distinguish between wave ripples and linguoid current ripples because of the lack of exposed bedding planes. Climbing ripple lamination is observed locally. The sand is interpreted as having been deposited by currents in an environment where interaction with or reworking by waves occurred intermittently, and it is suggested that the cross-laminated sand represents mouth-bar deposits resembling those described by Haszeldine (1984).

Large-scale cross-bedded sand

Solitary thick tabular sets of planar cross-strata, interpreted as transverse bars, and various types of trough cross-bedding generated by dunes are characteristic of the medium-grained sand which alternates with cross-laminated fine-grained sand (Fig. 4). Channel lags are rare, point-bar sequences have not been identified and fining-upward trends are not pronounced. The sand is interpreted as fluvial channel deposits.

Sedimentary sequence

The sedimentological logs compiled in Fig. 3 are almost identical and show that all the facies

FIG. 8. Shale interbedded with beds 1–20 cm thick of fine-grained sand with parallel or cross-lamination interpreted as distal mouth-bar deposits.

described above are laterally continuous. The shale abruptly overlies cross-bedded sand. A prominent level of sand streaks is seen in the lower half of the shale. Thin sand sheets occur at two levels which can be traced through most of the logs (Figs 2–4). They recur in the upper half of the shale and pass transitionally up into cross-laminated mouth-bar deposits overlain by fluvial sand. The shale sequence shows a distinct coarsening upwards, with an increasing amount of silt in the shale, an increasing sand grain size and an increasing sand-to-shale ratio upwards through the sequence.

Discussion of the depositional environment

Organic-rich shales with well-preserved plant fossils and bivalves in life position and almost without pyrite are described from lakes within coal-bearing Carboniferous sequences in NE England by Fielding (1984) and Haszeldine (1984). In contrast with the shales of the Upper Atanikerdluk Formation, the Carboniferous shales are thin (generally 1–2 m), overlie coal

seams and represent sediment-starved lakes. These were eventually filled by mouth-bar progradation (Fielding 1984, Fig. 2). A thick sequence of Tertiary lacustrine shales interfingering with sand deposited in peripheral fluvial-dominated deltas is described by Ayers (1986). Owing to the limited size and depth of the lake, basinal reworking of the delta front was slight, resulting in high constructive deltas and shoreline facies which are thin where present (Ayers 1986). Lake Lebo formed through rapid subsidence with intermittent phases of fluvial delta progradation. On these two points it deviates from the shale unit of the present study.

The studied shales are interpreted as lacustrine on the basis of palaeontological, mineralogical and sedimentological data. Koch (1959) interpreted the Naujât and Aussivik Members as non-marine owing to the presence of well-preserved plant remains, the lack of marine fossils and the absence of trace fossils. The lack of pyrite (and jarosite) is characteristic of non-marine deposits in the Skjernå delta (Postma 1982). Schiener & Leythaeuser (1978) claimed that TOC values of 10%–20% were characteristic of lacustrine Tertiary shales from the Nûgssuaq Embayment in

FIG. 9. Parallel beds of fine-grained mouth-bar sand separated by a few centimetres of sandy shale rich in plant debris. Cross-lamination is mainly current generated but wave ripples are seen locally. The normal graded bed of weak horizontal laminated sand is interpreted as a density flow deposit. The knife is 20 cm long.

'facies, where abundant autochthonous organic matter (partly liptinitic) was not sufficiently diluted with allochthonous organic and inorganic material'. The lower TOC values and the superabundance of type III kerogen (Figs 6 and 7) in the shale samples from section AB (Figs 1 and 3) can be explained as a result of dilution by allochthonous material. The samples analysed by Schiener & Leythaeuser (1978) represent distal prodelta shales, and the proximal prodelta shales are exposed in NE Disko (Fig. 6). Intermittent stratification of the lake is indicated by the sand-streaked shale. The high preservation of organic material indicates low dissolved oxygen contents in the bottom water, although the surficial but pervasive bioturbation of the shales excludes the existence of permanently anoxic bottom waters.

The sedimentary sequences are interpreted as deposition by a prograding delta. Few wave-generated structures are seen and indications of tidal redistribution of the sediments are absent. Basinal processes were apparently of little importance, which indicates a basin with limited fetch separated from marine tidal currents. The sharp lower boundary of the shale sequence suggests a rapid establishment of deep-water environments. The occurrence of shale on the floor of basins filled with subaqueously reworked volcanic sediments as breccias suggests that the size and depth of the lakes were controlled by volcanic activity. The shales are bounded to the S, and possibly to the E, by fluvial sand. The shales are *c.* 50 m thick after compaction, which suggests water depths in excess of 50 m during deposition. The depositional history is interpreted as follows: (i) volcanic eruptions and tectonic movements created barriers behind which river water was stemmed (Schiener & Leythaeuser 1978); (ii) the fluvial plain was rapidly drowned and succeeded by a lake more than 50 m deep in which mud was deposited; (iii) the river still supplied sediment

to the lake and gradually built a delta into it; (iv) lacustrine redistribution of the sediments was insignificant and the delta was therefore fluvially dominated.

Summarizing comments and conclusions

The Upper Atanikerdlek Formation on NE Disko comprises a 50 m thick sequence of dark grey shales passing through fine-grained cross-laminated sand up into medium-grained cross-bedded sand. The sequence is interpreted as lacustrine prodelta clay succeeded by mouth-bar sand and overlain by fluvial plain deposits, and represents the progradation of a fluvial-dominated delta into a relatively deep lake. Two similar cycles are represented by the Naujât and Aussivik Members on Nûgssuaq. Correlation between Disko and Nûgssuaq awaits establishment of the volcanic lithostratigraphy. The sedimentary sequences indicate that the lakes attained their greatest depth initially and afterwards became gradually more shallow. The distal lacustrine shales have high TOC contents and contain liptinitic organic matter, whereas the prodelta shales are dominated by allochthonous material.

ACKNOWLEDGMENTS: Grants j.nr. 11–5796 and 11–6350 from the Danish Natural Science Research Council and support from Arktisk Station, University of Copenhagen, and from the Geological Survey of Greenland are gratefully acknowledged. Earlier drafts of the manuscript were greatly improved by comments from Dr F. Surlyk, Dr A. Hurst and Dr A. Pulham. M. W. Jeppesen participated in the field work and P. Andersen, O. B. Berthelsen and R. Madsen helped in the preparation of the manuscript. I direct my sincere thanks to the persons and institutions mentioned above.

References

ALEXIADES, C. A. & JACKSON, M. L. 1966. Quantitative clay mineralogical analysis of soils and sediments. *Clays and Clay Minerals, Proceedings of the 14th Conference.* Pergamon, Oxford, 35–52.

ANDERSON, J. U. 1962. An improved pretreatment for mineralogical analysis of samples containing organic matter. *Clays and Clay Minerals, Proceedings of the 10th Conference.* Macmillan, New York, 380–388.

AYERS, W. B. JR. 1986. Lacustrine and fluvial–deltaic depositional systems, Fort Union Formation (Paleocene), Powder River Basin, Wyoming and Montana. *American Association of Petroleum Geologists Bulletin,* **70,** 1651–1673.

CLARKE, D. B. & PEDERSEN, A. K. 1976. Tertiary

volcanic province of West Greenland. *In:* ESCHER, A. & WATT, W. S. (eds) *Geology of Greenland.* Grønlands Geologiske Undersøgelse, Copenhagen, 364–385.

DEMAISON, G. J. & MOORE, G. T. 1980. Anoxic environments and oil source bed genesis. *American Association of Petroleum Geologists Bulletin,* **64,** 1179–1209.

ELLIOT, T. 1974. Interdistributary bay sequences and their genesis. *Sedimentology* **21,** 611–622.

FIELDING, C. R. 1984. Upper delta plain lacustrine and fluviolacustrine facies from the Westphalian of the Durham coalfield, NE England. *Sedimentology* **31,** 547–567.

FOLLETT, E. A. C., MCHARDY, W. J., MITCHELL, B. D.

& SMITH, B. F. L. 1965. Chemical dissolution techniques in the study of soil clays. *Clay Mineralogy* **6**, 23–34.

GUION, P. D. 1984. Crevasse splay deposits and roofrock quality in the Threequarters Seam (Carboniferous) in the East Midlands Coalfield, UK. *In:* RAHMANI, R. A. & FLORES, R. M. (eds) *Sedimentology of Coal and Coal-bearing Sequences*. International Association of Sedimentologists Special Publication **7**, 291–308.

HANSEN, J. M. 1980. *Stratigraphy and Structure of the Paleocene in Central West Greenland and Denmark.* Unpublished thesis, University of Copenhagen, 156 pp.

HASZELDINE, R. S. 1984. Muddy deltas in freshwater lakes, and tectonism in the Upper Carboniferous Coalfield of NE England. *Sedimentology* **31**, 811–822.

HENDERSON, G., ROSENKRANTZ, A. & SCHIENER, E. J. 1976. Cretaceous–Tertiary sedimentary rocks of West Greenland. *In:* ESCHER, A. & WATT, W. S. (eds) *Geology of Greenland*. Grønlands Geologiske Undersøgelse, Copenhagen, 341–362.

——, SCHIENER, E. J., RISUM, J. B., CROXTON, C. A. & ANDERSEN, B. B. 1981. The West Greenland basin. *In:* KERR, J. W. M. & FERGUSON, A. J. (eds) *Geology of the North Atlantic Borderlands*. Canadian Society of Petroleum Geologists Memoir **7**, 399–428.

KATZ, B. J. 1985. An organic geochemical comparison: clastic and carbonate lacustrine source-rock systems. *Abstracts, Lacustrine Petroleum Source Rocks Meeting 1985*. Geological Society, London.

KELTS, K. 1985. Depositional environments of lacustrine source rocks. *Abstracts, Lacustrine Petroleum Source Rocks Meeting 1985*. Geological Society, London.

KOCH, B. E. 1959. Contribution to the stratigraphy of the non-marine Tertiary deposits on the south coast of the Nûgssuaq peninsula, Northwest Greenland. *Meddr om Grønland* **162**, 100 pp.

—— 1963. Fossil plants from the Lower Paleocene of the Agatdalen (Angmârtussut) area, central Nûgssuaq peninsula, Northwest Greenland. *Meddr om Grønland* **172**, 120 pp.

MCCABE, P. J. 1984. Depositional environments of coal and coal-bearing strata. *In:* RAHMANI, R. A. & FLORES, R. M. (eds) *Sedimentology of Coal and Coal-bearing Sequences*. International Association of Sedimentologists Special Publication **7**, 13–42.

POSTMA, D. 1982. Pyrite and siderite formation in brackish and freshwater swamp sediments. *American Journal of Science* **282**, 1151–1183.

ROLLE, F. 1985. Late Cretaceous–Tertiary sediments offshore central West Greenland: lithostratigraphy, sedimentary evolution, and petroleum potential. *Canadian Journal of Earth Science* **22**, 1001–1019.

ROSENKRANTZ, A. & PULVERTAFT, T. C. R. 1969. Cretaceous–Tertiary stratigraphy and tectonics in northern West Greenland. *American Association of Petroleum Geologists Memoir* **12**, 883–898.

ROTH, C. B., JACKSON, M. L. & SYERS, J. K. 1969. Deferration effect on structural ferrous–ferric iron ratio and CEC of vermiculites and soils. *Clays and Clay Minerals* **17**, 253–264.

SCHIENER, E. J. 1975. Sedimentological notes on sandstones from Nûgssuaq, central West Greenland. *Rapports Grønlands Geologiske Undersøgelse* **69**, 35–44.

—— & LEYTHAEUSER, D. 1978. Petroleum potential off W. Greenland. *Oil and Gas Journal* October 2, 223–234.

SHEKHAR, S. C., FRANDSEN, N. & THOMSEN, E. 1982. *Coal on Nûgssuaq, West Greenland.* Geological Survey of Greenland, Copenhagen, 82 pp.

STURM, M. 1979. Origin and composition of clastic varves. *In:* SCHLÜCHTER, CH. (ed.) *Moraines and Varves: Origin, Genesis, Classification*. Balkema, Rotterdam, 281–285.

SURLYK, F. 1982. Kul på Nûgssuaq, Vestgrønland. *In:* SHEKHAR, S. C., FRANDSEN, N. & THOMSEN, E. *Coal on Nûgssuaq, West Greenland.* Geological Survey of Greenland, Copenhagen, 43–56.

TALBOT, M. R. 1985. Non-marine oil source rock accumulation—evidence from the lakes of tropical Africa. *Abstracts, Lacustrine Petroleum Source Rock Meeting 1985*. Geological Society, London.

G. K. PEDERSEN, Institute of General Geology, Øster Voldgade 10, DK-1350 Copenhagen K, Denmark.

The sedimentology of a Middle Jurassic lagoonal delta system: Elgol Formation (Great Estuarine Group), NW Scotland

J. P. Harris

SUMMARY: Deposition of N–S prograding lagoonal deltas took place in two separate extensional basins (Sea of the Hebrides and Inner Hebrides Basins) of different size and hence wave energy. Salinity of water in the basins varied spatially and temporally from marine to fresh–brackish. The reduced salinities were controlled by diminished connections with the open sea together with copious run-off from a hilly hinterland with a subtropical seasonal climate. Sections in N Trotternish, S Trotternish, Raasay (Sea of the Hebrides Basin) and Strathaird (Inner Hebrides Basin) represent three different styles of delta sedimentation corresponding to different salinities and basinal energy regimes (waves and tides). In N Trotternish, marine macrofaunas and microfloras occur in a wave–tide interaction delta system in which tidal currents controlled distributary-channel and mouth-bar hydrodynamics and high wave energies generated beach-ridge sediments at the delta shoreline. In S Trotternish and Raasay, probable brackish–marine salinities and a deeper basin are reflected in buoyant mouth-bar hydrodynamics. In Strathaird (Inner Hebrides Basin) probable fresh–brackish salinities correspond to a fluvial-dominated delta recording friction-dominated mouth-bar hydrodynamics. This lobate delta is characterized by low angle offshore inclined bar-front sandstones. In the Inner Hebrides Basin, the Elgol Formation delta reaches a depositional limit between Strathaird and Eigg. The N–S transition from marine to fresh–brackish salinities records the gradual establishment of a non-marine depositional system (which is maintained throughout the rest of the Great Estuarine Group) during progradation of the Elgol Formation delta.

The Inner Hebrides and Sea of the Hebrides Basins (Fig. 1) contain a thick marine Jurassic section in which the largely Bathonian Great Estuarine Group represents an important paralic episode (Hudson & Harris 1979). During the Bathonian, partial connections between the basins and the open sea together with freshwater run-off caused frequent salinity fluctuations (Fig. 2). In addition the two basins are of different size such that fetch and therefore wave energy varied widely.

The Elgol and underlying Cullaidh Formations occur at the base of the Group and also exhibit lateral and temporal salinity variations. The Elgol Formation has been interpreted as representing small river deltas prograding into enclosed water bodies (lagoons) represented by the underlying Cullaidh Formation (Hudson 1962, 1964; Hudson & Harris 1979; Harris 1984). It is the purpose of this paper to elaborate on this interpretation by describing Elgol Formation delta-front and distributary-channel deposits in the context of changing basin energy regimes and reducing basin water salinities together with basin water depth as the overriding controls.

The sediments are illustrated here by means of graphic logs representing three facies associations

corresponding to three distinct styles of delta progradation. These facies associations are interpreted as reflecting the distributary-mouth processes described by Wright (1977). The facies associations are (i) wave–tide-influenced sediments with marine faunas and palynofloras (N Trotternish, 12–12.5 m thick), (ii) fluvial-dominated successions with anoxic and marine to brackish–marine prodelta mudstones probably recording buoyant mouth-bar hydrodynamics (S Trotternish, up to 27 m thick; Raasay, 15.5 m thick) and (iii) fluvial-dominated successions with no marine indicators probably recording friction-dominated mouth-bar hydrodynamics (Strathaird, S Skye, up to 25 m thick).

The Elgol Formation sandstones are relatively coarse grained, locally pebbly and include granule conglomerates. The formation reaches a depositional limit in the Inner Hebrides Basin between S Skye (Strathaird) and Eigg, and palaeocurrent data demonstrate N–S progradation in both basins. Outcrops comprise sea cliff, scarp face and stream sections, and although the basal contact with the underlying Cullaidh Shale Formation (prodelta deposits) is well exposed the upper part of the formation and the contact with the overlying Lealt Formation is only rarely exposed.

From WHATELEY, M. K. G. & PICKERING, K. T. (eds), 1989, *Deltas: Sites and Traps for Fossil Fuels,* Geological Society Special Publication No. 41, pp. 147–166.

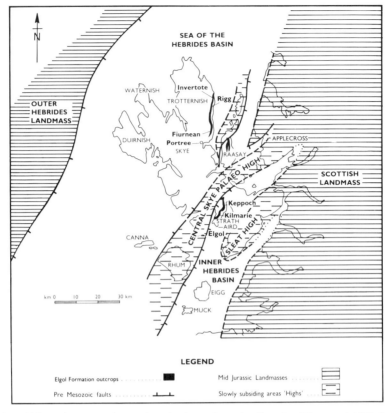

FIG. 1. Major mid-Jurassic structural controls on basin configuration (based on Binns *et al.* 1975; Steel 1976). Elgol Formation outcrops and localities are mentioned in the text.

Controls on the form and configuration of the Elgol Formation delta

Previous work on the rocks of the Great Estuarine Group has provided a detailed framework for the interpretation of the Elgol Formation in terms of the nature and climate of the hinterland (Hudson 1964; Tan *et al.* 1970; Tan & Hudson 1974; Hudson 1978) and the salinity of water in the receiving basins (Hudson 1962, 1963; Hudson & Morton 1969; Hudson & Harris 1979). In addition, the configuration of the basin margin faults (Fig. 1) can be defined by combining the structural map of Binns *et al.* (1975) with the Triassic palaeogeography of Steel *et al.* (1975) and Steel (1976).

Palaeogeography and characteristics of the hinterland

The palaeogeographic framework proposed by Hudson (1964) is supported by subsequent work

on the lithology of pebbles and heavy mineral assemblages (Harris 1984). Hudson (1964) proposes a Scottish land mass of variable relief in the E with a thick regolith derived from Old Red Sandstone sediments capping Moine and Dalradian rocks, and a less well-defined Outer Hebrides land area in the W composed largely of Lewisian gneiss. Deposition of Middle Jurassic rocks took place in a 'Minch Basin' between these two positive areas almost exactly corresponding to the present day Minch–Sea of Hebrides–Inner Hebrides area. Subsequent work on the structure of the area (Binns *et al.* 1975) and on Triassic palaeogeography (Steel *et al.* 1975; Steel 1976) supports this palaeogeographic scheme but suggests that, in the Skye area, the basin is divided into two NNE–SSW-trending troughs—the Sea of the Hebrides Basin in the W and the smaller Inner Hebrides Basin in the E. These two basins are separated by a fault-bounded palaeohigh which also trends NNE–SSW (*cf* Figs 1 and 15). Both basins are markedly asymmetrical with deep down-faulted western margins and more complex eastern margins, probably with numer-

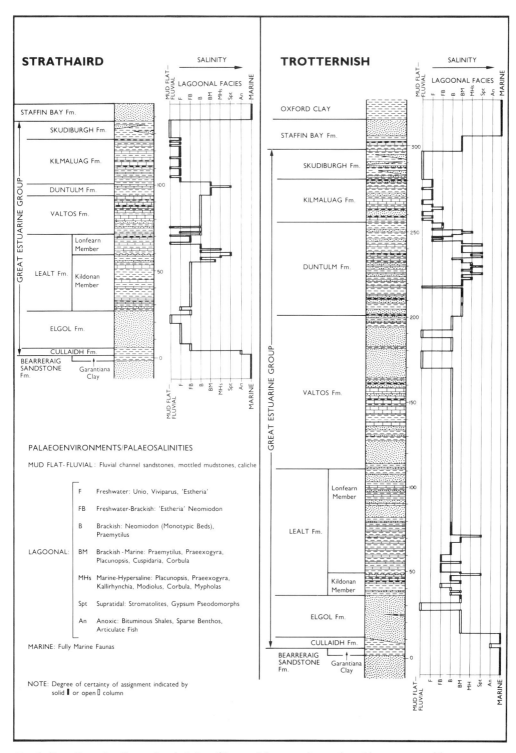

FIG. 2. Great Estuarine Group, Strathaird and Trotternish composite stratigraphic sequences with palaeoenvironmental–palaeosalinity plots (modified from Hudson & Harris 1979).

ous embayments. Differential subsidence, involving a progressive increase in the rate of subsidence southwestwards, is thought to have controlled the distribution of deltaic sediments in these basins (see discussion below).

The palaeolatitude was about 34°N (Smith *et al.* 1973), and the climate was warm (*c.* 17–22 °C) and humid with marked seasonality (Tan & Hudson 1974). These conditions led to periodic rapid run-off from a hilly hinterland which supplied large amounts of coarse clastic sediment to the basins.

Characteristics of the receiving basin

The base of the Elgol Formation either comprises a gradational passage from shale to siltstone to sandstone or is marked by sand–shale intercalations. This demonstrates the genetic relationship between the Elgol Formation (delta-front sandstones) and the upper Cullaidh Shale Formation (prodelta mudstones). However, the deposition of lignitic mudstones and oil shale in the lower part of the Cullaidh shale could have pre-dated delta progradation.

The characteristics of the receiving basin have been deduced from a combination of sources. Macrofaunas (Forsyth 1960; Hudson & Morton 1969) and microfloral data (W. Walton and J. P. G. Fenton, pers. comm.) from both the Cullaidh and Elgol Formations indicate palaeosalinities (*cf* Fig. 2). Palaeotemperatures based on oxygen isotope ratio data (Tan *et al.* 1970; Tan & Hudson 1974) from the overlying Lealt and Duntulm Formations demonstrate that the basin water was warm (17–22 °C). Wave and tidal processes are indicated by sedimentary structures within the Elgol Formation sandstones.

In N Trotternish, at Rigg and Invertote, the Cullaidh Shale Formation underlying the Elgol Formation comprises shales with a marine fauna (Hudson & Morton 1969), thin carbonaceous limestones and allochthonous lignitic shales. This formation records marine salinities in the receiving basin. In addition, thin (less than 1 cm) mudstones at the top of the Elgol Formation contain marine palynofloral assemblages (J. P. G. Fenton and W. Walton, pers. comm.). Sedimentary structures in the Elgol Formation include wave ripples and swash cross-lamination, demonstrating that the basin was relatively large. Bipolar palaeocurrent patterns indicate tidal currents in the basin and demonstrate the exchange of water with the open sea.

Further S in the Sea of the Hebrides Basin, at Fiurnean, the Cullaidh Formation includes an oil shale 2 m thick, representing anoxic bottom conditions, which is overlain by shales containing rare echinoids (*Diademposis cf woodwardi*; identification by C. Jeffries). On Raasay (also in the Sea of the Hebrides Basin), the formation also includes an oil shale and a bed with a marine fauna dominated by crinoid debris (Forsyth 1960). Palynofloral assemblages from both the Cullaidh and Elgol Formations in S Trotternish include acritarchs and dinoflagellates (W. Walton, pers. comm.). Species diversity is lower than in the underlying Garantiana clay, but the overall assemblage indicates near-normal marine salinities. In addition, leiospheres indicate 'stress' in the environment which is probably attributable to the proximity of nearby active distributaries. In Raasay, similar palynofloral assemblages also occur, but one sample from sands at the base of the Elgol Formation is dominated by derived freshwater palynomorphs. These macrofaunas and microfloras indicate brackish–marine salinities during the progradation of the Elgol Formation into the S Trotternish–Raasay area, although the basin had already been cut off from the open sea, allowing stratification of the water body and oil shale deposition.

In Strathaird in the Inner Hebrides Basin, the black shales of the Cullaidh Formation also represent an enclosed lagoon. However, the marine macrofauna found in the Sea of the Hebrides Basin is absent. Palynofloras in these shales have largely been destroyed by Tertiary thermal metamorphism (J. P. G. Fenton, pers. comm.) so that palaeosalinities remain uncertain.

Facies description and interpretation

The three facies associations referred to above, namely N Trotternish, which is wave–tide influenced (facies association 1), S Trotternish and Raasay, which is buoyancy dominated (facies association 2), and Strathaird, which is friction dominated (facies assocation 3), are described from N to S in the order of their deposition.

Facies association 1, N Trotternish: Invertote and Rigg Burn (Figs 3 and 4)

Description

At Invertote, the marine sediments of the Cullaidh Formation are overlain by a thin (1.5 m) burrowed sandstone and two complex fining-upward sequences with prominent scour surfaces and lags of quartz pebbles. Trough and tabular cross-stratification (5–30 cm) dominates the basal parts of these sequences and gives a complex

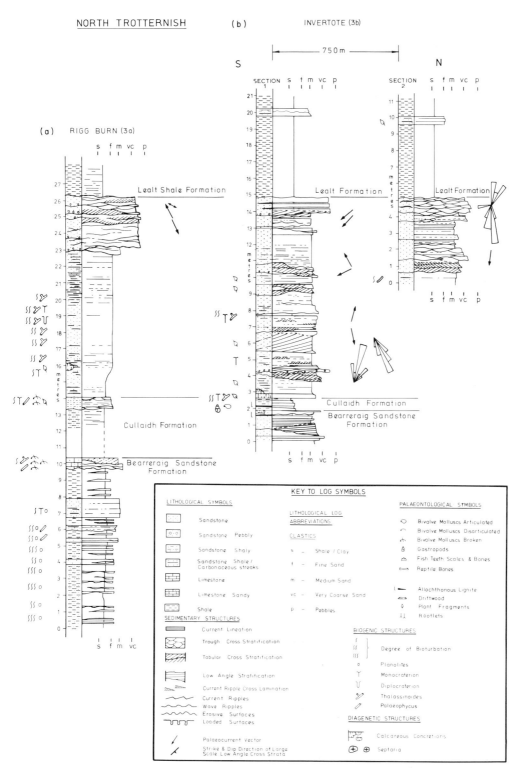

FIG. 3. Facies association 1: (a) log from the Rigg Burn, N Trotternish; (b) logs from near Invertote (section 1, sea cliffs close to the mouth of the Lealt River; section 2, from the foreshore 750 m to the N).

FIG. 4. Invertote, N Trotternish: facies association 1. The photograph corresponds to Fig. 3*b*, 0–11 m, viewed towards the W. The arrows mark the bases of sequences 1 and 2.

polymodal palaeocurrent pattern with both northerly and southerly modes. The sequences also include planar lamination and a single large-scale (1.50 m) tabular set (inclined towards 160°) with mud drapes on foreset surfaces (Figs 3 and 4). The two sequences are separated by a laterally persistent mudstone with *Monocraterion* and branching *Thalassinoides*? burrows.

At Rigg Burn the basal part of the formation comprises a gradational coarsening-upward sequence of fine-grained sands with *Diplocraterion*, *Monocraterion* and *Thalassinoides*? burrows overlain by argillaceous sands with wavy and bifurcating flasers.

The upper part of the Elgol Formation comprises sandstones with thin lenticular silty mudstones (maximum 7 cm) and wave ripples overlain by low angle planar-laminated sands with low angle discordances between sets (probably swash cross-stratification). These are inclined at up to 4° to the SW. The formation is capped by well-sorted granule conglomerates with low angle planar lamination dipping SW and thin (less than 2 cm) mudstone intercalations. Locally, trough and tabular cross-stratified pebbly sandstones with bimodal N–S palaeocurrent patterns cap the formation.

Interpretation

Macrofossils, microfloras and the diverse suite of trace fossils clearly indicate marine salinities in the basin during deposition of the Invertote–Rigg sequences. At Invertote the sequence of grain

sizes, structures and diametrically opposed palaeocurrent directions is comparable with the tidal channel and mouth-bar deposits described from the Recent Niger Delta (Allen 1965; Weber 1971) and the Quaternary Niger and Recent Netherlands coast (Oomkens 1974). Scour surfaces, numerous clay laminae and shale drapes are particularly characteristic of the tidal channel described by Oomkens (1974), and the occurrence of both large- and small-scale sedimentary structures in the Invertote section may also be characteristic of tidal-channel deposits (*cf* Barwis 1978).

It is proposed that the section represents tidal currents flowing in a distributary channel with subsequent progradation of the channel mouth by deposition from dominant basinward (southerly directed) ebb-tide and fl   rrents. This sequence of events resulted i   of channel and mouth-bar deposits an   ent deposition of fining-upward seque   fluctuating currents in laterally migr   distributary channels and channel-mo   s (*cf* Fig. 5).

The upper part of the f   at Invertote is dominated by wave-buil   tary structures (wave ripples and offs   inclined swash cross-stratification). Tl   ments represent the reworking by waves of the distributary sands to form a delta shoreline–shoreface–foreshore sequence (*cf* Clifton *et al.* 1971; Davidson-Arnott & Greenwood 1976; Hill & Hunter 1976). Sand was probably supplied directly to the shoreline by the distributaries, and wave reworking may have been responsible for the sealing of channel

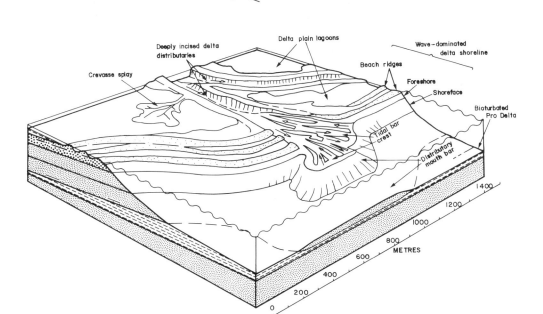

FIG. 5. Facies association 1: environmental reconstruction of the Elgol Formation delta around Invertote. The western part of the block is based on section 1, and the area around the distributary mouth is based on section 2. The delta is a fluvial–wave–tide interaction system.

mouths by beach sands as channel abandonment occurred. The granule conglomerates probably represent storm-generated backshore beach ridges separated by pools of standing water in which the dark shale intercalations could accumulate. Palaeocurrent directions in these sediments swing round from N–S to NE–SW in the Invertote area, probably indicating an arcuate configuration of the shoreline.

The Rigg Burn section (Fig. 3*a*, below 23 m) probably represents a wave-dominated part of the delta shoreline deposited away from active tidal distributaries. The coarse-grained pebbly sandstones at the top of the sequence probably represent the high wave energy delta foreshore–backshore.

The carbonaceous mudstones in the base of the overlying Lealt Formation probably represent small delta-plain lagoons developed landwards of the delta shoreline.

Facies Association 2, S Trotternish and Raasay: Fiurnean and N Dun Caan (Figs 6 and 7)

Description

The brackish–marine shales of the Cullaidh Formation are overlain gradationally by a se-

quence up to 24 m thick which coarsens upwards from very fine to fine-grained sandstone. The sandstones are structureless and moderately well sorted with minor amounts of detrital clay and occur in beds 2–20 cm thick. Bedding is defined by thin carbonaceous shale partings and horizons of small *Planolites* burrows. Low amplitude ripple forms (probably wave ripples) occur sporadically on bedding surfaces and are draped by mud laminae. The top of the sequence on Raasay comprises fine- to medium-grained planar-laminated sandstones without mud laminae.

This thick coarsening-upward sequence is truncated by trough and tabular cross-stratified, very coarse to medium-grained sandstones (sets 10–35 cm) with pebble accumulations. In S Trotternish these are overlain by trough cross-stratified cosets, also with basal pebble accumulations. Palaeocurrents are dominantly southerly and southwesterly. On Raasay (Fig. 6*b*, above 21 m) tabular cross-stratified sandstones with southerly palaeoflow directions are overlain by horizontal and low angle inclined planar-laminated sandstones.

On top of the trough cross-bedded cosets in S Trotternish (Fig. 6*a*, above 10.5 m, section 2) are fine medium-grained sands with thin medium

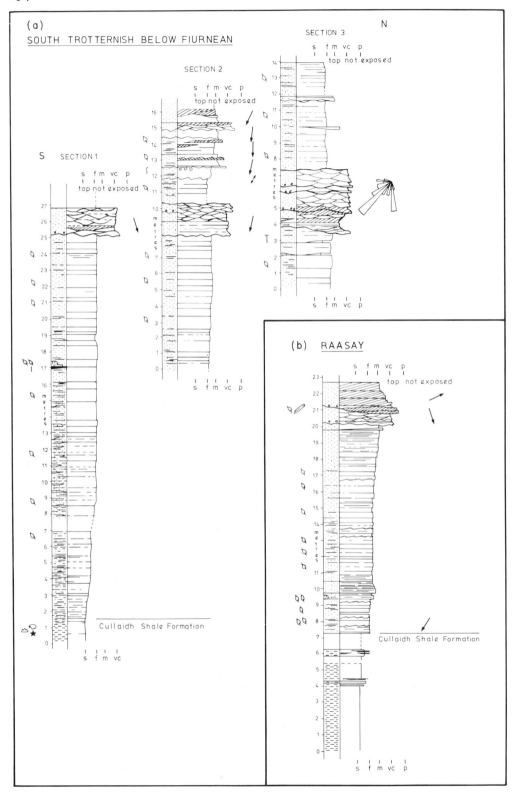

(a) SOUTH TROTTERNISH BELOW FIURNEAN

SECTION 3

SECTION 2

SECTION 1

S

N

(b) RAASAY

Cullaidh Shale Formation

coarse to very coarse grained cross-bedded sand sheets and lenses, all with southerly palaeoflow directions.

Interpretation

The underlying Cullaidh Formation in S Trotternish contains marine macrofossils in some beds and palynofloral assemblages which reflect stratification of the basin during deposition of the oil shale. These palynofloras record near-marine salinities in the upper Cullaidh Formation and continue into the Elgol Formation.

Water in the receiving basin was therefore more dense than river water entering the basin from Elgol Formation distributary channels. This would promote the generation of buoyantly supported plumes of sediment-laden river water offshore from the distributary mouths (hypopycnal outflow, *cf* Bates 1953). Deceleration and dispersal of these plumes in the manner described by Wright (1977) is taken as the process for the deposition of the structureless coarsening-upward sequence. This sequence is interpreted as representing a broad gently shelving fine-grained bar front. The wave ripples in this sequence represent post-flood reworking at the tops of sand beds, and the mud laminae and horizons of *Planolites* burrows record long periods between flood events. Structures recording traction-current processes are limited to low amplitude scour surfaces and planar lamination at the top of the Raasay sequence. These probably represent the mouth-bar crest and demonstrate that the saltwater wedge was displaced offshore during floods (*cf* Wright & Coleman 1974). Occasional poorly defined current ripples in the Raasay sequence possibly record the resedimentation of very fine grained sand by density currents flowing down the bar front.

This coarsening-upward sandstone has a sheet geometry. It probably represents a coalesced mouth bar formed from a series of closely spaced distributaries in the same manner as proposed for the pre-Recent 'shoal water' deltas of the Mississippi (Frazier 1967) and the modern Colorado River, Texas (Kanes 1970).

A problem with this interpretation is the importance that Wright (1977) places on deep distributary channels and deep water fronting the delta (and by inference fine-grained sediment load) in the promotion of buoyant outflow. The Elgol Formation distributary channels were shal-

FIG. 7. Cliffs below Fiurnean, S Trotternish: facies association 2. The photograph corresponds to Fig. 5a, section 1, 7–27 m, viewed towards the NE.

low, the water fronting the delta was relatively shallow (17–25 m deep measuring thickness from prodelta to delta plain) and the delta is a sand-dominated depositional system. Friction-dominated mouth-bar hydrodynamics might therefore be expected. However, traction-current structures are limited to the bar-crest sandstones in the Raasay sequence. This demonstrates that the basin was probably deep enough to maintain a pycnocline seaward of the mouth-bar crest during flood events and accounts for the contrast between the S Trotternish–Raasay facies association 2 and the Strathaird facies assocation 3. The closely spaced distributaries invoked above are represented by the medium- to coarse-grained sand sheet towards the top of the Elgol Formation. The basal erosion surface records incision of the distributaries and probably the removal of mouth-bar-crest sediments. These sandstones compare with the distributary channels described by Kanes (1970) and McGowan (1970) from modern deltas of the Texas Gulf Coast.

The sequence of sedimentary structures in the medium- to coarse-grained sand sheet (Fig. 6) records a distinct succession of bedforms. These represent channel incision and deposition of a pebble lag which was blanketed by in-channel bar bedforms (probably small-scale cross-channel bars). Subsequent in-channel aggradation is in the form of trough cross-stratified cosets repre-

---

FIG. 6. Facies association 2: (*a*) logs of sections 1–3, S–N from the cliffs below Fiurnean, S Trotternish; (*b*) logs from a stream section 2.5 km N of Dun Caan, Raasay.

senting stacked sinuous crested dunes. The
occurrence of shale drapes in the lower part of
the sequence represents fluctuations in discharge.
On Raasay, the planar-laminated and low angle
cross-laminated sandstones at the top of the
sequence are similar to the delta-plain sand-flat
deposits described by McGowan (1970).

The upper sequence in S Trotternish (Fig. 6*a*)
probably represents delta-plain sand flats with
flood-generated coarse-grained sand sheets. The
horizons of wave ripples demonstrate inundation
of sections of the delta plain or the ponding of
freshwater after floods.

Facies association 3, Strathaird: Elgol shore–Keppoch (Figs 8–12)

Description

In Strathaird hard silica-cemented sandstones
(cementation caused by Tertiary thermal metamorphism) form a distinct scarp traceable from
the type section of the Elgol Formation at the
Elgol shore (Harris & Hudson 1980) to Keppoch,
7.5 km to the NW (Figs 12 and 13). The scarp
exposes two distinct delta progradational sequences: a first phase comprising minor lenticular–lobate sand bodies enveloped by siltstone–
mudstone, and a second phase comprising thicker
more complex sandstones with a sheet geometry.

First progradational phase. A gradational coarsening-upward sequence 6 m thick (see Figs 10
and 13, logs 4 and 6) composed of structureless
and (towards the top) planar-laminated sandstones overlies shales of the Cullaidh Formation.
Locally, thin lignites and rootlet horizons together
with tabular cross-bedded and current-lineated
graded sandstones are preserved below a prominent scour surface at the top of the coarsening-
upward sequence. Palaeocurrents are bimodal,
with southerly and westerly flow directions.
Above the scour surface minor fining-upward
sandstones 1–3.5 m thick with tabular cross-bed
sets and scour and fill structures indicate southerly
palaeoflow directions. These are overlain by
rootlet-penetrated sandstones and capped by
mudstones with intensively bioturbated silt–sand
lenses. These sand bodies are lenticular. They
extend for a maximum of 400 m along the
escarpment and probably have an elongate–
lobate geometry with long axes oriented approximately SE.

Second progradational phase. On the Elgol shore
dark shales of the Cullaidh Formation have an
oscillatory transition to well-sorted fine-grained

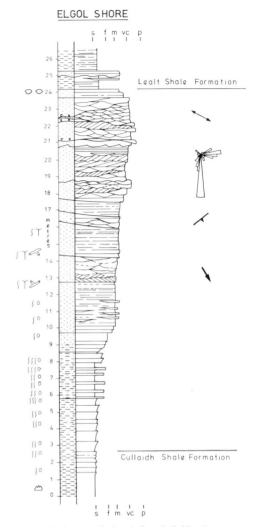

FIG. 8. Facies association 3: Strathaird log 1; type
section of the Elgol Formation, Elgol shore; sea cliff
and foreshore exposures.

sands at the base of the Elgol Formation (*cf* Fig.
8). Black micaceous shales are intercalated with
successively thicker shaly sandstones. These
sandstone units become progressively more intensively bioturbated upwards, culminating in the
lower part of the Formation in shaly sands with
pyrite nodules. *Planolites* burrows represent a
restricted assemblage of burrowing organisms.
The fine basal sandstones coarsen upwards to
well-sorted medium-grained sandstones with thin
lenticular coarse sand units. Above this are
moderately well sorted medium- to coarse-grained
sandstones with large-scale low angle inclined
surfaces (set height up to 5 m, see Fig. 9) dipping

FIG. 9. Elgol, Strathaird: facies association 3. The photograph corresponds to Fig. 6, log 1, 7–21 m, viewed towards the E. The arrows indicate the top and base of low angle inclined bar-front sandstones.

at between 4° and 7° to the SE (corrected for tectonic dip to the NW of 11°). These large-scale inclined surfaces commonly have weak basal scours and contain low amplitude (less than 10 cm) trough and tabular cross-stratification. Palaeocurrent flow directions are complex at this horizon but give a SE vector. *Monocraterion* and indistinct *Thalassinoides*? burrows occur intermittently.

Correlation of sections along the escarpment (Fig. 12) demonstrates how the deposits of the second phase thin markedly with the elimination of the low angle inclined surfaces wherever the elongate lobes of the first progradational phase are present. These two groups of strata have a mutually exclusive relationship everywhere, except possibly at Glasnakille Path (Figs 10 and 12, log 2) where the basal deposits are very poorly exposed.

The depositional dip direction of the large-scale low angle surfaces can be mapped at many localities along the scarp face. Their orientation swings round progressively from SE (142°–152°) at the Elgol shore to SW (234°) at N Elgol back to SSE (172°) at Drinan Road and W (260°) at Keppoch (Figs 12 and 13).

Coarse and very coarse grained sands with granules and pebbly lenses make up the top of the sequence. Distinct upper and lower units within this interval can be traced along the scarp with only minor lateral variation. The lower unit comprises trough and tabular cross-bed sets with planar lamination preserved between numerous scour surfaces. The upper unit comprises two trough cross-stratified cosets capped by medium grained sands with poorly defined planar lamination and the moulds of large bivalves (probably *Unio*; J. D. Hudson, pers. comm.). Palaeocurrents are bimodal in the lower unit and unimodal in the upper unit, but both give vector means which are subparallel to the dip direction of the underlying low angle surfaces.

At the top (base of the Lealt Formation) are 50 cm of carbonaceous silty shales capped by 10–20 cm of coarse erosively based sand with pyritic concretions. These are overlain by black silty shales.

At the northern end of the outcrop (Keppoch) the coarsening-upward sequence of the sandstones is truncated at the erosive base of a major composite fining-upward sequence (Figs 11 and 12). Pebble accumulations define the bases of these channel-form sand bodies which are dominated by large-scale tabular cross-stratification with occasional reactivation surfaces. Foreset surfaces in this material dip to the W and NW while trough cross-stratification in the finer-grained sands towards the top of the sequence gives a southwesterly palaeocurrent vector. At the top is approximately 1 m of highly carbonaceous argillaceous black sandstones.

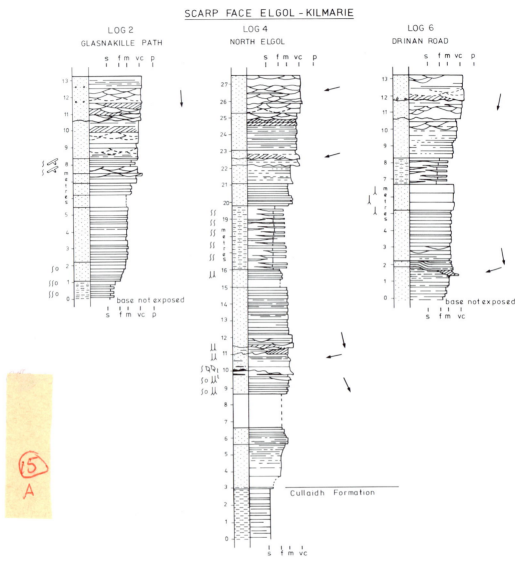

FIG. 10. Facies assocation 3: Strathaird logs 2, 4 and 6; scarp face sequences between the Elgol shore and Kilmarie.

Interpretation

First progradational phase. The shales of the underlying Cullaidh Formation represent an enclosed, probably stratified, water body but lack the marine macrofauna found in the Sea of the Hebrides Basin. Palynofloras have largely been destroyed by Tertiary thermal metamorphism (J. P. G. Fenton, pers. comm.) such that palaeosalinities in the receiving basin remain uncertain. The coarsening-upward sandstones clearly represent shoreline progradation in a shallow water body (rootlets occur 5.5 m above the base). The minor fining-upward sequences indicate sediment supply by small channels flowing southwards. The channel sandstones contain fine-grained sediment drapes and probably had fluctuating discharge regimes. These composite sand bodies have elongate–lobate geometries with long axes oriented SE subparallel to flow directions in the minor channel sandstones. The sequences are interpreted as minor fluvial-dominated elongate deltas similar to the Holocene Guadalupe delta described by Donaldson *et al.*

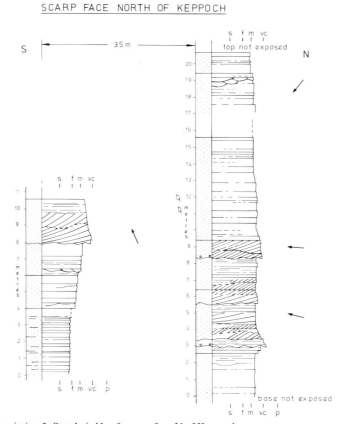

FIG. 11. Facies association 3: Strathaird log 9; scarp face N of Keppoch.

(1970). Similar sequences also occur in interdistributary bays, *eg* the minor mouth-bar–crevasse-channel couplet of Elliott (1974). Rootlets at the top of the sand bodies demonstrate plant colonization prior to abandonment and subsidence. The overlying mudstones with intensively bioturbated silt–sand lenses record abandonment and correlate with mudstones–siltstones in the base of the second progradational phase.

Second progradational phase. These sandstones contrast with the buoyancy-dominated sequences described from S Trotternish and Raasay. Instead of a gradational coarsening-upward sequence in structureless very fine to fine-grained sand, these sandstones include coarse-grained lenses together with trough and tabular cross-stratification. Traction currents capable of forming scour surfaces and transporting medium- and coarse-grained sand were therefore frequently active in the bar front. This is interpreted as indicating the friction-dominated outflow of Wright (1977) or the hyperpycnal flow (density-controlled underflow) of Bates (1953).

There is no direct evidence as to the salinity of the basin water after subsidence of the elongate delta lobes of the first progradational phase. Marine faunas are absent from the prodelta muds, but delta shoreline sands at the top of the formation contain moulds of probable Unionids demonstrating that by this time the basin probably contained freshwater.

The bar-front sandstones include low angle inclined surfaces (Fig. 9) which record the configuration and progradation direction of the delta front. These surfaces are mapped in Fig. 13, which demonstrates a broadly lobate geometry, probably formed by the coalescence of adjacent distributary mouth bars. Progradation was southwesterly at Keppoch (in the N) but changed to southerly as it continued. Palaeocur-

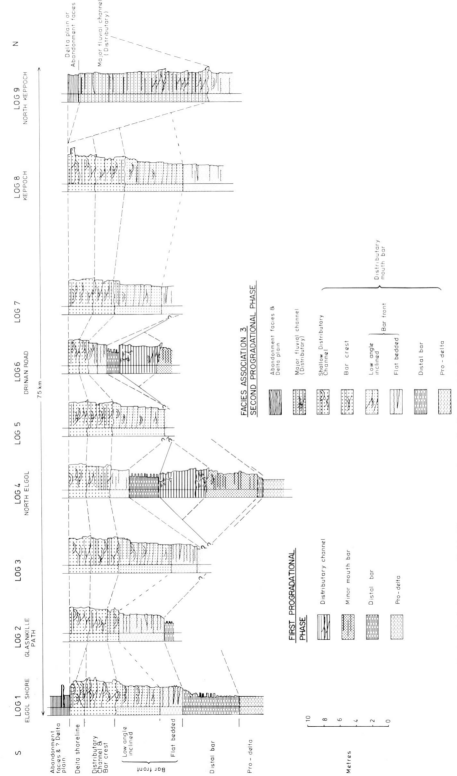

FIG. 12. Logs 1–9 from the Strathaird outcrop divided into groups of genetically related facies.

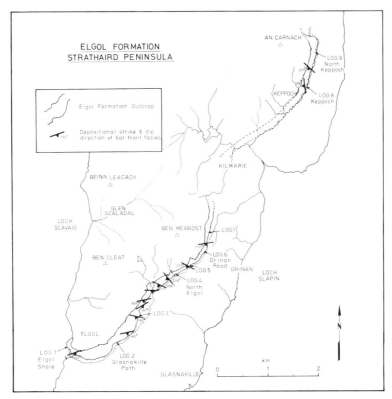

FIG. 13. Sketch map showing the distribution of the Elgol Formation outcrop in Strathaird, the distribution of the log localities and the depositional dip direction of units in the inclined bar-front facies.

rent measurements show that deposition took place from traction currents flowing down and oblique to the delta front. Deposition of sand in the delta front was episodic, allowing the development of burrows at some horizons, although mud drapes are absent.

The coarse-grained sandstones above 17.75 m in Fig. 8 include numerous scour surfaces and plane bedding with primary current lineation demonstrating high current energies. Palaeocurrent directions are strongly bimodal but give a vector parallel to the dip direction of the underlying low angle surfaces. These sandstones are interpreted as channelized bar-crest deposits. They also demonstrate friction-dominated hydrodynamics. The palaeocurrent pattern suggests channelized flow around features comparable with the 'middle-ground bars' described by Wright (1977, Fig. 5). These sandstones are truncated by pebble-lined erosion surfaces interpreted as the basal lag deposits of small bifurcating fluvial distributary channels. The palaeocurrent vector measured from trough axes is also subparallel to the dip direction of the underlying low angle surfaces.

At the northern end of the Strathaird outcrop (Keppoch), channel-form sand bodies representing major fluvial distributaries have eroded very deeply into the underlying mouth-bar deposits. These are interpreted as delta-plain channels. They were probably relatively straight and were filled predominantly with cross-channel and transverse-bar bedforms. It is proposed that these channels bifurcated repeatedly as they approached the delta shoreline in the manner emphasized by Elliott (1986). This repeated down-channel bifurcation during delta progradation is related to the formation of middle-ground bars in friction-dominated systems and would produce the shallow distributary channels at the delta shoreline which are invoked above.

Minor wave reworking of delta shoreline sands is probably recorded by the sands with moulds of *Unio* at the top of the formation at the Elgol shore. Delta-plain sediments are poorly exposed. The only outcrops which occur consist of carbonaceous sands and mudstones with a probable flood-generated sand sheet.

The configuration of the Elgol Formation delta in the Inner Hebrides Basin (reconstructed in

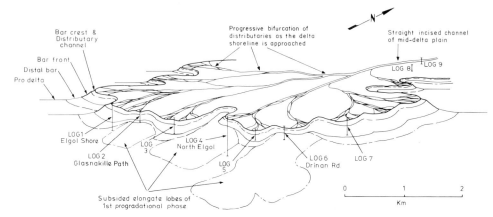

FIG. 14. Facies association 3: environmental reconstruction of the Elgol Formation delta in Strathaird, a lobate fluvial-dominated system with friction-dominated mouth bars.

Fig. 14) is dependent on friction-dominated hydrodynamics at a series of closely spaced shallow distributary mouths. Warm (*c.* 17–22 °C) freshwater in the receiving basin would promote this type of mouth-bar regime and would also allow the density-controlled underflow of colder sediment-laden river water which is evidenced by the coarse-grained sand lenses and cross-stratification in the lower bar-front deposits at the Elgol shore. However, water depth is also an important factor. This is indicated by the relationship between bar-front sands and the underlying abandoned lobes of the first progradational phase. The low angle inclined surfaces are eliminated whenever the first-phase lobes are present (Figs 13 and 14). Water depths greater than about 6 m were therefore necessary for the formation of the low angle surfaces. Because the basin was shallow (5.5 to 12–18 m), even if saltwater did occupy the basin during low discharge periods it could easily be displaced seaward of the delta front during flood periods and thereby allow the dominance of frictional processes at distributary mouths.

Conclusions and palaeogeography

Form and configuration of delta sand bodies

Lateral variation in the Elgol Formation delta sandstones, particularly in the bar front–bar crest, are attributed to variations in basin energy regime (waves and tides), basin water depth and salinity. These variations are particularly marked between the Sea of the Hebrides and Inner Hebrides Basins but are also apparent within the Sea of the Hebrides Basin, demonstrating that

the basin was partially cut off from the open sea during progradation of the delta, allowing a stratified anoxic basin to develop. This receiving basin is represented at least in part by the Cullaidh Shale Formation. The interaction of the above processes has resulted in three main facies associations within the Elgol Formation.

Facies association 1

In N Trotternish in the Sea of the Hebrides Basin the receiving basin was marine and both waves and tides were significant. Relatively thick channel sandstones with bipolar palaeoflow directions are interpreted as representing deep tidally influenced distributaries, and wave reworking is evidenced by the formation of beach-face sequences defining the delta shoreline. In N Trotternish the delta is defined as a fluvial–wave–tide interaction system (*cf* Elliott 1986).

Facies association 2

In S Trotternish and Raasay, also in the Sea of the Hebrides Basin, delta-front sandstones are very different. A thick gradational coarsening-upward sequence composed of structureless sands is interpreted as representing a broad gently shelving delta front. Deposition of very fine to fine-grained sand in this system is attributed to the deceleration and dispersal of buoyantly supported plumes of river water in a saline receiving basin (*cf* Wright 1977).

Facies association 3

In the Sea of the Hebrides Basin minor elongate delta lobes intervene between the Cullaidh

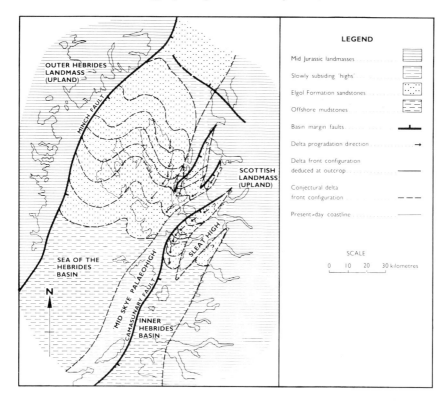

FIG. 15. Palaeogeography of the Inner Hebrides and Sea of the Hebrides Basins showing a series of tilted fault blocks as the major tectonic controls on sedimentation, delta progradation directions and the changing delta-front configuration of the Elgol Formation delta. Basin configuration based on Binns *et al.* (1975) and Steel (1976).

Formation and the main sandstones of the Elgol Formation. Thin coarse-grained sand lenses and cross-stratified sands in the lower bar front represent traction-current deposits. The lobate form of this bar front and the progradation direction of the delta are defined by low angle inclined surfaces. These sands are interpreted as representing the density-controlled underflow (hyperpycnal flow, *cf* Bates 1953) of cold sediment-laden river water beneath warmer fresh basin water within a friction-dominated shallow-water delta system (*cf* Wright 1977). Bar-crest and distributary-channel sandstones probably record the repeated bifurcation of channels around middle-ground bars as delta progradation took place.

Palaeogeography

Rapid progradation of this coarse-grained delta system supports the palaeogeographical scheme proposed by Hudson (1964) and the receiving basin and hinterland characteristics described by

Tan & Hudson (1974) and Hudson & Harris (1979). A hilly hinterland with a thick regolith and heavy seasonal rainfall is essential to the formation of this type of delta system. Evidence from these deltaic deposits indicates that the hinterland was uplifted, initiating a relatively rapid influx of clastic material. This caused copiously supplied deltas to prograde rapidly across the basins.

The palaeogeographic scheme shown in Fig. 12 is based on the present-day structural map of Binns *et al.* (1975) and the palaeogeographies of Hudson (1964), Steel *et al.* (1975) and Steel (1976). The modern structural framework was not established until the Late Jurassic or Early Cretaceous (Binns *et al.* 1975) but includes a number of major NNE–SSW-trending faults (principally the Great Glen, Camasunary and Minch Faults) which were in existence in the Triassic. The importance of these structures as controls on Triassic sedimentation has been demonstrated by Steel (1976) who proposed a system of elongate basins oriented NNE–SSW with shallow slowly subsiding NE margins. In the Great Estuarine Group (Strathaird in particular) subsidence-sensitive deposits of the Elgol Formation almost exactly mirror the pattern of sediment dispersal developed in the Triassic. It is therefore proposed that a system of differential subsidence or tilting and consequent basin fill by sediment transport subparallel to the basin margins (SSW), which was first established in the Triassic, also operated in the Middle Jurassic. The palaeographic map (Fig. 15) includes a central Skye palaeohigh, controlled at its southeastern margin by the Camasunary Fault (down-throw to the SE) and at its northwestern margin by a major fault exposed near Applecross (down-throw to the NW).

The central Skye palaeohigh appears on the regional structural map of Binns *et al.* (1975) and is a major element in Steel's (1976) Triassic palaeogeography. However, it was not included by Hudson (1964) and was dismissed as a Middle Jurassic feature by Harris & Hudson (1980), mainly because of the remarkable lateral continuity of facies within higher parts of the Great Estuarine Group, the individual argillaceous formations of which are identical to the N and S of the structure. It is therefore unlikely to have been anything except a very subdued feature during the Bathonian.

In this area of central Skye, marine siltstones of Liassic age overstep onto Durness Limestones and the Torridonian Sandstone basement (Hallam 1959) and interdigitate with locally derived ortho- and para-conglomerates (Steel *et al.* 1975). There was therefore considerable local relief on the southeastern margin of the structure in the Early Jurassic.

Middle Jurassic sediments in the area (Alt Strollamus) are difficult to interpret accurately because of poor exposure and severe Tertiary thermal metamorphism, but probably include rocks of the Bearreraig Sandstone Formation and the lower part of the Great Estuarine Group (Cullaidh, Elgol and Lealt Formations). The structure was therefore blanketed by Middle Jurassic sediments, at least at its southeastern margin. This provides additional evidence that it was not a major topographic feature in the Middle Jurassic. However, there is considerable lateral variation across the structure in the thickness and facies of all the Middle and Upper Jurassic sandstones. The Bajocian Bearreraig Sandstone Formation (immediately below the Great Estuarine Group) is up to 500 m thick in Strathaird, whereas it is a maximum of 210 m thick in Trotternish and Raasay N of the structure. Also, the palaeoflow directions of tidal currents in the Bearreraig Sandstone Formation parallel the

Camasunary, Applecross and Screapadal Faults (Morton 1983), demonstrating that the faults controlled the configuration of the Bajocian basins. Lateral variation in the facies of the Elgol Formation delta between the two basins is described above. The Valtos Sandstone Formation (Great Estuarine Group) is anomalously thin (21 m) in Strathaird, although this is on the down-throw side of the Camasunary Fault (Harris & Hudson 1980). The Callovian–Oxfordian section (immediately above the Great Estuarine Group) is predominantly argillaceous at Staffin, N of the structure, but consists almost entirely of coarse- and medium-grained sandstones in Strathaird (Sykes 1975). It seems probable therefore (although there is no direct evidence) that Steel's (1976) Triassic palaeohigh was still present during the Jurassic and controlled rates of subsidence in the basin and therefore the distribution and thickness of the Jurassic sandstones.

Steel's (1976) palaeogeography shows the Triassic basin margin (Inner Hebrides Basin) in N Strathaird swinging round from a SSW trend in the S to a WSW trend in the N (Fig. 13). This is paralleled almost exactly by a change in the progradation direction of the Elgol Formation delta (Figs 13–15) from WSW in the N to S in southern Strathaird. The progradation directions for the delta (controlled by basin configuration and style of subsidence) also match the SW palaeoflow direction for the Triassic braided-stream flood-plain phase for this part of the Inner Hebrides Basin (Steel *et al.* 1975). Therefore tilting or differential subsidence (a progressive increase in the rate of subsidence southwest-wards) which controlled this pattern of sedimentation was probably also responsible for the SW progradation direction and southerly depositional limit of the Elgol Formation delta in the Strathaird region and the southerly depositional limit of the delta to the S of Skye.

ACKNOWLEDGMENTS: This work was undertaken during the tenure of a National Environment Research Council Research Studentship at the University of Leicester. I am indebted to John Hudson for critically (and patiently) reviewing an earlier version of this paper and for introducing me to the Great Estuarine Group outcrops. I am also grateful to Andy Gardiner and two anonymous referees who critically read the manuscript and provided useful suggestions. William Walton (Sheffield) and Jim Fenton (Robertson Research) provided the very useful palynological data. Sue Button (Leicester), Tony Stephens and Roy Howgate (Robertson Research) draughted and helped in the design of most of the figures. The directors of Robertson Research, North Wales, are thanked for providing reprographic and photographic facilities.

References

ALLEN, J. R. L. 1965. Late Quaternary Niger Delta and adjacent areas; sedimentary environments and lithofacies. *American Association of Petroleum Geologists Bulletin* **49**, 547–600.

BARWIS, J. H. 1978. Sedimentology of some South Carolina tidal creek point bars, and a comparison with their fluvial counterparts. *In:* MIALL, A. D. (ed.) *Fluvial Sedimentology.* Memoir of the Canadian Society of Petroleum Geologists, Calgary, 129–160.

BATES, C. C. 1953. Rational theory of delta formation. *American Association of Petroleum Geologists Bulletin* **37**, 2119–2162.

BINNS, P. E., MCQUILLIN, R., FANNIN, N. G. T., KENOLTY, N. & ARDUS, D. A. 1975. Structure and stratigraphy of sedimentary basins in the Sea of the Hebrides and the Minches. *In:* WOODLAND, A. W. (ed.) *Petroleum and the Continental Shelf of Northwest Europe*, Vol. 1, *Geology.* Institute of Petroleum and Applied Science, Barking, Essex, 93–102.

CLIFTON, H. E., HUNTER, R. E. & PHILLIPS, R. L. 1971. Depositional structures and processes in the non-barred high-energy nearshore. *Journal of Sedimentary Petrology* **41**(3), 651–670.

DAVIDSON-ARNOTT, R. G. D. & GREENWOOD, B. 1976. Facies relationships on a barred coast, Kouchibouguac Bay, New Brunswick, Canada. *In:* DAVIS, R. A. JR & ETHINGTON, R. L (eds) *Beach and Nearshore Sedimentation.* Special Publication of the Society of Economic Paleontologists and Minerologists **24**, 149–168.

DONALDSON, A. L., MARTIN, R. H. & KANES, W. H. 1970. Holocene Guadalupe Delta of Texas Gulf Coast. *In:* MORGAN, J. P. & SHAVER, R. H. (eds) *Deltaic Sedimentation Modern and Ancient.* Special Publication of the Society of Economic Paleontologists and Minerologists **15**, 107–137.

ELLIOTT, T. 1974. Interdistributary bay sequences and their genesis. *Sedimentology* **21**, 611–622.

—— 1986. Deltas. *In:* READING, H. G. (ed.) *Sedimentary Environments and Facies* (2nd edn). Blackwell Scientific Publications, Oxford, 155–188.

FORSYTH, J. H. 1960. A marine shell-bed near the base of the Estuarine Series in Raasay. *Transactions of the Edinburgh Geological Society* **17**, 273–275.

FRAZIER, D. E. 1967. Recent deltaic deposits of the Mississippi delta: their development and chronology. *Transactions of the Gulf-Coast Association of Geological Societies* **17**, 287–315.

HALLAM, A. 1959. Stratigraphy of the Broadford Beds of Skye, Raasay and Applecross. *Proceedings of the Yorkshire Geological Society* **32**, 165–184.

HARRIS, J. P. 1984. *Environments of Deposition of Middle Jurassic Sandstones in the Great Estuarine Group, N.W. Scotland.* Unpublished PhD thesis, Leicester University.

—— & HUDSON, J. D. 1980. Lithostratigraphy of the Great Estuarine Group (Middle Jurassic), Inner Hebrides. *Scottish Journal of Geology* **16**, 231–250.

HILL, E. W. & HUNTER, R. E. 1976. Interaction of biological and geological processes in the beach and nearshore, northern Padre Island, Texas. *In:* DAVIS, R. A. JR & ETHINGTON, R. L. (eds) *Beach and Nearshore Sedimentation.* Special Publication of the Society of Economic Paleontologists and Minerologists **24**, 169–187.

HUDSON, J. D. 1962. The stratigraphy of the Great Estuarine Series (Middle Jurassic) of the Inner Hebrides. *Transactions of the Edinburgh Geological Society* **19**, 135–165.

—— 1963. The ecology and stratigraphical distribution of the invertebrate faunas of the Great Estuarine Series. *Palaeontology* **6**, 327–348.

—— 1964. The petrology of the sandstones of the Great Estuarine Series, and the Jurassic palaeogeography of Scotland. *Proceedings of the Geological Association* **75**, 499–528.

—— 1978. Concretions, isotopes, and the diagenetic history of the Oxford Clay (Jurassic) of central England. *Sedimentology* **25**, 339–370.

—— & HARRIS, J. P. 1979. Sedimentology of the Great Estuarine Group (Middle Jurassic) of North-West Scotland. *In: Symposium Sédimentation Jurassique, W. Européen, Paris, 9–10 May 1977.* Association des Sediments Française, Publication Speciale **1**, 1–13.

—— & MORTON, N. 1969. *Field Guide No. 4, Western Scotland.* International Field Symposium of British Jurassic, Keele University.

KANES, W. H. 1970. Facies and development of the Colorado River in Texas. *In:* MORGAN, J. P. & SHAVER, R. H. (eds) *Deltaic Sedimentation Modern and Ancient.* Special Publication of the Society of Economic Paleontologists and Minerologists **15**, 78–106.

MCGOWAN, J. H. 1970. Gum Hollow fan-delta, Neuces Bay, Texas. *Report of Investigation* **69**. Bureau of Economic Geology, University of Texas at Austin, 91 pp.

MORTON, N. 1983. Palaeocurrents and palaeo-environment of part of the Bearreraig Sandstone (Middle Jurassic) of Skye and Raasay, Inner Hebrides. *Scottish Journal of Geology* **19** (1), 87–95.

OOMKENS, E. 1974. Lithofacies relations in the Late Quaternary Niger delta complex. *Sedimentology* **21**, 145–222.

SMITH, A. G., BRIDEN, J. L. & DEWRY, G. E. 1973. Phanerozoic world maps. *Special Papers on Palaeontology* **12**, 1–42.

STEEL, R. J. 1976. Triassic rift basins of northwest Scotland—their configuration, infilling and development. *In: Mesozoic of the northern North Sea Symposium.* Norwegian Petroleum Society, Oslo, MNNSS/7, 1–18.

——, NICOLSON, R. & KALANDER, L. 1975. Triassic sedimentology and palaeogeography in central Skye. *Scottish Journal of Geology* **11**, 1–13.

SYKES, R. M. 1975. The stratigraphy of the Callovian and Oxfordian stages (Mid-Upper Jurassic) in northern Scotland. *Scottish Journal of Geology* **11**, 51–78.

TAN, F. C. & HUDSON, J. D. 1974. Isotopic studies on the palaeoecology and diagenesis of the Great Estuarine Series (Jurassic) of Scotland. *Scottish Journal of Geology* **10**, 91–128.

——, —— & KEITH, M. L. 1970. Jurassic (Callovian) palaeotemperatures from Scotland. *Earth and Planetary Science Letters* **9**, 421–426.

WEBER, K. J. 1971. Sedimentological aspects of oilfields of the Niger Delta. *Geologie en Mijnbouw* **50**, 559–576.

WRIGHT, L. D. 1977. Sediment transport and deposition at river mouths: a synthesis. *Geological Society of America Bulletin* **88**, 856–868.

—— & COLEMAN, J. M. 1974. Mississippi river mouth processes effluent dynamics and morphologic development. *Journal of Geology* **82**, 751–778.

J. P. HARRIS, Department of Geology, University of Leicester, Leicester LE1 7RH, UK.
Present address: Robertson Research International Ltd, Llandudno, Gwynedd LL30 1SA, UK.

Styles of soft-sediment deformation on a Namurian (Carboniferous) delta slope, Western Irish Namurian Basin, Ireland

O. J. Martinsen

SUMMARY: Soft-sediment deformation features from a delta-slope succession exposed in the Western Irish Namurian Basin, Ireland, are described and related to three main styles of deformation: (i) slumping, (ii) sliding and (iii) water-escape structures. While (i) and (ii) were formed by near-surface gravitational gliding of sediment, (iii) was formed by mainly vertical motion of sediment and escaping pore fluids. Deformation structures formed by slumping include slump folds and large- and small-scale faults, internal shear surfaces, deformation of early calcareous concretions and boudinage. The slumps were initially dominated by plastic deformation, whereas brittle deformation dominated the late stages. Deformation structures formed by sliding are mainly macroscopic faults, but microfaults and slide folds also formed locally. The slides were at all stages dominated by brittle deformation. Water-escape structures include mud and sand dykes and volcanoes, collapse depressions, mud diapirs and small-scale loading. These structures can be either related to or independent of slumping and sliding. While slumps are not restricted to any particular part of the delta slope, the occurrence of slides is, to a large degree, dependent upon grain size. Mud slides only occur on the muddy upper delta slope, while sand slides are only found on the lower delta slope. Water-escape structures tend to be more common on the lower delta slope.

Descriptions of soft-sediment deformation from deltaic settings have been made mainly from the Tertiary and more recent deposits of the Gulf Coast (eg Carver 1968; Morgan et al. 1968; Coleman et al. 1974, 1983; Coleman & Garrison 1977; Kosters 1985) and the Niger Delta (Weber & Daukoru 1975; Evamy et al. 1978). Relatively few accounts of deformation features from exposed ancient deltaic successions in intracratonic settings have been published. Notable exceptions include descriptions from Mesozoic deltaics in Svalbard (Edwards 1976; Nemec et al. 1988) and Carboniferous deltaic sequences in northern Europe (eg Rider 1978; Crans et al. 1980; Elliott & Lapido 1981). Most of these studies, however, have focused on large-scale deformation structures, particularly soft-sediment syndepositional growth faults, while smaller-scale structures have tended to be ignored (see, however, Maltman 1987). In addition, the geometry of the deformed sediment masses to which the individual structures are related is often omitted. The link between deformation structure and detailed depositional environment is often insufficiently discussed and interpreted.

Small-scale soft-sediment deformation features from an Upper Carboniferous delta-slope succession in the Western Irish Namurian Basin, Ireland, are described in this paper. The soft-sediment deformation features described as 'growth faults' and 'mud diapirs' which occur in the overlying proximal delta-front sediments are not included (see instead Rider 1978; Gill 1979; Crans et al. 1980). By relating the individual deformation structures to the deformation mechanism, which is itself largely dependent on the depositional environment, their environmental context is highlighted.

Regional geological setting

Upper Carboniferous (Namurian) outcrops in western Ireland lie in a N–S-trending zone divided by the Shannon Estuary (Fig. 1). Previous detailed sedimentological work in the area has concentrated mainly on the superbly exposed coastal cliff sections in County Clare (eg Rider 1974; Pulham, this volume) and the basin has commonly been named the West Clare Basin (eg Gill 1979). A recent study (Martinsen 1987) showed that all the Namurian rocks in western Ireland appear to have been deposited within the same basin, and a regionally more valid name, the Western Irish Namurian Basin, was suggested.

The basin was initiated during the Dinantian by NNW–SSE extension, probably associated with a minor right-lateral strike-slip component (Martinsen 1987). This transtensional regime was associated with shallow submarine and subaerial volcanism (Schultz & Sevastopulo 1965; Sevastopulo 1981). Subsequent rapid subsidence (thermally? and/or fault controlled) led to the development of a central basin trough 10–20 km wide with pronounced bathymetry, aligned roughly parallel to the present-day Shannon Estuary (Fig. 2). The earliest basin infill was

From WHATELEY, M. K. G. & PICKERING, K. T. (eds), 1989, Deltas: Sites and Traps for Fossil Fuels, Geological Society Special Publication No. 41, pp. 167–177.

168   O. J. Martinsen

FIG. 1. Location map.

derived from the subsiding trough margins as carbonate slumps and turbidites from surrounding carbonate platforms. Owing to differential subsidence these were transformed into regional slopes, perhaps in an analogous fashion to that

seen in the Bowland Basin of northern England (Gawthorpe 1986, 1987).

Delta systems advancing from the W and NW, together with decreasing differential subsidence rates, promoted shallowing and filling of the basin (Martinsen 1987). On a large scale, the infill (Fig. 2) shows a fairly symmetrical distribution with the thickest deposits in the basin centre. At several levels the basin infill is intersected by thin (maximum 50 cm) fossiliferous shales or marine bands which, among other fauna, contain goniatites. In the British Namurian, 64 marine bands have been recorded (Ramsbottom *et al.* 1979). Each marine band contains a characteristic goniatite species and occurs over large areas (some also extending into mainland Europe and Russia), and thus they are valuable chronostratigraphic markers for regional correlation. Furthermore, each stratigraphic interval encompassed by two succeeding marine bands can be estimated to have had an average duration of 170 000 years (based on the estimated duration of the Namurian of 11 Ma by Lippolt *et al.* (1984)). This provides an approximate method of estimating sediment accumulation rates (see below).

Sedimentation in the early Namurian was dominated by basinal mudstones and shales

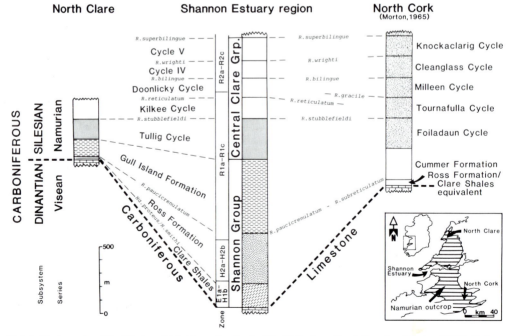

FIG. 2. Stratigraphy of the Western Irish Namurian Basin. The N Cork section is based on Morton (1965). Only the formation-bounding marine bands are included. The correlation with the N Cork section has not been confirmed and is thus only suggestive.

(Clare Shales). These deposits pass gradually upwards into the Ross Formation which, in the central parts of the basin, is represented by thick sandy turbidites with no apparent systematic vertical sequences (A. Kloster & B. Bakken, pers. comm.). On the basin margins, this formation is only represented by thin shales. Conformably above and separated from the Ross Formation by the *Reticuloceras paucicrenulatum* marine band lies the Gull Island Formation. In the lower part, this comprises extensively deformed slope deposits dominated by mud slumps and turbidites, while the upper part contains mud slumps and undeformed mudstone. A gradual coarsening-upward trend is seen from the top of the Gull Island Formation into the delta-front sediments of the overlying deltaic Tullig Cyclothem of the Central Clare Group. This group consists of five deltaic cyclothems, each produced by the progradation of a fluvial-dominated delta lobe (Rider 1974; Pulham, this volume).

The sediment accumulation rates were generally large within the Western Irish Namurian Basin and averaged 0.7 mm a^{-1} (Martinsen 1987). However, during shorter time intervals deposition was much more rapid. The Gull Island Formation and the Tullig Cyclothem accumulated between two marine bands (the *Reticuloceras paucicrenulatum* and the *Reticuloceras stubblefieldi* bands) which in the Namurian in Britain (where the most complete record of marine bands occurs) are only separated by four other such faunal bands. Ideally, this means that deposition took place in a time span of approximately 850 000 years. This yields an average sediment accumulation rate of close to 0.2 cm a^{-1}, which is extremely large. In comparison, the modern deltas of the Rhine and the Nile receive on average 0.6–0.7 mm a^{-1} (Schwarzacher 1979; Sestini, this volume).

Gull Island Formation

This study focuses on the deposits of the Gull Island Formation which were laid down on the slope in front of the earliest recorded deltaic progradation into the Western Irish Namurian Basin (the Tullig Cyclothem). The formation is a maximum of 550 m thick, and up to 75% of the sediments are deformed by soft-sediment deformation.

The Gull Island Formation can be divided into two: an upper part and a lower part. The lower part comprises mostly mud slumps, intercalated with heterolithic sand beds and interbedded mud, and randomly organized apparently tabular turbidite 'sheets' (maximum thickness 50 m) which

in a few cases were deformed by sand sliding. The upper part consists of dominantly undeformed mudstone with mud slumps, but mud slides are also present. Major channelling is very rare in both parts of the formation and only occurs where soft-sediment deformation is of minor importance.

The turbidity currents were apparently initiated at or close to the river mouths of the overlying delta lobe. On a small scale, the flow of turbidity currents was probably controlled by the topography created by frequent soft-sediment deformation, and the turbidite 'sheets' may actually represent interslump deposits (Hill 1984; Martinsen 1987). In addition, and on a larger scale, the relative positions of the river-mouth areas of the overlying delta lobe seem to have controlled turbidite sand deposition farther onto the slope. Areas that were situated axial to the active river mouths received more sand than the more lateral sites. It is possible that the frequent soft-sediment deformation may have prevented the development of large channels by rapid plugging of incipient channels and subsequent diversion of following turbidity currents to a new area.

The extensive soft-sediment deformation present within the Gull Island Formation was probably induced by a combination of differential subsidence in the basin causing delta-slope oversteepening, high sediment accumulation rates leading to underconsolidation and local rapid loading by deposition from turbidity currents.

Styles of soft-sediment deformation

The soft-sediment deformation in the Gull Island Formation can be classified as belonging to one of three diverse groups: (i) slumps (*sensu* Stow 1986), which are dominated by internal deformation, are primarily muddy and are formed by shallow gravitational gliding; (ii) slides (*sensu* Stow 1986), which have only minor internal deformation, are muddy or sandy and are also generated by relatively shallow gravitational gliding; (iii) water-escape structures, which are produced primarily *in situ* by liquefaction and fluidization.

Mud slumps and associated deformation structures

Mud slumps are the most common deformational style within the Gull Island Formation. They occur throughout the entire formation but tend to be more common and thicker in the lower part (Fig. 3). The slumps range in thickness from 0.5

170 O. J. Martinsen

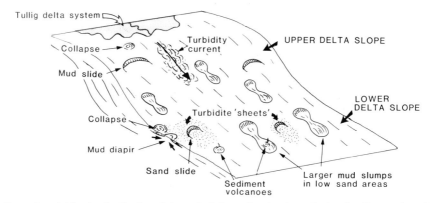

FIG. 3. General model for the distribution of the main deformational styles and related sedimentation within the Gull Island Formation (not to scale).

to 15 m, although units up to 50 m thick, which are possibly composite, have been observed. In small outcrops the slumps are apparently tabular, but in some large outcrops a lenticular form is evident. The real dimensions of the slumps are generally unknown, but it has been possible to correlate one slump 10 m thick over a distance of 5 km in a slope-parallel direction, suggesting a large width-to-thickness ratio.

Slump scars are rarely observed, but where present they are commonly small, with only three relatively large examples having been identified. The scars range in width from a few metres to several hundred metres, and in depth from 0.5 to 15 m. Complex geometries are seen in the smaller slump scars exposed on extensive bedding planes. 'Tributaries' feed into a main scar (Fig. 4). On a

slightly smaller scale, in plan view, slump scars resemble the scars of 'bottleneck slides' described from the Mississippi Delta (Prior & Coleman 1978). The lateral margins of slump scars can be transtensional, transpressional or purely strike-slip, depending on the relative sense of movement of the slump. It is often difficult to distinguish small slump scars from current scours, particularly when there are no deformed sediments in the scar. Evidence of contortion in the underlying beds and especially on the slump-scar surface (Fig. 4) is taken to indicate a slump-scar rather than a current-scour origin. However, this is questionable as shear stress at the base of large high concentration flows is probably also capable of inducing deformation in the sediment below.

Folds are the most common structures within

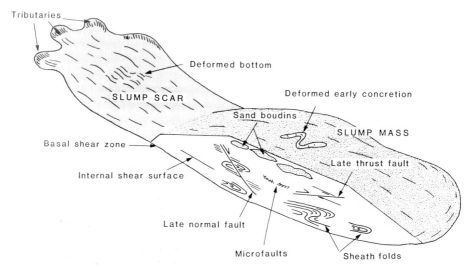

FIG. 4. Idealized model for slumps and associated deformation structures (not to scale). The most common positions of the individual structures are indicated.

mud slumps and show highly variable geometries, but are primarily sheath folds thought to have formed by simple shear (Fig. 5). Open and rounded folds are seen only locally, and then commonly adjacent to obstacles (*eg* sand boulders). These folds are apparently formed by compression relative to the obstacles. Slump folds are most commonly observed in the middle and lower parts of slump units (Fig. 4), probably because of higher shear in these regions.

Faults are generally developed as brittle shear zones, and usually occur internally within slumps (Fig. 4) but may also be bounding faults. Both normal and reverse senses of displacement are commonly seen. The faults are usually listric, but planar faults are also observed. Displacements on the faults are small (less than 5 m). The undeformed nature of most fault planes and the brittle mode of deformation suggest that faults are fairly late structures within slumps and are possibly related to antidislocation waves (Farrell 1984). Most slump faults observed are single and do not occur in groups or families, in contrast with faults associated with slides.

Internal shear surfaces show several features similar to faults but are different in that they are commonly fairly flat (Fig. 4) and seem to occur only within slumps. It is suggested that these surfaces represent internal shear within the slumps (see also Maltman 1987) during late movement stages.

Microfaults (*eg* Pickering 1983) seem only to have developed in very local zones (10–50 cm thick and less than 2 m wide) within slumps (Fig. 4). They are brittle in character, but also show signs of local plastic deformation along the fault planes. They have mainly normal senses of displacement. Microfaults, like larger faults, are considered to be late features, possibly developed as a result of local strain imposed as a result of pore-water escape. They do not seem to be restricted to any specific part of the slumps and may also occur on their tops and be draped by overlying sediment.

Deformed calcareous concretions (maximum 20 cm in diameter and 4 m in length) occur throughout the slumps. Their generally contorted nature and common alignment parallel to slump-fold axes and the fact that they are folded on top of bedding planes in some cases (Fig. 4) suggest that the concretions originated before soft-sediment deformation took place. Orientations of the long axes of the concretions have been used to supplement other data in establishing palaeo-slope trends (Martinsen 1987).

Boudins of sand, up to several metres in diameter, occur only locally within slumps, usually towards the top (Fig. 4), and demonstrate extension within the sediment. The stage in the deformation at which the boudins develop is unclear. There are some indications that they are relatively early, in that in one locality they are affected by slump faulting.

Plastic deformation (slump folding and contor-

FIG. 5. Sheath folds within a slump unit. Each 'belt' represents a fold closure. Note the highly variable orientation of the fold hinges. Lens cap (arrowed) for scale.

tion of early concretions) with a large degree of simple shear appears to have dominated the mud slumps in the early phases of movement, although local extensional stress fields also existed (boudinage). Brittle deformation patterns (faulting, microfaulting and internal shear surfaces) seem to have predominated during the late stages, probably as a response to increased sediment strength following pore-water escape.

Mud and sand slides and associated deformation structures

Slides occurring within the Gull Island Formation are of two types and are related to the dominant grain size. Sand slides occur exclusively in the lower part of the formation and only affect the turbidite 'sheets' (Fig. 3). These slides may be up to 20 m thick. Mud slides occur predominantly in the upper part of the formation (Fig. 3). These slides each affect a vertical sediment thickness of up to 50–60 m. The full three-dimensional geometries of both slide types have been difficult to document, but it seems that they approximate to the generally assumed 'spoon-shaped' morphology (Fig. 6) (eg Gawthorpe & Clemmey 1985).

Faults are the dominant structures in slides and occur mainly in groups, both as fault families dominated by normal faults (occasionally with

growth) (Fig. 7) and as imbricate thrust zones (Fig. 6). Although these fault groups may reflect the up-slope (head) or down-slope (toe) parts of a slide respectively, such groups may also occur within slides and record local extensional or compressional stress fields. Fault profiles are always listric and mostly concave upwards (Fig. 8). In the case of backthrusts related to thrust faults, they may also be convex upwards. The faults may be curved in plan view (Fig. 6); extensional faults are concave in inferred down-slope directions, while compressional faults are concave up-slope. The surface expression of compressional faults may appear as ridges which curve parallel to the fault traces (Figs 6 and 9). These ridges are analogous to the 'pressure ridges' described by Prior *et al.* (1984) from a modern submarine debris flow.

Displacements on the faults are generally less than 5 m but may reach several tens of metres. In contrast with faults generated by late-stage compressional and extensional strain waves in slumps (antidislocations) (Farrell 1984), similar structures in slides are early structures reflecting extension and compression in the sediment when the basal shear zone(s) cut up-section.

Microfaults and folds only affect small parts of slides. Microfaults are interpreted as late-stage structures, with occurrence and formation being

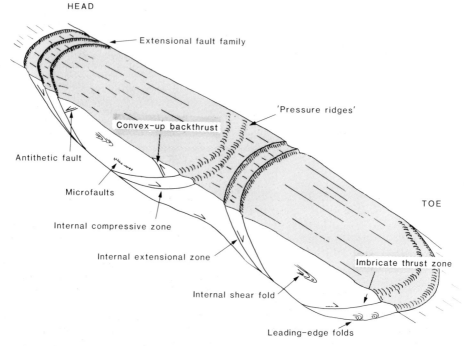

FIG. 6. Idealized model for slides and associated deformation structures (not to scale). Note the internal extensional and compressional fault zones and the internal shear folds.

FIG. 7. Extensional fault and associated antithetic fault (A) within a sand slide. Note how the unit marked X grows across the fault and how the fault dies out upwards within this unit. Lens cap (arrowed) for scale.

FIG. 8. Part of an extensional fault family within a sand slide. The major fault planes are arrowed. Note the rotation and erosion of the right fault block and that the slide is overlain by relatively flat-lying beds.

analogous to those of microfaults in slumps (Fig. 6). They may also offset earlier-formed folds in slides (Fig. 9). Slide folds may occur either associated with fault planes (most common, *eg* leading-edge folds related to thrust faults) or internally within the slides, where they indicate internal shear between otherwise undeformed parts on the slide (Figs 6 and 9). Whereas the commonly observed geometry of the former fold type suggests a compressional origin, the latter folds tend to be sheath folds indicating a simple shear origin.

In most cases it seems that the early stages of slide deformation were dominated by bedding-parallel slip, with the deformation concentrated in very thin shear or high strain zones (maximum 15 cm). Subsequently, the bedding-parallel slip zone cut up-section to produce extensional and compressional faults which are merely secondary

products of the bedding-parallel slip. Although the slides were dominated by brittle deformation at all times, the occurrence of folds shows that plastic deformation also took place locally.

Water-escape structures and vertical sediment motion

Water-escape structures occur as apparently late-stage phenomena in slumps and slides, but also affect sediment where there has been no gravitational gliding. They are found at all levels within the Gull Island Formation but tend to be more common low in the succession.

Sediment volcanoes and collapse depressions are occasionally found on the tops of slumps and mud diapirs, or on top of the immediately overlying beds (Fig. 3). Volcanoes can be either muddy or sandy cones and, in the Gull Island Formation, are up to 50 cm high and 1 m across. Substantially larger volcanoes are found in the underlying Ross Formation and were described by Gill & Kuenen (1958). Collapse depressions are circular to elliptical in plan view, up to 2 m deep and 20 m across. The rate of pore-water escape is considered an important factor in determining whether a volcano or a collapse forms. If escape is slow (*ie liquefaction*), a collapse may result because sediment is not entrained in the upward-moving water. The opposite (*ie fluidization*) applies to the formation of a sediment volcano. Collapse depressions in relation to mud diapirs seem to have formed by the lateral flow of mud to feed the rising diapirs.

Mud diapirs are relatively uncommon and are restricted to the lower part of the formation. They are of relatively small scale and grade into 'megaflames'. The largest diapirs affect 15–20 m of sediment, are a maximum of 25 m across and are conical to ridge like in three-dimensional form. These seem to have formed as a result of the progressive loading of turbidite sand in shallow channels, causing shallow subsurface lateral and vertical mud flow. Upward movement of mud caused extensive contortion of the beds adjacent to the diapirs and bent these upwards. Soft-sediment faults have also formed on the margins of the diapirs in an analogous fashion to the growth faults related to diapiric activity (*eg* Bruce 1973).

Sand and mud dykes are up to 30 cm wide, may be several metres long and penetrate up to 2–3 m of sediment. They are mostly found penetrating the beds on top of slumps. While water escape leading to the formation of sediment volcanoes and collapse structures was probably point sourced, dykes were apparently line sourced.

Loading occurs on the bases of turbidites. The

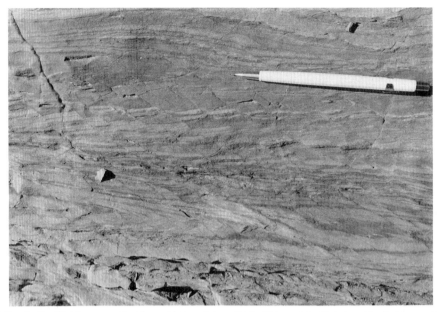

FIG. 9. Microfaults offsetting an earlier-formed internal shear fold within a slide. Note how the faults seem to die out upwards and their approximately listric character. The pen (for scale) is 15 cm long.

load features are mainly sand balls and pillows which, at the extreme, can be 1 m across but are more usually a few centimetres to 10 cm. Load features are very common but have not been studied in detail.

It seems likely that the water-escape structures described in this study can generally be related to either upward movement or sinking of sediment in response to the escaping pore water. Sediment volcanoes, mud diapirs and sediment dykes all formed by upward movement of sediment, while collapse depressions and loading formed by sinking of sediment. Collapse structures related to mud diapirs also formed by sinking of sediment, although the sinking was governed by lateral flow of mud to feed the rising diapirs.

Discussion

The abundance of soft-sediment deformation within the Gull Island Formation suggests that the delta slope on which it was deposited was generally unstable, with the conditions necessary to produce sediment failure almost always being present. The delta slope can thus be compared with the delta slope in front of the modern Mississippi Delta (*eg* Coleman *et al.* 1983) but with the apparent absence of the deeper penetrating structures (*eg* Coleman & Garrison 1977). This difference may be a result of the difference

in scale of the two systems, with the Mississippi system being an order of magnitude larger.

There are other important differences between the Gull Island Formation and the Mississippi Delta slope. Mud flows, which are of great importance in the Mississippi system, are not found in the Gull Island system. One possible reason for this may be the fundamental cause of deformation. In the Mississippi the main causes of the shallow soft-sediment deformation are rapid sedimentation and the presence of a large amount of organic material in the sediment leading to the production of biogenic methane (Coleman *et al.* 1983). These conditions, and particularly the presence of biogenic methane, may favour a high degree of plasticity in the sediment and cause flow rather than slip along discrete surfaces to produce sliding and slumping. A small (generally less than 0.5%) total organic carbon content suggests that generation of biogenic methane was not important in causing soft-sediment deformation in the Gull Island Formation. Another reason for the apparent absence of flows may be the fact that the slides and slumps observed in the present study do not generally appear to have moved very far (tens to hundreds of metres). Continued movement might have induced water mixing, remoulding and plastic deformation and transformed the masses into flows. Restricted amounts of transport are suggested by only minor rotation and modification of apparently early-formed deformation structures (*eg* slump folds) within the slumps and only minor displacements on most slide-bounding faults.

Turbidites are not involved in the deformation on the Mississippi Delta slope, but they appear to have played a major role in the soft-sediment deformation in the lower part of the Gull Island Formation. Their emplacement is thought to have acted as a triggering mechanism for mud slumping and sand sliding; they were deformed by sand sliding, and their flow patterns and deposition were probably controlled largely by the topography created by extensive sliding and slumping. In a qualitative way the Gull Island depositional system can thus be compared with the Nova Scotian continental slope described by Hill (1984), although there is no clear evidence on the Nova Scotian slope that the turbidites contributed directly to the soft-sediment deformation.

It is possible that the ENE-trending Silvermines–Navan fault system, shown to have been active during the Dinantian (Coller 1984; Sanderson 1984) and coinciding with the central axis of the basin, may also have been active during the Namurian and caused external triggering of soft-sediment deformation. The effect of seismicity as a triggering mechanism, however, is extremely difficult to show when deformation is so widespread.

Extensional and compressional faults may occur *within* slumps and slides and not necessarily at their respective up-slope and down-slope margins. Earlier models of slumps and slides (*eg* Lewis 1971; Dingle 1977) have caused some confusion in this respect owing to oversimplification. Gawthorpe & Clemmey (1985) invoked the existence of extensional faults within the overall compressional toe zone of their slide model and related them to 'extension in the transport direction of gravitationally unstable culminations produced by compressive ramp climb'. Farrell (1984) suggested that a slump halting at the up-slope margin would first cause extensional strain waves to propagate down-slope through slumps and make extensional structures overprint earlier-formed compressional structures. Farrell (1984) further suggested that slump stoppage at the down-slope margin would first cause a compressional strain wave to propagate up-slope through the slumps and cause overprinting of compressional structures on earlier-formed extensional structures. Both these studies, and the results presented here, illustrate the importance of considering the larger-scale context of soft-sediment deformation before extensional faults are interpreted as the head zone and compressional faults as the toe zone of slides or slumps.

The general distribution of gravity-driven soft-sediment deformation features in the Gull Island Formation seems, in part, to have been controlled by grain size. Sand slides only appear on the lower delta slope in relation to the sandy turbidite 'sheets', while slumps are restricted to dominantly muddy lithologies throughout the formation. It is nevertheless difficult to explain why mud slides and mud slumps coexist on the upper delta slope, and also why mud-sliding is uncommon on the lower delta slope. This may be related to the moving distance of the deformed sediment mass and therefore the time interval for strain imposition. It is probable that the longer the mass movement is, the greater will be the rate of internal shear deformation, all other factors being equal. In addition, it can be speculated that the degree of consolidation of the sediment at the time of failure may have influenced the rate of internal contortion of the moving sediment mass. However, as implied by Gawthorpe & Clemmey (1985), early consolidation will probably have a larger effect in carbonates than in clastics.

Water-escape structures formed most frequently near, or at, the sediment–water interface. The preferential occurrence of water-escape structures on the lower delta slope is probably a result of the occurrence of turbidite deposition at these sites. Rapid deposition of sand on mud may have induced density instabilities that led to relatively rapid pore-water escape (Lowe 1975). Nevertheless, the existence of, for example, collapse depressions on the muddy upper delta slope suggests that other factors were also important. Loading by slumps and slides moving down-slope and deposition of new sediment, even from suspension, on top of incompletely de-watered deformed units may have caused rapid pore-water escape.

ACKNOWLEDGMENTS: This work forms part of a Cand. Scient. thesis completed at the University of Bergen, Norway. The author would like to thank the two official referees, Dr Alex Maltman and Dr Kevin Pickering, for valuable comments leading to improvements in the manuscript. Professor J. D. Collinson supervised the project and suggested several improvements to the present manuscript. Dr Dag Nummedal also reviewed an early version of the manuscript. Norsk Hydro provided financial support for the field work in Ireland.

References

BRUCE, C. H. 1973. Pressured shale and related sediment deformation: mechanism for development of regional contemporaneous faults. *American Association of Petroleum Geologists Bulletin* **57**, 878–886.

CARVER, R. E. 1968. Differential compaction as a cause of regional contemporaneous faults. *American Association of Petroleum Geologists Bulletin* **52**, 414–419.

COLEMAN, J. M. & GARRISON, L. E. 1977. Geological aspects of marine slope stability, northwestern Gulf of Mexico. *Marine Geotechnology* **2**, 9–44.

——, PRIOR, D. B. & LINDSAY, J. F. 1983. Deltaic influences on shelfedge instability processes. *In:* STANLEY, D. J. & MOORE, G. T. (eds) *The Shelfbreak: Critical Interface on Continental Margins.* Special Publication of the Society of Economic Paleontologists and Minerologists **33**, 121–138.

——, SUHAYDA, J. N., WHELAN, T. & WRIGHT, L. D. 1974. Mass movement of Mississippi River delta sediments. *Transactions of the Gulf Coast Association of Geological Societies* **24**, 49–68.

COLLER, D. W. 1984. Variscan structures in the Upper Palaeozoic rocks of west central Ireland. *In:* HUTTON, D. H. W. & SANDERSON, D. J. (eds) *Variscan Tectonics of the North Atlantic Region.* Geological Society of London Special Publication **14**, 185–194.

CRANS, W., MANDL, G. & HAREMBOURE, J. 1980. On the theory of growth-faulting—a geomechanical delta model based on gravity sliding. *Journal of Petroleum Geology* **2**, 265–307.

DINGLE, R. V. 1977. The anatomy of a large submarine slump on a sheared continental margin (SE Africa). *Journal of the Geological Society, London* **134**, 293–310.

EDWARDS, M. B. 1976. Growth faults in Upper Triassic deltaic sediments, Svalbard. *American Association of Petroleum Geologists Bulletin* **60**, 341–355.

ELLIOTT, T. & LAPIDO, K. O. 1981. Syn-sedimentary gravity slides (growth faults) in the Coal Measures of South Wales. *Nature* **291**, 220–222.

EVAMY, B. D., HAREMBOURE, J., KAMERLING, P., KNAAP, W. A., MOLLOY, F. A. & ROWLANDS, P. H. 1978. Hydrocarbon habitat of the Tertiary Niger delta. *American Association of Petroleum Geologists Bulletin* **62**, 1–39.

FARRELL, S. G. 1984. A dislocation model applied to slump structures, Ainsa Basin, South Central Pyrenees. *Journal of Structural Geology* **6**, 727–736.

GAWTHORPE, R. L. 1986. Sedimentation during carbonate ramp to slope evolution in a tectonically active area: Bowland Basin (Dinantian), northern England. *Sedimentology* **33**, 185–206.

—— 1987. Tectono-sedimentary evolution of the Bowland Basin, N England, during the Dinantian. *Journal of the Geological Society, London* **144**, 59–71.

—— & CLEMMEY, H. 1985. Geometry of submarine slides in the Bowland Basin (Dinantian) and their relation to debris flows. *Journal of the Geological Society, London* **142**, 555–565.

GILL, W. D. 1979. Syndepositional sliding and slumping in the West Clare Namurian Basin, Ireland. *Geological Survey of Ireland Special Paper* **4**, 31 pp.

—— & KUENEN, P. H. 1958. Sand volcanoes on slumps in the Carboniferous of County Clare, Ireland. *Quarterly Journal of the Geological Society, London* **113**, 441–460.

HILL, P. R. 1984. Sedimentary facies of the Nova Scotian upper and middle continental slope, offshore eastern Canada. *Sedimentology* **31**, 293–309.

KOSTERS, E. C. 1985. Deeply buried Tertiary sand bodies in the northern Gulf of Mexico: examples from the Lower Hackberry (Oligocene) and Houma Embayments (Miocene). *Transactions of the Gulf Coast Association of Geological Societies* **35**, 161–170.

LEWIS, K. B. 1971. Slumping on continental slope inclined at 1°–4°. *Sedimentology* **16**, 97–110.

LIPPOTT, H. J., HESS, J. C. & BURGER, K. 1984. Isotopische alter pyroklastischen Sanidinen aus Kaolin–Kohlentonsteinen als Korrelationsmarken für das mitteleuropeische Oberkarbon. *Fortschritte in der Geologie von Rheinland und Westfalen* **32**, 119–150.

LOWE, D. R. 1975. Water escape structures in coarse-grained sediments. *Sedimentology* **22**, 157–204.

MALTMAN, A. J. 1987. Microstructures in deformed sediments, Denbigh Moors, North Wales. *Geological Journal* **22**, 87–94.

MARTINSEN, O. J. 1987. *Sedimentology and Syndepositional Deformation of the Gull Island Formation (Namurian R_1), Western Irish Namurian Basin, Ireland—With Notes on the Basin Evolution.* Unpublished Cand. Scient. thesis, University of Bergen, 327 pp.

MORGAN, J. P., COLEMAN, J. & GAGLIANO, S. M. 1968. Mudlumps: diapiric structures in Mississippian delta sediments. *In: Diapirism and Diapirs.* American Association of Petroleum Geologists Memoir **8**, 145–161.

MORTON, W. H. 1965. The Carboniferous stratigraphy of the area north-west of Newmarket, County Cork, Ireland. *Scientific Proceedings of the Royal Dublin Society* **2A**, 47–64.

NEMEC, W., STEEL, R. J., GJELBERG, J., COLLINSON, J. D., PRESTHOLM, E. & ØXNEVAD, I. E. 1988. Anatomy of a collapsed and re-established delta front in the Lower Cretaceous of eastern Spitsbergen: gravitational sliding and sedimentation processes. *American Association of Petroleum Geologists Bulletin* **72**, 454–476.

PICKERING, K. T. 1983. Small scale syn-sedimentary faults in the Upper Jurassic 'Boulder Beds'. *Scottish Journal of Geology* **19**, 169–181.

PRIOR, D. B. & COLEMAN, J. M. 1978. Submarine landslides on the Mississippi delta-front slope. *Louisiana State University School of Geosciences, Geoscience and Man* **19**, 41–53.

——, BORNHOLD, B. D. & JOHNS, M. W. 1984. Depositional characteristics of a submarine debris flow. *Journal of Geology* **92**, 707–727.

RAMSBOTTOM, W. H. C., HIGGINS, A. C. & OWENS, B. 1979. Palaeontological characterization of the Namurian of the stratotype area (a report of the Namurian Working Group). *8th International Congress on Carboniferous Stratigraphy and Geology, Moscow*, Vol. 3, 85–99.

RIDER, M. H. 1974. The Namurian of West County Clare. *Proceedings of the Royal Irish Academy* **74B**, 125–142.

—— 1978. Growth faults in Carboniferous of Western Ireland. *American Association of Petroleum Geologists Bulletin* **62**, 2191–2213.

SANDERSON, D. J. 1984. Structural variation across the northern margin of the Variscides in NW Europe. *In:* HUTTON, D. H. W. & SANDERSON, D. J. (eds) *Variscan Tectonics of the North Atlantic Region.*

Geological Society of London Special Publication **14**, 149–166.

SCHULTZ, R. W. & SEVASTOPULO, G. D. 1965. Lower Carboniferous volcanic rocks near Tulla, County Clare, Ireland. *Scientific Proceedings of the Royal Dublin Society* **2A**, 153–162.

SCHWARZACHER, W. 1979. *Sedimentation Models and Quantitative Stratigraphy.* Developments in Sedimentology **19**. Elsevier, Amsterdam.

SEVASTOPULO, G. D. 1981. Lower Carboniferous. *In:* HOLLAND, C. H. (ed.) *A Geology of Ireland.* Scottish Academic Press, Edinburgh, 147–172.

STOW, D. A. V. 1986. Deep clastic seas. *In:* READING, H. G. (ed.) *Sedimentary Environments and Facies.* Blackwell Scientific Publications, Oxford, 399–444.

WEBER, K. J. & DAUKORU, E. 1975. Petroleum geology of the Niger delta. *In: Proceedings of the 9th World Petroleum Conference*, 209–221.

O. J. MARTINSEN, Geologisk Institutt Avd. A, University of Bergen, Allégaten 41, 5007 Bergen, Norway.

Controls on internal structure and architecture of sandstone bodies within Upper Carboniferous fluvial-dominated deltas, County Clare, western Ireland

A. J. Pulham

SUMMARY: Deposition of Upper Carboniferous (Namurian R1–R2) deltas in the Clare Basin, western Ireland, was influenced by a combination of processes: fluvial, basinal (notably waves), synsedimentary deformation, and rapid subsidence rates within the receiving basin. Fluvial processes predominate; consequently the major deltaic sandstone bodies are fluvial channels and fluvial-dominated mouth bars. The internal structure of single channel sandstones reflects high energy tractive processes and aggradation of sands in relatively straight channel reaches. The structure of mouth-bar sandstone bodies reflects a variety of frictional and buoyant river-mouth processes. Depositional processes within the Clare Basin are often overprinted by syndepositional deformation. Delta-front sequences exhibit abundant evidence of synsedimentary faulting, slumping, clay diapirism and gravity sliding. Deformed packets of sediment are up to 60 m thick and over half of all delta-front sequences are disturbed. These processes led to a modification of mouth-bar structure and geometry. Some fault-bounded delta-front sandstone bodies fill fault-induced depocentres. Large, multistorey and laterally composite channel sandstones in the S of the basin may reflect major deformation across the delta plain, related to rapid subsidence and compaction rates. In contrast, the architecture and structure of delta-plain sequences in the N of the basin are consistent with a more stable fluvial-dominated setting.

Process-based classifications are widely used to describe the variability observed in modern and ancient deltas. The most popular scheme is the ternary diagram proposed by Galloway (1975) in which deltas are qualitatively plotted according to the relative balance between fluvial, wave and tide energy, particularly in the delta-front setting. This scheme and its variants have been adopted in many important texts (Coleman 1980; Miall 1984; Elliott 1986). These classifications have provided a foundation for evaluating newly discovered and previously documented deltas over the past decade or more.

The continuing documentation of deltas in terms of depositional processes is crucial to further progress in delta research. However, the focus of attention on fluvial, wave and, to a lesser extent, tidal processes has led to a perception that the prediction of variability within deltaic deposits and particularly the structure and geometry of sand bodies can be made by documenting depositional processes alone. The classification schemes based on depositional processes have neglected other important processes within deltas and have led to an oversimplification of observed variability. Indeed, many important deltas and deltaic deposits are sufficiently well documented and different from any other modern or known ancient deltas that they should be described as delta types in their own right (Elliott, this volume).

One important process that is well known in modern deltas and is documented in many ancient deltaic successions is penecontemporaneous deformation. For instance, the delta-front environment of the modern Mississippi Delta comprises a variety of shallow- and deep-seated deformation processes that are widespread and proceed at rapid rates (Coleman 1980). Despite this obviously important modifying process, the Mississippi Delta and most other modern deltas are classified according to their depositional processes alone. Little or no concession to the deformation processes in deltas is made in the modelling of sequences or the detail of basin fills.

Along the W coast of Ireland the Upper Carboniferous deltaic deposits of the Clare Basin are exposed in almost continuous coastal outcrop and provide an opportunity of documenting depositional and deformational processes and the relatively large-scale architecture of several stacked deltas. The observed geometry of sandstone bodies in parts of the Clare Basin is not that which might be predicted from an analysis of depositional processes alone.

Location and stratigraphy of the Clare Basin

Clastic sediments, up to 1600 m thick, of Namurian (E1–R2) age occur across a wide area of SW and central County Clare, western Ireland, and

From WHATELEY, M. K. G. & PICKERING, K. T. (eds), 1989, *Deltas: Sites and Traps for Fossil Fuels,* Geological Society Special Publication No. 41, pp. 179–203.

FIG. 1. Geological map and important locality names of SW County Clare, western Ireland.

overlie Visean (P1–P2) shelf limestones (Fig. 1). The stratigraphy of the Namurian sediments is based on important goniatite-bearing fossil bands (Fig. 2).

In the N of the basin, the basal unit of the Namurian comprises the Clare Shale Member, a condensed succession of anoxic marine mudstones approximately 12 m thick (Clarke 1966). The upper parts of the Clare Shale Member are a lateral equivalent of the Ross Sandstone Formation that occurs in the central and SW parts of the basin. These sandstone-dominated facies are at least 300 m thick and are interpreted as the deposits of turbidity currents derived from the W (Rider 1969, 1974). Overlying the Clare Shale Member in the N and the Ross Sandstone Formation in the S of the basin is the Gull Island Formation. This formation comprises a siltstone-dominated succession with abundant synsedi-

mentary deformation structures and packets of sandy turbidites. The formation is up to 395 m thick in the S of the Clare Basin but thins to 140 m in the N. It is interpreted as representing the progradation of an easterly facing deep-water slope (Rider 1969, 1974; Martinsen, this volume).

Above the slope deposit of the Gull Island Formation is a transition into the deltaic cyclothems of the Central Clare Group. The cyclothems range up to 200 m in thickness but are typically 100–150 m. The upper limit of each cyclothem is marked by a fossil band (Fig. 2). The fossil bands in the Central Clare Group accumulated during predominantly non-depositional transgressions that occurred between abandonments of the major delta complexes.

The lowest three cyclothems (Tullig, Kilkee and Doonlicky) have been examined in detail and form the basis for the following discussions.

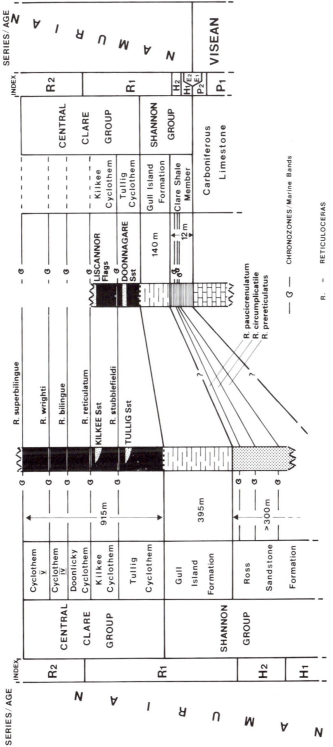

FIG. 2. Stratigraphy of the Namurian Clare Basin (modified after Rider 1974).

Deltaic sequences and depositional processes

Three principal depositional sequence types are recognized in the Central Clare Group (Fig. 3): (i) large-scale coarsening-upward sequences (26–120 m thick) that represent delta-front progradation; (ii) small-scale coarsening-upward sequences (2.5–14 m thick) that mostly represent deposition in interdistributary bays; (iii) erosively based sandstone-dominated channel sequences (up to 70 m thick).

The sandstones in all these sequence types are typically fine-grained, moderately to well sorted, lithic to sublithic arenites. Approximately 20%–25% of the Central Clare Group is sandstone.

Delta-front sequences

The thickest of the delta-front sequences represent progradations into relatively deep water (50–150 m). These sequences typically form the major components of each cyclothem. Their lower to middle parts display limited lateral variation and consist of sideritic mudstones that coarsen upwards into laminated siltstones. These fine-grained facies represent the deposition of clays and silts from suspension in prodeltaic and distal mouth-bar settings. The upper parts of the delta-front sequences vary considerably in structure from sandstone dominated to siltstone and mudstone dominated (Figs 4 and 5). This lateral complexity in upper delta-front structure is interpreted as reflecting progradations in contrasting axial and lateral river-mouth-bar settings (Fig. 6). Completely preserved delta-front sequences are characterized by a bioturbated upper surface commonly exhibiting *Zoophycos* and other marine ichnofauna. Rootlet facies are not observed.

The structures observed in the axial sandy mouth-bar settings indicate a mixture of fluvial and wave processes with fluvial currents dominant. A relatively rapid upward transition is commonly observed at the bases of mouth-bar

FIG. 3. Principal sequence types of the Central Clare Group, Tullig Cyclothem, Trusklieve. The lower part of the cliff is formed by the uppermost part of a major delta-front coarsening-upward sequence culminating in axial river-mouth-bar sandstones. These mouth-bar deposits are capped by a 2 m thick sandstone that was produced by basin-wave reworking during local abandonment of this part of the Tullig delta front. Above the mouth bar are at least three delta-plain coarsening-upward sequences, the lowest of which is 6 m thick and represents the progradation and aggradation of a subaqueous levee. The top of the cliff exposes the basal parts of a major erosively based multistorey channel that is up to 35 m thick in this area.

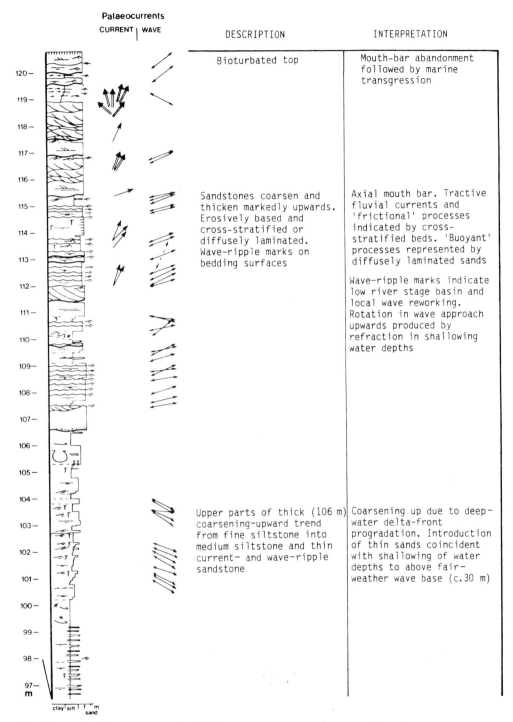

FIG. 4. Example of the upper parts of a delta-front sequence representing progradation in an axial fluvial-dominated mouth-bar setting, Tullig Cyclothem, Killard. Key overleaf.

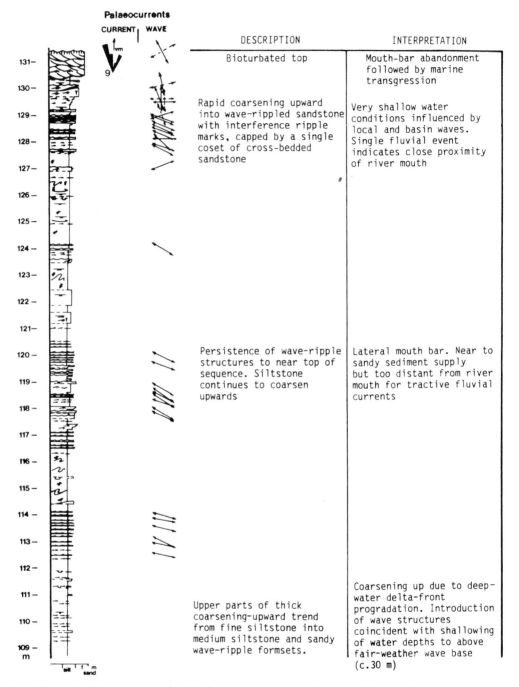

FIG. 5. Example of the upper parts of a delta-front sequence representing progradation in a lateral wave-influenced mouth-bar setting, Tullig Cyclothem, Killard.

KEY TO LOGS AND SECTIONS

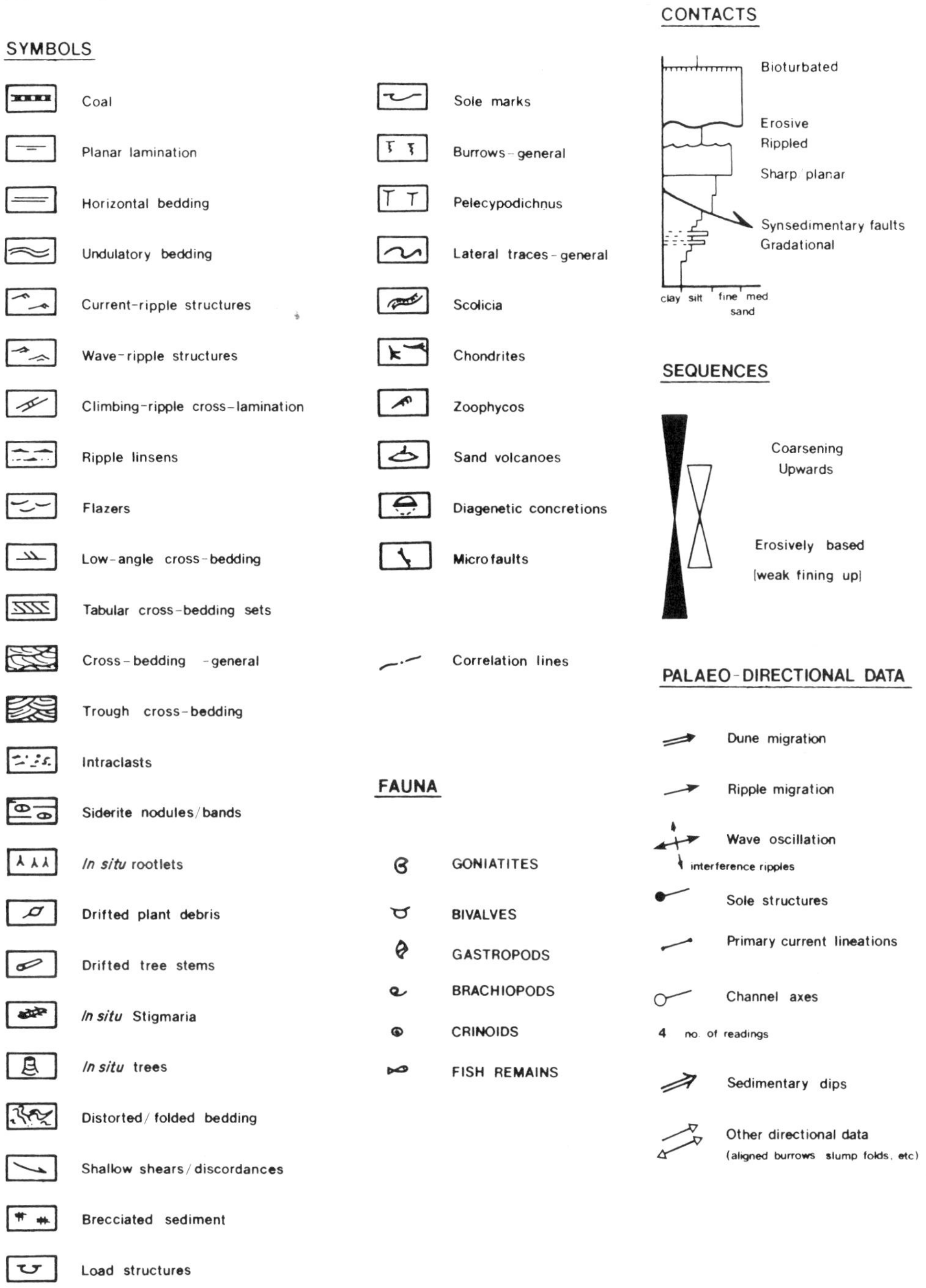

SYMBOLS

- Coal
- Planar lamination
- Horizontal bedding
- Undulatory bedding
- Current-ripple structures
- Wave-ripple structures
- Climbing-ripple cross-lamination
- Ripple linsens
- Flazers
- Low-angle cross-bedding
- Tabular cross-bedding sets
- Cross-bedding – general
- Trough cross-bedding
- Intraclasts
- Siderite nodules/bands
- *In situ* rootlets
- Drifted plant debris
- Drifted tree stems
- *In situ* Stigmaria
- *In situ* trees
- Distorted/folded bedding
- Shallow shears/discordances
- Brecciated sediment
- Load structures

- Sole marks
- Burrows – general
- Pelecypodichnus
- Lateral traces – general
- Scolicia
- Chondrites
- Zoophycos
- Sand volcanoes
- Diagenetic concretions
- Micro faults

- Correlation lines

FAUNA

- GONIATITES
- BIVALVES
- GASTROPODS
- BRACHIOPODS
- CRINOIDS
- FISH REMAINS

CONTACTS

- Bioturbated
- Erosive
- Rippled
- Sharp planar
- Synsedimentary faults
- Gradational

clay silt fine med sand

SEQUENCES

- Coarsening Upwards
- Erosively based [weak fining up]

PALAEO-DIRECTIONAL DATA

- Dune migration
- Ripple migration
- Wave oscillation interference ripples
- Sole structures
- Primary current lineations
- Channel axes
- 4   no. of readings
- Sedimentary dips
- Other directional data (aligned burrows slump folds. etc)

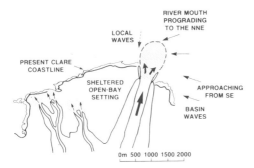

FIG. 6. Palaeogeography of a Tullig Cyclothem mouth bar that prograded in the Killard area.

sandstone bodies from siltstones with wave-ripple formsets into thickening-upward sandstone beds. The sandstones have two dominant stratification styles. In many cases beds are erosively or sharply based and comprise sets or cosets of unidirectional cross-stratification (Fig. 7a). These deposits are interpreted as the action of basinward-directed tractive fluvial currents and clearly represent frictional river-mouth processes. However, transitionally based sandstone beds comprising diffuse graded laminae and abundant *Scolicia* trace fossils are commonly observed and occasionally dominate the upper parts of some sequences (Fig. 7b). These sands are interpreted as the deposits from suspension, and indicate that buoyant river-mouth processes were also important in the proximal river-mouth bars of the Clare Basin.

Wave-ripple marks are observed on most sandstone bedding surfaces in the axial parts of the mouth-bar sequences (Fig. 7c). These indicate that, during the low river stage, basin waves were able to rework the upper parts of the fluvially deposited sandstones. In many instances the orientations of the wave-ripple marks rotate vertically through the mouth-bar sequence, suggesting an increasing degree of basin-wave refraction in the shallow water depths developed during basinward progradation of the mouth bars (Fig. 4a). In addition, a single mouth-bar sandstone body in the Tullig Cyclothem displays a unit of wave-reworked sandstone at its top (Fig. 3). This structure at the mouth-bar top indicates that in the absence of fluvial processes abandoned river mouths were liable to significant basin-wave reworking prior to the re-establishment of sediment supply.

In lateral river-mouth settings, the delta-front sequences show limited evidence of tractive fluvial currents. Sandstone beds are thin and infrequent and comprise wave-produced structures. Typically, wave-ripple formsets and micro-

linsens appear some 50 m from the tops of these sequences in the delta-front siltstones (Fig. 5). The position and context of these structures suggest a basin-wave base in excess of 50 m. Upwards, the wave-ripple linsens become more abundant and interconnected, and eventually thin sandstone beds of bundled wave-ripple formsets are observed (Fig. 7d). This gradual upward development in wave structures is interpreted as a shallowing of water depths in a predominantly wave-influenced setting.

The uppermost parts of the lateral river-mouth-bar progradations occasionally comprise some erosively based cross-stratified sandstones that are the products of unidirectional fluvial currents (Fig. 5). In some instances, these represent fluvial processes on the margins of the major river mouths. However, reconstructions of delta-front palaeogeographies suggest that they more commonly represent the onset of overbank, crevasse-splay or minor-mouth-bar processes during establishment of open-bay environments between the basinward-migrating major river mouths (Fig. 6).

Interdistributary bay sequences

Small-scale coarsening-upward sequences are observed in the upper parts of the Clare Basin cyclothems and are typically associated with channel deposits. These sequences vary considerably in structure but most comprise an upward transition from laminated mudstone to laminated and cross-laminated siltstone and cross-laminated sandstone. Unidirectional current-generated structures are dominant in most sequences, but wave-ripple structures are important in some instances (Fig. 8). The upper surfaces of these sequences are typically marked by bioturbated surfaces, rootlet facies and/or very thin coals.

The variability in these sequences can be readily classified according to the scheme proposed by Elliott (1974). They represent the fills of shallow-water bays produced by overbank flooding, crevasse and minor-mouth-bar progradations. The dominance of unidirectional current processes indicates that the interdistributary bays in the Clare Basin were fluvially dominated. However, open-bay settings that were exposed to the basin were significantly influenced by basin waves, resulting in the development of bay-mouth sequences (Fig. 8b).

The occurrence of very thin coals above only some of the bay sequences and the significant influence of wave processes indicate that only lower delta-plain bays are present in the Central Clare Group. This interpretation is supported by commonly observed marine bioturbation in the

FIG. 7. (*a*) Unidirectional cross-bedded sandstones from an axial mouth-bar sequence, Tullig Cyclothem, near Killard (scale is 15 cm long); (*b*) diffusely laminated sandstones from an axial mouth-bar sequence, Tullig Cyclothem, near Killard; (*c*) wave-ripple marks on the top surfaces of sandstone beds in an axial and proximal river-mouth setting, Tullig Cyclothem, near Trusklieve (the wavelength of ripple marks is *c.* 7 cm); (*d*) bundled wave-ripple linsens from the upper part of a lateral delta-front progradation, Kilkee Cyclothem, near Tullig (the direction of wave approach is from right to left).

188   *A. J. Pulham*

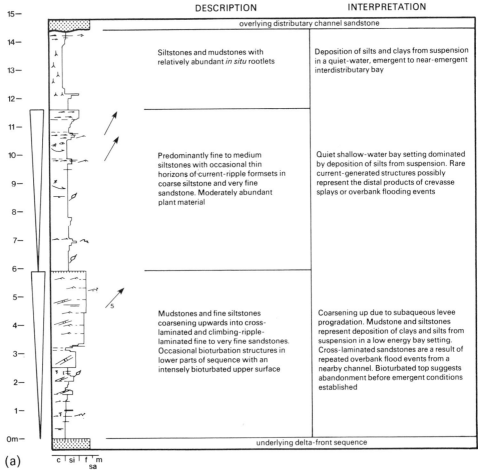

FIG. 8(*a*). (*Caption opposite*).

tops of other delta-plain sequence types, including channels, and suggests that relatively high subsidence rates persisted throughout the evolution of the Clare Basin deltas.

Channel sequences

The sandstone-dominated channel sequences of the Clare Basin are dominated by high energy unidirectional current processes. Planar lamination with primary current lineation on bedding surfaces and decimetre- to metre-scale cross-bedding are the dominant structures. There is no evidence of reversing flows that might indicate tidal processes, and therefore the channels are all considered fluvial in origin.

The major channel sequences are typically multistorey comprising up to six separate erosively based vertically stacked channel fills (Fig.

9). The erosive bases of the channels are typically irregular across palaeoflow, but commonly dip gently downcurrent parallel to palaeoflow. Local erosive relief on the bases of individual storeys can be up to several metres.

The internal structures of the major channel sequences are dominated by bounding surfaces that mark cosets of cross-bedding, and they are flat to gently inclined, suggesting predominantly vertical accretion within the channel fills. In only two examples are there significantly dipping coset boundaries that might be attributed to lateral accretion of facies owing to lateral channel migration. In the upper parts of sand bodies, within multistorey sequences, large dune bedforms draped by intrachannel mudstone and siltstone are commonly observed. The preservation of these structures is interpreted as indicating that channels altered their position on the delta

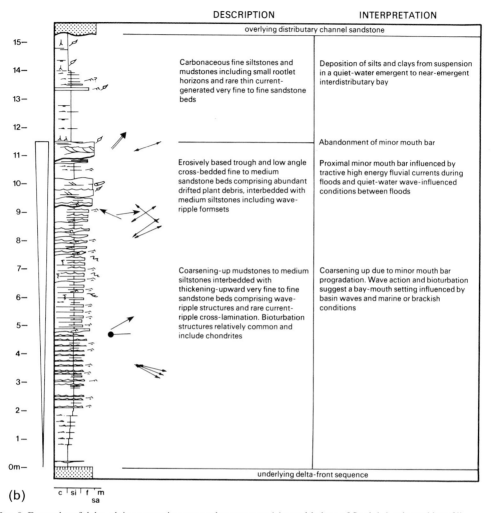

FIG. 8. Examples of delta-plain coarsening-upward sequences: (*a*) graphic logs of fluvial-dominated bay-fill sequences (these are also illustrated in Fig. 3); (*b*) detailed graphic log of wave-influenced bay-mouth sequence, Tullig Cyclothem, Killard.

plain by rapid avulsion rather than by gradual migration.

Despite the frequent observation of intra-channel fine-grained material in many multistorey sequences, the lateral persistence and proportion of these facies are both low. The sandstone content of channel sequences is greater than 95% and lenses of siltstone and mudstone rarely extend laterally for more than a few tens of metres.

The uppermost parts of channel sequences usually exhibit several decimetre-scale cosets of current-generated cross-lamination. This decline in energy marks the final abandonment of channel

processes within delta-plain successions of the Clare Basin but is not normally associated with any significant fining-upward grain-size trend. The tops of channel sequences are often intensely bioturbated and include *Zoophycos*. Rootlet facies and thin coals are also observed, however.

Summary of depositional processes

The structure of the various sequence types in the Central Clare Group of the Namurian Clare Basin leads to a number of conclusions about the important depositional deltaic processes.

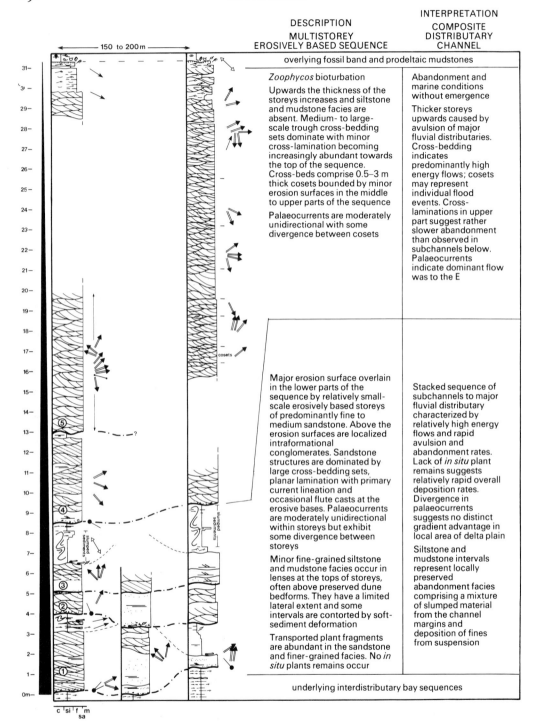

FIG. 9. Example of a detailed graphic log through a major multistorey channel sandstone sequence, Tullig Cyclothem, Trusklieve.

(1) The major deep-water delta fronts were characterized by isolated basinward-prograding river mouths. These were fluvially dominated but wave influenced. There are no structures observed that indicate the action of significant tidal processes.

(2) Axial river-mouth processes were mixed. Both frictional and buoyant processes were important in the sandy proximal river-mouth deposits of the Clare Basin. The evidence for buoyant expansion of river discharge suggests that delta-front waters were relatively saline even during high river stage conditions. This assumption is partly supported by the observed ichnofacies in the mouth bars (*Scolicia*).

(3) The refraction of basin waves observed in many delta-front sequences indicates protruding river-mouth bars and an irregular indented deltaic shoreline.

(4) Lateral river-mouth areas were typically non-sandy open-bay settings influenced predominantly by basin-wave processes.

(5) The overall dominance of mudstone and siltstone facies in the delta-front sequences indicates that the river discharge into the Clare Basin was dominated by a fine-grained suspended sediment load.

(6) Subsidence rates in the Clare Basin were fast (*c.* 0.18–0.29 mm a^{-1}), possibly partly as a consequence of the fine-grained sediment load. Fast subsidence contributed to the development of only lower delta-plain environments during maximum progradation of the deltas. These subaerial settings in the Clare Basin were fluvially dominated but were significantly wave influenced around their fringes. Fluvial channels were probably low sinuosity and bifurcating, and migrated by avulsion.

A reasonable comparison between the Clare Basin deltas and modern fluvial-dominated deltas can be based on this analysis (Fig. 10). The apparent lack of tidal processes and the significance of wave processes suggests that the deltas could be plotted on the classic ternary classification scheme along the upper part of the fluvial–wave axis (Fig. 11). This suggests that the Clare Basin deltas may bear some comparison with modern deltas such as those of the Danube or Mississippi. In terms of reservoir analysis, the combination of depositional processes might therefore suggest that bar-finger or shoestring sandstone bodies oriented normal to the shoreline will be the most prominent features (Fig. 11). An analysis of these Clare Basin deltas in terms of fluvial and wave processes alone would ignore a

further process that is observed in many of the major delta-front sequences, *ie* synsedimentary deformation.

Deformational processes

Relatively large-scale deformation structures that can be attributed to penecontemporaneous processes with sedimentation occur throughout the Namurian of the Clare Basin and have been described in detail (Rider 1969, 1978; Gill 1979; Martinsen, this volume). Within the Central Clare Group, large-scale synsedimentary deformation is confined to delta-front sequences. Previously described structures, and many that have not been documented, have been examined during a more recent study (Pulham 1987).

The principal structures recognized in the delta-front sequences are fault-bounded slides, slumps and mudstone diapirs (Fig. 12). Packets of delta-front sediments involved in these deformation features are up to 60 m thick. Over half of all the delta-front sequences examined in detail in the Clare Basin are disturbed by one or more of these structures. For instance, deformation processes are ubiquitous within the major deep-water delta-front progradation of the Doonlicky Cyclothem.

The causes of such deformation, particularly within modern and recent deltaic successions, have been discussed by many workers (Morgan 1961; Morgan *et al.* 1963; Dailly 1976; Coleman 1980; Coleman *et al.* 1980; Crans *et al.* 1980; Prior & Coleman 1982). Chief amongst most lists of synsedimentary deformation processes are shallow- and deep-seated instabilities generated by undercompaction and overpressuring of rapidly deposited fine-grained sediments. Sediment failures are commonly triggered by oversteepened slopes, loading by greater density material or high energy weather systems such as hurricanes. Within ancient successions such as the Clare Basin precise causes of sediment failures usually can only be broadly deduced. However, the fine-grained nature of the Namurian succession is considered perhaps the most important reason why the deltas are so intensely deformed.

The effects of the deformation on facies distribution and structure within the Clare Basin delta-front sequences are twofold and depend to a large extent on the depositional setting in which the deformation took place.

(1) In lateral mouth-bar and distal delta-front settings, relatively large-scale down-slope displacements of sediment packets are commonly observed. In some instances this

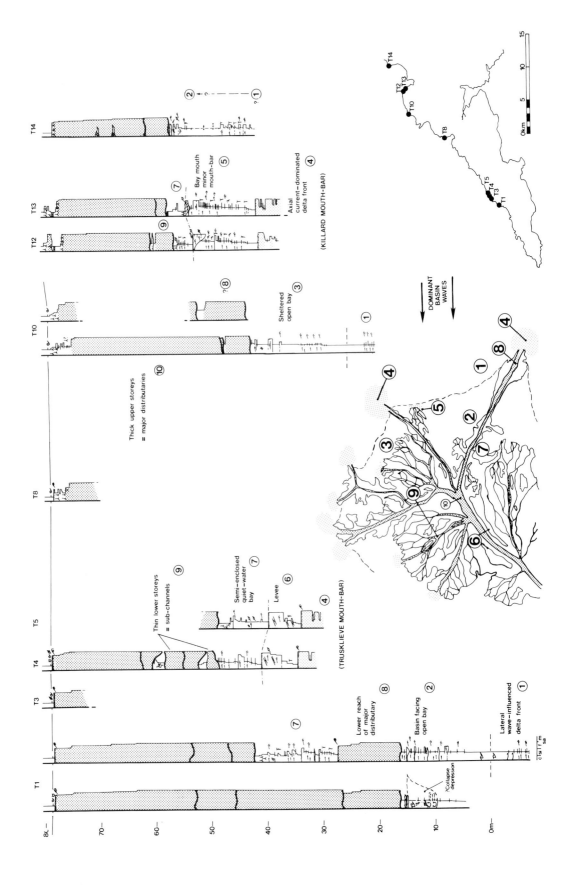

T14

T13 T12 (KILLARD MOUTH-BAR)

② ← ♀ ---- ♂ ①

⑦

⑨

Bay mouth minor mouth-bar ⑤

Axial current-dominated delta front ④

T10

⑧?

Sheltered open bay ③

①

Thick upper storeys = major distributaries ⑩

T8

T5 T4 (TRUSKLIEVE MOUTH-BAR)

⑨

Thin lower storeys = sub-channels

Semi-enclosed quiet-water bay ⑦

⑥ Levee

④

T3

T1

⑦

Lower reach of major distributary ⑧

Basin facing open bay ②

Lateral wave-influenced delta front ①

?Collapse depression

DOMINANT BASIN WAVES

④
⑧ ①
⑤ ② ⑦
③
⑨
⑩
⑥

T14
T12 T13
T10
T8
T5 T4
T3 T1

0km 5 10 15

8L 70 60 50 40 30 20 10 0m

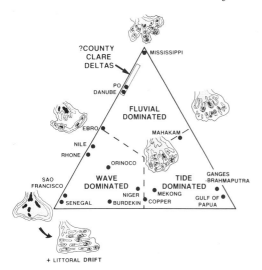

FIG. 11. Interpreted position of the Clare Basin deltas on a ternary diagram of delta types. Based purely on depositional processes derived from interpretations of vertical facies sequences. (Modified after Coleman & Wright 1975; Galloway 1975.)

process can result in the displacement of sheets of shallow-water upper delta-front facies against deeper-water lower delta-front facies. Toe structures such as synsedimentary thrust faults are recognized in these settings (Fig. 12c). Ambient sediment accumulation rates in these settings are relatively slow, and therefore 'growth' of facies across deformation structures is typically restricted to passive draping. However, thin sandy delta-front turbidites are usually observed associated with deformation in the Clare Basin and punctuate the delta-front siltstones whenever faults, slides and slumps are present.

(2) In contrast with the above environments, deformation in axial and marginal mouth-bar environments in the Clare Basin, where depositional rates are fast, commonly results in significant 'growth' structures. Growth of facies in these settings may simply be reflected in an increased thickness of mouth-bar sandstones on the flanks of mudstone diapirs (Fig. 12b) or on the down-thrown sides of extensional growth faults. In many instances discrete sandy depocentres generated by extensional faulting are also recognized (Fig.

13). These entirely fault-bounded sandstone units are commonly observed in marginal mouth-bar settings, but some do occur in lower delta-front settings of the Tullig Cyclothem associated with delta-front turbidite deposits. The thickening of successive sandstone beds into the faults indicates their origin as growth-fault fills rather than redeposited (mass flow) packets. The internal structure of the growth-fault fills may comprise unidirectional cross-stratification directed down-slope and normal to the bounding fault. In one example lateral accretion surfaces are observed within a growth-fault fill, possibly reflecting a tilting of the sediment surface during deformation.

The principal effects of deformation processes in the Clare Basin that can be directly observed show a redistribution of delta-front facies downslope, a modification of the geometry of sandy mouth bars and the creation of fault-bounded depocentres. These effects are significant enough in the delta fronts to encourage a modelling of vertical facies sequences and sandstone-body geometry that accounts for deformation. A purely depositional model similar to those proposed for many modern deltas would clearly fail to predict many of the features of the overall delta-front structure in the Clare Basin.

The distribution of sandstone in the Central Clare Group is largely concentrated in the delta plains rather than in mouth bars of the delta front. An understanding of controls on sandstone-body geometry in all environments is important for modelling the overall architecture of the Clare Basin deltaic succession.

Before the lateral and vertical distribution of sandstone bodies in the Central Clare Group is described the palaeogeography of the deltas is briefly discussed. This is necessary in order that cross-sections through the sequences are oriented relative to the progradation direction and shoreline trends.

Palaeogeography of the Central Clare Group

The palaeogeography of the deltas in the Central Clare Group is derived from a summation of all

FIG. 10. Subenvironments of the Tullig Cyclothem in the S of the Clare Basin. Model based on the modern Mississippi River Delta and drawn to the same scale as the location map. Section numbers (T1–T14) refer to principal Tullig Cyclothem exposures (*opposite*).

(a)   (b)

(c)

FIG. 12. (*a*) Large-scale synsedimentary deformation structures in the mid-parts of a delta-front sequence, Doonlicky Cyclothem, Foohagh Point. To the right is a listric extensional fault in well-bedded siltstone facies with relatively minor dip-slip displacement. The fault scarp has been filled and the footwall draped by more homogeneous siltstones. To the left is a similar structure in which the hanging wall has been displaced completely out of section. The resultant 'gully' feature has also been filled with homogeneous siltstone facies. The lenticular sandstone beds are products of delta-front density currents and commonly occur in association with deformation structures in the Central Clare Group. Cliffs are *c*. 50 m high. (*b*) Mudstone diapir in an axial mouth-bar sequence, Tullig Cyclothem, near Kilkee. Sandstones deposited by tractive fluvial currents are thickest on the flanks of the diapir. Cliffs are *c*. 30 m high. (*c*) Large synsedimentary listric fault in the mid-parts of a delta-front sequence, Kilkee Cyclothem, near Kilkee. The dip-slip displacement on the fault declines upwards and indicates an upwards sense of movement in the hanging wall. The orientation and structure of this fault leads to the interpretation that it is an oblique lateral ramp at the down-slope end of a major gravity slide. Cliffs are *c*. 30 m high.

palaeocurrent data in channels and mouth bars and from basin wave-ripple data (Fig. 14).

In single fluvial-channel sequences, cross-bedding and primary current lineation structures indicate a broadly W–E palaeocurrent pattern. The channel data indicate some differences in the channel orientations between the Tullig and Kilkee Cyclothems. However, the Tullig Cyclothem data are dominated by channels from the N of the basin which flow towards the NE. The data from major mouth bars, particularly the frictionally dominated types, are in broad agreement with the data from the fluvial channels. They indicate river mouths prograding towards

the E. Data collected from wave ripples in the delta-front sequences indicate a consistent trend throughout the Central Clare Group. The wave approach was from the SE, possibly slightly oblique to the palaeo-shoreline of the deltas.

These data suggest that the principal sediment input to the Clare Basin during the deposition of the Central Clare Group was from the W. The Tullig, Kilkee and Doonlicky deltas prograded towards the present County Clare coastline. The basin-wave approach further suggests that the basin was open towards the SE. Therefore cross-sections compiled from the excellent coastal exposure along the western County Clare coast-

(a)

(b)

FIG. 13. (*a*) Listric growth fault in an upper delta-front setting, Doonlicky Cyclothem, Foohagh Point. The progradation direction is broadly to the left. Cliffs are 50 m high; cows for scale. (*b*) Listric growth fault in an upper delta-front setting, Tullig Cyclothem, near Trusklieve. This structure occurs on the margin of the mouth-bar sandstones illustrated in Fig. 3 and the progradation direction is to the left and into the cliff face. The sandstone fill of the growth fault is *c.* 20 m thick.

A. J. Pulham

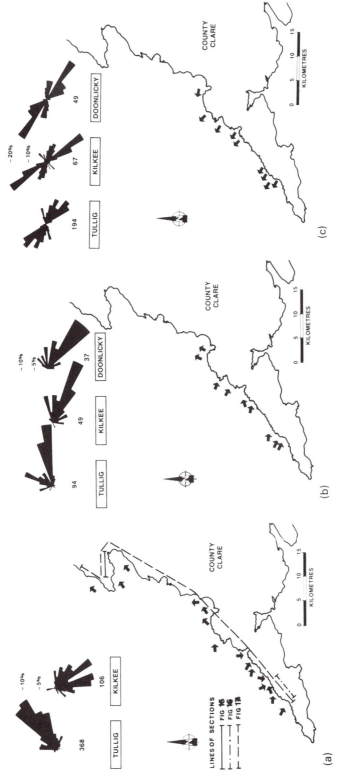

FIG. 14. Fluvial-channel, major-mouth-bar and basin-wave palaeocurrents of the Central Clare Group: (a) fluvial channel palaeocurrents derived from cross-bedding and primary current lineation structures (single arrows represent vector mean fluvial currents from localities/sequences with more than 10 readings); (b) mouth-bar palaeocurrents derived from cross-stratification structures (single arrows represent vector mean fluvial currents from localities/sequences with more than 10 readings); (c) basin-wave palaeocurrents derived from wave-ripple marks in the middle to upper parts of delta-front sequences and plotted as palaeo-oscillations (single arrows represent vector mean wave approach from localities/sequences with more than 10 readings).

line are mostly oriented across the deltas and palaeoflow directions.

Sandstone bodies: type and geometry

Variation in the abundance, scale and structure of sandstone bodies within the Central Clare Group occurs across the Clare Basin. The most significant differences occur between the N and the S of the basin.

N Clare

In N Clare, the thicknesses of the Gull Island Formation and the Tullig Cyclothem are significantly less than in the S of the basin (Fig. 2). The structure of the Tullig Cyclothem comprises a relatively thin delta-front coarsening-upward trend without major deformation structures. This is overlain by a delta-plain succession dominated by relatively thin single-storey channel sandstones (up to 15 m thick) (Fig. 15). Channel margins can readily be identified and the sandstone bodies are limited in lateral extent. The other delta-plain sequences are thin fluvial-dominated bay fills. Abandonment facies associated with these sequences comprise rootlets and thin coals.

The above features suggest that the progradation of the Tullig Cyclothem in N Clare was relatively stable. The thin nature of the delta-plain sequences and the cyclothem, together with the dominance of emergent abandonment facies, are interpreted as evidence for relatively slow subsidence rates during progradation. The isolated channel sandstones observed in sections normal to palaeoflow suggest that linear sandstone bodies are present. This would be consistent with a fluvial-dominated deltaic setting producing 'bar-finger' or shoreline-normal shoestring sandstones.

S Clare

The overall architecture of the Tullig Cyclothem changes markedly in S Clare. The thickness of the cyclothem increases southwards, and this is reflected in all sequence types from delta front to delta plain. In addition, deformation becomes an important feature in many of the delta-front sequences. The construction of cross-sections over relatively short intervals of the County Clare coastline indicates that growth-faulted sandstones are relatively unimportant and generally isolated features (Fig. 16). The mouth-bar sandstone bodies are also isolated and generally encased in delta-front siltstones. These sandstone bodies are typically a few kilometres across, suggesting river mouths and channels 0.5-1 km wide. This is approximately similar in scale to modern river-mouth settings of the Mississippi River Delta.

Despite the effects of deformation in the delta-front environment, the major mouth-bar sandstones have geometries consistent with the fluvial-dominated processes in the deltas. However, when the fluvial channel sequences in the S of the Clare Basin are correlated and mapped a very different sandstone distribution is observed.

In the Tullig and Kilkee Cyclothems the channel sequences are virtually all multistorey and reach thicknesses of up to 70 m. There are major contrasts with the channel sequences observed further N. Reconstruction of the entire cyclothems from S to N illustrates that the channel sandstone bodies in the S of the basin are laterally composite features (Fig. 17). In both the Tullig and Kilkee Cyclothems, the scale of these sandstones is in excess of 25 km wide and generally greater than 30 m thick. They form such prominent features in the cyclothems and along the SW County Clare Coast (Fig. 18) that they have been given member status: the Tullig Sandstone Member and the Kilkee Sandstone Member respectively (Rider 1969).

These large multistorey and laterally composite channel sandstones are clearly not 'bar-finger' or shoestring features. Despite the limited lateral control parallel to palaeoflow, they suggest that a sheet-like geometry, possibly elongate along the deltaic shoreline trend, is most likely. This pattern is interpreted as resulting from rapid and repeated channel avulsion over a wide area of the delta plain. The establishment of this channel behaviour apparently excluded the preservation of interdistributary bay sequences, which in the Tullig and Kilkee Cyclothems are only preserved below and lateral to the large channel sandstones (Figs 16 and 17).

Further contrasts in the structure of the Tullig Cyclothem between the N and S of the basin are also observed in the distribution of abandonment facies. Rootlet facies become increasingly rare southwards. On the top of the Tullig Sandstone Member, *Zoophycos* bioturbation is the most common abandonment style, and bioturbation, not rootlets, occurs in the uppermost parts of most bay-fill sequences. Wave processes are also most common in the delta-plain sequences in the S of the Clare Basin. These features are interpreted as indicating relatively rapid subsidence rates in the S of the Clare Basin.

The marked contrasts in channel sandstone geometry from the S to the N of the Clare Basin are not mimicked by changes in depositional

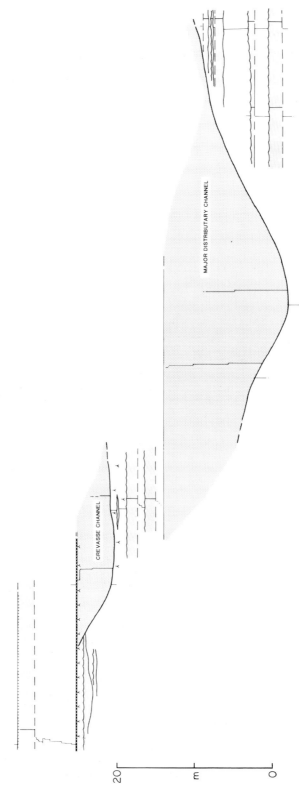

Fig. 15. Cross-section of the upper part of the Tullig Cyclothem exposed along the northern coast of Liscannor Bay. The section is oriented W–E and oblique to the principal NE palaeocurrent direction in the area. Horizontal scale is c. 5 km. See Fig. 14 for the location of the section.

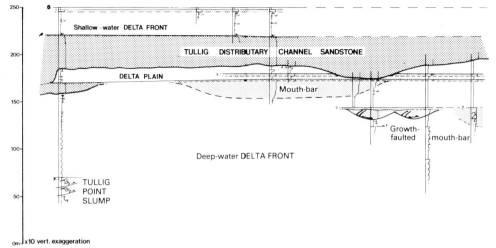

Fig. 16. Cross-section of the Tullig Cyclothem from Tullig to the Trusklieve area. The section is oriented SSW–NNE and normal to the principal E and SE palaeocurrent directions in the area. See Fig. 14 for the location of the section.

processes. Across the basin, the deltas are fluvially dominated and wave influenced. The isolated mouth-bar sandstones in the S of the basin also appear to be inconsistent with rapidly switching channels on the delta plain. The major contrasts in process that can be observed are variations in subsidence rates, reflected in thicknesses and abandonment facies, and degree of delta-front deformation.

A model of delta formation in the Clare Basin is therefore proposed which has deformation and basin subsidence as integral processes combined with the important fluvial and wave controls on deposition.

Model and discussion

Principal depositional processes in the deltas were fluvial with a significant influence of basin waves. Tidal processes were unimportant. A Clare Basin delta model must take into account the following observed features: (i) the deltas include a significant component of delta-front deformation that was active during deposition and was more important in the S of the basin than in the N; (ii) subsidence rates were fast during deposition of the deltas, and these also varied across the basin with most rapid apparent subsidence rates in the S; (iii) fluvial-channel sandstone bodies in the S of the basin are anomalously thick and laterally continuous given the depositional processes in the deltas. The scale and shape of these sandstones are apparently inconsistent with underlying and associated mouth-bar sandstones and contrast with the smaller-scale channel sandstone bodies in the N of the basin.

Faster subsidence rates and more abundant deformation in the S of the basin may be at least partly responsible for the observed differences in overall sandstone-body architecture from S to N across the basin. Synsedimentary deformation increased in importance during each successive progradation.

The observed structure of the cyclothems in the S of the basin records relatively unstable delta-front progradations. The delta fronts that prograded in this area were characterized by several widely spaced major river mouths, typical of modern fluvial-dominated deltas. The coarsening-upward trends produced by progradation of these delta fronts record sandy axial and silty lateral mouth-bar settings, indicating relatively fixed channel positions on the nearby delta plain. Some modification in the geometry of resultant delta-front sandstone bodies occurred as a result of clay diapirism, growth faulting, slumping and sliding. At the outcrop scale the deformation appears to be entirely confined to the delta-front environment.

Continued progradation eventually resulted in the establishment of lower delta-plain environments. This is best illustrated in the structure of the Tullig Cyclothem (Fig. 16), where progradation of the deltas reveals a marked change in basinal conditions here in the stratigraphy. The fluvial channel behaviour alters and rapid avulsion rates produced a widespread deposit that is

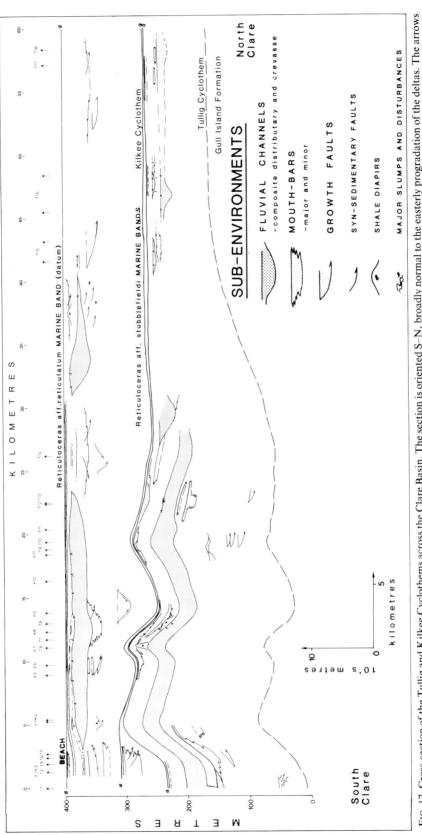

FIG. 17. Cross-section of the Tullig and Kilkee Cyclothems across the Clare Basin. The section is oriented S–N, broadly normal to the easterly progradation of the deltas. The arrows mark the location of detailed sedimentary logging. See Fig. 14 for the location of the section.

FIG. 18. View of SW County Clare looking SW from Kilkee. Atlantic Ocean to the right. The large 'whale-back' topographic features are the major distributary channel sandstone bodies outcropping on the limbs of gentle folds. Distance in view *c.* 15 km.

almost entirely channel sandstone facies. Similar processes did not occur in the N of the basin where delta-plain successions comprise interdistributary bay sequences preserved between and adjacent to single-storey channel sandstones.

The cause of the change in channel behaviour is interpreted as widespread delta collapse, possibly caused by the loading of already unstable delta-front deposits by higher-density delta-plain deposition. A result may have been an order-of-magnitude change in the scale of deformation, perhaps on large growth faults (Fig. 19). The effects of a widespread collapse across a delta plain could have significant consequences: gradient advantages would be developed into the deforming area encouraging avulsion of nearby distributaries, repeated deformation would discourage the previously fixed nature of the channels and, as an increasing proportion of sand deposition became involved, incision into previous channel sands would favour the continued stability of increased channel avulsion rates.

The apparent repetition of this sequence of processes in two successive cyclothems of the Central Clare Group (Tullig and Kilkee) argues against other controls such as hinterland changes. External controls would also be expected to be observed across the basin, not only in specific parts. In addition, the ubiquitous deformation observed in the Doonlicky Cyclothem is associated with widespread sandy deposition in the upper delta-front environment. This style of delta-front progradation may reflect the types of

delta-plain processes that occurred during deposition of the earlier Tullig and Kilkee Sandstone Members.

The link between deformation and sandstone distribution in deforming deltas has been discussed in several relatively recent publications (Edwards 1980, 1981; Winker & Edwards 1983). Many of the observations in these studies also seem true of these Clare Basin deltas, although the outcrop and seismic scales are several orders of magnitude greater. Edwards cites shoreline-parallel sandstone distribution patterns as being typical of subsurface deformed deltaic successions, despite a variety of depositional processes. The main controlling structures in Edwards's examples are synsedimentary faults that would be seen rarely, if at all, on limited strike sections. Therefore it may be that detailed overall architecture of deltaic successions will be the only way to infer higher-order deformation processes in surface studies of ancient deformed deltas such as those in the Clare Basin.

Conclusions

(1) Syndepositional deformation was an important process during progradation of fluvial-dominated deltas in the Namurian Clare Basin.

(2) Rates of deformation are interpreted as increasing during each successive progradation into the basin. In parts, this led to the

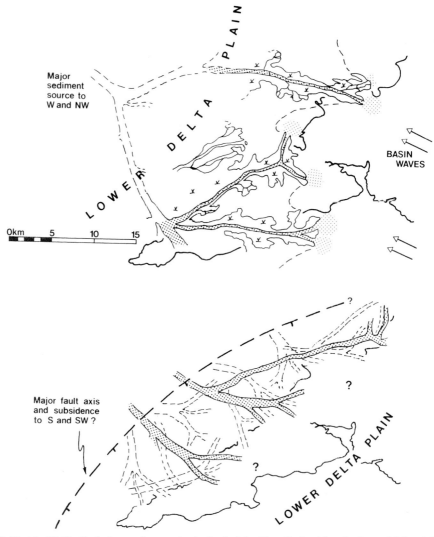

FIG. 19. Model of Tullig Cyclothem sedimentation in the S of the Clare Basin: (*a*) early phase of delta-plain progradation characterized by relatively fixed channels and isolate mouth bars along the delta front (this style of sedimentation is typical of the entire Tullig Cyclothem in the N of the Clare Basin); (*b*) later phase of delta-plain deposition after widespread collapse of the delta such that channels are now encouraged to avulse rapidly in a ?fault-controlled depocentre. Contrasts between the resultant sandstone-body geometries that are produced in each type of setting are illustrated in Fig. 17.

development of anomalously thick lateral-continuous shoreline-parallel fluvial sandstones not normally associated with fine-grained fluvial-dominated deltas.

(3) It is speculated that the presence of relatively small scale (outcrop sized) deformation structures in deltaic successions may indicate the presence of much larger deformation structures within the basin fill, with the converse also being true.

ACKNOWLEDGMENTS: I would like to thank Trevor Elliott for his valuable support during the study from which this work was derived, and other colleagues too numerous to name. This work comprises part of a 3 year postgraduate research project undertaken at University College Swansea and supported by a National Environment Research Council (NERC) Award. Thanks are also due to BP Petroleum Development for providing facilities and technical support during the writing of this paper, and to Manuel Di Lucia of Kilkee, County Clare, for providing Fig. 18.

References

CLARKE, M. J. 1966. *Orionastrea* in Ireland. *Irish Nature* **15**, 177–178.

COLEMAN, J. M. 1980. *Deltas: Processes of Deposition and Models for Exploration* (2nd edn). Burgess, Minneapolis, MN.

—— & WRIGHT, L. D. 1975. Modern river deltas: variability of processes and sand bodies. In: BROUSSARD, M. L. (ed.) *Deltas, Models for Exploration*. Houston Geological Society, Houston, TX.

——, PRIOR, D. B. & GARRISON, L. E. 1980. Subaqueous sediment instability in offshore Mississippi River delta. *Bureau of Land Management Open File Report* **80–01**, 63 pp.

CRANS, W., MANDL, G. & HAREMBOURE, T. 1980. On the theory of growth faulting: a geomechanical delta model based on gravity sliding. *Journal of Petroleum Geology* **2**, 265–307.

DAILLY, G. C. 1976. A possible mechanism relating progradation, growth-faulting, clay diapirism and overthrusting in a regressive sequence of sediments. *Canadian Society of Petroleum Geologists* **24**, 92–116.

EDWARDS, M. B. 1980. The Live Oak delta complex: an unstable shelf-edge delta in the deep Wilcox trend of South Texas. *Transactions of the Gulf Coast Association of Geological Societies* **30**, 71–79.

—— 1981. Upper Wilcox Rosita Delta System of South Texas: growth faulted shelf-edge deltas. *Bulletin of the American Association of Petroleum Geologists* **65**, 54–73.

ELLIOTT, T. 1974. Interdistributary bay sequences and their genesis. *Sedimentology* **21**, 611–622.

—— 1986. Deltas. In: READING, H. G. (ed.) *Sedimentary Environments and Facies* (2nd edn). Blackwell Scientific Publications, Oxford, 113–154.

GALLOWAY, W. E. 1975. Process framework for describing the morphologic and stratigraphic evolution of deltaic depositional systems. In: BROUSSARD, M. L. (ed.) *Deltas, Models for Exploration*. Houston Geological Society, Houston, TX.

GILL, W. D. 1979. Syn-depositional sliding and slumping in the West Clare Namurian Basin, Ireland. *Geological Survey of Ireland Special Paper* **4**.

MIALL, A. D. 1984. Deltas. In: WALKER, R. G. (ed.) *Facies Models* (2nd edn). Geoscience Canada, Reprint Series 1, 105–118.

MORGAN, J. P. 1961. Mudlumps at the mouth of the Mississippi river. In: *Genesis and Paleontology of the Mississippi River Mudlumps*. Bulletin of the Louisiana Department of Conservation Geology **35** (1), 116 pp.

——, COLEMAN, J. M. & GAGLIANO, S. M. 1963. *Mudlumps at the Mouth of South Pass, Mississippi River; Sedimentology, Paleontology, Structure, Origin and Relation to Deltaic Processes*. Louisiana State University Press, Baton Rouge, LA, 50 pp.

PRIOR, D. B. & COLEMAN, J. M. 1982. Active slides and flows in underconsolidated marine sediments on the slopes of the Mississippi delta. In: SAXOV, S. & NIEUWENHUIS, J. K. (eds) *Marine Slides and Other Mass Movements*. NATO Conference Series **IV:6**, Plenum, New York, 21–49.

PULHAM, A. J. 1987. *Depositional and Syn-sedimentary Deformation Processes in Namurian Deltaic Sequences of West County Clare, Ireland*. Unpublished PhD thesis, University of Wales, Swansea.

RIDER, M. H. 1969. *Sedimentological Studies in the West Clare Namurian and the Mississippi River Delta*. Unpublished PhD thesis, Imperial College, London.

—— 1974. The Namurian of West County Clare. *Proceedings of the Royal Irish Academy* **74B** (9), 125–143.

—— 1978. Growth faults in the Carboniferous of western Ireland. *Bulletin of the American Association of Petroleum Geologists* **62**, 2191–2213.

WINKER, C. D. & EDWARDS, M. B. 1983. Unstable progradational clastic shelf margins. In: STANLEY, D. J. & MOORE, G. T. (eds) *The Shelfbreak: A Critical Interface on Continental Margins*. Special Publication of the Society of Economic Paleontologists and Mineralogists **33**, 139–157.

A. J. PULHAM, Geology Department, University College, Singleton Park, Swansea SA2 8PP, UK. *Present address:* Stratigraphic Laboratory, BP Petroleum Development (North West Europe) Ltd, Britannic House, Moor Lane, London EC2Y 9BU, UK.

Upper Proterozoic passive margin deltaic complex, Finnmark, N Norway

A. Siedlecka, K. T. Pickering & M. B. Edwards

SUMMARY: The Upper Proterozoic Båsnaering Formation in the 9000 m thick Barents Sea Group is a succession of deltaic siliciclastics, 2500–3500 m thick, with an exposed extent covering an area of 3000 km². The Båsnaering Formation is underlain by up to 3200 m of deep-marine submarine fan deposits, while above the deltaics there are up to 1500 m of shallow- and marginal-marine (intertidal and supratidal) carbonates and siliciclastics. The lower part of the Båsnaering Formation records the change from deep- to shallow-marine, slope and prodelta environments to delta-front and delta-top environments. The middle and upper parts of the formation comprise delta-top, fluvial, marginal-marine, interdistributary bay and related environments. The Båsnaering Formation appears to record the northeastward progradation of a fluvially dominated wave-modified deltaic system. Tidal processes influenced the delta development, but the importance and magnitude of such tides remains uncertain. Gravity-controlled sediment instability was prevalent in the prodelta and delta-front environments.

The Båsnaering Formation is extensively exposed in coastal sections, stream profiles and inland cliffs of the northeastern parts of the Varanger Peninsula, N Norway (Fig. 1). The formation was defined and briefly described by Siedlecka & Siedlecki (1967). Subsequent work by Siedlecka & Edwards (1980) resulted in a subdivision of the formation into distinctive members, and in the interpretation of the Båsnaering Formation as a major delta system. The Båsnaering delta complex is probably one of the world's largest well-preserved Upper Proterozoic coastal and shallow-marine systems, with a preserved areal extent of about 3000 km².

The Båsnaering Formation is about 2500–3500 m thick and is one of four formations of the Barents Sea Group, which is a predominantly siliciclastic succession 9000 m thick underlying most of the Barents Sea region of the Varanger Peninsula (Fig. 1). The group (base unknown) starts with the Kongsfjord Formation, a succession of turbidites and other sediment gravity-flow deposits c. 3000 m thick interpreted as a submarine fan system (Siedlecka 1972; Pickering 1981, 1985). The basin slope and deltaic system of the overlying Båsnaering Formation is succeeded by c. 1500 m of coastal-marine siliciclastic and carbonate deposits of the Båtsfjord Formation (Siedlecka 1978). The uppermost part of the Barents Sea Group is the c. 1500 m thick Tyvjofjell Formation, interpreted as shallow-marine and fluviatile deposits (Siedlecka & Siedlecki 1967, 1972; Siedlecka 1978).

The Barents Sea Group is truncated by a c. 12° angular unconformity, above which there is the predominantly shallow-marine Løkvikfjell Group (Siedlecki & Levell 1978). A relatively small area in the NW part of the Varanger Peninsula comprises part of the Kalak Nappe Complex of the N Norwegian Caledonides. The Barents Sea region is separated from the northeastern margin of the Fennoscandian Shield and its Upper Proterozoic cover by the NW–SE-oriented Trollfjord–Komagelv Fault Zone, interpreted as a dextral transform fault with 500–1000 km displacement prior to the Caledonian deformation (Kjøde et al. 1978). Thus the Barents Sea region represents a suspect or displaced terrane as a dissected remnant of a sedimentary basin that developed farther W prior to its juxtaposition with Fennoscandia.

The Barents Sea Group forms an overall regressive sedimentary succession interpreted as having accumulated in an extensional fault-controlled passive continental margin which may continue southeastwards along the Fennoscandian craton towards the Timanian aulacogen (Siedlecka 1975). In this context, the Trollfjord–Komagelv Fault Zone is interpreted as an ancient lineament that was active during the evolution of the Barents Sea Group and the Løkvikfjell Group as a major basin-margin feature, and which was later reactivated as a dextral strike-slip fault to emplace the Barents Sea region in its present position (Kjøde et al. 1978; Siedlecka 1985). Restored balanced cross-sections through the Caledonides of N Norway, combined with the palaeomagnetic data of Kjøde et al. (1978), suggest that the Barents Sea region was part of the Baltoscandian Plate (Baltica) that rifted away and subsequently was translated to its present position during collision in the Finnmarkian Orogeny (Gayer et al. 1987); the calculated maximum of 626 km of shortening in the Cale-

From WHATELEY, M. K. G. & PICKERING, K. T. (eds), 1989, *Deltas: Sites and Traps for Fossil Fuels*, Geological Society Special Publication No. 41, pp. 205–219.

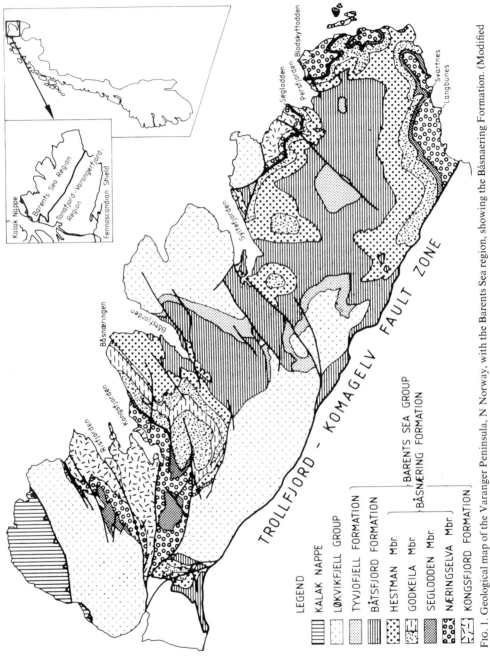

FIG. 1. Geological map of the Varanger Peninsula, N Norway, with the Barents Sea region, showing the Båsnæring Formation. (Modified from Vidal & Siedlecka 1983; Siedlecka 1985.)

donian nappes (Gayer *et al.* 1987) is compatible with a 500–1000 km dextral displacement for the Barents Sea block prior to shortening in the W.

The intensity of the Caledonian deformation increases northwestwards and metamorphism is mostly in the lower greenschist facies (Roberts 1972, 1975). Swarms of dolerite dykes dissect the rocks of the Barents Sea region (Beckinsale 1975). Rb–Sr whole-rock age determinations on cleaved mudrocks from the Kongsfjord Formation suggest that the folding and cleavage formation in the Barents Sea Group occurred at 520 ± 47 Ma (Taylor & Pickering 1981).

The Riphean–Vendian acritarchs found in the lower Båtsfjord Formation (Vidal & Siedlecka 1983) and the K–Ar age of *c*. 640 Ma for dolerite dyke intrusion (Beckinsale *et al.* 1975), together with lithostratigraphic correlations, all suggest a Late Proterozoic age for the Barents Sea Group and the Løkvikfjell Group.

Båsnaering delta complex

The Båsnaering Formation comprises four lithostratigraphic members (Figs 1 and 2) from base to top: (i) the Naeringselva Member, greenish-grey mainly mudrocks in the lower part, siltstones and fine-grained sandstones (increasing in abundance upwards) interpreted as basin slope, pro-delta and delta front; (ii) the Seglodden Member, red-violet sandstones; (iii) the Godkeila Member, greenish-grey mudrocks, siltstones and sandstones; (iv) the Hestman Member, red-violet sandstones. The Seglodden, Godkeila and Hestman Members are interpreted as stacked coarsening-upward sequences, mainly representing delta-front, lower-delta-plain, including interdistributary bay, and associated environments. Waves and tides appear to have reworked parts of the delta system. Siedlecka & Edwards (1980) published detailed sedimentary logs from the Naeringselva Member, SE of Seglodden, and from the Godkeila Member, Godkeila (Fig. 3).

The thickness of the Båsnaering Formation, and its members, shows considerable lateral variations (Fig. 2). The maximum 3500 m thickness of this succession occurs on the peninsula of Båsnaeringen, whereas farther E in the Persfjord area the formation is approximately 2700 m thick (Fig. 2). The lower two members of the Båsnaering Formation, the Naeringselva and Seglodden Members, tend to thicken eastwards, while the upper Godkeila and Hestman Members tend to thin towards the E.

Petrographically, the Båsnaering Formation sandstones are quartzo-feldspathic arenites or subarkoses containing 10%–20% feldspar (Siedlecka & Edwards 1980). Muscovite, chlorite, tourmaline, zircon, opaque minerals (mainly haematite) and chert fragments form the accessory minerals, with cements of quartz, feldspar and carbonates. The predominance of plagioclase feldspar, together with either an untwinned plagioclase or an alkali feldspar with low potassium (identified using staining techniques), suggests a non-granitic (igneous) and/or metamorphic source area. The lithic fragments such as chert indicate sedimentary rocks also in the source area.

Naeringselva Member

The Naeringselva Member shows considerable thickness variation, reaching 1000–1200 m in the E (Figs 2 and 3*a*). The Naeringselva Member is divided into two parts, the lower mainly comprising siltstones and mudrocks, and the upper part mainly comprising medium to very thick bedded sandstones. The Lower Naeringselva Member varies from 60 m to 540 m thick and comprises four main facies: parallel-laminated and structureless mudstone (Figs 4*a*, *b*); current-rippled and oscillation or combined-flow rippled fine-grained sandstone (Fig. 4*c*, *d*) (Pickering 1982, 1984). Locally, mudflake breccias occur in sheet-like to very irregular layers (Fig. 4*e*). Wet-sediment deformation is abundant, as slide, slump and *in situ* liquefaction and fluidization structures (Fig. 5*a*). There are erosional gully-shaped depressions with widths of metres to tens of metres and depths of up to approximately 5 m, infilled with slide deposits and draped by normal bedding (Figs 5*b*, *c*). Erosional discordances define packets of beds on a decimetre to metre scale.

Figure 6, from the Lower Naeringselva Member, shows the distribution of wet-sediment deformed horizons, interpreted as slope-failure deposits. Sandstones in the upper part of the member are cross-stratified, commonly as low angle tangential, multistorey sets (Fig. 7). Locally, there are deformed horizons of muddy siltstone and sandstone, described by Siedlecka & Edwards (1980) as large-scale ball-and-pillow structures (Fig. 8) but which may be incipient mud diapirs or mud lumps analogous to those described from the Mississippi delta front and prodelta (Prior & Coleman 1982).

Palaeocurrent data, together with a limited number of slide-fold readings, suggest flow to the ENE (Fig. 9*a*). These palaeocurrents show a similar orientation to those recorded from the underlying Kongsfjord submarine fan system (Pickering 1981, 1985).

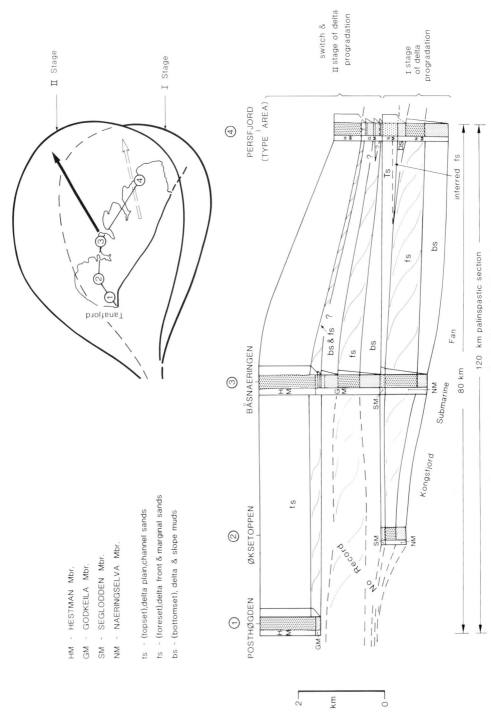

Fig. 2. Lithostratigraphic correlation and main sedimentary environments for the Båsnæring Formation.

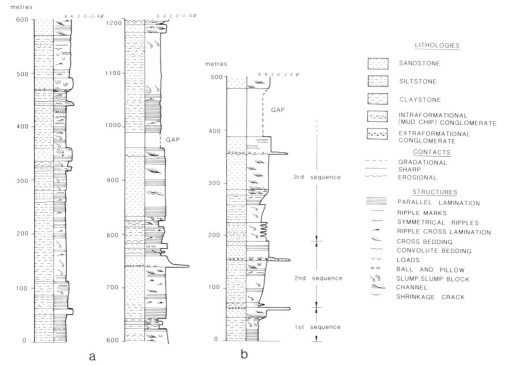

FIG. 3. Summary sedimentary logs: in the Båsnaering Formation; (*a*) the Naeringselva Member, NW of Seglodden, and (*b*) the Godkeila Member, Godkeila (from Siedlecka & Edwards 1980).

Seglodden Member

The Seglodden Member is up to approximately 350 m thick in the type area (northwestern and southeastern sides of Persfjorden) and thins westwards to about 100 m (Fig. 2); SW of Kongsfjorden, thicknesses tend towards the maximum 350 m observed further E in the type area.

The main lithologies in the Seglodden Member are fine-grained sandstones, mainly red-violet to variegated greenish-grey and red-violet. Beds typically range from 50 to 200 cm thick, and vary from laterally discontinuous to slightly lenticular in shape. Tabular to wedging and planar to tangential cross-stratification is common (Fig. 10), and in some cases the stratification is defined by concentrations of opaque heavy minerals such as ilmenite and magnetite. Reactivation surfaces are common, together with examples of oversteepened cross-bedding. Bounding surfaces to beds are generally flat and sharp, but with local erosional scours on a decimetre scale. Mudstone drapes, up to about 20 cm thick, occur above many sandstones. Mudflake breccias typically occur in the sand-rich lithologies immediately overlying such mudstone horizons.

Channels up to 5 m deep occur, filled with pebble conglomerates of carbonate-cemented sandstone and red mudflakes in a sandy matrix. Palaeocurrents, from cross-stratification, show predominant flow towards the SE and E (Fig. 9*b*). In general, the Seglodden Member has a much greater proportion of sand-grade sediment compared with the upper part of the Naeringselva Member, not only in terms of individual sandstones but also with fewer mudstone intercalations between the sandstone packets. The Seglodden Member shows a paucity of shallow-marine structures such as wave ripples.

Godkeila Member

The Godkeila Member varies from approximately 500 m thick in the type section on the southeastern side of Persfjord, immediately SW of the headland of Godkeila, to about 1450 m thick on the Båsnaering Peninsula (Fig. 2).

In the type section, the Godkeila Member comprises three stacked coarsening-upward sequences (Fig. 3*b*). The lowest sequence is 64 m thick and shows an upward increase in both the frequency and thickness of thin-bedded lami-

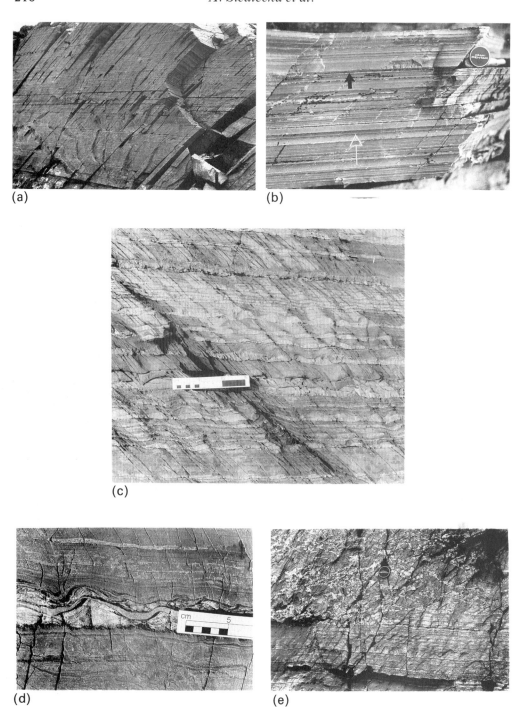

FIG. 4. Typical facies in the Lower Naeringselva Member, Båsnaering Formation: (*a, b*) parallel-laminated and structureless mudstone, SE Persfjord (camera lens cap for scale to right of centre); (*c*) current-rippled siltstone and oscillation-rippled fine-grained sandstone, SE Persfjord (15 cm scale); (*d*) close-up of fine-grained asymmetric combined-flow ripple with silty mudstone drape, SE Kongsfjord; (*e*) mudflake breccia above rippled siltstones, SE Persfjord (lens cap for scale).

Fig. 5. Typical wet-sediment deformation in the Lower Naeringselva Member, Båsnaering Formation: (*a*) multiple erosional surfaces infilled by sediment slides, SE side of Kongsfjord (15 cm scale arrowed); (*b*) gully-shaped erosional depression (outlined), infilled by sediment slide, SE Persfjord (30 cm high shoulder bag (arrowed) for scale); (*c*) close-up of right-hand (western) side of the gully infilled by sediment slide shown in (*b*) (30 cm high shoulder bag (arrowed) for scale).

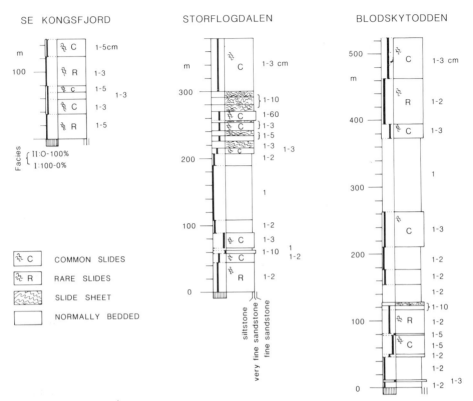

FIG. 6. Sedimentary logs through the Lower Naeringselva Member to show the distribution of facies associations and the relative abundance of wet-sediment deformation, mainly sediment slides: facies I, parallel-laminated and structureless mudstone; facies II, rippled siltstones/sandstones. Estimated average bed thicknesses are shown to the right of the logs. The grain-size and per cent facies scales are similar for each log. (From Pickering 1982, 1984.)

nated siltstones. Higher up in the sequence, lenticular thin-bedded parallel- and ripple-laminated sandstones with erosive bases appear, some of which show wet-sediment slide structures. The uppermost part of this sequence consists of trough cross-stratified medium- to thick-bedded sandstones (Fig. 10) with some oversteepened cross-bedding. The top of the sequence is marked by a mudflake conglomerate associated with deep scours into the underlying sandstones (Fig. 11), and locally the conglomerate passes laterally into these sandstones.

The start of the second coarsening-upward sequence is defined by 10 m of grey-black parallel-laminated 'paper shale' with some intercalated lenticular erosively based sandstones. Above the shales, there is a gradual change into grey-green laminated mudstones. Throughout this 80 m thick unit, there is an overall increase in mean grain size, with sandstones becoming more common (Fig. 12). Abundant straight-crested asymmetric ripples occur, together with parallel-

and wavy-laminated lenticular sandstones with features that resemble parts of the sequence of hummocky cross-stratification (Dott & Bourgeois 1982). A few steep-sided channels were observed, together with wet-sediment deformed horizons including ball-and-pillow structures. The second coarsening-upward sequence is capped by a 35 m thick erosively based sandstone unit that varies from being locally structureless to ball-and-pillow structure.

The uppermost coarsening-upward sequence is 198 m thick and starts with laminated mudstones and intercalations of medium- to thick-bedded structureless to wet-sediment-deformed sandstones. A unit of greenish-grey laminated mudstones with intercalated structureless or internally deformed siltstones and mudstones occurs above the sandstones. At about 260 m from the base of the measured type section, abundant straight-crested sandy and silty ripples appear, grading up into parallel-laminated and cross-stratified sandstones with silty partings. The uppermost

FIG. 7. Cross-stratified fine- to medium-grained sandstone body in the Upper Naeringselva Member, with 30 cm high hammer for scale (arrowed).

FIG. 8. Large-scale deformation structures in the Upper Naeringselva Member. The dome-shaped structures may represent incipient mud diapirs or mud domes. See text for details.

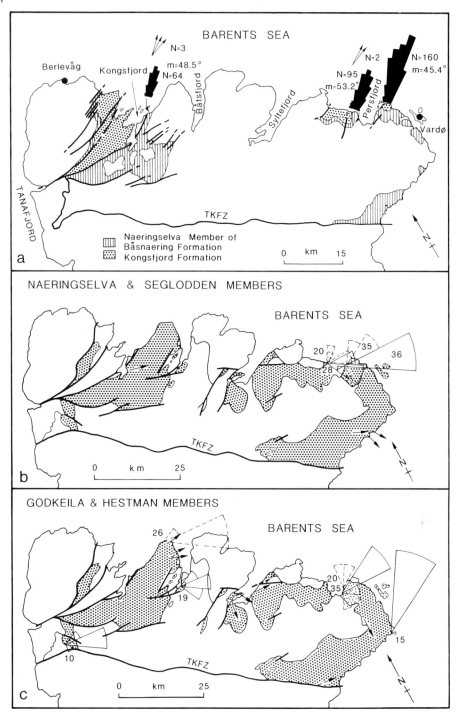

FIG. 9. Maps to show the outcrop pattern of the members of the Båsnaering Formation with rose diagrams showing cross-bedding and current-ripple and combined-flow-ripple directions (TKFZ, Trollfjord–Komagelv fault zone). The numbers of readings are shown, with only an arrow denoting the average current flow direction for less than 10 readings: (a) Lower Naeringselva Member data; (b) Naeringselva Member (— — —) and Seglodden Member (———) data; (c) Godkeila Member (— — —) and Hestman Member (———) data. (Data collated from Siedlecka & Edwards 1980; Pickering 1982.)

FIG. 10. Thick-bedded sheet-like sandstone beds with solitary large-scale cross-stratification in the Seglodden Member, Seglodden.

FIG. 11. Mudflake breccia infilling an erosional hollow: top of the lowest coarsening-upward sequence, Godkeila Member. 30 cm hammer for scale on the left.

100 m of the coarsening-upward sequence comprises 70 m of medium-bedded mainly trough cross-stratified sandstones, overlain by 30 m of thin- to thick-bedded mainly structureless sandstones. The base of the 30 m thick unit is an erosional surface overlain by a mudflake conglomerate. Palaeocurrents from the Godkeila Member indicate flow towards the NE, E and SE (Fig. 9c).

Above the three coarsening-upward sequences, there is no exposure until the overlying Hestman Member in the type section. However, by extrapolation from other sections, the gap appears to represent mainly fine-grained thin and very thin bedded deposits.

The coastal section at Båsnaeringen is considerably thicker and contains a major sandstone body in the middle (Fig. 2). The sandstone

FIG. 12. Thick-bedded sandstones, including a solitary set of trough cross-stratified sandstone from the middle coarsening-upward sequence, Godkeila Member. 30 cm hammer for scale.

comprises thick to very thick beds that have a ball-and-pillow structure, with overturned cross-stratification containing heavy mineral concentrations. Some channelling and breccias of red mudstone clasts occur in the lower middle part of the sandstone body. Parallel-laminated and rippled mudstone with graded sandstone beds constitute the upper and lower parts of this section.

Hestman Member

The Hestman Member thickens from approximately 600 m in the type section on the northwestern slopes of Hestman to an estimated 1300 m at Båsnaeringen and Posthøgden (Fig. 2).

The main lithologies in the Hestman Member are medium- to thick-bedded red-violet to pinkish-grey fine-grained sandstones and subordinate intercalated mudstone units from 5 cm to 25 m thick (Kiberg area). Tabular tangential cross-stratification is the most common bedform in the sandstones. Ripple lamination occurs in the finer-grained lithologies, for example in the transition from sandstones to mudstones as part of fining-upward sequences.

Reactivation surfaces and deep scours occur in sets of cross-stratified sandstone. There is some foreset truncation by structureless sandstone layers dipping in the same direction as the foresets but at a lower angle, below which oversteepened cross-stratification may occur. Such reactivation surfaces were first described by Siedlecka & Edwards (1980). Some of the deep scours are infilled by conglomerates. For example,

a 3 m deep scour, E of Kiberg, occurs as a rounded depression with steep sides and contains a breccia of large siltstone slabs in a matrix of fine-grained sandstone with scattered pebbles; 15 m of mainly structureless sandstone with lenses of mudflake conglomerate occur above the scour, and are in turn overlain by about 25 m of mudstones.

Depositional environments in the Båsnaering Formation

The Båsnaering Formation occurs above a deep-marine turbidite system of the Kongsfjord Formation, more than 3000 m thick, interpreted as a submarine fan system (Pickering 1979, 1981, 1985), and below shallow-marine intertidal and supratidal mixed siliciclastics and carbonates of the Båtsfjord Formation (Siedlecka 1978). The sedimentology of the Båsnaering Formation, together with its regional context, suggest (i) a large-scale deltaic complex and associated sedimentary environments that developed as part of (ii) an evolving or mature passive continental margin.

In the Lower Naeringselva Member, the common occurrence of slide sheets and other types of wet-sediment deformation, the thin-bedded and fine-grained nature of the deposits, and the stratigraphic position above a thick submarine fan system and below shallow-marine/nearshore deposits suggests sedimentation on a continental slope to prodelta. The slope was constructed by terrigenous sediments fed towards the NE quadrant by a variety of processes including (i) fine-grained turbidity currents (current-rippled siltstone and parallel-laminated mudstone), (ii) distal storm events (oscillation and combined-flow rippled fine-grained sandstone), (iii) possible annual phytoplankton/algal blooms (some parallel-laminated mudstone) and (iv) hemipelagic deposition from nepheloid layers (structureless mudstone).

The upward coarsening and transition into fluvio-deltaics (see below) suggests that the Lower Naeringselva Member is, at least towards the top, part of a prodelta (Siedlecka & Edwards 1980; Pickering 1982) (Fig. 2). Gullies and channels, typically with depths on a metre scale and widths greater than tens of metres, were excavated into this slope. Slope instability appears to have been an inherent feature of the Lower Naeringselva Member, producing structures similar to those observed on the Mississippi River Delta such as bottleneck slides, mudflow gullies and coalescing mudflow noses (Prior & Coleman 1982).

The dramatic thickening of the slope succession by more than 400 m (compacted thickness) from W to E on the Varanger Peninsula, over an estimated down-slope distance of 5–10 km (Pickering 1982), suggests a NNW–SSE-striking basin margin. Growth faulting may account for this thickening, but field observations to support this hypothesis have not been made.

The Upper Naeringselva Member is characterized by medium- to thick-bedded mainly fine-grained sandstones, some sand-filled channels up to metres in depth, slide deposits, coarsening-upward sequences and an upward increase in the frequency of wave ripples and combined flow ripples, all of which support prodelta/delta-front and some delta-top distributary channel or mouth-bar deposits (*cf* Elliot 1976, cycles 2, 7 and 8 lateral to a distributary channel axis). Indeed, the coarsening upward from the prodelta to delta-front deposits is accompanied by increased wave activity.

The Seglodden Member is significantly more sand rich than the Naeringselva Member and is predominantly cross-stratified, and there is no clear evidence supporting a marine environment. Thus the Seglodden Member is tentatively interpreted as fluvial–fluvio-lacustrine deposits, particularly distributary channel sandstones with subordinate amounts of overbank, interdistributary bay and delta-associated coastline deposits. However, it is possible that some of the cross-stratification, particularly the heavy mineral laminated sets, represents transgressive reworking of delta-top sands into shallow-marine sandwave complexes.

The Godkeila Member marks a return to marine conditions. The vertically stacked regressive sequences of the Godkeila Member, together with their wave and combined-flow ripples, suggest successive progradation of wave-influenced delta environments (probably delta-front to lower-delta-plain distributary channel sandstones and mouth-bar deposits). The unusually fine-grained sediments in the lower part of the second coarsening-upward sequence may represent interdistributary bay or lagoonal sediments (*cf* Tye & Kosters 1986). The lenticular and flaser bedding in the second sequence suggests the presence of tidal currents.

The comparatively thin nature of the upper two sequences in the Godkeila Member compared with the older members, on a scale from 10 m to a few tens of metres, favours delta-lobe switching rather than major prodelta to delta-top progradations. However, towards the W, on Båsnaeringen (Figs 1 and 2), the member thickens considerably. Such a dramatic thickness change would necessitate increased subsidence in the W, with a major switch in delta progradation (Fig. 2, inset on top right). Two phases of delta progradation, separated by the major depocentre switch (above), are tentatively proposed. The first phase is represented by the Naeringselva–Seglodden Member regressive sequence that shows ENE progradation. The second phase was initiated prior to the pre-Godkeila Member transgression, followed by the Godkeila–Hestman Member regressive sequence showing progradation towards the NE.

The Hestman Member marks the final episode of fluvio-deltaic sedimentation. The absence of sedimentary structures attributable to marine conditions, together with the channelized nature of some of the sandstone bodies between packets of finer-grained deposits, can be interpreted as indicating lower-delta-plain fluvial sedimentation.

Post-deltaic sedimentation

There is a sharp contact between the fining-up red-violet sandstone–mudstone succession of the uppermost Hestman Member and the greenish-grey, grey and dark grey terrigenous and carbonate rocks of the lower Båtsfjord Formation (Siedlecka 1978). Sedimentary structures include herring-bone cross-stratification, oscillation ripples, ripple-drift, lenticular, flaser and parallel lamination. Desiccation cracks are ubiquitous in the claystones and dolomite beds, together with intraformational breccias. Some limestone beds contain stromatolites. Mixed terrigenous–carbonate rocks are typical and are arranged in sequences that show an upward increase in the proportion of carbonate relative to terrigenous siliciclastics. These deposits have been interpreted mainly as marginal marine, reflecting shallowing-up intertidal to supratidal environments (Siedlecka 1978).

The development of tidal flats above the abandoned delta-plain deposits of the Båsnaering Formation (Hestman Member) suggest (i) a transgression associated with a eustatic rise in sea level or (ii) a more local relative change in sea level as a result of delta-lobe switching or tectonically controlled subsidence.

Plate-tectonic setting

The overall stratigraphy and thickness of the Barents Sea Group and Løkvikfjell Group demonstrate the late Precambrian development of a passive continental margin stratigraphy, juxtaposed by dextral strike-slip faulting against the rocks of the southern Varanger Peninsula in early Cambrian times (Kjøde *et al.* 1978; Gayer *et al.* 1987). Restored balanced cross-sections through the Caledonides of N Norway, combined with the palaeomagnetic data of Kjøde *et al.* (1978), suggest that the Barents Sea Group may have developed on the NW margins of the Baltoscandian Plate rather than on other plates which are now separated (Gayer *et al.* 1987).

Modern analogues

Modern analogues for the Barents Sea Group, and in particular the Båsnaering Formation delta system, may in part be the Mississippi, Niger or Amazon delta systems. Although no one modern delta may be entirely comparable with the Båsnaering Formation, various aspects of the above modern deltas provide useful frameworks for understanding this ancient deltaic system.

The abundant wet-sediment deformation, as slides, slumps and other features, which is particularly well developed in the Naeringselva Member, may be comparable with many of the structures described from the offshore Mississippi Delta (*cf* Prior & Coleman 1982). However, different modern deltas may provide better analogues for other aspects of the Båsnaering Formation delta system.

The Amazon Delta and its associated environments provides an interesting analogue because it is a good modern example of a coastal and shelf system with important storm and tide reworking of the delta-plain and delta-front sands, silts and muds. Despite an annual river discharge of more than 1 billion tonnes of sediment, the mouth of the Amazon River has little subaerial expression and is therefore not a classic delta (Nittrouer *et al.* 1986). Strong tidal currents near the river mouth (tidal range more than 6 m) up to about 100 cm s^{-1}, coastal currents (the N Brazilian coastal current flows N with surface speeds of 50 cm s^{-1}) and surface waves serve to disperse the fluvial discharge. The subaqueous Amazon deltaic system comprises very-large-scale sigmoidal clinoform structures stretching for hundreds of kilometres offshore and divisible into three regions: (i) topset strata of muds and sands less than 40 m deep that dip gently seaward and thin laterally northward (downcurrent); (ii) muddy and silty foreset strata approximately 40–60 m deep dipping seaward at less than 1°; (iii) muddy bottomset strata approximately 60–100 m deep that are now burying a basal transgressive sand layer (Nittrouer *et al.* 1986). The outer shelf bedforms include sand waves 3–6 m high and 100–200 m in wavelength which are perpendicular to the shoreline; farther N, the sand shoals are 20–30 m high and 6000–8000 m in wavelength, and it is these bedforms that are currently being covered by muds of the bottomset.

The particular appeal of the Amazon coastal system as a modern analogue to the Båsnaering Formation is that the latter deposits show little evidence for a specifically fluvially dominated delta, and also that the Seglodden and Hestman Members show considerable evidence for reworking of probable delta-front/delta-plain sediments. Moreover, the Amazon system is strongly influenced by both waves and tides, again something that is believed to have been important for the Båsnaering Formation. The scale of the Amazon submarine fan, slope and delta system is comparable with that suggested for the palaeo-environments of the Kongsfjord and Båsnaering Formations.

Although the scale and plate-tectonic setting of the Mississippi Delta provides a possible modern analogue for the Båsnaering Formation, the Gulf of Mexico is microtidal and wave activity is insignificant relative to fluvial distributary processes. The Niger Delta may be more promising as a modern counterpart because it is significantly modified and influenced by both tide and storm processes. Unfortunately there is still little published data on this system to allow detailed comparisons with the Båsnaering Formation.

References

BECKINSALE, R. D., READING, H. G. & REX, D. C. 1975. Potassium–argon ages for basic dykes from East Finnmark: stratigraphical and structural implications. *Scottish Journal of Geology* **12**, 51–65.

DOTT, R. H. JR & BOURGEOIS, J. 1982. Hummocky stratification: significance of its variable bedding sequences. *Bulletin of the Geological Society of America* **93**, 663–680.

ELLIOTT, T. 1976. Upper Carboniferous sedimentary cycles produced by river-dominated, elongate deltas. *Journal of the Geological Society, London* **132**, 199–208.

GAYER, R. A., RICE, A. H. N., ROBERTS, D., TOWNSEND, C. & WELBON, A. 1987. Restoration of the Caledonian Baltoscandian margin from balanced cross-sections: the problem of excess continental crust. *Philosophical Transactions of the Royal Society of Edinburgh, Earth Sciences* **78**, 197–217.

KJØDE, J., STORETVEDT, K. H., ROBERTS, D. & GIDSKEHAUG, A. 1978. Palaeomagnetic evidence for large-scale dextral displacement along the Trollfjord–Komagelv Fault, Finnmark, North Norway. *Physics of the Earth and Planetary Interiors* **16**, 132–144.

NITTROUER, C. A., KUEHL, S. A., DEMASTER, D. J. & KOWSMANN, R. O. 1986. The deltaic nature of the Amazon shelf sedimentation. *Bulletin of the Geological Society of America* **97**, 444–458.

PICKERING, K. T. 1979. Possible retrogressive flow slide deposits from the Kongsfjord Formation: a Precambrian submarine fan, Finnmark, N. Norway. *Sedimentology* **26**, 295–305.

—— 1981. The Kongsfjord Formation—a late Precambrian submarine fan in north-east Finnmark, North Norway. *Norges Geologiske Undersøkelse* **367**, 77–104.

—— 1982. A Precambrian upper basin-slope and prodelta in northeast Finnmark, North Norway—a possible ancient upper continental slope. *Journal of Sedimentary Petrology* **52**, 171–186.

—— 1984. Facies, facies-associations and sediment transport/deposition processes in a late Precambrian upper basin-slope/pro-delta, Finnmark, N. Norway. *In:* STOW, D. A. V. & PIPER, D. J. W. (eds) *Fine-grained Sediments: Deep Water Processes and Facies.* Geological Society of London Special Publication **15**, 343–362.

—— 1985. Kongsfjord Turbidite System, Norway. *In:* BOUMA, A. H., NORMARK, W. R. & BARNES, N. E. (eds) *Submarine Fans and Related Turbidite Systems.* Springer, New York, 237–244.

PRIOR, D. B. & COLEMAN, J. M. 1982. Active slides and flows in underconsolidated marine sediments on the slopes of the Mississippi Delta. *In:* SAXOV, S. & NIEWENHUIS, J. K. (eds) *Marine Slides and Other Mass Movements.* NATO Conference Series, Series IV: Marine Sciences. Plenum, New York, 21–49.

ROBERTS, D. 1972. Tectonic deformation in the Barents Sea Region of Varanger Peninsula, Finnmark. *Norges Geologiske Undersøkelse* **282**, 1–39.

—— 1975. Geochemistry of dolerite and metadolerite dykes from Varanger Peninsula, Finnmark, North Norway. *Norges Geologiske Undersøkelse* **322**, 55–72.

SIEDLECKA, A. 1972. Kongsfjord Formation—a Late Precambrian flysch sequence from the Varanger Peninsula, Finnmark. *Norges Geologiske Undersøkelse* **278**, 41–80.

—— 1975. Late Precambrian stratigraphy and structure of the north-eastern margin of the Fennoscandian Shield (East Finnmark–Timan Region). *Norges Geologiske Undersøkelse* **316**, 313–348.

—— 1978. Late Precambrian tidal-flat deposits and algal stromatolites in the Båtsfjord Formation, East Finnmark, North Norway. *Sedimentary Geology* **21**, 277–310.

—— 1985. Development of the Upper Proterozoic sedimentary basins of the Varanger Peninsula, East Finnmark, North Norway. *Geological Survey of Finland, Bulletin* **331**, 175–185.

—— & EDWARDS, M. B. 1980. Lithostratigraphy and sedimentation of the Riphean Båsnaering Formation, Varanger Peninsula, North Norway. *Norges Geologiske Undersøkelse* **355**, 27–47.

—— & SIEDLECKI, S. 1967. Some new aspects of the geology of Varanger peninsula (Northern Norway). *Norges Geologiske Undersøkelse* **247**, 288–306.

—— & —— 1972. Lithostratigraphical correlation and sedimentology of the Late Precambrian of Varanger Peninsula and neighbouring areas of East Finnmark, Northern Norway. *Proceedings of the 21st International Geological Congress, Nice.* IAS, Section 6, 349–358.

SIEDLECKI, S. & LEVELL, B. K. 1978. Lithostratigraphy of the Late Precambrian Løkvikfjell Group on Varanger Peninsula, East Finnmark, North Norway. *Norges Geologiske Undersøkelse* **343**, 73–85.

TAYLOR, P. N. & PICKERING, K. T. 1981. Rb–Sr isotopic age determination on the late Precambrian Kongsfjord Formation, and the timing of compressional deformation in the Barents Sea Group, East Finnmark. *Norges Geologiske Undersøkelse* **367**, 105–110.

TYE, R. S. & KOSTERS, E. C. 1986. Styles of interdistributary basin sedimentation: Mississippi delta plain, Louisiana. *Transactions of the Gulf Coast Association of Geological Societies* **36**, 575–588.

VIDAL, G. & SIEDLECKA, A. 1983. Planktonic, acid-resistant microfossils from the Upper Proterozoic strata of the Barents Sea Region of Varanger Peninsula, East Finnmark, Northern Norway. *Norges Geologiske Undersøkelse* **382**, 45–79.

A. SIEDLECKA, Norges Geologiske Undersøkelse, Postboks 3006, 7001 Trondheim, Norway.

K. T. PICKERING, Department of Geology, University of Leicester, Leicester LE1 7RH, UK.

M. B. EDWARDS, 42 Eagle Court, Spring, TX 77380, USA.

Petroleum- and Gas-related Case Histories

Reservoir geology of the Sirikit oilfield, Thailand: lacustrine deltaic sedimentation in a Tertiary intermontane basin

S. Flint, D. J. Stewart & E. D. van Riessen

SUMMARY: The Sirikit oilfield, situated some 400 km N of Bangkok, Thailand, is a fault-bounded structure within a major half-graben basin, comprising a relatively undeformed eastern sector and a western area characterized by a complex pattern of isolated fault blocks. The 8 km thick basin-fill succession includes alluvial fan–fluvial deposits overlain by the fluvio-deltaic Lan Krabu Formation, which contains two main oil reservoirs.

A reservoir geological model has been produced based on cores and wireline log suites from some 45 wells. The geology is characterized by cycles of regressive coarsening-upward sequences, up to 15 m thick, which reflect cyclic progradation of fluvial-dominated deltas into a large relatively shallow lake in a humid tropical climatic regime. The cycles are commonly topped by subaerial flood-plain claystones. Distributary channel sandstones show evidence of stage fluctuation and intermittent abandonment. Individual mouth-bar reservoir sandstones range in thickness from 1 to 5 m and form complex sheet-like bodies of lateral extent up to 20 km². Thus the Sirikit deltas have little inferred geometrical similarity with contemporary bar-finger models postulated for river-dominated lacustrine deltas. The sheet-delta geometry may be due to a combination of high frictional forces related to similar river and lake water densities, shallow lake depth and a large but variable fluvial sediment load. This combination of factors resulted in rapid dumping of sediment, the plugging of channel mouths and subsequent channel bifurcation.

Existing models for lacustrine deltaic sedimentation are based on idealized 'bed-load' (Gilbertian) or modern Mississippi Delta morphologies. Bed-load deltas are relatively uncommon, and are usually only found in glacial and arid regions. The modern Mississippi 'bird's foot' delta geometry is generally accepted as a model for sand-body development in fluvial-dominated delta systems, as seen in some recent shallow lakes (Hyne et al. 1979). A model for lacustrine deltaic sedimentation in which the deltas are considered to be lobate with sandstones of sheet-like geometries is presented in this paper. The controlling factors on delta growth and implications for hydrocarbon reservoir development are discussed.

Location and geological setting of the Sirikit oilfield

The Sirikit oilfield is located some 400 km N of Bangkok in the central plains of Thailand. It was discovered in 1981 within the Phitsanulok Basin, which is one of a series of large Tertiary basins in central Thailand. The Phitsanulok Basin is a major Oligocene extensional structure overlying Mesozoic basement, which is complexly folded, partially metamorphosed and block faulted owing to the Indosinian orogeny and later deformation (Workman 1975; Knox & Wakefield 1983).

The basin is situated in a triangular zone at the intersection of two regional strike-slip fault zones, the NW–SE-trending Mae Ping Fault Zone and the NE–SW-trending Uttaradit Fault Zone (Fig. 1). The western basin margin is represented by a normal fault (western boundary fault), giving an overall half-graben basin geometry (Fig. 2) with a maximum depth of 8 km adjacent to the western boundary fault. The Sirikit field is situated on the southern flank of the graben on a local basement high and is a fault-bounded structure with a relatively undeformed eastern half and a complexly block-faulted western sector. The main faults are listric in style, trending roughly N–S, and appear to have been reactivated subsequently by left-lateral strike-slip movement.

Tertiary stratigraphy and basin development

The oil-bearing succession in the Sirikit field consists of two major units, the fluvio-deltaic Lan Krabu Formation and the open lacustrine Chum Saeng Formation (Fig. 2). The interdigitation of these two Miocene formations has resulted in a subdivision of the sequence into the L sandstones at the base, the upper intermediate seal, the K sandstones and the main seal at the top (Plate 1).

The Lan Krabu–Chum Saeng package unconformably overlies coarse-grained clastic sedi-

From WHATELEY, M. K. G. & PICKERING, K. T. (eds), 1989, *Deltas: Sites and Traps for Fossil Fuels*, Geological Society Special Publication No. 41, pp. 223–237.

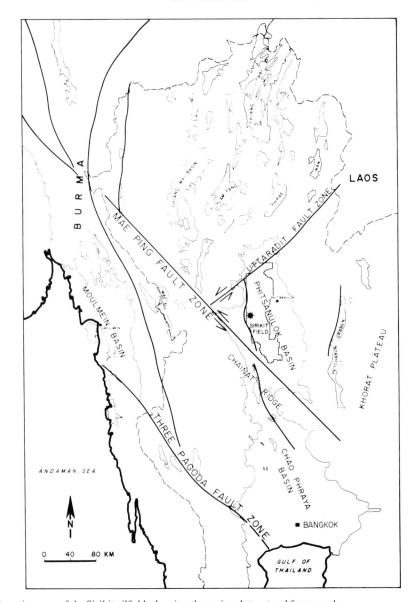

FIG. 1. Location map of the Sirikit oilfield, showing the regional structural framework.

ments of the Khom Formation and is itself overlain by fluvio-alluvial sediments of the Pratu Tao Formation.

Methods of study

The Phitsanulok Basin succession is completely covered by Recent alluvium. Data have thus been acquired through a three-dimensional seismic survey and the drilling of some 45 production wells. Facies were defined on the basis of cores taken through the reservoir sections in seven wells and have been substantiated by the presence of macro and micro palaeontological and palynological data. The core-derived facies associations were matched with their corresponding log responses to enable recognition in uncored wells. During this exercise it was found that the gamma-ray and sonic logs were most useful. Correlation panels were then constructed to facilitate bed-by-bed geological correlations based on the sonic

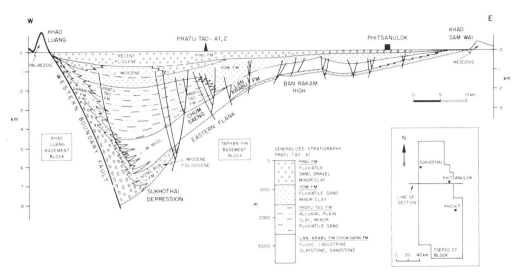

Fig. 2. W–E cross-section of the Phitsanulok Basin showing the structural configuration and basin-fill stratigraphy.

log's response to field-wide acoustically soft transgressive lacustrine claystone marker beds.

Facies analysis

Based on core analyses, a series of lithofacies was delineated (Table 1) on colour, sedimentary structures (where present), grain size and the presence of body/trace fossils. These lithofacies include a series of three claystones, a heterolithic facies, a pelletal siltstone and three sandstone facies.

Facies C1: mottled claystone

Description. This facies consists of pale green to greenish-grey claystones commonly including laminae and patches of greenish siltstone. A characteristically mottled appearance is due to calcite and siderite nodules (Fig. 3). Rootlets and plant fragments are rare and the overall organic carbon content is low. However, rare non-correlatable coaly beds have been recognized in the Greater Sirikit area. Bioturbation varies from moderate to intense, but body fossils are absent. This facies ranges in thickness from a few centimetres to more than 5 m and is distinguished from other claystone facies by its relative hardness and corresponding sonic log response.

Interpretation. The colour and bioturbation indicate subaqueous depositional conditions. The pedogenic mottling and rootlets, however, suggest

periods of emergence. C1 is thus interpreted as a humid flood-plain deposit representing vertical aggradation. Seasonal flooding resulted in clay deposition by simple settling out of suspended sediment. Drying out of the temporary lakes resulted in the initiation of oxidative pedogenetic processes.

Facies C2: carbonaceous claystone

Description. This facies is characterized by dark brown to grey claystones with rare bioturbation, no benthonic fossils and a high content of terrigenous organic material ($C_{org} = 17\%$–40%). Beds never exceed 20 cm in thickness and are of source-rock potential.

Interpretation. The high organic content and paucity of benthonic fossils indicate low oxygen levels at the sediment–water interface. C2 facies may represent the bottom deposits of an anoxic lake or flood-plain swamp sediments. Precise interpretations depend on association with other facies as discussed below.

Facies C3: grey-green molluscan claystone

Description. These sediments consist of structure-less to laminated claystones characterized by orange siderite nodules and freshwater gastropod and bivalve shells (Figs 4 and 5). Trochospiral gastropods are found floating in the claystones. Pyritized plant fragments are common. Organic carbon contents range from effectively zero to

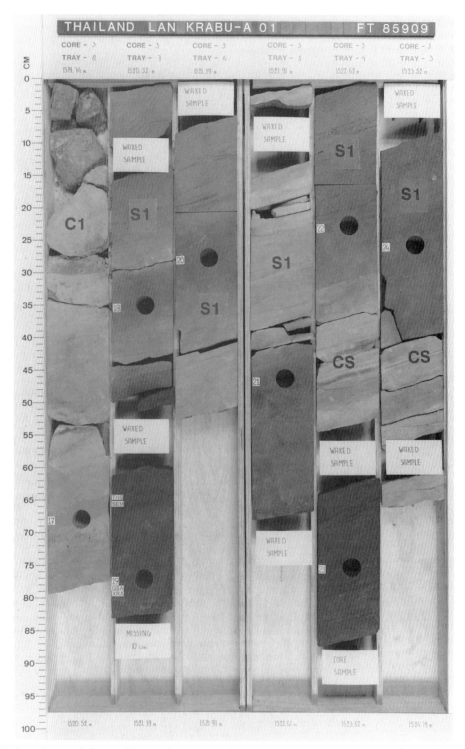

Fig. 3. Core photograph from well Lan Krabu A-01 showing a complex mouth-bar unit (alternating CS and S1 facies) overlain by facies C1 flood-plain claystone.

TABLE 1. *Summarized facies types and characteristics in the reservoir succession*

Facies	Lithology	Colour	Bioturbation	Fossils	Thickness	Depositional environment
C1	Mottled claystone + siltstone laminae + siderite nodules	Pale green	Moderate	Plant fragments	0.2–5.0 m	Well-drained flood plain
C2	Organic-rich claystone	Dark grey	No	Fish and plant fragments	< 0.3 m	Poorly drained flood plain or anoxic lake bottom
C3	Claystone ± siderite nodules	Green-grey	Heavy	Molluscs and fish fragments	0.2–15 m	Open lacustrine; organic content related to lake-floor oxicity
CS	Interbedded claystone–siltstone	Green-grey	Variable	Rare	0.3–2.5 m	Delta front/lower mouth bar
P	Pelletal siltstone and sandstone	Dark green-grey	No	Fish and mollusc fragments	< 0.2 m	Transgressive lag
S1	Fine sandstone; coarsens upwards; cross-bedded; sheet geometry	Brown-grey	Rare	Plant fragments	0.5–5 m	Mouth bar
S2	Fine- to coarse-grained sandstone; fining-upward trend; cross-bedded	Brown-grey	No	Rare plant fragments	0.3–8 m	Active channel fill
S3	Fine-grained sandstone; flat based	Brown-grey	No	Rootlets and plant fragments	< 2 m	Crevasse splay/crevasse delta

20%, and in the latter case are transitional into facies C2. Siltstone and fine-grained sandstone laminae are common. The organic content has an inverse relationship with bioturbation and gastropod shells. Fish fragments are the only fossil material found in the organic-rich claystones. Bed thicknesses vary from 20 cm to 10–15 m. Both facies C2 and C3 are acoustically soft owing to their organic content and are thus distinguished from C1 on the sonic log (Fig. 6).

Interpretation. The oxidized nature of much of the C3 claystone, as evidenced by the siderite nodules, bioturbation and abundant benthonic fauna, indicates oxygenated bottom conditions. Coupled with palaeontological–palynological data, the evidence is consistent with a shallow-lake-bottom environment. Thin siltstone and sandstone laminae may represent storm or flood deposits or reworked edges of crevasse deltas. The organic-rich claystones represent locally deeper or restricted areas of the lake, as they are characterized only by nectonic fauna.

Heterolithic facies CS

Description. This facies consists of interlaminated greenish-grey claystones, pale coloured siltstone and fine-grained sandstone (Figs 3 and 4). The arenaceous material contains sedimentary structures including wavy lamination and trough cross-lamination. Individual trough sets commonly have millimetre-scale clay drapes. Bioturbation is restricted to isolated vertical burrows. The claystones contain gastropod shells, and plant material is concentrated along bedding planes. Wet-sediment deformation is widespread and siderite nodules are common. This facies is difficult to distinguish from facies C1 on the log response (Fig. 6), although its sequential position (see below) is normally diagnostic.

Interpretation. The claystones represent deposition from suspension in a low energy lacustrine environment which was periodically subject to the input of high energy traction currents, leading to sand and silt deposition. The absence of rootlets and soil profiles preclude a subaerial setting. Coupled with its association with other facies types, facies CS is interpreted as prodelta–lower-mouth-bar deposits.

Facies P: pelletal siltstone–sandstone

Description. This facies is characterized by an abundance of poorly sorted dark greenish–grey pellets, fish fragments and shell fragments. Facies P beds reach maximum thicknesses of 25–30 cm,

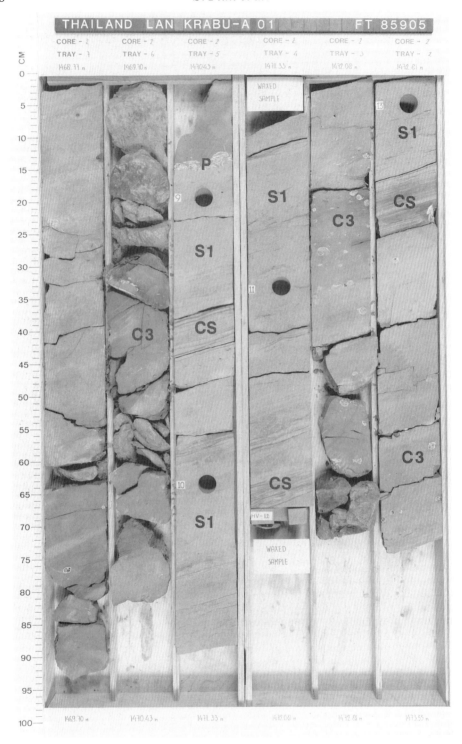

FIG. 4. Core photograph from well Lan Krabu A-01 showing the upward passage from facies C3 lacustrine claystones through CS heterolithics into mouth-bar sandstones (facies S1). Note the erosive contact with the transgressive facies P shell lag which passes upwards into open lacustrine claystones.

FIG. 5. Section of core from Lan Krabu E-02 showing a channel sandstone (facies S2) directly overlying lacustrine claystones. This unit may have cut down through its mouth bar. See Fig. 8 for log responses and the inferred position within the delta model.

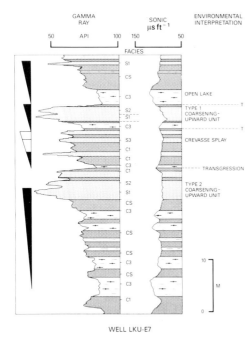

WELL LKU-E7

FIG. 6. Wireline logged section of well Lan Krabu E-07 showing vertical stacking of regressive units and flood-plain sequences, interbedded with transgressive lacustrine claystones.

invariably erosively overlie facies S1–S2 sandstones or C1 claystones and always pass gradationally upwards into lacustrine claystones (Fig. 4).

Interpretation. Based on its sequential position, fossil content and indicators of moderately high energy sedimentation, this facies is interpreted as a lacustrine transgressive deposit.

Sandstone S1

Description. Facies S1 comprises very fine to fine grained, moderately well to very well sorted sandstones (Figs 3 and 4). A well-developed coarsening-upward trend is normally associated with the upward passage from parallel lamination into cross-lamination and cross-bedding. Bioturbation decreases towards the tops of beds, which range in thickness from 0.5 to 5 m. Plant fragments are common, and wet-sediment deformation structures are widespread. The units normally overlie facies CS heterolithics and are characterized by a distinctive coarsening-upward gamma-ray response. Claystone intercalations occasionally result in a more complex serrate log shape (Fig. 6).

Interpretation. The trends in grain size and sedimentary structures seen in this facies are

consistent with an increase in depositional energy. The overall coarsening-upward sequence of the S1–CS packages is indicative of progradational processes consistent with a mouth-bar environment.

Sandstone S2

Description. This facies consists of pebbly to fine-grained sandstones which exhibit well-developed fining-upward trends. Common features include erosive bases and the upward passage from cross-bedding to cross-lamination and parallel lamination. Bed thicknesses range from 0.5 to 8 m. Gamma-ray shapes vary from blocky to bell shaped, depending on grain-size trends. S2 sandstones erode into lacustrine facies C3 claystones (Figs 5 and 6), flood-plain sediments or S1 sandstones. Thin intercalations of claystones within the sandstones are common, giving a modified gamma-ray trace (Fig. 6).

Interpretation. The features of this facies in both core and log response are consistent with an active channel-fill sandstone.

Sandstone S3

Description. This facies comprises thin sandstones (less than 2 m thick) which have flat to erosive bases and contain claystone rip-up clasts and plant fragments. Beds are ungraded to normally graded and vary between structureless and cross-bedded. S3 sandstones are usually well cemented and occasionally show evidence of rootlets in the upper few centimetres.

Interpretation. Based on a combination of the above individually inconclusive features and the sequential position of C3 sandstones within C1 flood-plain claystones, these sandstones are tentatively interpreted as crevasse-splay deposits. They exhibit a typical serrated gamma-ray log response (Fig. 6).

Facies associations and depositional model

The integration of core with log data has allowed the recognition and interpretation of gross facies associations, but not necessarily individual facies, in the dominantly uncored wells (Fig. 7). The detailed correlation of field-wide acoustically soft transgressive lacustrine claystone marker beds has made it possible to ascertain the lateral continuity and resultant geometries of sand bodies. Extending this technique to both the K and upper L reservoirs (some 300–400 m of section) has allowed constraints to be put on the dimensions and morphology of the Sirikit deltaic successions.

The most common facies association in the Lan Krabu Formation is a coarsening-upward association 1.5–15 m thick (Fig. 6 and Plate 2). These associations represent cyclical progradation of deltas into a large relatively shallow freshwater lake. Two subdivisions of the regressive sequences can be made, based on log character and supported by core data.

Type 1 coarsening-upward sequences (Figs 4 and 6) grade from open lacustrine facies C3 claystones at the base through delta-front heterolithics (CS) into mouth-bar–channel sandstones (facies S1–S2) (Fig. 6 and Plate 2). The reservoir units are overlain sharply by transgressive open lacustrine claystones (facies P–C3) which form field-wide barriers to vertical fluid communication between mouth-bar sandstones. Type 1 sequences are thought to represent relatively distal locations within the delta.

Type 2 sequences are more common than type 1 sequences, and are also more complex (Fig. 3). They differ from the former in having a fining-upward vertical aggradational flood-plain unit or abandonment channel fill on top of the mouth-

bar–channel reservoir sandstones (Fig. 6 and Plate 2). In the type 2 sequences many of the mouth-bar and channel units are composite, with individual sandstones separated by minor claystone intercalations; these may represent periods of temporary abandonment related to a variable-discharge fluvial system. Type 2 sequences are considered to be proximal equivalents to type 1 units.

Two scales of the above regressive sequences are seen in the Sirikit field. Those regressive sequences which represent the progradation of deltas into the main lake average 4–8 m in thickness (Plate 2). Smaller-scale (1–3 m thick) coarsening-upward sequences (commonly with little or no sandstone) represent either the distal equivalents of the sandier sequences or crevasse deltas, building into flood-plain lakes or into lake bays between the larger delta lobes. However, well data and detailed log correlations clearly show that the Sirikit mouth bars have an extensive lateral continuity (Fig. 7 and Plate 2). Furthermore, production data suggest that many of these sands are interconnected, forming drainage areas of up to 20 km².

Relatively thick (up to 30 m) flood-plain sequences are developed in the middle K (Plate 1) and lower L reservoirs. This facies association is characterized by facies C1 claystones interbedded with thin S3 crevasse-splay sandstones (Fig. 6). Such sequences represent periods of net flood-plain aggradation. They can be interpreted either as upper-delta-plain facies, deposited in areas where differential subsidence has promoted the accumulation of delta-top sediments, or as the result of rapid rises in lake level, which led to the trapping of sediment in the flood plain.

Frequent lacustrine transgressions are marked by facies P siltstone–shell debris, overlain by acoustically soft C3 claystones. A related facies association comprises a 1–4 m thick sequence which passes upward from a thin C2 organic-rich claystone, sometimes through CS heterolithics into an S3 sandstone. This association may represent local crevasse–delta deposition in minor flood-plain lakes.

The data presented above indicate that a depositional model based on high-constructive lobate (rather than elongate) deltas is applicable to Sirikit (Fig. 8). Such deltas typically prograde into very shallow non-stratified water bodies with relatively low wave energy and high sediment load. An example is the modern Atchafalaya Delta (Fig. 9*A*) (Van Heerden & Roberts 1980). The lobate pattern is caused by complex bifurcation of distributary channels at the point where they enter a shallow open-water body. During initial delta formation, a sand bar is deposited at

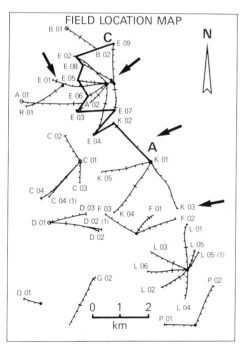

FIG. 7. Field map showing well positions and line of section of Plate 2 (A–C). Arrows indicate directions of lacustrine delta progradation during deposition of the K and upper L reservoirs.

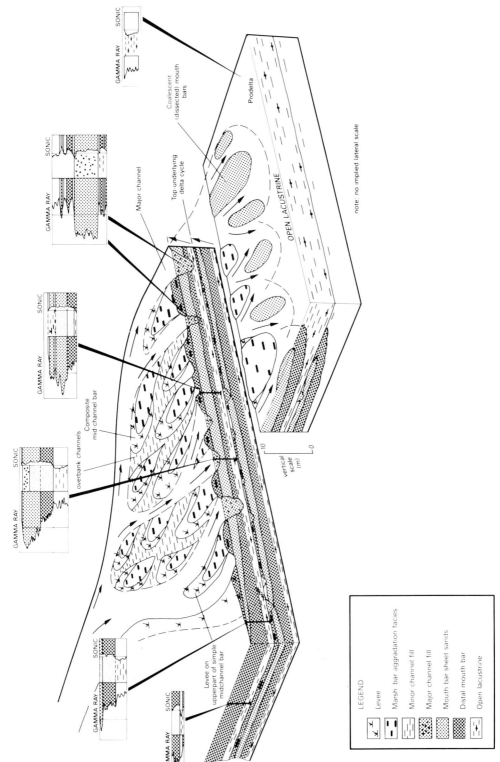

Fig. 8. Depositional model for the highly constructive lobate deltas of the Lan Krabu Formation.

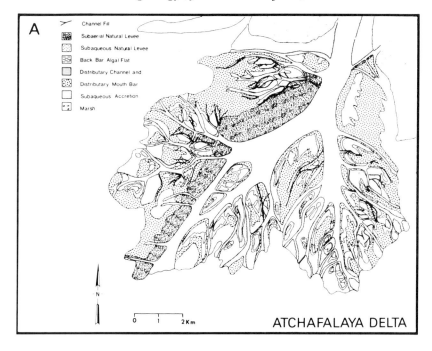

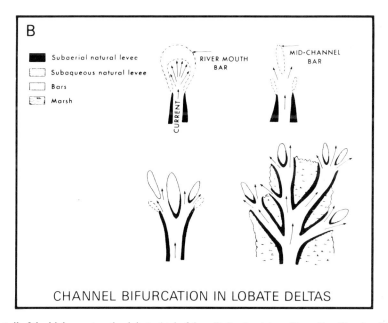

FIG. 9. (*A*) Detail of the high-constructive lobate Atchafalaya Delta, Louisiana (from Van Heerden & Roberts 1981); (*B*) mechanism for the growth of lobate deltas through continued channel plugging and upstream bifurcation.

the river mouth (Fig. 9B). As the delta develops, this bar is quickly dissected to form mid-channel bars which may subsequently amalgamate to form composite sand bars (Figs 8 and 9B). Coalescence of mid-channel bars, channel-fill sequences and mouth bars produces a composite sand body (or lobe) of highly variable thickness.

The rapid reduction in flow velocity and subsequent dumping of sediment at a river mouth is a function of exceptionally high frictional forces. The most important factors controlling such depositional processes are highly variable river discharges and a receiving basin which both has a higher base level than the actual river channels and contains non-saline standing water. The Atchafalaya Basin is open to the sea but is almost completely diluted by freshwater input. This means that there are no sediment buoyancy forces and therefore frictional forces dominate, ie the homopycnal flow of Bates (1953). The net result is that channel mouths are plugged by rapid sedimentation which results in upstream river bifurcation and the backfilling of the original channel by abandonment deposits.

Discussion

Despite the comparative paucity of detailed studies on lacustrine deltas compared with their marine counterparts, it is known that their morphology and resulting sand distribution are controlled by many factors common to both settings. The majority of documented modern and ancient deltas in shallow lakes are of the high-constructive elongate type, with many workers drawing analogies with the modern Mississippi Delta. Examples include the Catatumbo Delta of Lake Maracaibo, Venezuela (Hyne et al. 1979), and the St Clair Delta in the North American Great Lakes (Pezzetta 1973). Ancient elongate lacustrine deltaic successions have been documented from the Powder River Basin, Wyoming (Ayers 1986), and Antarctica (Farquharson 1982).

Deltas with lobate morphologies are less common; modern examples include the Atchafalaya Delta as discussed above, the Lafourche lobe of the pre-Recent Mississippi Delta (Frazier 1967) and the Tonle Sap Delta in the Grand Lac, Kampuchea. We have studied the latter delta geomorphologically (unpublished data) using satellite data and topographic maps. In the ancient record, lobate deltas are usually of the bed-load-dominated Gilbertian type with high angle foresets (Stanley & Surdam 1978; Farquharson 1982). Low gradient lobate deltas are known

from flood-plain lake–crevasse delta sequences in the Westphalian of northern England (Fielding 1984; Haszeldine 1984) and from the present-day Mississippi system (Roberts et al. 1980). In these examples, the lakes are inferred to have had depths of a few metres and extents of tens of kilometres. To our knowledge, no ancient lobate deltaic successions associated with large lakes have been documented. Some poorly exposed deposits of possible lobate deltas have recently been summarized from the marine to brackish Central Pennine Basin, England (Collinson 1987). Lobate marine deltaic successions described from the subsurface of the Gulf Coast, USA (Fisher 1969; Fisher et al. 1969) differ generally from lacustrine deposits in showing evidence of considerable wave reworking. However, our investigations of the Tonle Sap Delta suggest that reworking of sediment by longshore currents may be locally important. Unfortunately we have no data on actual facies distributions in this delta.

Factors known to be important in delta development (Elliott 1986) include basinal energy (wave and tidal processes), type of inflow (hyperpycnal, homopycnal or hypopycnal (Bates 1953)), type of sediment load (suspended, mixed or bed load), climatic regime (rate and variability of inflow and presence of bank-stabilizing vegetation), depth of the ambient water and rates of subsidence. The conditions prevailing during construction of the Sirikit deltas are inferred to have included (i) a shallow large lake, (ii) mixed-load rivers, (iii) homopycnal flow, (iv) minor wave energy, (v) highly variable river discharge rates and (vi) poorly developed levees.

The first four points are common to both elongate and lobate deltas. Sedimentological evidence suggests that the Sirikit deltas were fed by mixed-load streams with a markedly variable discharge rate. The inferred periodic fast sediment accumulation rates contrast with the steady stream flow of recent and ancient elongate deltas (Pezzetta 1973; Ayers 1986). As discussed above, erosional processes related to periodic high energy sediment-laden flood water resulted in overdeepening of distributary channels, plugging of channel mouths and upstream channel bifurcation. The latter process is aided by the absence of permanent levees, which tend to confine flood waters to the main distributary channel and are characteristic of elongate deltas (Ayers 1986). However, the lobate Tonle Sap Delta is subject to catastrophic seasonal flooding and complete submergence during the monsoon, when most of the sedimentation takes place. This highly seasonal pattern of sedimentation tends to inhibit the formation of stable distributary channels.

Conclusions

Core facies analysis, coupled with the correlation of wells, provides good evidence that deltaic sandstones in the Sirikit field overall are of sheet rather than bar-finger geometries. In terms of reservoir development and continuity, this observation represents a significant departure from commonly applied elongate models for low gradient deltas formed in large shallow lakes. Correct identification of the morphology of ancient lacustrine deltaic successions in the subsurface is critical in predicting reservoir sandbody geometries and thus has application in oilfield development planning. Until facies distribution models for lobate deltas associated with large shallow tropical lakes become available, models developed from marine deltas should be applied with caution to ancient and particularly subsurface lacustrine deltaic deposits.

ACKNOWLEDGMENTS: The data presented here represent the efforts of Shell geologists over the last 5 years. In particular, the core facies scheme is based on the work of L. van Geuns and H. Bürgisser. This paper is published with the permission of Shell Internationale Petroleum Maatschappij and Thai-Shell Exploration and Production Company Ltd.

References

AYERS, W. B. JR 1986. Lacustrine and fluvial-deltaic depositional systems, Fort Union Formation (Paleocene), Powder River Basin, Wyoming and Montana. *American Association of Petroleum Geologists Bulletin* **70**, 1651–1673.

BATES, C. C. 1953. Rational theory of delta formation. *American Association of Petroleum Geologists Bulletin* **37**, 2119–2162.

COLLINSON, J. D. 1987. Controls on Namurian sedimentation in the central basins of northern England. *In:* BESLY, B. & KELLING, G. (eds) *Sedimentation in a Synorogenic Basin Complex: The Upper Carboniferous of NW Europe.* Blackwell Scientific Publications, Oxford, 85–101.

ELLIOTT, T. 1986. Deltas. *In:* READING, H. G. (ed.) *Sedimentary Environments and Facies* (2nd edn). Blackwell Scientific Publications, Oxford, 113–154.

FARQUHARSON, G. W. 1982. Lacustrine deltas in a Mesozoic alluvial sequence from Camp Hill, Antarctica. *Sedimentology* **29**, 717–725.

FIELDING, C. R. 1984. Upper delta plain lacustrine and fluviolacustrine facies from the Westphalian of the Durham coalfield, NE England. *Sedimentology* **31**, 547–567.

FISHER, W. L. 1969. Facies characterization of Gulf Coast basin delta systems, with some Holocene analogues. *Transactions of the Gulf Coast Association of Geological Societies* **19**, 239–261.

——, BROWN, L. F., SCOTT, A. J. & MCGOWAN, J. H. 1969. *Delta Systems in the Exploration for Oil and Gas.* Bureau of Economic Geology, University of Texas, Austin, 78 pp.

FRAZIER, D. E. 1967. Recent deltaic deposits of the Mississippi delta: their development and chronol-ogy. *Transactions of the Gulf Coast Association of Geological Societies* **17**, 287–315.

HASZELDINE, R. S. 1984. Muddy deltas in freshwater lakes, and tectonism in the Upper Carboniferous Coalfield of NE England. *Sedimentology* **31**, 811–822.

HYNE, N. J., COOPER, W. A. & DICKEY, P. A. 1979. Stratigraphy of intermontane lacustrine delta, Catatumbo river, Lake Maracaibo, Venezuela. *American Association of Petroleum Geologists Bulletin* **63**, 2042–2057.

KNOX, G. & WAKEFIELD, L. L. 1983. An introduction to the geology of the Phitsanulok Basin. *Proceedings of the Conference on Geology and Mineral Resources of Thailand, Bangkok, Thailand, 1983.* Geological Society of Thailand, Bangkok.

PEZZETTA, J. M. 1973. The St. Clair river delta: sedimentary characteristics and depositional environments. *Journal of Sedimentary Petrology* **43**, 168–187.

ROBERTS, H. H., ADAMS, R. D. & CUNNINGHAM, R. H. W. 1980. Evolution of sand-dominant subaerial phase, Atchafalaya delta, Louisiana. *American Association of Petroleum Geologists Bulletin* **64**, 264–279.

STANLEY, K. O. & SURDAM, R. C. 1978. Sedimentation on the front of Eocene Gilbert-type deltas, Washakie Basin, Wyoming. *Journal of Sedimentary Petrology* **48**, 557–573.

VAN HEERDEN, I. L. & ROBERTS, H. H. 1980. The Atchafalaya Delta—Louisiana's new prograding coast. *Transactions of the Gulf Coast Association of Geological Societies* **30**, 497–506.

WORKMAN, D. R. 1975. Tectonic evolution of Indochina. *Journal of the Geological Society of Thailand* **1**, 3–9.

S. FLINT & D. J. STEWART, Koninklijke/Shell Exploratie en Produktie Laboratorium, Postbus 60, 2280AB, Rijswijk Z-H, The Netherlands.
E. D. VAN RIESSEN, Thai-Shell Exploration and Production Company Ltd, 10 Soonthornkosa Road, Klontoey, Bangkok, Thailand.

Review and computer modelling of the Brent Group stratigraphy

W. Helland-Hansen, R. Steel, K. Nakayama & C. G. St. C. Kendall

SUMMARY: The depositional history of the Brent Delta system can be analysed on three different levels: (i) the individual formations and their facies distribution, used for the interpretation of the depositional processes; (ii) the vertical succession of formations in each well, indicating the temporal changes in environment; (iii) the two- or three-dimensional array of sequences illustrating both the temporal and spatial variations within the delta system. The Brent Group is essentially one regressive–transgressive megacycle. The regressive part comprises the Rannoch and Etive Formations (delta front) and the lower Ness Formation (delta plain), and the transgressive system contains the Tarbert Formation (delta front) and the (upper) Ness Formation (delta plain). With the aid of a computer program, the succession geometry can be analysed in terms of the relative importance of sediment supply, eustasy and tectonics. The change from regressive to transgressive conditions was probably initiated by an increase in subsidence rates, whereas regressive pulses in the overall transgressive Tarbert Formation can be explained in part by eustatic sea-level falls.

The Brent Group and its lateral equivalents, a Middle Jurassic succession of alternating sandstones, siltstones, mudstones and coals up to 600 m thick, is best known for its prolific hydrocarbon reservoirs in both the British and Norwegian sectors of the northern North Sea. It has long been recognized as largely deltaic in origin (eg Bowen 1975; Deegan & Scull 1977), but only since the early 1980s have details of sedimentation and palaeogeography begun to emerge. Important papers by Eynon (1981) and Skarpnes et al. (1980) were the first published attempt to predict the regional development of the Brent Group, and thus the first palaeogeographic maps of this Jurassic deltaic complex. The early work also established that there had been an important early Middle Jurassic uplift in the area around and to the S of the present southern Viking Graben, and that this uplift, together with a consequent relative fall in sea level, led to the northward progradation of the Brent Group deltaic system (Fig. 1).

Further sedimentological and biostratigraphical studies have provided a better understanding of the large- and small-scale stratigraphic relationship and time framework *within* the Brent Group succession. Critical elements in this latest work have been the following:

(1) Mapping of the sandy alluvial fan–deltaic lobes which prograded into a marine northern North Sea Basin from both the W (Broom Formation) and the E (Oseberg Formation) while the Brent Delta proper was establishing itself farther S in the Viking Graben. Studies by Brown et al. (1987), Graue et al. (1987) and Brown & Richards (this volume) document this earliest (Aalenian–early Bajocian)

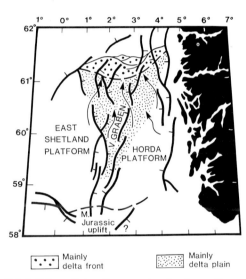

FIG. 1. Schematic location map showing the position of the Viking Graben relative to the Bajocian Brent Delta system.

lateral infill phase of the northern North Sea basin, and imply early Jurassic uplift along the flanks of the embryonic Viking Graben in addition to uplift in the S.

(2) Understanding and mapping of the broadly contemporaneous S–N facies changes within the deltaic 'package', eg in Bajocian times from fluvial through lagoonal and lower-delta-plain deposits to barrier and other shoreline sandstones and eventually to prodelta mudstones in the northern reaches of the East Shetland Basin (Budding & Inglin

1981; Johnson & Stewart 1985; Brown et al. 1987; Brown & Richards, this volume).

(3) Better definition of the northernmost termination of the Brent Group sandstones (see also Graue et al. 1987). This is important for our understanding of the temporal and spatial relationships between the Rannoch–Etive and Tarbert Formations at a time when the deltaic system reverted from being dominantly progradational to being dominantly retrogradational.

(4) Improved understanding of the nature, extent and origin of the Tarbert and Hugin Formations, a succession believed to be the sandy product of the 'transgressive' phase of delta development (Graue et al. 1987). Local work reported by Rønning & Steel (1987) and regional work reported here strongly suggest that the Tarbert Formation (taken as the marine shoreline sandstones deposited during the gradual drowning of the Brent Delta) is stratigraphically complex. The Tarbert and Hugin Formations can be regarded as analogous to the Etive–Rannoch Formations. They represent the shoreline deposits from the dominantly retrogradational phase of delta building and therefore tend to interfinger with and *overlie* the Ness–Sleipner delta-plain deposits. In contrast, the Etive–Rannoch Formations are shoreline deposits from the delta's regressive phase and therefore tend to *underlie* the Ness Formation. We therefore suggest that it is inappropriate to refer to the Tarbert Formation as a 'transgressive sheet sand' (*eg* Johnson & Stewart 1985), because the formation represents a series of marine sand bodies which are highly diachronous in their development and vary in age from Bajocian to Callovian (see Fig. 13).

The main aims of this paper are to review briefly some of the lateral variability of the Brent Group stratigraphy and to present the preliminary results of computer modelling of the deltaic advance and retreat, as interpreted from the Brent Group stratigraphy.

Brent Group succession

The dynamic stratigraphy of the Brent Group and the internal characteristics and lateral variability of the Brent Delta system can be illustrated on three different levels: (i) the individual formations with their facies distribution used to interpret the depositional environments and the processes which built the delta system; (ii) the vertical succession of formations at an individual locality that reflect the temporal change in environments; (iii) the two- or three-dimensional vertical array of sequences that illustrates the overall geometry of the deltaic system as well as the details of internal variability. Interpretation at this level potentially allows the basin-infill history to be unravelled.

The lithostratigraphic components

The *Broom and Oseberg Formations* are the oldest lithostratigraphic units in the Brent Group, and represent 'early' sands laterally dispersed into the northern North Sea Basin. These deposits are not components of the main Brent Delta complex, which resulted from longitudinal progradation essentially northwards along the present Viking Graben, but are nevertheless an important part of the basin development history.

The Oseberg Formation (up to 60 m thick), located along the eastern edge of the Viking Graben, accumulated as fault-controlled fan deltas, with a sediment source near the present Norwegian mainland, and advanced across the Horda Platform (Graue et al. 1987). The Broom Formation, which is more than 25 m thick in places (Brown et al. 1987), has been variously interpreted (Chauvin & Valachi 1980; Eynon 1981) but is probably a western analogue to the Oseberg Formation, although the deposition of the two formations is not necessarily synchronous. Facies sequence signatures in both formations emphasize flood-generated sedimentation, often by highly concentrated partly turbulent subaqueous flows and commonly within Gilbert-type fan-delta bodies (Graue et al. 1987).

The *Rannoch Formation* most commonly consists of one or several coarsening-upward sequences with thicknesses in the range from 5 m to several tens of metres. The lower part of the formation contains interbedded shales, siltstones and micaceous very fine-grained sandstones which are often bioturbated and organized into flat-laminated and wave-ripple-laminated units. The upper part of the formation is dominated by fairly well sorted micaceous very fine and fine-grained sandstones which show hummocky cross-stratification. The types of strata and facies sequences are typically those generated by storm-wave processes (Brown et al. 1987; Brown & Richards, this volume) and are summarized in Fig. 2. The Rannoch Formation represents the lower to middle reaches of a prograding wave-influenced delta front, whereas the Etive Formation represents the upper part.

The *Etive Formation* is dominated by low angle laminated, trough cross-stratified or structureless medium- to coarse-grained sandstones. It either

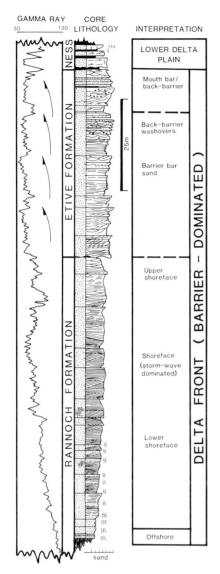

GAMMA RAY   CORE
30 ___ 130 LITHOLOGY   INTERPRETATION

LOWER DELTA PLAIN

Mouth bar/ back–barrier

Back–barrier washovers

Barrier bar sand

Upper shoreface

Shoreface (storm-wave dominated)

Lower shoreface

Offshore

sand

NESS

ETIVE FORMATION

RANNOCH FORMATION

DELTA FRONT (BARRIER – DOMINATED)

FIG. 2. Representative facies sequences in the Rannoch and Etive Formations, Tampen Spur area.

sharply overlies or grades up from the underlying Rannoch Formation. Despite the formation's 'blocky' or overall coarsening-upward character, smaller scale (0.5–3 m) fining-upward units are common. These units often have pebble-lined erosive bases and are probably of lensoid geometry, and they are interpreted in terms of a barrier-bar and mouth-bar setting (Fig. 2). Thus the Etive Formation represents the upper part of the Brent delta front. This marine sand belt was established during the advance of the Brent Delta and appears to have achieved a more uniform and wider distribution than the Tarbert Formation sand belt which developed during the delta retreat.

The *Ness Formation* comprises interbedded shales, coals, siltstones and sandstones with sequences typical of fluvial channel, overbank, interdistributary bay and lagoonal environments. Facies sequences are characteristically flood generated and commonly capped by coal layers (Fig. 3). Wave-influenced bay-fill sequences are particularly common in the Ness Formation in the northern areas. The Ness Formation is a delta plain time-equivalent to the Rannoch–Etive delta front during the period of Brent Delta advance. Moreover, the formation continued to be deposited during the retreat of the delta systems. It is also likely that much of the Sleipner Formation (farther S) accumulated during the transgressive phase of the delta evolution (Graue *et al.* 1987).

The *Tarbert Formation*, variably developed over the whole region as the Brent Delta retreated, consists of marine sandstones, siltstones and shales. A landward interfingering of these marine deposits with lagoonal and other lower-delta-plain deposits results in coals, non-marine shales and flood-generated sandstones interfingering with the Tarbert Formation marine facies, as illustrated in Fig. 4.

Time trends of sedimentation

The overall temporal trend of sedimentation represented by the Brent Group is usually portrayed to show a succession such as that in Fig. 5, with an initial classic regressive phase (Rannoch–Etive–Ness Formations) followed by a destructive or 'transgressive' phase (Ness–Tarbert–Heather Formations). This picture is essentially correct for the development of the Brent Group in the northern reaches of the East Shetland Basin (especially N of 61°N), but becomes severely modified in (i) areas of 'proximal' deltaic development, where either there has not been any significant delta front developed or where such deposits are lacking, and (ii) areas which have been subject to unusually fast or unusually slow rates of subsidence.

Figure 6 illustrates (i) above, where the Rannoch and Etive Formations are absent and alluvial and delta-plain deposits (the Sleipner Formation) rest directly on older marine deposits. This is well illustrated in the stratigraphic model shown in Fig. 8. The latter situation, where variation in subsidence rates markedly affects the stratigraphy, has been noted and documented by Johnson & Stewart (1985). Figure 6 shows a greatly thickened Sleipner Formation, here a product of fast subsidence rates in the region of

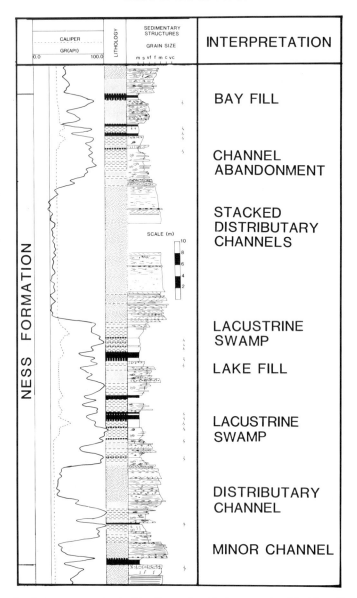

FIG. 3. Representative facies sequences in the Ness Formation, Oseberg area (modified from Ryseth 1987).

the present Viking Graben during the transgressive phase of the Brent Delta. Figure 7, showing a succession from the edge of the Horda Platform, illustrates the effects of overall slow subsidence rates with a consequently thin or condensed Brent Group.

This variation in succession thickness and the time trend of sedimentation is important in illustrating the lateral variability which exists in the Brent Group, as is discussed further below.

Lateral variability in basin fill

Two correlation panels illustrate the lateral variability within the Brent deltaic succession. Figure 8 shows an essentially N–S (dip) section, whereas Fig. 9 is somewhat more oblique. Figure 8 forms the basis of the computer modelling discussed below.

The following features should be noted from Figs 8 and 9 as an introduction to the stratigraphic

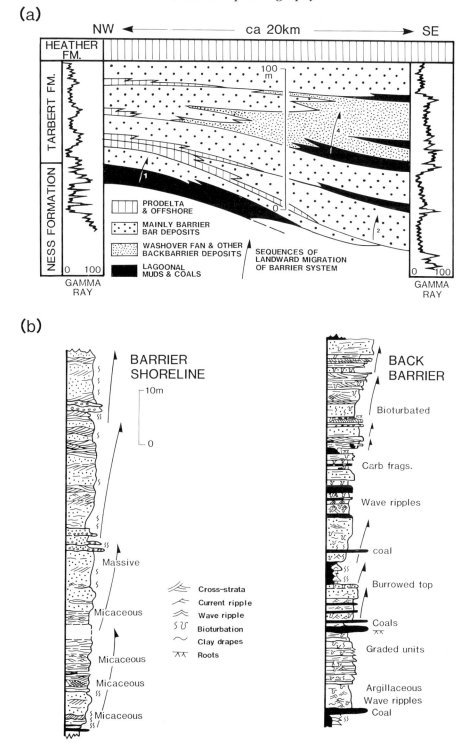

FIG. 4. Representative sequences in the Tarbert Formation: (*a*) lateral facies changes from marine deposits to lower-delta-plain deposits (W margin of the Viking Graben); (*b*) representative facies sequences.

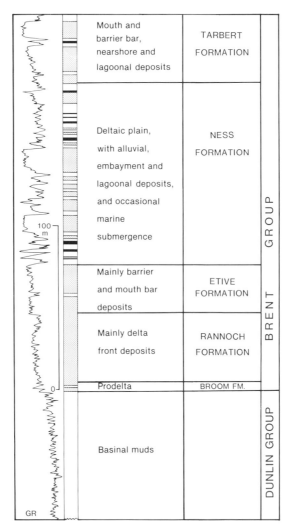

Fɪɢ. 5. Example of a typical vertical Brent Group succession from the Tampen Spur area.

modelling: (i) the basinward thickening of the Rannoch and Etive Formations; (ii) the overall 'transgressive' offset-stacked Tarbert and Hugin Formations; (iii) the time equivalence of the Ness Formation with the Rannoch and Etive Formations, as well as with the Tarbert and Hugin Formations.

Computer testing of Brent Group stratigraphy

The stratigraphic relations proposed above for the Brent Group have been quantified in a computer program that generates two-dimen-

sional representations of the time lines within sediment bodies as a response to input parameters of basin topography, sediment supply, subsidence (or uplift) and eustasy. The simulation is based on the concept that important gross features of depositional geometries, averaged over long periods of time, are controlled by the above processes. The program is briefly described before its application to the Brent Group is demonstrated. A more detailed description of the program is presented by Helland-Hansen *et al.* (in press).

Input parameters (Fig. 10)

The main input parameters are (i) the configura-

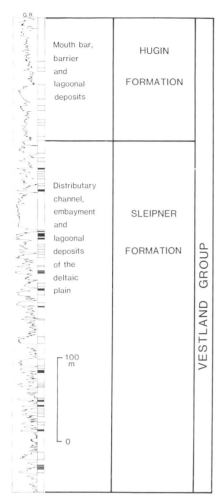

FIG. 6. Example of a 'proximal' development of the Brent Group from the central Viking Graben area.

and generates a depositional surface or time line (Helland-Hansen *et al.*, in press).

Sequence of operations within each time step (Fig. 11)

The basic sequence of processes followed in each time step is as follows.

(1) Define depositional triangles whose area expresses the amount of sand and shale to be deposited for that particular time step.
(2) Locate the shoreline according to the given eustatic position of the sea level.
(3) Subside all surfaces according to the given tectonic subsidence for that time step.
(4) Locate the shoreline (position modified after subsidence).
(5) Erode sediments that are elevated above the 'alluvial base line' to a depth determined by the erosion formula (the 'alluvial base line' is defined by a line from the shore landwards with an inclination that is given by the user).
(6) Deposit sediments, including those eroded above and those to be deposited for that time step (as defined by the triangles).
(7) Compact all sediment layers to equilibrium according to the weight of the overburden.
(8) Subside all surfaces isostatically in response to sediment and water loading on the crust.

This sequence of subsidence, erosion, deposition, compaction and loading is carried out for

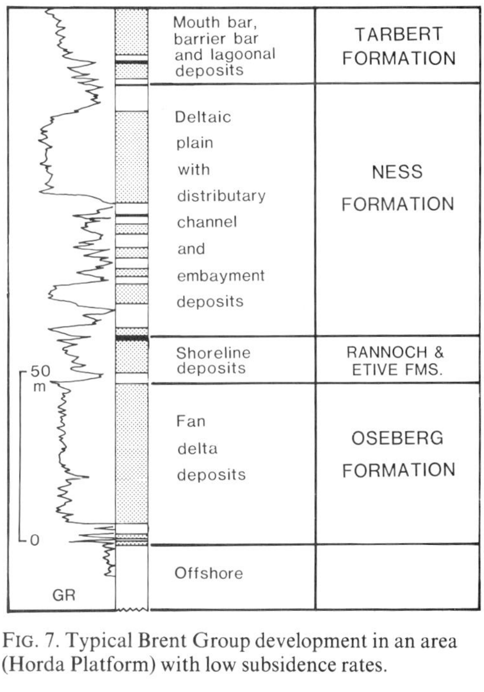

tion of the initial depositional surface (Fig. 10*a*), (ii) sediment supply versus time (for sand and shale respectively) (Fig. 10*b*), (iii) subsidence (or uplift) versus time (which can be defined for several positions in the basin) (Fig. 10*c*) and (iv) eustasy versus time (Fig. 10*d*). Moreover, sediments are dispersed according to a given gradient of the alluvial plain and a given distance from the shoreline to sand and shale pinch-out basinwards. The sediments are compacted according to empirical relations of overburden versus porosity, and the level of the basin floor is adjusted vertically as a response to the sediment and water load. The time span over which the simulation is defined is divided into time steps (maximum of 45). For each time step, the program reads off a value for each of the input parameters

FIG. 7. Typical Brent Group development in an area (Horda Platform) with low subsidence rates.

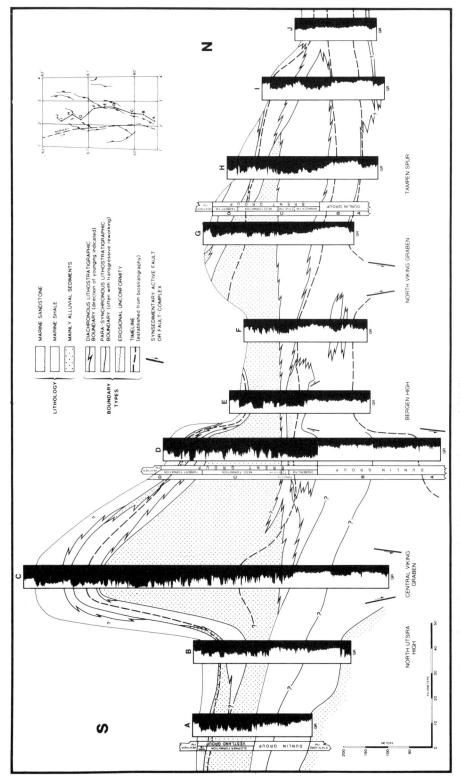

Fig. 8. N–S stratigraphic correlation through the Brent Group (from Graue *et al.* 1987).

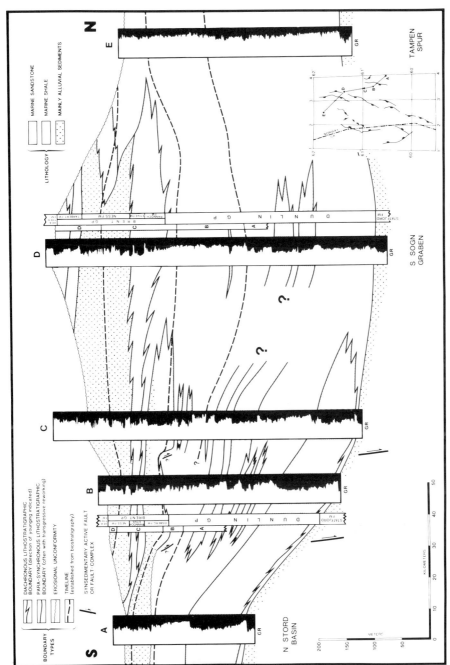

FIG. 9. NNW–SSE stratigraphic correlation through the Brent Group (along the eastern basin margin).

(a) INITIAL BASIN SURFACE

Provide horizontal and vertical
position of vertices (linear
interpolation between)

(b) SEDIMENT ACCUMULATION RATE

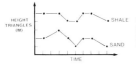

Provide triangle heights against
time–steps for sand and shale
(linear interpolation between)
Lengths of triangles are constant

(c) SUBSIDENCE RATE

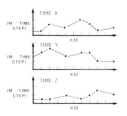

Provide subsidence rate against
horizontal location in basin for
designated time - steps (here x,y,z).
Program will interpolate linearly in
both space and time dimension

(d) EUSTASY

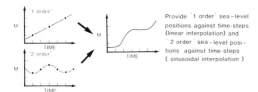

Provide '1 order' sea - level
positions against time-steps
(linear interpolation) and
'2 order' sea - level positions against time-steps
(sinusoidal interpolation)

FIG. 10. Main controls in the simulation for the
configuration of the time lines within the sediment fill
(modified from Helland-Hansen et al., in press).

each time step. Deposition involves the creation
of new surfaces above the previous one. Both
erosion and compaction involve reductions in the
layer thicknesses. Tectonic subsidence and loading cause the surfaces to drop vertically (Helland-
Hansen et al., in press).

Output plots

Figure 12 shows four examples of output plots.
The input data do not relate to actual examples.
Each of the lines within the sequences represents
a time line. The asterisks indicate the shoreline
position for each of the time lines. In these four
examples both the initial basin topography, which
has a simple concave-up profile, and the sediment
supply (constant through time) are the same.

However, different values are given for eustasy
and subsidence.

In Fig. 12a the eustatic sea level is constant
with no tectonic subsidence and the basin floor
only adjusts to maintain isostatic equilibrium. In
Fig. 12b the basin floor is also responding to
tectonic subsidence (being constant in time and
space), giving a progradational sequence with a
stronger component of vertical aggradation than
in the previous example. Figure 12c exemplifies
how a sinusoidally varying eustatic sea level
together with no tectonic subsidence affects the
sequence geometry. Finally, Fig. 12d shows the
influence of pulses of high tectonic subsidence.
The eustatic sea level is constant here. Figures
12c and 12d both show examples of transgressive–
regressive cycles but with quite different internal
geometries.

Application to Brent Group stratigraphy

When simulating the Brent Group stratigraphy,
we have input a set of parameters for which the
depositional sequence generated has its main
geometrical and genetical features in common

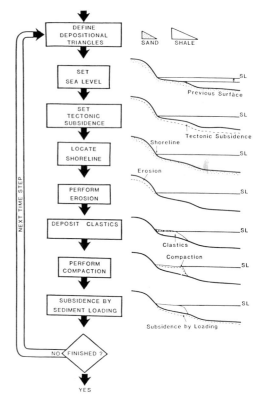

FIG. 11. Flow chart of simulation sequence (from
Helland-Hansen et al., in press).

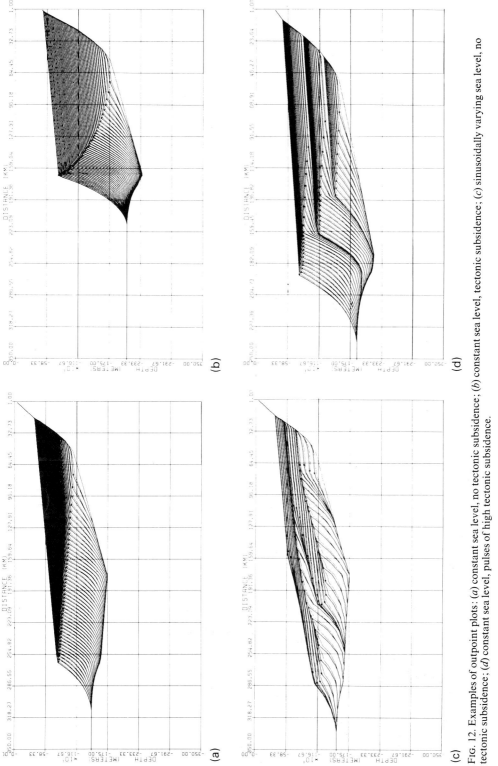

FIG. 12. Examples of outpoint plots: (*a*) constant sea level, no tectonic subsidence; (*b*) constant sea level, tectonic subsidence; (*c*) sinusoidally varying sea level, no tectonic subsidence; (*d*) constant sea level, pulses of high tectonic subsidence.

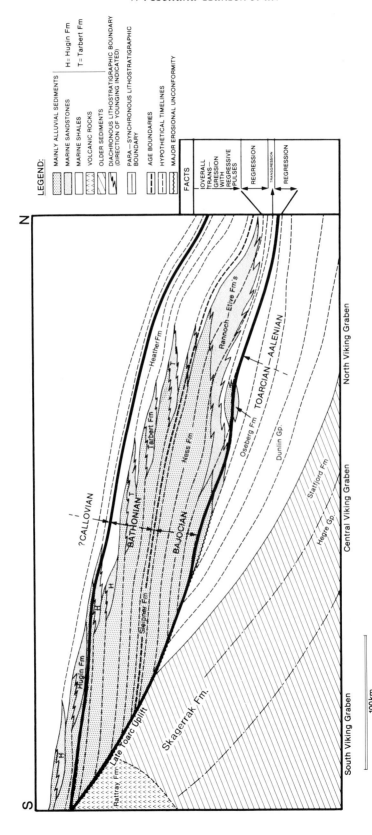

FIG. 13. Conceptual model of the Brent Group stratigraphy with the simulated sequence enveloped by the bold line (modified from Graue et al. 1987).

TABLE 1. *Input data for simulation of the Brent Group*

Start time (Ma)	0.0							
End time (Ma)	14.0							
Number of time steps (maximum, 45)	45							
Start, horizontal position (km)	1							
End, horizontal position (km)	350							
Length of sand deposition triangle (km)	5							
Length of shale deposition triangle (km)	60							
Angle for alluvial deposition (deg)	0.005							
Time steps	1	18	24	45				
Height of sand deposition triangle (m)	180	180	180	180				
Height of shale deposition triangle (m)	45	45	45	45				
Density of shale matrix (g cm^{-3})	2.65							
Density of sand matrix (g cm^{-3})	2.65							
Density of mantle (g cm^{-3})	3.40							
Locations (km)	1	350						
Time step	1							
Tectonic subsidence (m per time step)	1.0	1.0						
Time step	18							
Tectonic subsidence (m per time step)	1.0	1.0						
Time step	24							
Tectonic subsidence (m per time step)	8.0	8.0						
Time step	45							
Tectonic subsidence (m per time step)	8.0	8.0						
Time steps for average sea level	1	45						
Amplitude	− 50.0	− 86.0						
Time steps per quarter cycle	1	2	3	8	12	24	30	33
	36	37	38	41	45			
Amplitude of sea level (from average) (m)	0.0	5.0	0.0	− 10.0				
	0.0	30.0	0.0	− 10.0				
	0.0	1.0	0.0	− 28.0				
	0.0							
Maximum burial for sediment (m)	0.300E + 04							
Horizontal position of vertices (initial surface) (km)	1	50	100	350				
Vertical position of vertices (initial surface) (m)	150.0	65.0	0.000E + 00	− 200.0				

with the depositional model in terms of areal cross-section, geometry of regressive and transgressive cycles *etc.* Moreover, the input parameters have been critically evaluated in terms of their geological validity (see below).

The stratigraphic cross-section in Fig. 8 has been used as a basis for the modelling. The main elements of this N–S dip section are the basinward-thickening Rannoch–Etive succession backed by the alluvial sediments of the (lower) Ness Formation and overlain by the overall transgressive offset-stacked progradational units of the Tarbert Formation. The shorelines of the Tarbert Formation also fronted an extensive alluvial plain (upper Ness Formation). Southwards, the Ness Formation grades into the Sleipner Formation and the Tarbert Formation grades into the Hugin Formation.

The advancing stage of this delta development took place mainly within Bajocian–Aalenian times, whereas the retreat took place mainly in the Bathonian and Callovian. In the modelling, we have focused on the Bajocian–Aalenian to Bathonian phase of the sequence development (Fig. 13). In the simulation, 14 Ma has been used as the input value for the time span covered by the sequence.

The thickness distribution of the simulated sequence is not intended to account for complex thickness variations across faults; instead it shows an 'average' thickness distribution as suggested by a regional grid of well data. Sediment supply has been input as constant through time and has been determined by dividing the area of the sediments present (decompacted) within the actual sequence by the number of time steps. The sand-to-shale ratio (of uncompacted sediments) has been input as 1:3.

For the distribution of sediment, the distance from the shore to the basinwards pinch-out has been input as 5 km and 60 km for sand and shale respectively, and the alluvial gradient has been given as 0.005° (in accordance with recent low gradient alluvial systems (*cf* Collinson 1978)). Of importance in the simulation of the succession is the absence of clear sedimentological evidence

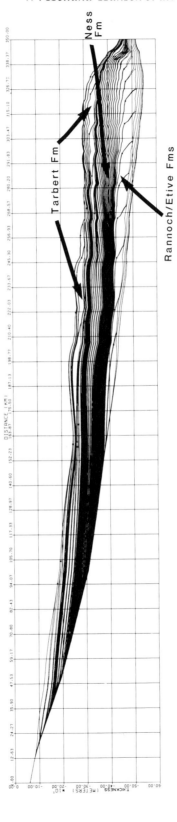

FIG. 14. Output plot of the computer simulation of the Brent Group.

for major erosional truncations within the Brent Group. Eustatic sea-level variations have been input in accordance with the latest global sea-level curves from the Exxon group (Haq *et al.* 1987) for the time period 157–171 Ma.

In addition to the compaction performed according to the weight of the intrasection overburden, an additional 3000 m of sediments were laid down above the Brent Group and account for the compaction resulting from post-Brent Group overburden.

Finally, the magnitude of the tectonic subsidence has been used as a trial-and-error factor to adjust the simulated geometries to the stratigraphic model. The input data are listed in Table 1.

Results

Figure 14 shows the simulated succession obtained using the input data outlined above. Figure 15 is a graphical illustration of the variation of eustasy and tectonic subsidence. Here, the gradient of the cumulative subsidence curve increases considerably after 5.5 Ma (when 157 Ma is considered to be zero time); this change is an expression of the fact that subsidence rates have to be increased to generate the transgression leading to the overall retreat of the Brent Delta. The two eustatic falls from 7 to 10 Ma and from 12 to 13 Ma respectively are responsible for the two regressive packages in the upper part of the output geometry corresponding to the Tarbert Formation (*cf* Fig. 15). It should be noted that these two eustatic falls do not give *relative* sea-level falls owing to the high subsidence rate.

However, the accommodation is temporarily decreased so that progradation can take place. The Tarbert Formation may contain more than two regressive phases (Graue *et al.* 1987). These, if present, have to be attributed to mechanisms other than eustasy (assuming that the Exxon sea-level curve is valid).

Important constraints on the above modelling are the assumptions of constant sediment supply and the correctness of the eustatic curve, both of which are highly speculative. As a result of limited biostratigraphical control the timing of deposition for the succession is also a matter of debate. It is of interest to note from the output geometry that each transgressive episode, here a product of increased accommodation (increased subsidence or eustatic rise), also produces expanded sections in the alluvial counterpart (*cf* Fig. 14). This is because the alluvial sediments have to adjust to new, and higher, base levels.

Alternatively, a similar succession could be produced by decreasing the sediment supply and keeping eustasy and tectonic subsidence constant. In such a case, the feature of expansion of the time lines in the alluvial plain would not be present since the alluvial sediments would not have to adjust to a higher base level. Future research on the Ness and Sleipner Formations may lead to improved understanding of these matters.

Conclusions

Several hypotheses can be put forward concerning the cause of the change from an advancing to a

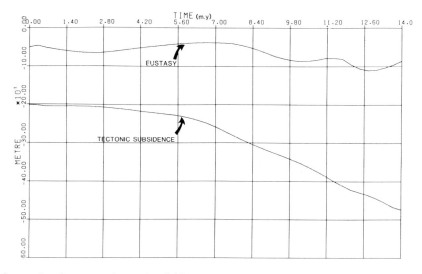

FIG. 15. Output plot of eustasy and tectonic subsidence.

retreating stage of the Brent Delta development. Increased accommodation from eustasy seems less likely when the timing of the delta retreat is compared with the latest global sea-level charts. A reduction of sediment supply could explain this change. A general levelling of the source area with time would contribute to such an effect. Moreover, an expansion of the alluvial plain as the (Rannoch–Etive) progradation took place would give an increasing 'alluvial trapping' of sediment, thus reducing the amount of sediments supplied to the shore.

Alternatively, if it is assumed that sediment supply was constant through time and that the global sea-level chart is valid, an increase in tectonic subsidence has to be invoked. This scenario has been computer tested. More or less coeval transgressive events across the mid-Norwegian and northern Norwegian shelves support the notion of a regional increase in tectonic subsidence rates.

The offset backstepping progradational prisms in the overall transgressive Tarbert and Hugin Formations could be explained in terms of eustasy, tectonic subsidence and sediment supply. Eustasy can only account for two regressive packages in the actual period because only two eustatic falls are present in this time span. A variable sediment supply, however, could be attributed to either lateral switching of the sediment source or an irregular sediment supply on a regional basis.

References

BOWEN, J. M. 1975. The Brent oil-field. *In:* WOODLAND, A. W. (ed.) *Petroleum and the Continental Shelf of North-West Europe.* Applied Science Publishers, Barking, 353–362.

BROWN, S., RICHARDS, P. C. & THOMSON, A. R. 1987. Patterns in the deposition of the Brent Group (Middle Jurassic), The North Sea. *In:* BROOKS, J. & GLENNIE, K. (eds) *Petroleum Geology of North West Europe.* Graham & Trotman, London, 899–913.

BUDDING, M. C. & INGLIN, H. F. 1981. A reservoir geological model of the Brent sands in Southern Cormorant. *In:* ILLING, L. V. & HOBSON, G. D. (eds) *Petroleum Geology of the Continental Shelf of North-West Europe.* Heyden, London, 326–334.

CHAUVIN, A. & VALACHI, L. 1980. Sedimentology of the Brent and Statfjord Formations of Statfjord Field. *In: The Sedimentation of the North Sea Reservoir Rocks.* Norwegian Petroleum Society, Geilo, 16/1–17.

COLLINSON, J. D. 1978. Alluvial sediments. *In:* READING, H. G. (ed.) *Sedimentary Environments and Facies.* Blackwell Scientific Publications, Oxford, 15–60.

DEEGAN, C. E. & SCULL, B. J. 1977. A proposed standard lithostratigraphic nomenclature for the Central and Northern North Sea. *Report of the Institute of Geological Sciences* **77/25**, 36 pp.

EYNON, G. 1981. Basin development and sedimentation in the Middle Jurassic of the Northern North Sea. *In:* ILLING, L. V. & HOBSON, G. D. (eds) *Petroleum Geology of the Continental Shelf of North-West Europe.* Heyden, London, 196–204.

GRAUE, E., HELLAND-HANSEN, W., JOHNSEN, J., LØMO, L., NØTTVEDT, A., RØNNING, K., RYSETH, A. & STEEL, R. 1987. Advance and retreat of Brent Delta System, Norwegian North Sea. *In:* BROOKS, J. & GLENNIE, K. (eds) *Petroleum Geology of North West Europe.* Graham & Trotman, London, 915–937.

HAQ, B. U., HARDENBOL, J. & VAIL, P. R. 1987. Chronology of fluctuating sea levels since the Triassic. *Science* **235**, 1156–1167.

HELLAND-HANSEN, W., KENDALL, C. G. ST. C., LERCHE, I. & NAKAYAMA, K. A simulation of continental basin margin sedimentation in response to crustal movements, eustatic sea-level change and sediment accumulation rates, in press.

JOHNSON, H. D. & STEWART, D. J. 1985. Role of clastic sedimentology in the exploration and production of oil and gas in the North Sea. *In:* BRENCHLEY, P. J. & WILLIAMS, B. P. J. (eds) *Sedimentology: Recent Developments and Applied Aspects.* Geological Society of London Special Publication **18**, 249–310.

RØNNING, K. & STEEL, R. 1987. Depositional sequences within a 'transgressive' reservoir sandstone unit: the Middle Jurassic Tarbert Formation, Hild area, northern North Sea. *In:* KLEPPE, J., BERG, E. W., BULLER, A. T., HJELMELAND, O. & TORSÆTER, O. (eds) *North Sea Oil and Gas Reservoirs.* Norwegian Institute of Technology, Trondheim, 169–176.

RYSETH, A. 1987. *Facies and Distribution of Coals and Associated Clastic Deposits of the Ness Formation (Middle Jurassic), Oseberg Area.* Unpublished cand. scient. thesis, University of Bergen.

SKARPNES, O., HAMAR, G. P., JACOBSSON, K. H. & ORMAASEN, D. E. 1980. Regional Jurassic setting of the North Sea north of the central highs. *In: The Sedimentation of the North Sea Reservoir Rocks.* Norwegian Petroleum Society, Geilo, 13/1–8.

W. HELLAND-HANSEN, Norsk Hydro, PO Box 4313, 5028 Bergen, Norway.

R. STEEL, Norsk Hydro, PO Box 200, 1321 Stabekk, Norway.

K. NAKAYAMA, Japex, Tokyo 107, Japan.

C. G. ST. C. KENDALL, University of South Carolina, Columbia, SC 29208, USA.

Facies and development of the Middle Jurassic Brent Delta near the northern limit of its progradation, UK North Sea

S. Brown & P. C. Richards

SUMMARY: The lithofacies patterns in the Middle Jurassic Brent Group, the principal hydrocarbon reservoir in the UK northern North Sea, are examined in seven wells. The sections, from the area of the Don and Murchison oil fields, record the nature of the Brent Delta near the northern limit of its progradation. Here the deltaic deposits of the Brent Group, which regionally constitute a major regressive–transgressive clastic wedge, progressively lose the interbedded sands and muds of the delta plain which characterize the middle part of the succession further S. This leads locally to a maximum sand-to-mud ratio for the Brent Group reservoir interval.

A regressive storm-wave-influenced marine succession, from offshore transition-zone silts and muds up to hummocky cross-stratified nearshore sand, is overlain in turn by stacked distributary channel sands. Distributary mouth deposits are absent. The interbedded delta-plain succession, where it occurs, exhibits some marine influence and records wave reworking of interdistributary bay or lagoonal sediments. The Brent Group succession is capped by transgressive sands which were deposited in a storm-wave-influenced nearshore environment. Delta-plain drowning is also evident in the gradual marine incursions into distributary channels at the northern limit of the study area.

The character of the delta locally is compared with the nature of the wave-dominated Brent Delta elsewhere on its progradational path. High energy nearshore deposits in both regressive and transgressive phases are widely distributed, but there is a contrast between the dominance of fluvial processes in the study area and the development elsewhere of a barrier island complex in front of a coastal plain.

The Brent Group (Aalenian to Bathonian) is the principal hydrocarbon reservoir in the UK northern North Sea. It consists predominantly of deltaic deposits which belong to a regressive–transgressive clastic wedge up to 300 m thick.

In this account, the Brent Delta is examined close to its limit of northward progradation using core and gamma-ray logs from seven wells from around the Don (UK Blocks 211/13 and 211/18) and Murchison (UK Block 211/19) fields (Figs 1 and 2). Patterns of sedimentation are used to determine the local character of the delta, and this is compared with the nature of the delta elsewhere on its progradational path.

Regional setting

The Brent Group occurs in the northern part of the northern Viking Graben; in the UK sector it is generally found below *c.* 3000 m depth and is distributed over block-faulted terraces (the East Shetland Basin) which lie between the Viking Graben axial zone and the Graben margin in the W. The Brent Group occurs above early Jurassic marine shales (Dunlin Group) and is succeeded by Bathonian to late Jurassic marine shales (Humber Group).

The Brent Group is subdivided into five formations by Deegan & Scull (1977), but Brown et al. (1987) postulate that the basal Broom Formation is genetically unrelated to the Brent Delta. It is thin or absent in the study wells. The remaining formations, ie Rannoch, Etive, Ness and, at the top, Tarbert, record delta advance and retreat. Vail & Todd (1981) note a small marine hiatus at the Broom–Rannoch boundary caused by a rise in sea level which outstripped sediment supply. The upper boundary of the Brent Group is commonly an unconformity. The Brent Delta is considered by Johnson & Stewart (1985) and Graue *et al.* (1987) to have prograded northwards, broadly parallel to the trend of the Viking Graben, but in the east Shetland Basin Richards *et al.* (in press) favour transverse drainage with a component of easterly progradation.

Description and interpretation of facies

Five main facies associations are recognized from core examination and gamma-ray log response.

Regressive offshore to shoreface deposits

These deposits consist of muddy heterolithic deposits and micaceous sandstones (Figs 3–5) and are present in all the study wells.

From WHATELEY, M. K. G. & PICKERING, K. T. (eds), 1989, *Deltas: Sites and Traps for Fossil Fuels,* Geological Society Special Publication No. 41, pp. 253–267.

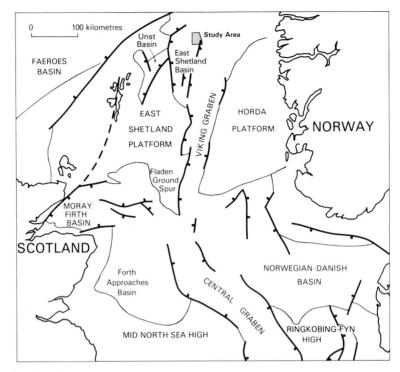

FIG. 1. Location of study area.

The heterolithic beds are dark grey silty micaceous and carbonaceous mudstone, locally structureless but more commonly with lenticular bedding or with very fine, even, parallel silty laminations. Ripple cross-lamination, some of which is wave formed, is present, as are very fine-grained sandstone layers with laminar concentrations of mica. The coarser-grained layers commonly have a sharp base and either fine upwards

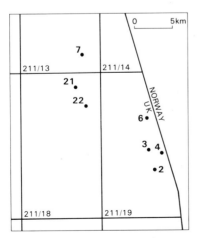

FIG. 2. Location of wells used in the study.

or have sharp, typically uneven, upper boundaries. Minor low angle concave-up scours and low angle planar erosion surfaces are recorded. Biogenic reworking is locally intense. Palynofacies analysis of the 211/19-4 deposits by Parry *et al.* (1981) points to a marine environment, but with the freshwater alga *Botryococcus* locally abundant.

The boundary with the overlying micaceous sandstones is gradational. These sandstones are highly micaceous, mostly very fine grained, subarkosic, and locally have a calcite cement. The alternation of mica-rich and mica-poor laminae, on a scale of less than 3 mm, gives a bipartite character to the even parallel lamination. This fabric is usually well defined, although structureless intervals occur in places. Low angle (less than 15°) wedge-shaped cross-laminated sets occur, with laminae parallel, or almost so, to subjacent truncation surfaces. Angle-of-repose laminae in medium-scale structures are absent. Steep scours and low angle concave-up scours with near-concordant bipartite laminated fills are recorded, as are examples of subtle convex-up surfaces mantled by bipartite laminated sand. Thin wave-rippled zones and burrowed zones occur towards the base of the sandstone succession. Burrowed beds have erosional tops and

UK 211/19 – 4 (CONOCO)

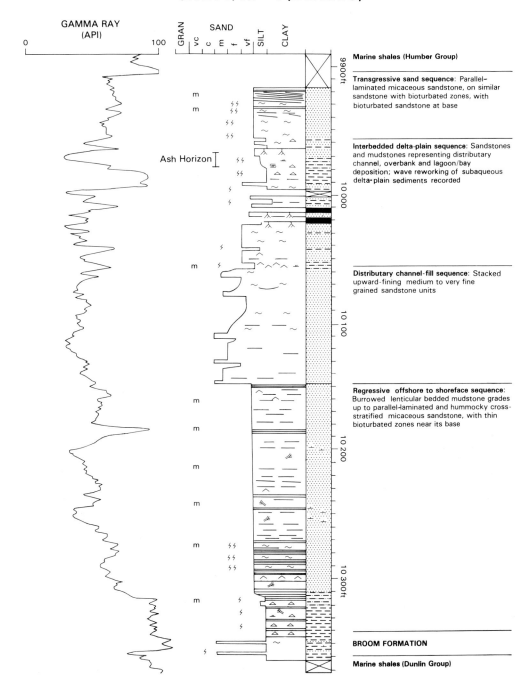

GAMMA RAY (API)

0   100

GRAN vc c m f vf SILT CLAY
SAND

Marine shales (Humber Group)

9900ft

Transgressive sand sequence: Parallel-laminated micaceous sandstone, on similar sandstone with bioturbated zones, with bioturbated sandstone at base

Ash Horizon

Interbedded delta-plain sequence: Sandstones and mudstones representing distributary channel, overbank and lagoon/bay deposition; wave reworking of subaqueous delta-plain sediments recorded

10 000

Distributary channel-fill sequence: Stacked upward-fining medium to very fine grained sandstone units

10 100

Regressive offshore to shoreface sequence: Burrowed lenticular bedded mudstone grades up to parallel-laminated and hummocky cross-stratified micaceous sandstone, with thin bioturbated zones near its base

10 200

10 300ft

BROOM FORMATION

Marine shales (Dunlin Group)

FIG. 3. Gamma-ray and core log for the Brent Group in well 211/19-4.

UK 211/19 – 3 (CONOCO)

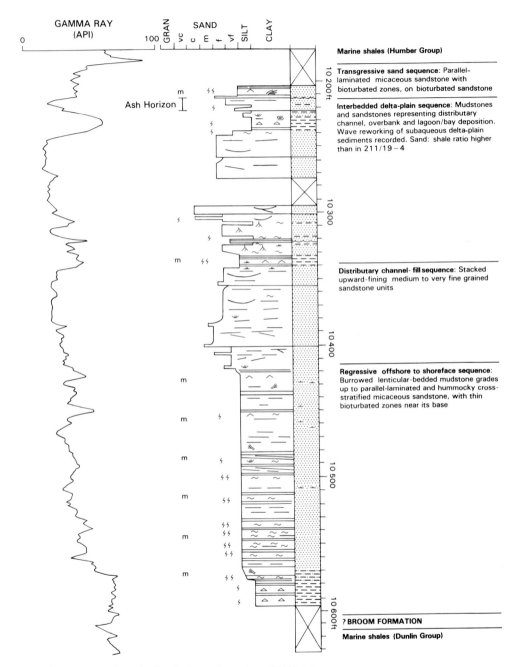

Marine shales (Humber Group)

Transgressive sand sequence: Parallel-laminated micaceous sandstone with bioturbated zones, on bioturbated sandstone

Interbedded delta-plain sequence: Mudstones and sandstones representing distributary channel, overbank and lagoon/bay deposition. Wave reworking of subaqueous delta-plain sediments recorded. Sand: shale ratio higher than in 211/19 – 4

Distributary channel-fill sequence: Stacked upward-fining medium to very fine grained sandstone units

Regressive offshore to shoreface sequence: Burrowed lenticular-bedded mudstone grades up to parallel-laminated and hummocky cross-stratified micaceous sandstone, with thin bioturbated zones near its base

? BROOM FORMATION

Marine shales (Dunlin Group)

FIG. 4. Gamma-ray and core log for the Brent Group in well 211/19-3.

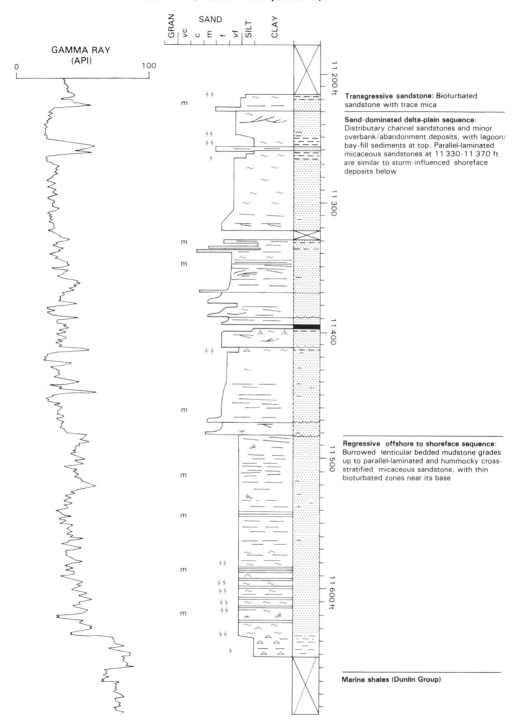

UK 211/18a – 22 (BNOC)

GAMMA RAY
(API)

0   100

GRAN | SAND | SILT | CLAY
vc c m f vf

11 200 ft

11 300

11 400

11 500

11 600 ft

Transgressive sandstone: Bioturbated
sandstone with trace mica

Sand-dominated delta-plain sequence:
Distributary channel sandstones and minor
overbank/abandonment deposits, with lagoon/
bay-fill sediments at top. Parallel-laminated
micaceous sandstones at 11 330-11 370 ft
are similar to storm-influenced shoreface
deposits below

Regressive offshore to shoreface sequence:
Burrowed lenticular bedded mudstone grades
up to parallel-laminated and hummocky cross-
stratified micaceous sandstone, with thin
bioturbated zones near its base

Marine shales (Dunlin Group)

FIG. 5. Gamma-ray and core log for the Brent Group in well 211/18a-22.

transitional bases, passing in both cases to laminated micaceous sandstone. In well 211/18a-21 thin erosively based and rapidly upward-fining bands of very coarse sandstone occur within the sequence (Figs 6 and 7). The section is overlain abruptly, and in well 211/18a-22 (Fig. 5) clearly erosively, by less micaceous coarser-grained deposits.

Interpretation

The regressive offshore to shoreface succession is assigned to the Rannoch Formation. The heterolithic facies is considered to represent deposition at the transition between marine offshore and shoreface zones, close to the limit of effective wave action on bottom sediments. Low energy mud deposition was intermittently interrupted by the emplacement of silt and very fine sand layers. In the context of the storm-wave-influenced sands deposited to landward during delta progradation (see later), the coarser-grained sediment was probably generated and emplaced by storm-related processes. Storm waves touched bottom on occasions with sufficient energy to rework and scour the substrate. The heterolithic facies can be compared with the shallow-marine muddy platform facies (M_1 and M_2) of de Raaf *et al.* (1977).

Similar deposits of micaceous sandstone in an equivalent stratigraphic position have been described from the Northwest Hutton oil field, about 40 km to the S, by Richards & Brown (1986). A storm-wave-influenced nearshore environment of deposition was postulated. The bipartite laminated micaceous sandstones satisfy the criteria for the recognition of hummocky cross-stratification (HCS) in the core (Walker 1984). In the present wells, the micaceous sandstones mostly conform to the amalgamated HCS of Dott & Bourgeois (1982) but with the development of wave-rippled zones in the lower sands pointing to lower wave intensity. The character of the sandstones is comparable with the micaceous Cape Sebastian Sandstone (late Cretaceous) described by Hunter & Clifton (1982). They argued that the mica-rich–mica-poor couplets may be deposited from flows with a significant oscillatory component and that each couplet may be the product of an individual wave.

The coarse-grained sand layers in 211/18a-21 (Figs 6 and 7) are uncommon in our experience of the Rannoch Formation. Given their association with the micaceous sandstones, they are considered to have been transported to, and reworked in, a nearshore environment by storm processes. Alternative explanations such as transgressive lags or coarse-grained concentrates at the foreshore–shoreface transition are considered less likely. The effects of fluvial effluent processes during flood together with offshore-directed storm surges may have aided transport. The lateral extent of these coarse-grained layers and the probable nature of the transporting agent, *ie* whether confined or not, has not been determined. Clifton (1981) records thin layers and lenses of coarse sand within comparable nearshore micaceous sands of Miocene age, and interprets them as post-storm lags left after fair-weather winnowing on a high energy coast. Independent positive evidence of fair-weather processes is absent from the nearshore deposits of the Brent Group.

Distributary channel-fill deposits

The distributary channel-fill deposits, which rest directly on micaceous sandstone of the Rannoch Formation, consist of fine to rarely coarse grained, relatively well sorted quartzose sandstone. Characterized by stacked fining-upward units, some with erosive bases, the sediments display an indistinct even parallel lamination as well as traces of medium-scale cross-bedding and ripple cross-lamination. Structureless sandstone is also common. There are no significant argillaceous intercalations, and this results in the rather uniform gamma-ray log response of the section.

These sediments are present in wells 211/19-4 (Fig. 3) and 211/19-3 (Fig. 4), and probably also in 211/19-2 (Fig. 10) and 211/18a-21 (Fig. 6).

Interpretation

These sandstone deposits are assigned to the Etive Formation on the basis of lithology and gamma-ray log patterns. Parry *et al.* (1981) interpret the sandstones in 211/19-4 as stacked distributary channel deposits; this is confirmed and extended to the equivalent strata in 211/19-3, 211/18a-21 and the uncored 211/19-2.

Morton & Humphreys (1983) record the leaching of apatite grains at the top of this distributary sandstone section in wells 211/19-3 and 211/19-4. They attribute the leaching to emergence and subaerial weathering processes (see also Morton 1986).

Sand-dominated delta-plain deposits

This succession lies above the regressive shoreface sandstones in wells 211/19-6 (Fig. 8), 211/18a-22 (Fig. 5) and 211/13-7 (Fig. 9), and is distinguished from the distributary channel fills by a more variable ('ratty') gamma-ray log response. Core examination confirms a greater lithological variability. In all cases, the succession remains predominantly sandy, varying from

211/18a – 21 (BNOC)

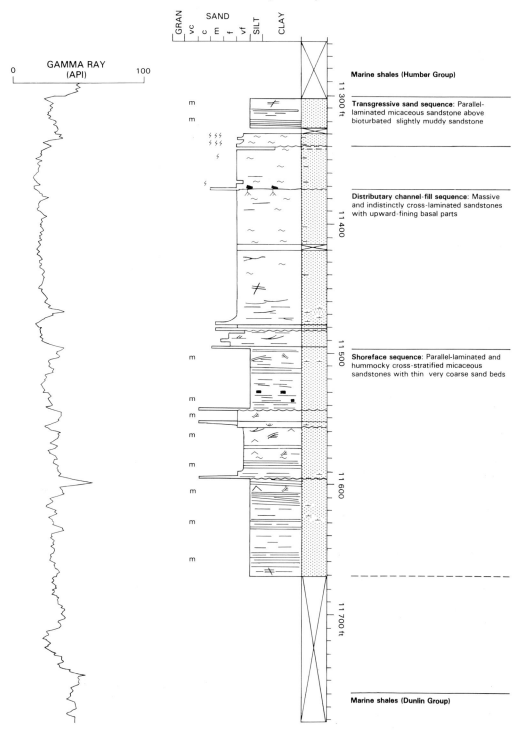

GAMMA RAY
(API)
0   100

Marine shales (Humber Group)

Transgressive sand sequence: Parallel-
laminated micaceous sandstone above
bioturbated slightly muddy sandstone

Distributary channel-fill sequence: Massive
and indistinctly cross-laminated sandstones
with upward-fining basal parts

Shoreface sequence: Parallel-laminated and
hummocky cross-stratified micaceous
sandstones with thin very coarse sand beds

Marine shales (Dunlin Group)

FIG. 6. Gamma-ray and core log for the Brent Group in well 211/18a-21.

FIG. 7. Example of sharp-based coarse-grained sandstones truncating micaceous laminae in the Rannoch Formation of well 211/18a-21. (Continuous length of core; base of section at bottom right.)

mostly fine to rarely coarse grained, and is again characterized by stacked fining-upward units.

In well 211/19-6, these units are mostly thin in the lower half of the succession. The indistinctly cross-laminated sands are intercalated with rare mica-rich partings near the base and with a thin *in situ* coal. A prominent plane-parallel-laminated and low angle cross-laminated very fine-grained micaceous sandstone, with a bipartite laminar fabric similar to that of the Rannoch sands, is found below 3232 m (10 600 ft) (see Fig. 8 and the 'mica zone' in Fig. 10). The fining-upward units above this are thicker than those below. Thin wispy mud seams also contribute to the rather 'ratty' gamma-ray log motif. In well 211/18a-22 the succession is similar (Fig. 5), with a micaceous sandstone unit containing hummocky cross-stratified beds below 3456 m (11 335 ft).

In well 211/13-7 equivalent strata are referred to distributary channel to marginal-marine/embayment deposits (Fig. 9). Fining-upward medium- to very fine-grained sandstones are marked by abundant wispy mud intercalations and pervasive bioturbation (Fig. 11), in contrast with the deposits described above. The basal parts of the fining-upward units tend to be structureless

or indistinctly laminated, and relatively free from mud. Mud wisps and the degree of biogenic reworking increase upwards as the host sand changes from medium or fine grained up to very fine grained. Thin bioturbated silty mudstone layers occur at the top of some fining-upward units. The biogenic structures present include the oblique cylindrical *Scoyenia* spp and the horizontal cylindrical *Planolites* spp. Both tend to be found in Brent Group deposits which are known from palynofacies analysis to have a marine influence. The appearance of the intensely burrowed sands in 211/13-7 resembles the bioturbated sands commonly found at the base of the uppermost transgressive sands (Tarbert Formation) in many Brent Group successions.

Interpretation

These deposits in wells 211/19-6 and 211/18a-22 are interpreted as largely of stacked distributary channel fills, following the discussion of broadly laterally equivalent strata in 211/19-4 and 211/18a-21. The main contrast is in the greater representation of fine-grained channel-abandonment deposits and intercalated muds which point to less uniformly energetic current activity in the channels. The micaceous sandstones in 211/19-6 and 211/18a-22, found within stacked fining-upward units, form part of a mica-rich zone in the middle of the Brent Group which can be traced into wells 211/19-4 and 211/19-3 (see later).

The bioturbated deposits in 211/13-7, which retain a fining-upward grain-size motif, are interpreted tentatively as distributary channel sands that were gradually drowned by marine incursions. Here, at the northern limit of progradation, the distributaries of an over-extended fluvial system may have been drowned as rises in sea level intermittently outpaced sediment supply.

One reason for distinguishing the sand-dominated delta-plain deposits from the distributary channel-fill sediments (Etive Formation) described previously is to note and account for the lateral variability in gamma-ray log response in these largely arenaceous successions and to clarify their lithostratigraphy. The 'ratty' character and associated mica and mud within the sand-dominated delta-plain section is atypical of the Etive Formation regionally. However, as Fig. 10 shows, it is possible to extend the Etive Formation from its typical development, *eg* in 211/19-4, to include the genetically related but less uniform sand-dominated delta-plain deposits. In wells 211/19-6 and 211/18a-22 the upper boundary of the Etive Formation is placed below the mica zone. In 211/18a-21 the amalgamated distributary

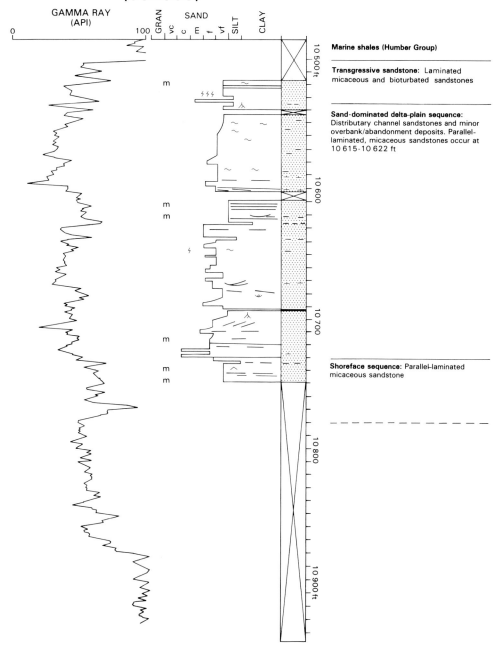

FIG. 8. Gamma-ray and core log for the Brent Group in well 211/19-6.

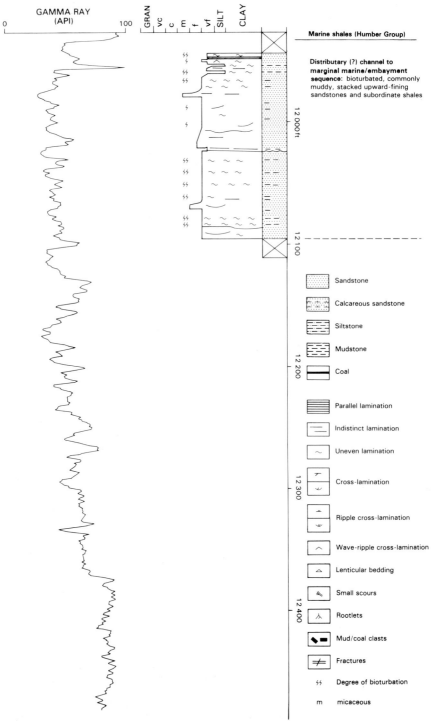

FIG. 9. Gamma-ray and core log for the Brent Group in well 211/13-7.

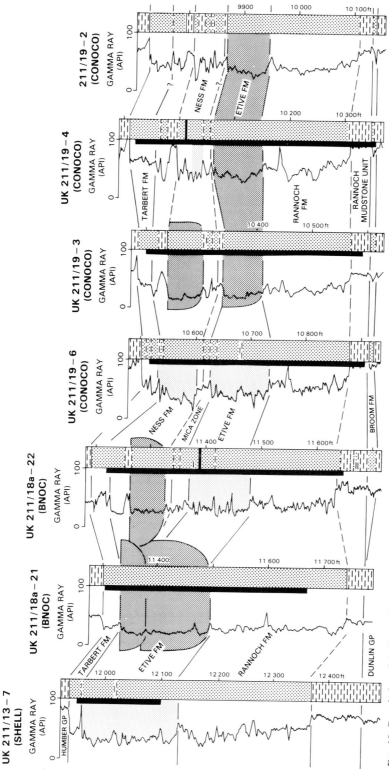

FIG. 10. Correlation of the Brent Group (see Fig. 2 for well locations) showing lithostratigraphic boundaries (full lines). The shaded areas between the graphic lithology columns indicate sections characterized by fining-upward distributary channel deposits. The denser stipple represents diagrammatically the thicker more uniform channel sands. Their occurrence above the Rannoch Formation in wells 211/19-2, 211/19-4 and 211/19-3 gives a gamma-ray response typical of the Etive Formation regionally. Identification of the Etive Formation in wells 211/19-6, 211/18a-22 and 211/13-7 is less clear on gamma-ray log pattern alone. The black bands to the left of the lithology columns indicate the positions of cores shown elsewhere in the paper.

FIG. 11. Bioturbation at the top of the fining-upward cycle in well 211/13-17. Such sections are interpreted as representing marine drowning of rivers near the northern limit of delta progradation. (Continuous length of core; base of section at bottom right.)

FIG. 12. Wave-rippled siltstone from the basal part of the interbedded delta-plain deposits in well 211/19-3.

channel sand succession, which replaces the mixed Ness deposits, is assigned to the Etive Formation. This broadening of the lithostratigraphic classification, however, places sediments of rather different reservoir properties in the Etive Formation.

Interbedded delta-plain deposits

An interbedded sandstone, mudstone and minor coal succession is developed in wells 211/19-4 (Fig. 3), 211/19-3 (Fig. 4) and 211/19-2 (Fig. 10). The 211/19-4 succession has been interpreted by Parry *et al.* (1981) and Simpson & Whitley (1981) as deposits of a delta plain with distributary channel, crevasse-splay, swamp and lagoonal sediments. We agree with this interpretation and in the present account concentrate on several specific horizons.

Firstly, the basal beds of the section in 211/19-4 and 211/19-3 are characterized by wave-ripple cross-laminated highly micaceous siltstone (Fig. 12) and very fine-grained sandstone. In 211/19-3, this gradually coarsens up to fine-grained sandstone with rootlets at its top.

In 211/19-3, the thinly interbedded deposits are partly replaced by a thick development of stacked fining-upward fine- to coarse-grained sandstones. These have a gamma-ray response similar to that of the distributary channel-fill deposits below. Morton & Humphreys (1983) again record evidence of apatite leaching at the top of this sandstone. They also find similar evidence at *c.* 3050 m (10 000 ft) in well 211/19-4.

Immediately above these leached horizons is a ripple cross-laminated and burrowed silty mudstone which coarsens up overall to fine-grained sandstone. In 211/19-4, rootlets are found at the top of the sandstone. The mudstone was deposited in an environment with marine influence, at least at its base (J. B. Riding, pers. comm.; see also Morton & Humphreys 1983).

Interpretation

This mixed succession is characteristic of the Ness Formation regionally. The basal mica-rich beds can be correlated with the even, parallel-laminated and hummocky cross-stratified mica-ceous sandstones recorded in the sand-dominated delta-plain deposits of 211/19-6 and 211/18a-22 (mica zone in Fig. 10). These sediments represent the temporary drowning of the distributary channel system and either shoreline retreat to form an embayment or the development of a lagoon with sufficient fetch to sustain wave reworking of bottom sediment. The highest wave energy is recorded in the hummocky cross-stratified sandstones of 211/18a-22.

The reservoir character of the interbedded succession is strongly influenced by the reappearance of thicker distributary channel sands in well 211/19-3. These are, in part, laterally equivalent to the thin-channel and overbank deposits in 211/19-4 and to the less uniform distributary channel deposits above 3232 m (10 600 ft) in 211/19-6 (Fig. 10). The amalgamation of distributary channel sandstones across the study area results in the loss of a distinct interbedded character and the loss of the Ness Formation (*sensu* Deegan & Scull 1977).

The mudstones in 211/19-4 and 211/19-3 directly below the volcanic ash horizon (Figs 3 and 4) recorded by Morton & Humphreys (1983) were probably deposited in an embayment or lagoon and are succeeded by a prograding minor mouth bar–channel sandstone unit.

Transgressive sand deposits

Except in well 211/13-7, where the distinction is less clear, the various delta-plain deposits are overlain abruptly by bioturbated sandstone. In 211/19-4 (Fig. 3), 211/19-3 (Fig. 4), 211/19-6 (Fig. 8) and 211/18a-21 (Fig. 6) the bioturbated sandstones are succeeded by micaceous, even, parallel-laminated very fine-grained sandstone, with evidence of HCS. Biogenic structures diminish rapidly in abundance upwards. In 211/19-3 and 211/18a-22 a thin medium-grained sandstone bed, with a sharp and uneven base, lies at the base of the bioturbated sandstone. The bioturbated sandstone has a marine character, with the burrow forms *Scoyenia*, *Palaeophycus* and *Planolites* recognized (Fig. 13). Evidence of bioturbation occurs throughout much of the cored section in well 211/13-7 (Fig. 9) and no distinct sharp-based transgressive sand unit is identified in the core.

Interpretation

These deposits are assigned to the Tarbert Formation on the basis of lithology and stratigraphic position. Comparison with Brown *et al.*

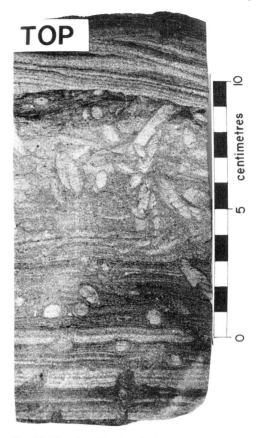

FIG. 13. Bioturbated horizon in the transgressive Tarbert Formation of well 211/19-3 with an upper truncation surface and gradual transition down to less intensely burrowed, even, parallel and bipartite laminated micaceous sandstone.

(1987) and Graue *et al.* (1987) demonstrates the variable nature of Tarbert Formation gamma-ray log patterns regionally. The deposits studied here lack the well-developed coarse-grained sand to pebble lags commonly found elsewhere in the Tarbert Formation above ravinement surfaces. The thin medium-grained sandstones at the base of the section in 211/19-3 and 211/18a-22 may be genetic equivalents.

Diverse and abundant biogenic structures in the basal sandstones, the presence of micaceous sandstones comparable with those of the Rannoch Formation shoreface deposits and comparison with the Brent Group regionally suggest that the deposits represent the final drowning of the Brent Delta plain. Shoreline retreat was followed by the re-establishment of nearshore marine conditions influenced by storm processes, similar to those during the regressive phase of delta development.

Conclusions

The Brent Group deltaic strata behave as a strongly layered hydrocarbon reservoir in northern North Sea oil fields; a first-order zonation closely reflects the internal stratigraphy and evolving depositional environments of the deltaic deposits. The succession was constructed by the progradation of a storm-wave-influenced delta front, the aggradation of a delta plain which had fluvial and bay or lagoonal deposits, and finally by the migration of a storm-wave-influenced transgressive coast. In this study of wells close to the northern limit of progradation of the delta in the UK sector, interbedded delta-plain deposits (Ness Formation) sandwiched between regressive and transgressive sands further S are progressively lost. They are replaced by distributary channel sands. The sand-to-mud ratio locally reaches its maximum value for the Brent Group reservoir interval and vertical continuity of sands is improved.

High energy shoreface deposits are succeeded by distributary channel sands without evidence of an intervening mouth-bar development. Although the age relationships between distributary channel sands and the subjacent shoreface deposits are only known from the superposition in vertical sections, it is possible, given the occurrence of freshwater palynomorphs at least locally in the Rannoch Formation, that rivers did debouch directly at the coast where wave energy redistributed any mouth-bar sediment (*cf* van Straaten (1959) on the Rhone Delta).

The aggrading delta- or coastal-plain succession, to which the distributary channel-fill and delta-plain deposits belong, varies considerably both vertically and laterally, depending on the maintenance of distributary channels. The devel-opment of bays or lagoons terminated major distributary channels on at least two occasions within several of the sections studied. Terrestrial environments were again established by the outbuilding of minor delta lobes or by the return of distributary channel sedimentation. The Brent Group shows a degree of symmetry in its vertical succession of facies, with the occurrence of Rannoch-type facies in the upper transgressive sand sequence.

Elsewhere on its progradational path, the Brent Delta has been characterized as a wave-dominated delta with a coastal barrier complex recorded by the Rannoch and Etive Formations (Budding & Inglin 1981; see also Brown *et al.* 1987). In this model, the Etive Formation belongs to the laterally migrating or progradational part of the Brent Group regressive–transgressive wedge. In the study area, the dominance of fluvial processes is evident and sediments assigned to the Etive Formation belong to an aggradational phase of Brent Group sedimentation.

In the rather anomalous succession in well 211/13-7, where fining-upward sandstones are stacked above the shoreface deposits, mud is deposited in intimate association with sands, and bioturbation is common throughout. This leads to the tentative hypothesis that gradual drowning of rivers may have occurred locally near the distal limits of the delta. This contrasts with the mode of delta drowning characterized by the development of extensive lagoons (Mid Ness Shale lagoonal deposit of Eynon 1981), the temporary submergence of the delta plain reported in this paper and the transgression associated with prominent shoreface erosion surfaces reported by Rønning & Steel (1985) and Brown *et al.* (1987).

ACKNOWLEDGMENTS: The authors wish to thank Mr J. M. Dean for his comments on an early draft of this paper. The paper is published with the permission of the UK Department of Energy and the Director, British Geological Survey (Natural Environment Research Council).

References

BROWN, S., RICHARDS, P. C. & THOMSON, A. R. 1987. Patterns in the deposition of the Brent Group (Middle Jurassic) UK North Sea. *In:* BROOKS, J. & GLENNIE, K. (eds) *Petroleum Geology of North West Europe.* Graham & Trotman, London, 899–913.

BUDDING, M. C. & INGLIN, H. F. 1981. A reservoir geological model of the Brent Sands in Southern Cormorant. *In:* ILLING, L. V. & HOBSON, G. D. (eds) *Petroleum Geology of the Continental Shelf of North-West Europe.* Heyden, London, 326–334.

CLIFTON, H. E. 1981. Progradational sequences in

Miocene shoreline deposits, southeastern Caliente Range, California. *Journal of Sedimentary Petrology* **51**, 165–184.

DEEGAN, C. E. & SCULL, B. J. 1977. A standard lithostratigraphic nomenclature of the Central and Northern North Sea. *Report of the Institute of Geological Sciences* **77/25.**

DOTT, R. H. & BOURGEOIS, J. 1982. Hummocky stratification; significance of its variable bedding sequences. *Bulletin of the Geological Society of America* **93**, 663–680.

EYNON, G. 1981. Basin development and sedimentation

EYNON, G. 1981. Basin development and sedimentation in the Middle Jurassic of the North Sea. *In:* ILLING, L. V. & HOBSON, G. D. (eds) *Petroleum Geology of the Continental Shelf of North-West Europe.* Heyden, London, 196–204.

GRAUE, E., HELLAND-HANSEN, W., JOHNSEN, J., LØMO, L., NØTTVEDT, A., RØNNING, K., RYSETH, A. & STEEL, R. 1987. Advance and retreat of Brent delta system, Norwegian North Sea. *In:* BROOKS, J. & GLENNIE, K. (eds) *Petroleum Geology of North West Europe.* Graham & Trotman, London, 915–937.

HUNTER, R. E. & CLIFTON, H. E. 1982. Cyclic deposits and hummocky cross stratification of probable storm origin in Upper Cretaceous rocks of the Cape Sebastian area, southwestern Oregon. *Journal of Sedimentary Petrology* **52**, 127–143.

JOHNSON, H. D. & STEWART, D. J. 1985. Role of clastic sedimentology in the exploration and production of oil and gas in the North Sea. *In:* BRENCHLEY, P. J. & WILLIAMS, B. P. J. (eds) *Sedimentology: Recent Developments and Applied Aspects.* Geological Society of London Special Publication **18**, 249–310.

MORTON, A. C. 1986. Dissolution of apatite in North Sea Jurassic sandstones; implications for the generation of secondary porosity. *Clay Minerals* **21**, 711–733.

—— & HUMPHREYS, B. 1983. The petrology of the Middle Jurassic sandstones from the Murchison field, North Sea. *Journal of Petroleum Geology* **5**, 245–260.

PARRY, C. C., WHITLEY, P. K. J. & SIMPSON, R. D. H. 1981. Integration of palynological and sedimentological methods in facies analysis of the Brent Formation. *In:* ILLING, L. V. & HOBSON, G. D. (eds) *Petroleum Geology of the Continental Shelf of North-West Europe.* Heyden, London, 205–215.

DE RAAF, J. F. M., BOERSMA, J. R. & VAN GELDER, A. 1977. Wave generated structures and sequences from a shallow marine succession, Lower Carboniferous, County Cork, Ireland. *Sedimentology* **24**, 451–483.

RICHARDS, P. C. & BROWN, S. 1986. Shoreface storm deposits in the Rannoch Formation (Middle Jurassic), North-West Hutton oilfield. *Scottish Journal of Geology* **22**, 367–375.

——, ——, DEAN, J. M. & ANDERTON, R. A new palaeogeographic reconstruction for the Middle Jurassic of the northern North Sea. *Journal of the Geological Society of London* **145,** in press.

RØNNING, K. & STEEL, R. 1985. Transgressive reservoir sands and architecture in the Tarbert Formation, northern North Sea. *International Seminar, North Sea Oil and Gas Reservoirs.* University of Trondheim, Norway, 30–31 (abstract only).

SIMPSON, R. D. H. & WHITLEY, P. K. J. 1981. Geological input to reservoir simulation of the Brent Formation. *In:* ILLING, L. V. & HOBSON, G. D. (eds) *Petroleum Geology of the Continental Shelf of North-West Europe.* Heyden, London, 310–314.

VAN STRAATEN, L. M. J. U. 1959. Littoral and submarine morphology of the Rhône delta. *In:* RUSSEL, R. J. (ed.) *Proceedings of the 2nd Coastal Geographers Conference, Baton Rouge, LA.* National Academy of Science, National Research Council, Washington, DC, 233–264.

VAIL, P. R. & TODD, R G. 1981. Northern North Sea Jurassic unconformities, chronostratigraphy and sea level changes from seismic stratigraphy. *In:* ILLING, L. V. & HOBSON, G. D. (eds) *Petroleum Geology of the Continental Shelf of North-West Europe.* Heyden, London, 216–235.

WALKER, R. G. 1984. Shelf and shallow marine sands. *In:* WALKER, R. G. (ed.) *Facies Models.* Geoscience Canada, Reprint Series 1, 141–170.

S. BROWN & P. C. RICHARDS, Hydrocarbons Research Programme, British Geological Survey, Grange Terrace, Edinburgh EH9 2LF, UK.

Facies associations and sand-body geometries in the Ness Formation of the Brent Group, Brent Field

S. E. Livera

SUMMARY: The Brent Group in the Brent Field contains over 1000×10^6 bbl of recoverable oil in a simple gently dipping half-graben structure. It is 243 m (800 ft) thick and represents a sequence amplified by synsedimentary fault-block subsidence close to the axis of the Viking Graben. On a regional scale the depositional model is of a major shoreface and coastal sequence, the Rannoch and Etive Formations, underlying the partly time-equivalent shallow lagoonal to emergent delta-plain deposits of the Ness Formation. However, more localized and detailed sedimentary modelling is required to predict reservoir geometries and production behaviour, particularly in the Ness Formation.

The Ness Formation is the largest reservoir-bearing unit in the Brent Field, being up to 160 m (525 ft) thick. Five main facies associations form reservoir-quality rock: varied channel sandstones, fluvial sheet sandstones, major mouth bars, lagoonal shoals (which are usually associated with, and capped by, thin beaches) and lagoonal sheet sandstones. These associations make up 50%–60% of the sand-rich sequence and are embedded in the non-reservoir facies associations: lagoonal mudstones, emergent flood-plain mudstones and allochthonous and autochthonous coals. The coals are often field wide in extent. Within the amplified Ness Formation the most important environments were the lagoonal deltas which were wave and fluvial dominated with minor tidal influence. As a result of this importance the sand bodies produced by these deltas are subjected to a more detailed description in this paper. Wave energies were sufficient to redistribute large amounts of sediment and indicate significant fetch in the brackish lagoons diluted by major freshwater input.

Sand-body geometries are very variable, reflecting both localized controls on sediment distribution, particularly immediately above the compacting shoreface pile, and more regional effects including coarse-sediment supply, basinal processes and fault-block subsidence. The dominant controls on sedimentation resulted in regular drowning events on the delta plains and had a significant impact by producing a strongly layered reservoir.

The Brent Field is located in licensing block 211/29 at the eastern margin of the UK offshore sector. The field was discovered in 1971 when well 211/29-1 proved oil-bearing Jurassic sediments. The Brent Field proved to be the largest of a number of similar accumulations in the UK area of the East Shetland Basin (Fig. 1). Over 60% of the Brent Field hydrocarbon reserves are in the Bajocian and Bathonian Brent Group, which contains in excess of 1000×10^6 bbl of recoverable oil; the remaining 40% of reserves are found in the underlying Statfjord Formation (Bowen 1975; Johnson & Krol 1984). Since the first oil was produced in 1976, development has proceeded rapidly and 98 wells now penetrate the Brent Group in the field (Fig. 2).

Detailed description of the Ness Formation is required for two reasons: it is particularly variable, and it contains most of the oil in the Brent Group reservoir. The layered nature of these sediments and their lateral variability are of particular importance in the field. Because synsedimentary fault-block subsidence favoured the deposition of lagoonal delta-front sediments

in the area these environments are the dominant features of the Ness Formation.

Primary development drilling is now almost complete (Diehl 1978), and as the field has gone into a period of plateau production detailed sedimentological characterization is required to explain its performance and to predict its future behaviour.

Structure

The Brent Field occurs in one of a number of rotated fault blocks that are found between the Shetland Platform and the Viking Graben (Fig. 1). This graben was formed in a period of lithospheric extension during which time marginal hinterlands shed considerable volumes of coarse clastic sediments into the developing basin, especially during the Bajocian (Leeder 1983). Rifting and fault-block rotation is thought to have occurred in two phases, during the Triassic and between the Bathonian and the Berriasian (Eynon 1981; Badley et al. 1984).

From WHATELEY, M. K. G. & PICKERING, K. T. (eds), 1989, *Deltas: Sites and Traps for Fossil Fuels*, Geological Society Special Publication No. 41, pp. 269–286.

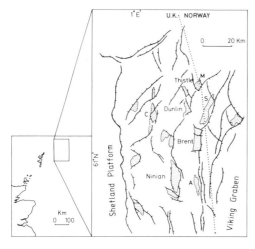

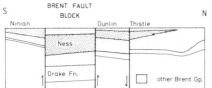

FIG. 1. Regional location map showing the position of the Brent and other oilfields (stippled) and a schematic S–N cross-section across the basin (based on Johnson & Stewart 1985): A, Alwyn; C, Cormorant; M, Murchison; S, Statfjord. The Ness Formation is amplified in the Brent fault block by synsedimentary thickening across the Ninian–Dunlin fault zone.

However, the major faults in the area also have pronounced synsedimentary thickening across them during the Bajocian when the majority of the Brent Group was deposited. Brent Group sediments become progressively thicker eastwards towards the graben axis, and as a result the Brent fault block preserves an amplified sequence, particularly in the Ness Formation, compared with the rest of the UK sector (Fig. 1).

The Brent Field is structurally rather simple and is bordered to the E by a major N–S-trending graben-forming fault near the crest of the Statfjord Formation (Fig. 2). The Brent Group hydrocarbon accumulation is a linear feature approximately 4 km × 16 km in areal extent with a single gas-on-oil column delineated to the N and S by W–E- or NW–SE-trending faults (Fig. 2). The gentle westerly dipping (8°) fault block is cut by two cross-faults with a maximum throw of 70 m at the Brent Group level. A combination of fault-block rotation and erosion produced shallow listric failure surfaces at the crest of the structure that sole out in the underlying Dunlin Group

(Fig. 2) (Livera 1986). These features are mostly restricted to the gas zone, and their effect on oil production is not thought to be great. The hydrocarbon column is trapped against a series of unconformities and condensed sequences at the basinward side of the fault block, some of which are of regional significance (Rawson & Riley 1983). This simple overall structure contrasts with other oilfields in the area, particularly those overlying NE–SW-oriented strike-slip Caledonian lineaments. For example, such a lineament has produced a complex fault pattern in the Cormorant Field (Watters et al. 1985). Because structural complications in the Brent Field are minimal, the dominant influences on the behaviour of the reservoir are the primary depositional patterns of the Brent Group sediments. Therefore the Brent Field is also an ideal analogue for other Ness Formation accumulations where structural complexity makes sand-body geometries difficult to quantify.

Regional sedimentology

The Brent Group (Deegan & Scull 1977) represents a limited period of mostly Bajocian (approximately 6 Ma) (van Hinte 1976) sand-rich clastic sedimentation and overlies shallow-marine Sinemurian to Aalenian mudstones and siltstones of the Dunlin Group. The uppermost parts of the reservoir in Brent, including the Tarbert Formation, are of early Bathonian age, as are the overlying mudstones of the Heather Formation of the Humber Group.

A regional sedimentological model for the deposition of the Brent Group is summarized by Johnson & Stewart (1985). This model infers an initial northerly prograding wave-dominated delta that deposited the Rannoch and Etive Formations (Deegan & Scull 1977), overlain by coastal- and delta-plain Ness Formation sediments and finally capped by a transgressive sand sheet of the Tarbert Formation. The basal Broom Formation is now regarded as a separate depositional event from sedimentological and petrographic considerations (Morton 1985). Formation boundaries transgress time lines and, according to the regional model, the Ness Formation was deposited in an area partly protected by contemporaneous coastal barrier sediments of the Etive Formation to the N. However, more detailed studies of individual fields, eg Cormorant (Budding & Inglin 1981) and Murchison and Statfjord (Parry et al. 1981), have shown that regional variability exists, particularly in the Ness Formation.

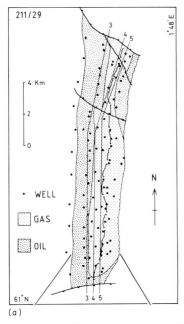

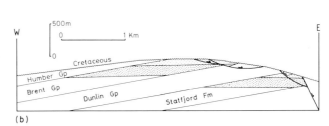

(a)   (b)

FIG. 2. (*a*) Top Brent Group map and (*b*) structural cross-section through the Brent Field. The Brent Group crest is a complex of listric faults and failure surfaces that sole out in the Dunlin Group, whilst the main graben-forming fault lies E of the Statfjord Formation crest. The lines marked 3, 4 and 5 on the map indicate the locations of Figs 3, 4 and 5 respectively.

Brent Field reservoir subdivision

All 98 wells that penetrate the Brent Group have a full suite of wireline logs. Approximately 1250 m of core has been taken and described in detail. By careful correlation of core to wireline log response the sedimentological interpretations have been extrapolated throughout the reservoir and enabled a rigorous subdivision to be made. This subdivision has had to account for, and be tested against, a wealth of well production and fluid pressure data.

The original reservoir subdivision of Bowen (1975) and the stratigraphic subdivision of Deegan & Scull (1977) based on the Brent type well 211/29-3 have, not surprisingly, proved insufficient to model the production behaviour of the field adequately. As more *in situ* fluid pressure data have been collected, reservoir units have been redefined because the four original drainage units (cycles I–IV) were found to contain a number of subreservoirs (Johnson & Stewart 1985). The reason for this is that extensive shales and abandonment coals exist, particularly in the Ness Formation, which form horizontal barriers to vertical flow during oil production. Sand bodies between these barriers behave as separate subres-

ervoirs with their own pressure regimes. The extensive coals are easily recognized on wireline logs and help to define reservoir units (Table 1) that are now selectively drained. This layer-cake framework of coals, which has been given an alphabetical coding for identification (Figs 3–6), also delimits very distinctive and different sedimentological sequences within the Ness Formation.

Facies associations of the Ness Formation

Two groups of facies association occur: those interpreted to be of upper-delta–coastal-plain origin and those considered to represent lower-delta–coastal-plain and lagoonal delta-front environments (Table 2). These two interdigitate within the reservoir and illustrate a complex history of upper-delta-plain progradation and contraction throughout Brent times. The three youngest reservoir cycles of Table 1 are illustrated in Figs 3–5 which are near-field-length S–N panels drawn from core and wireline log data.

TABLE 1. *Subdivision of the Brent reservoir in the Brent Field*

FORMATION	CYCLE	RESERVOIR UNIT	THICKNESS (RANGE) m	COALS (N–S)	DESCRIPTION
TARBERT	I	1.1	11 (0-21)		Coarsening-upward very fine to medium grained, bioturbated marine sandstone. Deposition as shelf sand thinning to nothing at southern end of field.
		1.2	10 (2·5-17)		Crossbedded shoreface and bioturbated thin washover sands, interdigitating with lagoonal mudstones to the south. Thin coals.
		1.3	15 (12-20)		Bioturbated coarsening-upward lagoonal delta and sheet sandstones, capped by mostly drifted coals.
		1.4	17 (11-23)		Dark lagoonal mudstones and mostly drifted coals. Small wave-reworked shoals and beaches; one lagoonal delta.
NESS	II	2.1	11·5 (3-24)		Channel and sheetflood sandstones close to the transition from lower to upper delta plains. Floodplain, lagoonal mudstones and coals.
		2.2	13 (8-22)		Stacked channel sandstones in the south, major lagoonal delta in the north. Floodplain and lagoonal mudstones and coals.
		2.3	9 (6-13)		Floodplain mudstones, pedogenic ironstones, thick splitting coal seams and rare isolated fluvial channel sandstones.
		2.4	12 (6-20)		Coarsening-upward mouth bar and distributary channel sandstones overlying lagoonal mudstones and overlain by floodplain mudstones.
		2.5	17 (13-23)		Deep, lagoonal mudstone coarsening upward rapidly to lagoonal beach and minor-mouth-bar sandstones, deeply incised by later major distributary sandstones.
	III	3.1	14 (7·5-20)		Sand-rich sequence of large coalesced wave-modified mouth bars. Basal sheet and minor-mouth-bar sandstone.
		3.2	22 (11-32)		Floodplain mudstones, variety of isolated channel and crevasse sandstones. Laminated mudstones of small lakes in upper-delta-plain environment.
		3.3	20 (14-27)		Lagoonal mudstones and small lagoonal delta and sheet sandstones. Minor deltas rapidly **abandoned**.
ETIVE		3.4	5 (0-14)		Variety of poorly correlatable massive dune sandstones, **abandoned** channel and interdune swale sandstones and mudstones. Thin coal at top.
	IV	4.1	22 (5-49)		Medium-grained cross-bedded sandstones of upper shoreface and complex channel origin, thin coals, roots at top.
RANNOCH		4.2	37 (27-45)		Hummocky and low angle cross stratified middle and lower shoreface fine-grained sandstones of wave-dominated Brent delta-front regressive phase. Relatively uniform thickness.
Broom		4.3	13 (9-18)		Offshore marine shales interdigitating with coarse-grained offshore marine sand patches and sheets of Broom Formation.

(All units belong to the BRENT GROUP)

Shaded units = Upper Delta Plain Ness Formation Sequences

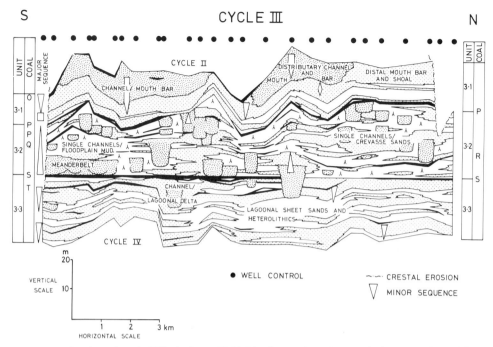

FIG. 3. S–N section through cycle III in the Brent Field. The diagram has 100 × vertical exaggeration and is based on well control points projected onto the section. The key is given in Fig. 6.

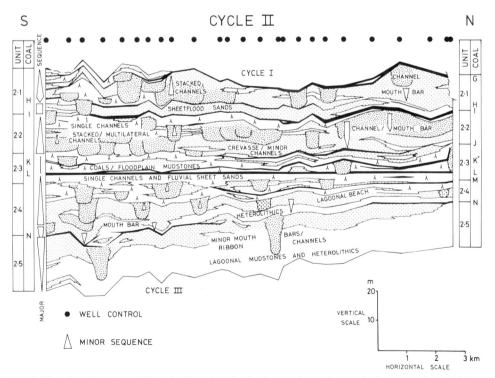

FIG. 4. S–N section through cycle II in the Brent Field. The diagram has 100 × vertical exaggeration and is based on well control points projected onto the section. The key is given in Fig. 6.

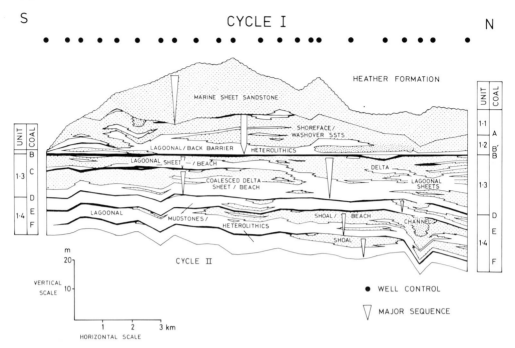

FIG. 5. S–N section through cycle I in the Brent Field. The diagram has 100 × vertical exaggeration and is based on well control points projected onto the section. The key is given in Fig. 6.

Depositional history of the Brent Group in the Brent Field

Wave-dominated shoreface of Brent progradation

The Rannoch and Etive Formations consist of 60–100 m of predominantly sandstone forming a single coarsening-upward sequence, which over-

FIG. 6. Key for all figures.

lies more open shallow-marine sediments including the interbedded Rannoch shale and Broom Formation. The Rannoch and Etive Formations represent a wave-dominated distributary-fed shoreface where large amounts of sediment were supplied and reworked by marine processes resulting in rapid northerly progradation (Johnson & Stewart 1985). The scale of deposition and the facies associations developed during this period are very different from those seen in the overlying Ness Formation. However, this sequence both allowed the development of the overlying coastal-plain environments and, if the regional model is correct, protected them from the wave-dominated marine environments that existed in the N of the basin.

Coastal-plain establishment

The Ness Formation sequence immediately overlying the Etive and Rannoch deposits is an extension of the initial progradational phase of the Brent system. Unit 3.3 contains a number of thin (3–10 m) coarsening-upward minor deltas set in a matrix of lagoonal mudstones (Fig. 3) and fed by smaller channels than found in the Etive deposit below. These minor deltas are unique in the Brent Field because they form laterally discontinuous sand bodies with poorly developed

TABLE 2. *Facies associations of the Ness Formation in the Brent Field*

	Facies	Description	Thickness and lateral continuity	Porosity/Permeability (average)
Upper Delta Plain	Fluvial channel sandstones	Very variable channel types not usually correlatable at present well spacings. Grain size varies from fine sand to granule. Clean and dominated by dm scale planar cross-stratification. Marked fining-upward sequences, overlain by discontinuous coals. Includes mixed load straight channels and one meander belt in unit 3.2.	2–10 m thick, few hundred metres wide 1-5 km wide meander belt in unit 3.2	21-32 (26.5) % / 200-4000 (950) mD
	Fluvial sheet sandstones (sheet flood and crevasse)	Fine to very fine grained sandstone with ripple cross-lamination, but usually intensely modified by soft-sediment deformation, or well developed roots. Thin crevasse sandstones. One stacked sheet - flood sequence of PCL plane bedded sandstone in unit 2.1 is 4 m thick.	< 0.3 m by few hundred metres 4 m stacked sheet flood 40 sq km in unit 2.1	15-29 (23) % / 10 - 3000 (180) mD
	Floodplain mudstones	Pale grey 'massive' mudstones with slickensides, soft-sediment deformation, roots and sphaerosiderite. Desiccation cracks later deformed. Well-laminated pale siltstones with organic debris. Emergent flood-plain and small shallow flood-basin standing water bodies.	sequences up to 10 m thick field wide 16 × 4.5 km extent	non reservoir
	Coals I	Discontinuous coals of very varied thickness. May coalesce, split or thin, can form continuous sequences. Well-developed roots at base, mostly autochthonous organic debris. Poorly recovered in cores but recognized on logs.	up to 2-3 m, usually <1 m < 1 km to field wide where coalesced	—
	Coals II	Simple single thin pyritic black coaly shales and poor coals. Thin roots below. Coals well laminated, mostly allochthonous material. Poor core recovery, but recognized on logs.	< 1 m Field wide	—
Lower Delta Plain and Delta Front	Lagoonal Deltas (mouthbars and distributary channels)	Very fine - medium grained small-scale cross-laminated to through/planar cross-bedded sandstones. Characteristic dm - cm scale graded beds with basal scours. Rare bioturbation. Sharp scours and grain size changes mark channels. Two sand-body scales, 'major' and 'minor'.	minor 1-3 m / coalesced up to several km major 3-10 m / 5-20 + km	20-30 (25.5)%/30-2000 (250) mD
	Lagoonal shoal sandstones (associated beaches)	Wave-ripple cross laminated and hummocky cross-stratified fine sandstone with gradational bases to flaser/wave-ripple formsets. Tops have low angle and horizontal laminated sandstone. Roots and rare bioturbation. Coalesce with minor mouth bars.	2-7 m / several km	15-29 (22.5)% / 2-800 (42) mD
	Lagoonal sheet sandstones	Very fine grained sandstones, thin sharply defined beds, small-scale current ripple cross-laminated or homogeneous. Bioturbated, may contain roots. Interbedded with lagoonal mudstones. Volumetrically unimportant, formed by variety of mechanisms.	<2 m / hundreds of metres	14-29 (23.5)% 1-2000 (100) mD
	Lagoonal mudstones*	Variable lithofacies and colour, where more arenaceous contain cm graded beds, wave-generated ripple form sets, synaeresis cracks and some bioturbation. Contain acritarchs (brackish conditions) but no full marine indicators.	generally 2-10 m and field wide	non reservoir

* For detailed lithofacies characterization see Budding & Inglin (1981).

abandonment coals. A possible explanation for this discontinuity is that the underlying shoreface sediments and the mudstones of the Dunlin Group were compacting actively and that this led to the rapid abandonment of the minor deltas whilst promoting and prolonging the existence of interbedded lagoons. Continued progradation of the Brent Group system allowed the development of the upper-delta-plain facies associations found in unit 3.2, which consists of single and stacked channel systems and flood-plain mudstones. The boundary between the two units is taken at the top of the S Coal (Fig. 3) which is overlain by a multilateral fining-upward channel sand body, thought to represent a meander belt, in the southern part of the field. The top of unit 3.2 contains relatively few channel sands, well-developed root-bearing flood-plain shales and a thick coal, which all indicate waning of coarse-grained sediment supply as the first phase of upper-delta-plain expansion gradually diminished. Unit 3.2 is a complex reservoir unit with variable lateral sand continuity, and many of the channel sand bodies cannot be correlated from well to well (Fig. 3).

The middle Ness sequences

A major coastal-plain drowning event is recorded at the base of unit 3.1, and the succeeding units (3.1, 2.5 and 2.4, informally termed here the middle Ness) (Figs 3 and 4) contain contrasting Ness lagoonal delta-front sequences outlined in detail below. The mudstones of unit 2.5 are equivalent to the mid-Ness shale of other fields (Budding & Inglin 1981). These deltaic coarsening-upward sequences are less than 20 m thick, much thinner than the underlying Rannoch and Etive shoreface (Table 1). Units 3.1 and 2.5 terminate in thin laterally extensive coals which formed prior to the quite rapid transgressions that deposited the lagoonal mudstones of the succeeding units. By considering the decompacted thickness (about twice the present thickness) of these coarsening-upward sequences (the silty mudstones of unit 2.5 are now some 10 m thick) we can estimate that the drowning events resulted in water depths up to a maximum of 20 m. However, detailed facies analysis suggests that the deepening of the lagoons was gradual and involved some sedimentation prior to their maximum development (Budding & Inglin 1981). Although the prograding delta infills did not represent single events, the overall pattern of aggradation to a base level represented by the coals results in the characteristic laterally extensive sheet-like morphology of the sand bodies, which can be traced in excess of 15 km (Figs 3

and 4). This sheet-like geometry contrasts with that seen in unit 3.3 where the individual sand bodies do not coalesce. A different control must have been responsible for the geometry of the middle Ness sequences, which present much simpler reservoir units for production purposes.

Coastal-plain re-establishment

From the upper part of unit 2.4 to the top of unit 2.1 the Ness Formation is dominated by flood-plain depositional systems, particularly in the southern part of the field (Fig. 4). This period represents the second phase of upper-delta-plain expansion in the Ness Formation. Throughout cycle II coarse sediment supply was variable. Unit 2.3 was formed during a period of reduction in sandy sediment supply when thick coals and root-bearing mudstones were deposited. In contrast, units 2.2 and 2.1 were periods of major sand influx separated and underlain by thin lagoonal mudstones that mark minor transgressive events over the coastal plain. These units are represented by single mouth-bar and distributary channel sequences in the N of the field, separated by the I Coal. Core data indicate that these northerly sediments are very similar to those seen in underlying unit 3.1, which is one of the units described in detail below. In the southern part of the field the transgressive events cannot readily be identified, and both units are represented by stacked or isolated channel sandstones and root-bearing flood-plain mudstones where the I Coal is missing. Correlation is still maintained over a large part of the field by the H and I Coals (Fig. 4).

Initial failure of the Brent Group terrestrial sediment supply

Units 1.3 and 1.4 differ from the underlying units because of their low sand-to-shale ratios and facies association development. There appears to have been a fall in sediment supply and the units are characterized by very dark organic-rich lagoonal mudstones. This horizon contains chamositic ooids in the neighbouring Statfjord Field (Parry et al. 1981) but these have not been recorded in the Brent Field cores. Small-scale thin (less than 7.5 m) coarsening-upward sequences which terminate abruptly in simple, but field-wide, coals are a distinctive feature (Fig. 5). A gradual increase in bioturbation upwards (both in amount and diversity) indicates more saline conditions, but no fully marine indicators are found in the mudstones. The sheet-like geometry of the sand bodies together with field-wide mudstones and coals means that correlation in these upper parts of the Ness Formation is very easy.

Complete failure of the Brent Group terrestrial sediment supply

The upper two units (1.1 and 1.2) of the Brent Group reflect the final destruction of the Brent system and belong to the Tarbert Formation. Unit 1.2 represents a complex facies association of shoreface sediments in the N of the field. Thinner sands, interpreted partly as washover sediments, interdigitate with heavily bioturbated lagoonal mudstones to the S (Fig. 5). This unit appears to be much thinner than the Tarbert sediments in oilfields to the S (Hild and Alwyn), although Rønning & Steel (1986) interpret the latter successions as having formed in similar depositional environments. Unit 1.1 is unique and consists of an entirely bioturbated gradually coarsening-upward sandstone of Bathonian age containing marine microplankton. This sandstone has a marked lateral thickness variation (0–20 m) and has more affinity to Upper Jurassic shelf sandstones of the Central Graben (Johnson *et al.* 1985) than to the underlying Ness sediments or the Tarbert Formation in other oilfields. The sand is interpreted as a shelf ridge, formed after complete transgression over the Brent coastal-plain sediments, and probably represents a relict feature similar to those found off the present E coast of the USA (Swift 1976). Lower Bathonian Heather Formation mudstones conformably overlie this unit, but the contact is marked by a pyritized bioturbated interdigitation of the two sediment types.

The informal sixfold stratigraphic subdivision of the Brent Group outlined above is very broadly similar to that which Eynon (1981) proposed for the Statfjord, Hutton and Ninian fields that lie to the N and W of Brent. However, his idea that this subdivision represents individual deltaic lobes is probably misleading when we consider the possible controls on sedimentation within the Ness Formation.

Examples of Ness Formation sequences

Three of the reservoir units are described in more detail to illustrate the variability in delta-front and lower-delta-plain deposits, and the sequence generated in one of the main phases of upper-delta-plain progradation. These units differ, reflecting local variation in water depths produced by drowning events, sediment supply and basinal processes.

Unit 3.1

This unit overlies root-bearing emergent flood-plain mudstones and coal of the first phase of upper-delta-plain development (Table 1). A major drowning event submerged the underlying sequences over a considerable area (well beyond the field boundaries). Unit 3.1 consists of a rapidly coarsening-upward lower sand body 0–2.5 m thick overlain by a more complexly coarsening-upward main sand body 15 m thick (Fig. 7). The isopach map and cross-section of Fig. 7 show two depocentres, each some 3 km wide, separated by a thinner sequence which overlies a stacked channel sand body in unit 3.2. The main coarsening-upward sequence can be subdivided and is illustrated in Plate 1. The lower part is dominated by fine-grained sandstone with hummocky cross-stratification (HCS) and planar low angle cross-stratification overlying and partly capped by wave ripples, which in one instance are preserved as formsets draped with muscovite mica. This micaceous horizon appears correlatable over a large area (Fig. 7). The upper part is medium- to coarse-grained trough and planar cross-stratified sandstone separated from the lower part by scours (wells A, C and D) or with a gradational contact (well B). Sharp-based centimetre to decimetre thick graded beds, medium grained at the base and fine grained and wave rippled at the top, are common in the transition to the upper part (Plate 1). This upper part of the sand body also contains a coal in well C (Fig. 7). Towards the N (well E) three coarsening-upward units are found on the fringe of the northernmost of the depocentres. Bioturbation is rare and is only found in the lower part of the main sand body. The unit is capped by a thin coal in the central and southern parts of the field (Fig. 7).

Unit 3.1 is interpreted as a sand-rich lagoonal mouth-bar and distributary channel complex where the channel bases are marked by the major scour surfaces and sharp grain-size changes. However, the distinction between the channels and the upper mouth-bar sandstone is not possible from wireline log data alone as both represent very high quality reservoir rock with the same petrophysical response. During deposition of the unit, a combination of gradual subsidence and continued coarse-sediment supply resulted in a sheet of sand with a lateral extent of over 16 km. The sequence is interpreted as a complex of lobate deltas deposited by two closely spaced distributaries entering a lagoon in which there was enough fetch for wave reworking of the

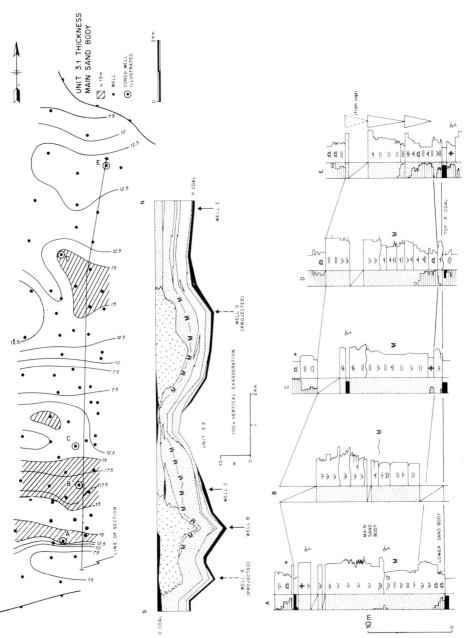

FIG. 7. Map, S–N cross-section and examples of cored sequences from reservoir unit 3.1 in the Brent Field. The sand body is divided into two parts, an upper main body and a lower section. See text for details.

sediment. The wave reworking produced a sand apron fronting the distributary mouth bars and resulted in amalgamated HCS units which are very characteristic of the Brent lagoons. The lobate shape is interpreted from the net sand map and inferred from the wave influence on the delta front (Coleman & Wright 1975). The sharp-based graded beds of the mouth bar in well B suggest that sediment was fed directly into the lagoons by traction currents from the distributaries in a friction-dominated delta front. This, and the general low diversity of bioturbation, suggests that the lagoon was dominated by freshwater influx from the distributaries. Further in front of the delta, and marginal to the distributaries, finer-grained sediment was reworked into shallow shoals and laterally to possible emergent (roots and coal) beach ridges. The coal in well C (Fig. 7) and the three coarsening-upward units of well E imply that the distributaries were not constantly supplying sediment and suggest that some autocyclic control allowed the distributaries to prograde and wane during deposition and that the whole unit formed during a period of gradual subsidence. It is also possible that only one of the distributaries was active at any one time, a situation seen in similar modern fluvial and wave-modified deltas such as that of the Danube (Coleman & Wright 1975). Abandonment was eventually rapid but the distributaries were choked by sediment and have sharp tops without a fining-upward profile.

Unit 2.5

Unit 2.5 (Fig. 8) is probably equivalent to the mid-Ness shale of the Cormorant Field (Budding & Inglin 1981) and Dunlin Field (Johnson & Stewart 1985) and represents another major drowning event. Lagoonal mudstones dominate the sequence. Water depths in this lagoon appear to have been significantly deeper than in any of the other sequences in the Ness in Brent. The presence of wave-generated ripples and thin graded laminae within the mudstones suggests that it was always close to or above the storm-wave base. Synaeresis cracks are common in the mudstones, possibly indicating variations in salinity during deposition. The lagoon was eventually filled by a series of fine-grained low angle and trough cross-bedded and ripple cross-laminated sandstones and capped by a thin coal. The sandstone forms a sheet of coalesced sand bodies which are interpreted as including minor mouth bars and lagoonal beaches (Fig. 8). The sandstones have a maximum thickness of 7 m, which contrasts with the much thicker unit 3.1 underneath. The thick amalgamated sequences of HCS

found below the major mouth bars in the rest of the Ness are not present in unit 2.5. A striking feature of the unit are two channel bodies that scour deeply into the underlying lagoonal mudstones. Their E–W orientation is defined by well correlation and they are thick (up to 15 m) but narrow (less than 500 m) sand bodies (Fig. 8). Unfortunately, no core is available for detailed internal characterization of these channels.

Unit 2.5 formed after a major transgression produced an extensive (2100 km^2; Brown *et al.* 1987) lagoon that was gradually filled by a series of minor distributaries where coarse-sediment supply was limited and extensively redistributed along the shoreface by along-shore currents. After the lagoon was nearly filled in the Brent area, two major distributaries switched into the area by avulsion and cut deeply through the earlier sediments but accreted to similar base levels. These autocyclic events re-introduced significant fluvially transported sediment through the Brent area, and it is interesting to note that there was insufficient basinal energy to choke the channels with reworked sand when abandoned and they have a fining-upward partial abandonment profile. These large distributaries probably fed mouth bars of the same size as those seen in unit 3.1. There is no direct evidence in the unit of the gradual subsidence during deposition recognized in unit 3.1, and although actual water depths are difficult to estimate they probably did not exceed 20 m. The mudstone lithofacies in unit 2.5 are similar throughout the vertical section and a combination of gradual subsidence and sedimentation may account for this observation.

Unit 2.4

Unit 2.4 forms the last of the lagoonal-dominated sequences in the middle of the Ness Formation and marks the second phase of upper-delta-plain progradation. The unit consists of a thin coarsening-upward sand body with HCS (Fig. 9, well A). This sand body gives way northward to a thinner sequence cut by small channels, as in well C. Overlying this sand sheet are emergent flood-plain mudstones with roots, slickensides and coal seams. In this matrix a series of channels up to 5 m thick with associated crevasse sandstones form good but disconnected reservoir-quality rock. Iron-rich horizons (dominantly sphaerosiderite) and common slickensides suggest pedogenesis, and the iron carbonate forms high matrix density horizons that can be correlated from wireline log responses for up to 5 km. At the top of unit 2.4 the thick L and K coals mark a field-wide abandonment surface (Fig. 9). This upper-delta-plain coal series splits and thins, probably

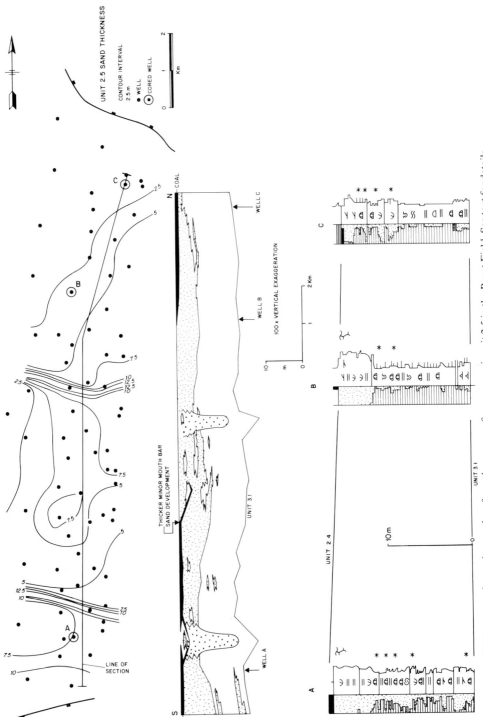

FIG. 8. Map, S–N cross-section and examples of cored sequences from reservoir unit 2.5 in the Brent Field. See text for details.

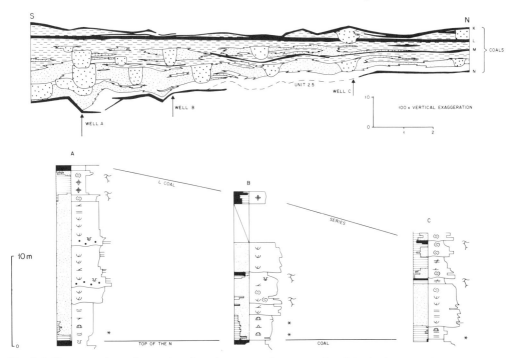

FIG. 9. S–N cross-section and examples of cored sequences from reservoir unit 2.4 in the Brent Field. See text for details.

owing to differential compactional effects during prolonged accumulation, in contrast with the simpler lower-delta-plain coals of unit 1.4. Similar features of lower- and upper-delta-plain coals have been recognized from the Carboniferous of the USA (Howell & Ferm 1980). When these coals split and thin, correlations are maintained by considering all the sedimentological features seen in cores and on wireline log responses.

Unit 2.4 represents a major expansion of upper-delta-plain conditions in the Brent Group and also varies in thickness and sand content from S to N, giving the wedge shape seen in Fig. 9. In the southern part of the field the thick HCS unit is interpreted as part of a major mouth bar which grades and thins northwards into a lagoonal beach. The thinning is perpendicular to the E–W channel sand orientations deduced from well correlations. If the unit represents the advancement and abandonment of a single delta lobe, the observed thinning (from a depocentre in the S to a sheet-like wing in the N) can be used to estimate the minimum geometry of a shallow-water Ness delta. The asymmetry suggests that the lobe was at least twice the observed width of the unit in the Brent Field, giving a minimum figure of about 30 km for the diameter of an individual lagoonal delta.

This unit represents a shallow-water delta which prograded rapidly through the Brent Field and was abandoned to marshy conditions. The abandonment of the unit was by autocyclic processes—a decrease in sediment supply from distributaries—rather than by rapid drowning events as seen in the two units below (3.1 and 2.5). Unit 2.4 is very similar to the overlying units 2.2 and 2.1, where small-scale drowning events preceded delta building into shallow-water lagoons.

A general model for lagoonal deltas in the Ness Formation

Although the three reservoir units described above differ in detail it is possible to make some generalizations from them about the Ness lagoonal deltas (Fig. 10).

The major Ness deltas were fed by thick (up to 15 m) but narrow (less than 500 m) straight distributary channels that led into large brackish lagoons. The rarity of marine fauna and the low amounts and diversity of bioturbation suggest that the lagoons were significantly diluted by freshwater input, particularly adjacent to the

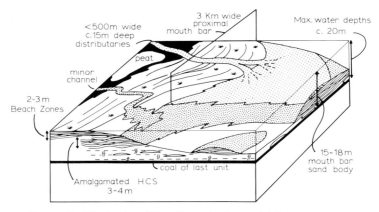

FIG. 10. Summary diagram of a fully developed Ness Formation lagoonal delta.

large distributaries. These lagoons had variable but significant water depths (up to *c.* 20 m?), sufficient in some instances to accommodate the unimpeded development of mouth bars such as in unit 3.1. Where water depths were initially shallower, rapid progradation resulted in a thin sheet-like sand body and a sudden transition to flood-plain marsh conditions as in unit 2.4. Minor distributary channels (up to 5 m thick) also transported sand to the lagoons, forming thinner minor mouth bars that coalesced with other shoreface sediments. The fault-block-wide drowning events produced low offshore slopes adjacent to the lagoonal shorefaces. Both these conditions meant that the heavily sediment-laden distributaries formed friction-dominated mouth bars (Coleman & Wright 1975) where flood events transported sediment directly into the lagoons. These floods formed thin medium- to coarse-grained sandstone beds which were subsequently partially reworked by wave action. The amalgamated upper-mouth-bar and channel depocentres approach a width of 3 km, and when the HCS units are included the sand bodies are up to 18 m thick.

HCS is found in association with the major mouth bars but is only poorly developed in the thinner lagoonal shoreface sand bodies. The exceptions to this are the sheet sandstones of unit 1.4, which were deposited during the initial abandonment phase of the Brent system. The HCS sequence in unit 3.1 is over 3.5 m thick (Plate 1) and is of the amalgamated hummocky only (HHH) type (Dott & Bourgeois 1982). In general these sequences are dominated by concave-upward scour surfaces and have a fairly sharp base overlying wave-rippled sandstones, and the HCS is favourably preserved over fair-weather sedimentary structures. Although de-

tailed interpretations from data limited to a 102 mm (4 inch) core are dangerous, there is also evidence that the HCS includes the accretionary type described by Brenchley (1985) which indicates that it formed from the vertical growth of true bedforms. From its position adjacent to the major mouth bars there is a temptation to suggest that the HCS formation was favoured by high sediment influx where distributary effects modified wave-produced currents. This would explain why the HCS is not well developed in unit 2.5. Its presence indicates significant wave modification of the lagoonal delta fronts in a deeper-water environment seaward of the distributaries.

Transport parallel to the shoreline allowed the development of marginal delta-front shoals which formed wings up to 6 km away from the distributary source in unit 1.4. These shoals became attached to the shoreface upon progradation of the lagoonal shorelines and are capped by thin beaches and eventually by coals. There are no indications of tidal drapes in the channel sandstones, and tidal influences on the lagoons or delta fronts are thought to have been relatively insignificant compared with wave modifications.

The general model for the Ness deltas suggests that they were lobate in form and both fluvial and wave influenced. Sand distribution patterns resemble those of the Rhone Delta (Oomkens 1967). However, the Rhone Delta is not a very suitable analogue for the Brent lagoonal deltas because water depths allow coarsening-upward sequences of up to 70 m to develop and the delta has a larger areal extent and forms in a marine setting. No suitable modern analogue resembles the Brent examples. The most fully documented lagoonal deltas are those of the Colorado River (Kanes 1970) and the Guadalupe River (Donaldson *et al.* 1970) which have formed behind barriers on the Gulf Coast of the USA. These deltas formed in very shallow water where the effects of along-shore waves and storms are relatively minor owing to the limited fetch of the narrow lagoons. The general observations from the Brent Field suggest some similarity with the Eocene wave-modified deltas described by Chan & Dott (1986). However, in Brent the deltas were formed in broad shallow shelving lagoons which would not favour intense wave modification of the delta fronts. In the examples given by Chan & Dott a narrow shelf and large fetch would have contributed considerably to wave influences. If the regional Brent model is correct and the lagoons were protected by a contemporaneous Rannoch–Etive barrier system to the N, the wave energy could only come from the lagoons themselves which would have had a fetch of the order of 50 km. This would infer that either the Bajocian climate was storm dominated or that a strong prevailing wind was in existence during Brent times. Certainly relatively low wave energy can still produce wave-modified deltas, such as the modern Ebro (Maldonado 1975), provided that fluvial input is low or fluctuates substantially; thus the presence of thick HCS sequences in the Ness does not necessarily mitigate against the regional model.

Palaeotrends in the Ness Formation

The progradation direction of the Brent Group system and Bajocian palaeogeography is well established from regional mapping of facies belts (Johnson & Stewart 1985). In the Brent Field, heavy mineral (zircon) layers that probably represent time lines have been mapped in the Rannoch Formation shoreface. These layers dip due N at between 0.4° and 0.6° and give a localized indication of the Rannoch–Etive barrier progradation that verifies the regional model. However, correlation of closely spaced wells and sandstone isopach mapping in the Ness Formation indicate that the regional trend is not reflected in the lower-delta-plain channels. The majority of the correlatable channels have an E–W orientation, as in unit 2.5 (Fig. 8). No dipmeter information is available but a few palaeocurrent data on oriented cores indicate a westerly direction, which agrees with the trend of sandstone isopach thinning in units 2.5 and 3.1. This palaeocurrent divergence from the regional Brent model is not unexpected because the small shallow-water Ness lagoonal deltas would be influenced by localized basinal effects, such as compaction in underlying sediments or possibly the creation of localized slopes within individual

synsedimentary fault blocks. The regional model cannot be used to predict preferential sand-body orientations in the complex Ness Formation, even though it was deposited in a confined rift basin.

Controls on sedimentation

The vertical and lateral variability of the Ness Formation must reflect the balance between sediment supply, basinal processes (related to water depths and fetch), tectonic activity and eustasy in the basin. Within the Ness sequence it is possible to recognize at least four orders of scale of control on sedimentation that contribute to this variability:

(1) Within reservoir units. The controls within the reservoir units are manifested in two forms. Deposition is seen to have avoided areas of thick sandy accumulations in underlying units, as in unit 3.1 in the central part of the field (Fig. 3), suggesting a compaction-dominated control, even in the lower delta plain, similar to that built into simulation models of alluvial architecture (Bridge & Leeder 1979). This control is particularly evident in unit 3.3 above the thick Rannoch and Etive sequences. A second and more regionally significant feature is the evidence of large-scale distributary avulsion within the framework of the reservoir units. In unit 2.5 this autocyclic control is most clearly illustrated where two major distributaries cut through earlier sand sheets formed at the periphery of a lagoonal delta.

(2) Between individual reservoir units. The layer-cake stratigraphy seen in the Ness Formation is easily defined by the field-wide coal horizons. In the middle and upper parts of the Ness this pattern resulted from rapid lagoonal transgressions (drowning events) followed by sediment aggradation during relative still-stand giving asymmetric vertical sequences. However, the coals are always preserved, indicating that the rises in base level were sudden and that little erosion occurred during the transgressions. This contrasts with some Tertiary marine-wave-dominated deltas on the Gulf Coast of the USA (Weise 1979), where extensive erosion and bioturbation is found on the abandoned delta top sands. This scale of control has had a significant impact on the production characteristics of the Ness Formation.

(3) Forming the gross 'stratigraphic' subdivision with repeated periods of upper- and lower-

delta-plain deposition. The development of extensive lagoonal systems in the middle Ness with thick emergent flood-plain sequences above and below is on a scale larger than the drowning events outlined above.

(4) On the scale of the whole Brent Group system. The development and abandonment of the Brent Group spanned a period of about 6 Ma and was a rather abrupt coarse-grained event in relation to the finer-grained clastics of the Dunlin and Heather sediments above and below.

The small-scale autocyclic processes are easily understood when we consider the size of the lagoonal deltas in relation to the extent of the basin. The control forming the field-wide reservoir units does not appear to have an autocyclic origin and we must look elsewhere to understand the mechanism that created such rapid but periodic rises in base level.

Tectonic influences within the Viking Graben rift basin must have played an important part in sedimentation, and they are directly reflected in the major synsedimentary thickening within the Ness Formation across active fault zones (Fig. 1). It would be reasonable to suggest that the subsidence of whole fault blocks was responsible for this vertically repeated pattern of sedimentation. However, where coarse sediment supply was low, in units 1.3 and 1.4 (Fig. 5), the drowning intervals begin to appear fairly regularly spaced. The mudstone lithofacies in these two units are the same, and variations in thickness between coals is mainly due to differences in sand content. If the mudstone sedimentation rates are roughly constant and dominant over sand deposition rates, and as there are no indications of hiatuses in the sequence, a uniform periodicity in drowning events becomes plausible. This suggests that a more regular sea-level control was the dominant influence in forming these asymmetric cycles, as recently inferred for some Carboniferous deltaic sequences (Holdsworth & Collinson 1987). However, the actual control must remain rather conjectural until detailed field-to-field correlations have been carried out and a better estimate of sedimentation rates has been obtained. Insufficient biostratigraphic data are available in the Ness Formation to establish the possible periodicity of these cycles. In most of the units a combination of several controls was probably active at the same time, including gradual compactional subsidence of the coastal plain.

The 'stratigraphic' level ((3) above) represents periods where the Brent fault block was not acting as a selective trap for the lagoonal deltas in the Brent province and upper-delta-plain

sediments were deposited. These major prograding and regressive phases may reflect changes in sediment supply basin wide, possibly related to rift development and rates of marginal uplift and rejuvenation (Leeder 1983). The Ness Formation accumulated during a period of gradually rising sea level punctuated by one third-order eustatic cycle (Vail & Todd 1981). However, there is little evidence of such a scale of sea-level control within the Brent Group in Brent Field and these effects were probably overwhelmed by the development of the rift basin.

Conclusions

The Ness Formation in the Brent Field is a complex of deltaic and coastal-plain sediments with strong internal vertical ordering. In the Brent area this sequence was amplified by synsedimentary fault-block subsidence which created a preferential depocentre for lagoonal deltas. These deltas are on a much smaller scale than the coarsening-upward progradational phase of the Brent Group system which provided a shallow but extensive and protected coastal-plain setting. Within this setting two periods of upper-delta-plain expansion and contraction occurred, separated by volumetrically more important lower-delta-plain and delta-front sequences. The Ness lagoonal deltas differ in detail and a general model can only be applied with caution as each one reflects both localized and regional controls on sediment supply and distribution. Where drowning events provided water depths (c. 20 m?) sufficient to accommodate a freely growing delta and sediment supply was high, the deltas had a lobate form (as in unit 3.1). Fluvial and wave effects controlled this geometry as the tidal regime was ineffectual at transporting large amounts of sediment. The deltas were friction dominated and the lagoons were diluted by fluvially supplied freshwater. The lagoons had sufficient fetch to promote wave modifications of the delta fronts, and broad frontal and lateral zones of HCS are a feature of these deposits. At the margins of the deltas thin but extensive sheet sandstones accreted by shoreline transport of sand, resulting in lagoonal shoals which became attached to the shoreline on progradation to form beach systems.

The main control on facies distributions was a vertical one caused by external factors rather than internal autocyclic mechanisms. The result is a strong layer-cake pattern which has a considerable impact on production behaviour in the reservoir. The external factors related to the development of the rift basin also manifest themselves in the variable sand-body trends in the Ness which bear no relation to the regional model of a northerly prograding system.

Although it must be realized that the Brent Group in Brent Field is only a small sample of a much larger basin, its unfaulted character and dense well coverage allow a detailed analysis of the Ness Formation. However, a thorough understanding of the whole basin will be required to unravel and quantify the controls on sediment dispersal in the Brent Group system as a whole.

ACKNOWLEDGMENTS: I would like to thank my colleagues at Shell, particularly Alan Heward, John Marshall and John Kantorowicz, for their criticisms of early versions of this manuscript, and Chris Deegan and Harold Reading for their comments as referees. The permission of Shell UK Exploration and Production, Shell International Expro and Esso UK to publish this paper is gratefully acknowledged.

References

BADLEY, M. E., EGEBERG, T. & NIPEN, O. 1984. Development of rift basins illustrated by the structural evolution of the Oseberg Feature, Block 30/6, Offshore Norway. *Journal of the Geological Society, London* **141**, 639–649.

BOWEN, J. M. 1975. The Brent Oilfield. *In:* WOODLAND, A. W. (ed.) *Petroleum and the Continental Shelf of North-West Europe*, Vol. 1, *Geology*. Applied Science Publishers, Barking, 353–362.

BRENCHLEY, P. J. 1985. Storm influenced sandstone beds. *Modern Geology* **9**, 369–396.

BRIDGE, J. S. & LEEDER, M. R. 1979. A simulation model of alluvial stratigraphy. *Sedimentology* **26**, 617–644.

BROWN, S., RICHARDS, P. C. & THOMSON, C. 1987. Patterns in the deposition of the Brent Group (Middle Jurassic), U.K. North Sea. *In:* BROOKS, J. & GLENNIE, K. (eds) *Petroleum Geology of North West Europe*. Graham & Trotman, London, 899–913.

BUDDING, M. B. & INGLIN, H. F. 1981. A reservoir geological model of the Brent Sands in Southern Cormorant. *In:* ILLING, L. V. & HOBSON, G. D. (eds) *Petroleum Geology of the Continental Shelf of North-West Europe*. Heyden, London, 326–334.

CHAN, M. A. & DOTT, R. H. JR 1986. Depositional facies and progradational sequences in Eocene wave-dominated deltaic complexes, southwestern Oregon. *Bulletin of the American Association of Petroleum Geologists* **70**, 415–429.

COLEMAN, J. M. & WRIGHT, L. D. 1975. Modern river deltas: variability of processes and sand bodies. *In:* BROUSSARD, M. L. (ed.) *Deltas*. Houston Geological Society, Houston, TX, 99–150.

DEEGAN, C. E. & SCULL, B. J. 1977. A standard lithostratigraphic nomenclature for the Central and Northern North Sea. *Report of the Institute of Geological Sciences* **77/25**, 36 pp.

DIEHL, A. L. 1978. The development of the Brent Field—a complex of projects. *European Offshore Petroleum Conference and Exhibition*, London, 24–27. October Proceedings Paper EUR108, 397–406.

DONALDSON, A. C., MARTIN, R. H. & KANES, W. H. 1970. Holocene Guadalupe delta of Texas Gulf Coast. *In:* MORGAN, J. P. & SHAVER, R. H. (eds) *Deltaic Sedimentation, Modern and Ancient*. Special Publication of the Society of Economic Paleontologists and Mineralogists **15**, 107–137.

DOTT, R. H. JR & BOURGEOIS, J. 1982. Hummocky

stratification: significance of its variable bedding sequences. *Bulletin of the Geological Society of America* **93**, 663–680.

EYNON, G. 1981. Basin development and sedimentation in the Middle Jurassic of the northern North Sea. *In:* ILLING, L. V. & HOBSON, G. D. (eds) *Petroleum Geology of the Continental Shelf of North-West Europe*. Heyden, London, 196–204.

VAN HINTE, J. E. 1976. A Jurassic time scale. *Bulletin of the American Association of Petroleum Geologists* **60**, 489–497.

HOLDSWORTH, D. K. & COLLINSON, J. D. 1987. Millstone Grit cyclicity revisited. *In:* KELLING, G. & BESLEY, B. (eds) *Sedimentation in a Synorogenic Basin Complex*. Blackie, Glasgow, 132–152.

HOWELL, D. J. & FERM, J. C. 1980. Exploration model for Pennsylvanian upper delta plain coals, southwest West Virginia. *Bulletin of the Geological Society of America* **64**, 938–941.

JOHNSON, H. D. & KROL, D. E. 1984. Geological modeling of a heterogeneous sandstone reservoir: Lower Jurassic Statfjord Formation, Brent Field. *Society of Petroleum Engineers 59th Annual Technical Conference and Exhibition, Houston, TX, 16–19 September 1984*. Society of Petroleum Engineers, Paper 13050.

—— & STEWART, D. J. 1985. Role of clastic sedimentology in the exploration and production of oil and gas in the North Sea. *In:* BRENCHLEY, P. J. & WILLIAMS, B. P. J. (eds) *Sedimentology: Recent Developments and Applied Aspects*. Geological Society of London Special Publication **18**, 249–310.

——, MACKAY, T. A. & STEWART, D. J. 1985. The Fulmar oil-field (Central North Sea): geological aspects of its discovery, appraisal and development. *Marine and Petroleum Geology* **3**, 99–126.

KANES, W. H. 1970. Facies and development of the Colorado river delta in Texas. *In:* MORGAN, J. P. & SHAVER, R. H. (eds) *Deltaic Sedimentation, Modern and Ancient*. Special Publication of the Society of Economic Paleontologists and Mineralogists **15**, 78–106.

LEEDER, M. R. 1983. Lithospheric stretching and North Sea Jurassic clastic sourcelands. *Nature* **305**, 510–514.

LIVERA, S. E. 1986. Geological controls on heterogeneous reservoir behaviour in complex deltaic sediments, Brent Field, northern North Sea.

Bulletin of the American Association of Petroleum Geologists **70**, 613 (abstract only).

MALDONADO, A. 1975. Sedimentation, stratigraphy, and development of the Ebro Delta, Spain. *In:* BROUSSARD, M. L. (ed.) *Deltas.* Houston Geological Society, Houston, TX, 311–338.

MORTON, A. C. 1985. A new approach to provenance studies: electron microprobe analysis of detrital garnets from Middle Jurassic sandstones of the northern North Sea. *Sedimentology* **32**, 553–566.

OOMKENS, E. 1967. Depositional sequences and sand distribution in a deltaic complex. *Geologie en Mijnbouw* **46**, 265–278.

PARRY, C. C., WHITLEY, P. K. J. & SIMPSON, R. D. H. 1981. Integration of palynological and sedimentological methods in facies analysis of the Brent Formation. *In:* ILLING, L. V. & HOBSON, G. D. (eds) *Petroleum Geology of the Continental Shelf of North-West Europe.* Heyden, London, 205–215.

RAWSON, P. F. & RILEY, L. A. 1983. Latest Jurassic–early Cretaceous events and the 'Late Cimmerian Unconformity' in the North Sea area. *Bulletin of the Geological Society of America* **66**, 2628–2648.

RØNNING, K. & STEEL, R. 1985. Transgressive reservoir sands and architecture in the Tarbert Fm., north-ern North Sea. *International Seminar, North Sea Oil and Gas Reservoirs.* University of Trondheim, Norway, 30–31 (abstract only).

SWIFT, D. J. P. 1976. Coastal sedimentation. *In:* STANLEY, D. J. & SWIFT, D. J. P. (eds) *Marine Sediment Transport and Environmental Management.* Wiley, New York, 255–310.

VAIL, P. R. & TODD, R. G. 1981. Northern North Sea Jurassic unconformities, chronostratigraphy and sea-level changes from seismic stratigraphy. *In:* ILLING, L. V. & HOBSON, G. D. (eds) *Petroleum Geology of the Continental Shelf of North-West Europe.* Heyden, London, 216–235.

WATTERS, D. G., PAGAN, M. C. T. & PROVAN, D. M. J. 1985. Cormorant Field, United Kingdom North Sea—synergistic approach to optimum development of a complex reservoir. *Bulletin of the Geological Society of America* **69**, 315 (abstract only).

WEISE, B. R. 1979. Wave-dominated deltaic systems of the upper Cretaceous San Miguel Formation, Maverick Basin, South Texas. *Transactions of the Gulf Coast Association of Geological Societies* **29**, 202–214.

S. E. LIVERA, Koninklijke/Shell Exploratie en Produktie Laboratorium, Volmerlaan 6, 2288 GD Rijswijk ZH, The Netherlands.

Coal-related Case Histories

Coal reviewed: depositional controls, modern analogues and ancient climates

R. S. Haszeldine

SUMMARY: Ancient economic coals usually contain more than 90% organic matter and less than 1% sulphur. Sedimentological reconstructions show the interbedded clastics to be fluvial-plain, deltaic, barrier coastline, estuarine or alluvial-fan deposits. However, bases of coal beds unconformably transgress these various facies, and coals only rarely pass laterally into clastic facies. Thus coal-forming peats *post-dated* the underlying clastic sedimentation and most coals did not form in deltas. Local coal thicknesses can be controlled by compaction of underlying sediment, syndepositional faulting, or basement structure.

Coals with more than 1%–2% sulphur are usually overlain within 10 m by marine sediments which provided SO_4^{2-} by diffusion. However, most coals contain less sulphur and must be freshwater in origin. Lithotype and spore profiles vertically through coals indicate a characteristic change of peat floras. This resulted from equisites-dominated seat earths, followed by diverse tree-sized vegetation which formed bright coal and lastly by low diversity stunted vegetation which formed dull coal. These features compare well with peats formed in modern ombrogenous blanket or raised bogs. The ancient mires formed in rainwaters (low sulphur), excluded clastic sediments (low ash), had a lateral floral zonation, usually post-dated clastic deposition and occurred inland from the shoreline.

Modern ombrogenous mires are abundant, especially in humid temperate climates. These peats are forming today during a period of long-term climatic cooling in areas with a frequent reliable rainfall. Ancient coals may also have formed during long-term climatic cooling epochs when volcanic CO_2 was being removed from the atmosphere. This could be regarded as a Gaian biosphere maintaining Earth temperatures similar to those of the present day.

Coal has been exploited for centuries because of its combustible qualities, and our knowledge of this rock type must equal or surpass many others. Much of the scientific study of coal in the late 1800s and up to the 1950s was concerned with a coal's calorific content, chemistry, detailed petrography or macrofloristic remains. Surprisingly, there are still many enigmas concerning the geology and formation of this organic sediment, arising from the lack of integrated studies and communication between chemists, petrographers, geologists, sedimentologists, engineers and palaeo- or neo-botanists. It is clear from this review, that of McCabe (1984) and those in Scott (1987) that interdiscipline exchange must increase, that the present can tell us something about ancient coals and also that ancient coals may tell us something about how the present planetary biosphere functions.

In this paper we examine aspects of the earth science factors that coals record, ranging from palaeo-environments, controls of sulphur content and local tectonics to mire types and climatic controls.

Coal depositional settings

Deltas

The inclusion of this paper in a publication devoted to deltas is testimony that coals are widely regarded as 'deltaic'. Such settings have long been implied by palaeographical reconstructions (Wills 1954) but cannot now be regarded as correct—in contrast many coals formed after clastic sedimentation in fluvial settings.

Initially, the new discipline of sedimentology rapidly added credibility to the analogies which could be made with modern deltaic processes (Ferm 1970; Wanless 1970). According to McCabe (1984), coal-forming deltaic processes unfortunately and misleadingly became identified with the modern Mississippi Delta (*eg* Frazier & Osanik 1969), following Fisk's (1960) study of organic-rich backswamps adjacent to distributary channels. The American work eventually culminated in a benchmark paper (Horne *et al.* 1978) which modelled coal deposition in the Appalachians as a fluvially dominated marine delta system with minor wave-produced barriers (Fig. 1). This model predicted optimum coal accumulation in the transitional zone between the lower and upper delta plains. Generalized coal seam geometries could also be predicted. This model formalized a need, has worked well for many years and can still be applied today (Ferm 1984). However, it suffers from several weaknesses, in that it considers only the clastic sedimentation (rather than the coal), and syndepositional tectonics must also have had complicating effects (Ferm & Staub 1984). With further sedimentological work in other coalfields, it has rapidly become apparent that coals formerly interpreted as 'deltaic' actually accumulated in many settings other than marine deltas (see below).

Deltaic models of coal deposition are still applicable to many coal-bearing sequences in the USA, UK and Australia although, as McCabe (1984) points out, fluvially dominated freshwater settings are now being recognized in many cases. These may lie up-river of the true lacustrine or marine deltas. Many workers may have been too enthusiastic in applying the model from Horne *et al.* (1978) to their own coalfields and re-interpretations may appear (*eg* Breyer 1987). Even in coalfields regarded as classically deltaic, the areal delineation of individual delta systems by sedimentological argument alone appears to be difficult. For example, in the Pennine coalfields of N and E England, Fielding (1984*a, b*) interprets progradation of one Mississippi-scale delta with coal growth on lower-delta-plain to upper-delta-plain environments. In contrast, the context of European palaeogeography, in the same study area, suggested to Haszeldine (1984) that coal deposition was on a fluvially dominated coastal plain which fed down-river to several coalescing small deltas, analogous to the Cenozoic Gulf Coast (Galloway 1981).

Coal-bearing deltas and fluvial plains entering 'lacustrine' water bodies are now recognized in ensialic foreland basins (*eg* Ayers & Kaiser 1984) and strike-slip fault basins (Hacquebard & Donaldson 1969; Gersib & McCabe 1981).

Coal not in deltas

Although most economic coals occur in fluvial-plain settings often associated with deltas, it appears that coal has also accumulated in association with any humid siliciclastic environment where plant colonization took place. Sedimentologically based work has assigned coals to back-barrier settings (*eg* Hobday & Horne 1977; Stavrakis 1987), alluvial fans and fan deltas (Heward 1978; Whateley & Jordan, this volume), and estuarine settings (Breyer 1987). Most of the present-day peat-forming areas do not occur in deltas or any active clastic environments (see below) but on tidal coasts, abandoned coastal plains and inactive fluvial plains, which are predominantly *inland* and completely removed from significant coexisting sedimentation or shoreline influence.

Neither does coal accumulate in all fluvial plains or deltas, for it is a *subaerial* sediment, recording the climatic influences acting on an

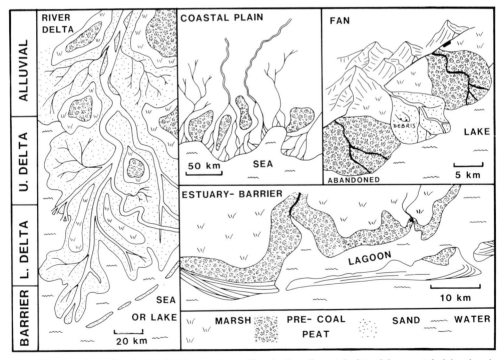

FIG. 1. Sketches of coal-forming environments proposed by clastic sedimentologists: delta, coastal plain, chernier or barrier coastline, estuary and alluvial fan. Modern peats in these settings usually have high sulphur and ash contents.

area. Deltas and fluvial plains occur today in climates ranging from arctic to temperate to tropical to arid, with consequent variation in organic accumulation on their flood plains.

The relationship between coal seams and clastics must be emphasized here. Coals *overlie* the clastic sediments deposited by rivers, deltas, marine barriers or lakes. The coal base *unconformably transgresses* many underlying facies, and yet the coal is usually overlain by a lake or marine transgression or a channel which has eroded through such deposits. In many basins, coals are laterally persistent for tens of kilometres and only rarely pass laterally into coeval fluvial sedimentation (*eg* Elliott 1965), although they often split and pass laterally into areas of more rapid subsidence. Thus these facies geometries show that the pre-coal peat (see Fig. 7) *post-dated* the clastics. Clastic deposition occurred, was abandoned and was colonized by plants to accumulate clastic-free peat some distance inland from the contemporary shoreline.

Thus coal is neither restricted to deltas nor indicative of deltas; it is a common accessory sediment, reflecting the subaerial climate of the varied environments across which it has formed.

Syndepositional controls on coal thickness

Compactional subsidence

Following the great enthusiasm for coal-forming analogues based on the Mississippi Delta, much has been made of autocyclical (sedimentological) mechanisms of producing cyclicity within deltaic sediments and hence of producing thick coals (*eg* Ferm & Staub 1984). The concept is that muds and silts compact more than sands in the preceding cycle. This subsidence will be greatest, and coals thickest, in areas laterally offset from underlying sands (Fig. 2).

However, published maps do not demonstrate these compactional controls of coal thickness. A thick coal may overlie a sand-poor interval, and 5–10 km laterally the same coal may overlie a sand-rich interval (Fig. 2) (Eggert 1984; Ferm & Staub 1984, Fig. 10). Neither do published cross-sections provide a clear demonstration of compactional effects. For example Flores (1981, Fig. 20) invoked compaction and showed that the Palaeocene Pawnee coal thickens laterally away from immediately underlying less compactible sandstone. However, Ayers & Kaiser (1984, Figs 17–19) suggest that thick coals in this same basin grew in flood basins lateral to coeval small deltas which had presumably been attracted into the fastest-subsiding areas. Likewise cross-sections through the fluvio-deltaic New Mexico coals (Cavaroc & Flores 1984, Figs 9 and 12) fail to demonstrate any lateral stacking of coals on a 1 km scale laterally adjacent to earlier channels.

One explanation for this overlap of thick coal above an 'uncompactible' sandstone could be lack of stratigraphic resolution, as such sandstones may be older parts of the pre-coal cycle. Alternatively the compaction of a much thicker sequence of underlying sediments may have controlled the coal thickness—such effects will require computer modelling. Thus I conclude that compaction of immediately underlying sediments

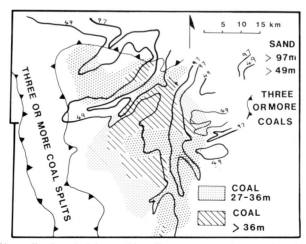

FIG. 2. Compaction of immediately underlying muddy sediment can sometimes produce thick coals. Here the coal splits into the basin, and a thick NE–SW sandstone (slow compacting?) has thin coal over it and thick coal adjacent. A thick N–S sandstone (earlier deposited, compacted?) has thick coal overlying it. Data from the Palaeocene Powder River Basin (Ayers & Kaiser 1984).

may sometimes have controlled coal thickness on a lateral scale of a few kilometres but that this effect usually did not act alone and may often have been influenced by local faulting or compaction of much thicker underlying sediments.

Faulting and basement subsidence

Coals occur where subsidence conditions are suitable in any particular basin. If subsidence was too slow, then the pre-coal peat would oxidize away. If subsidence was too fast, the pre-coal peat would be drowned. Successive drownings at the basinward end of a cross-section will cause a seam to split into the basin. Such a situation was demonstrated in the Pennine coalfield of England (Rayner 1981, Fig. 48) and can also be seen in the Palaeocene Powder River Basin (Ayers & Kaiser 1984).

Faulting during deposition does not necessarily occur in foreland basins, although basement weaknesses may reactivate (Allen *et al.* 1986). Transcurrent fault basins contain such thick sediment sequences that controls inherited from the basement are probably weak compared with the prominent syndepositional faults at basin edges (Heward 1978; Ballance & Reading 1980). In contrast, rifted basins and their subsequent thermal subsidence phase are certainly significantly affected by faulting and basement reactivation. Examples of these effects are cited below. In such basins any sedimentary facies model or coal exploration model may be totally dominated by structural effects. It is surprising, then, that the vast majority of ancient river, delta or coal studies (and most studies of modern environments) completely ignore the tectonic and structural framework of the deposits being considered.

The Tertiary rift basin brown coals of N Germany have long been known to split and thin into the deeper parts of the basin on a 20 km scale (Teichmuller & Teichmuller 1968, Fig. 5). On a scale of hundreds of metres, syndepositional faulting has infrequently caused rapid thickening of these seams on the down-throw side of faults (Fig. 3) (Teichmuller & Teichmuller 1968, Fig. 8). Faulting usually reduced seam thicknesses on its down-throw side; Li Sitian *et al.* (1984, Fig. 26) show that, in the late Jurassic transcurrent fault basins of NE China, thick coals occur on a 1 km scale over the slower-subsiding areas on the up-throw side of faults.

In the Carboniferous of the Black Warrior Basin, Alabama, Weisenfluh & Ferm (1984, Figs 4 and 6) demonstrate that thick coals are localized on the up-throw sides of NW–SE normal faults with 30 m throws (Fig. 3). In the E Kentucky Carboniferous, Trinkle & Hower (1984) show

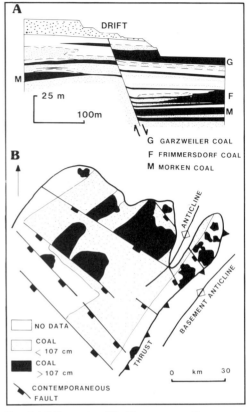

G GARZWEILER COAL
F FRIMMERSDORF COAL
M MORKEN COAL

NO DATA

COAL
< 107 cm

COAL
> 107 cm

CONTEMPORANEOUS FAULT

FIG. 3. (*A*) Slow rates of faulting may sometimes enable thick coals to develop on the down-throw side (from Teichmüller & Teichmüller 1968); (*B*) thick coals often occur on the up-throw side of syndepositional faults, and coal is split on the down-throw side (from Weisenfluh & Ferm 1984).

that syndepositional basement highs were active on a 20 km scale and caused low vitrinite contents in the coals. The same highs acted as high heat flow areas, so that higher coal ranks later developed.

In the Upper Carboniferous rift basin of NE England Haszeldine (1981) demonstrated that thick coal on a 100 km² scale occurred above a slow-subsiding high underlain by buoyant granite basement. Fluvial sedimentation was attracted into the adjacent faster-subsiding areas (Fig. 4). Fielding (1986, Fig. 8) observed similar relationships in this area. To the N of this granitic basement, Haszeldine (1984, Fig. 9) demonstrated that the siting and orientation of intracoalfield deltas were controlled by syndepositional faults of 20–50 m throw.

It therefore appears that thick coals may be localized by both syndepositional faulting effects

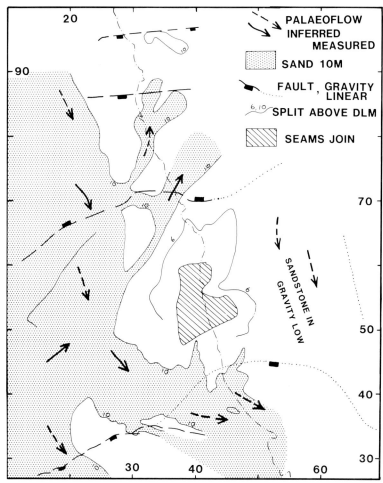

FIG. 4. Deep basement may subside slowly and permit thick coals to develop by exclusion of clastics. Example from NE England Westphalian after Haszeldine (1981). Data from above Durham Low Main (DLM) coal and below Maudlin–Bensham coal, which merge over the granitic basement of the Alston block. Compiled from 1:10 000 British Geological Survey (BGS) maps, borehole data and BGS regional memoirs.

and subsidence effects due to the underlying deep basement. Consequently, river channel deposits may be more abundant or may be vertically stacked on the down-throw sides of faults with small throws or over faster-subsiding basement areas (cf Horne et al. 1978). Thick coal deposits are often localized over slower-subsiding footwall blocks or basement highs. However, this relationship may be reversed if the rate of faulting was too slow. It is possible that fault and basement controls of facies locations could be delineated in many more coalfield basins, regardless of whether they are of rift, foreland, transcurrent or thermal subsidence type.

We must therefore adapt our tectonically anodyne sedimentological facies models to ac-commodate the structural conditions within each basin.

Sulphur content and marine influence

A positive relationship between marine influence and high sulphur content in a coal has been noted by many workers. In the Appalachian Carboniferous, Williams & Keith (1963) demonstrated higher sulphur contents in coals with roofs of marine mudrock than in coals with roofs of freshwater mudrock. Likewise Horne et al. (1978), in a similar area, show that marine deposits immediately overlying a coal can raise the sulphur content from less than 1% to more than 4%; such changes can occur laterally on a 5 km scale. In Germany, the northern end of the Miocene Rhine Graben contains high sulphur coals and interbedded marine faunas, whereas 20 km southwards coals are thick with non-marine sediments. Shimoyama (1984) illustrates that the great majority of Eocene and Oligocene coals mined in Japan have sulphur contents of less than 1% and that this low sulphur level occurs where any overlying brackish or marine deposits are at least 10 m above the coal. Conversely, any coal overlain within 10 m by marine deposits has a much higher sulphur content. Love et al. (1983) used a detailed isotopic study in the English Westphalian to suggest that marine sulphate can diffuse down 10 m through muds during the first stages of burial. Casagrande (1987) provides a review of sulphur in coal and concludes that high sulphur contents of organically bound sulphur or framboidal pyrite are linked to marine influence during peat deposition or during burial. Cleat pyrite formed later.

Results obtained from studies of modern peats on the Fraser River Delta show that marine sulphur is the source of sulphur in pre-coal peat (Styan & Bustin 1983; Bustin & Lowe 1987). Here the Lulu Island peat is forming on the transition from the lower to the upper delta plain and contains between 6% and 0.4% sulphur. In contrast, the Pitt Meadows peat is forming on the freshwater fluvial plain at the head of the delta (Fig. 5) and contains 0.1%–0.5% sulphur.

The great majority (87.5%) of this sulphur is bound to the organic molecules within peat debris rather than being in the form of discrete metallic sulphides. In contrast, the sulphur in coals forms predominantly as metallic framboidal FeS_2 or by replacing plant debris (during early diagenesis?) or along cleat veins (Horne et al. 1978; Averetti 1984). Cross-plots of sulphur types and total percentage sulphur in ancient coals (Fig. 6) can be used to discriminate broadly between lower- and upper-delta-plain environments (Hunt & Hobday 1984).

Most ancient economic coals contain very little sulphur: 0.5%–1.5% in the USA, 0.4%–2.5% in Britain, 0.2%–0.9% in Australia, 0.5%–1.7% in South Africa (Barnsley 1984) and 0.2%–1% in Japan (Shimoyama 1984). Qualitative information shows that most appraised coals in China, the USSR and worldwide have low sulphur contents (International Energy Agency 1983). The exceptions are coals with marine roofs, as is the case in New Zealand (Suggate 1959, Table 6) where many coals have sulphur contents below 1% but some basins contain coals with up to 9% sulphur. Although these data may well be biased in favour of the more economic low sulphur coals, the implication is clearly that the majority of ancient coals have low sulphur contents. This implies settings without access to marine sulphate. We must therefore envisage ancient mires which developed in freshwater (low sulphate) settings. These mires were then transgressed and

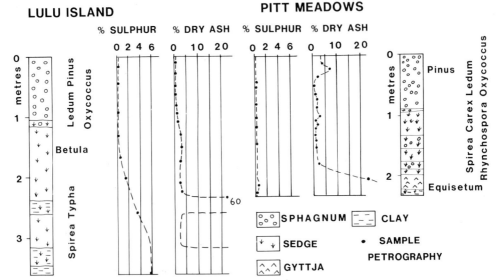

FIG. 5. Setting of modern temperate raised mires on the Fraser River Delta. The sulphur and ash contents of the Pitt Meadows peat are most analogous to ancient economic coals. Note that the Lulu Island peat in the transitional setting from the lower to the upper delta plain has a high sulphur content. (From Styan & Bustin 1983; Bustin & Lowe 1987.)

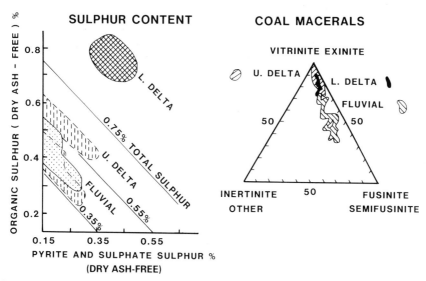

FIG. 6. Cross-plots of sulphur content in coal can distinguish between coal overlying different environmental positions over a delta. In addition the different coal maceral groups can broadly distinguish environmental positions over an abandoned delta. (Redrawn from Hunt & Hobday 1984.)

buried beneath freshwater deposits. This could perhaps have occurred either in fluvial-plain settings or in deltaic and coastal settings with such high ancient rainfalls that marine influence was literally diluted or, lastly (and most probably), because the pre-coal mires had formed across abandoned sediments on an older fluvial or coastal plain in a position inland from the palaeoshore (Fig. 7). The present-day Okefenokee Swamp (Cohen 1984) is formed across abandoned barrier coastlines, and present-day thick peat-forming mires in England develop inland across abandoned estuarine and fluvial deposits (Barber 1981). The coal-forming mire was not coeval with sedimentation of the underlying clastics but *post-dated* them, possibly by thousands of years.

Spore and lithotype profiles in ancient coals

Lithotypes

Coal is not a homogeneous rock but is composed of bedded layers like any laminated sediment. On a hand-specimen scale, these layers can be described as one of four lithotypes (Stopes 1919; Davis 1984). Clarain and vitrain are brightly reflecting lithotypes; durain and fusain are dull lithotypes. Upon microscopic examination, it

appears that in general terms vitrain represents little-oxidized bark shells from tree-sized vegetation, durain results from slower preservation rates, with periods of shallow submergence alternating with periods of good drainage and oxidation, clarain is composed of smaller layers of durain and vitrain which may have been partially oxidized and fusain results from extreme oxidation by exposure to either palaeoweathering or palaeofire (Davis 1984). The vertical sequence of lithotypes through a coal is relatively easily recorded and numerous examples have been published. In general terms, very many coals commence with bright lithotypes and merge upwards into dull lithotypes. This occurs in the UK Carboniferous (Smith 1962, 1968; Smith & Butterworth 1967), the German Carboniferous (Stach 1982, Fig. 115), the USA Carboniferous (Ting & Spackman 1975), the E Canada Carboniferous (Cameron 1971) and the Australian Permian (Smyth 1967, 1970). The vertical sequence then becomes complicated in detail, usually with interbedding of bright and dull lithotypes, and may terminate with either dull or bright lithotypes. These profiles suggest that a characteristic sequence of conditions occurred during development of a pre-coal mire (Fig. 8). Initial rapid burial of organic debris or poorly drained conditions resulted in the formation of unoxidized vitrinite. As the peat became thicker, burial of organic debris was slower or the mire

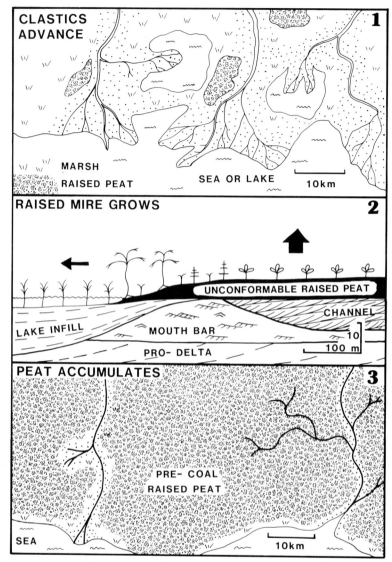

FIG. 7. Coal seams unconformably transgress underlying facies and rarely pass laterally into clastic facies. Therefore pre-coal peats developed *later* than the clastics, overlying them, and were *not* contemporaneous with clastic sedimentation. Studies of modern environments should drill through peaty mires into sediment, rather than examining active deltas. Ancient coals are low in sulphur and must have accumulated inland, away from marine influence, on the surfaces of abandoned clastics.

was better drained, resulting in dull coals. These conditions alternated during most of the peat accumulation. The mire was terminated either by abrupt drowning which did not give time for bright-coal floras to re-establish (dull top) or by gradual drowning giving time for the bright-coal water-edge floras to re-establish (bright top). Consequently (Fig. 9), lower-delta-plain coals or rapid-subsidence areas contain more vitrinite than the duller coals of the slower-subsiding

upper delta plain (Butterworth 1964, and in Smith & Butterworth 1967; Hunt & Hobday 1984).

Spores

The actual plant floras which existed during accumulation of a coal seam are recorded by the macerals within a coal (Davis 1984) and more precisely by the spores deposited by the plants. Smith (1962, 1968) and Smith & Butterworth

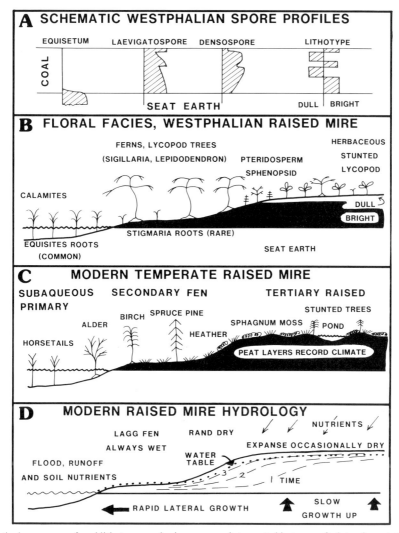

A SCHEMATIC WESTPHALIAN SPORE PROFILES

EQUISETUM   LAEVIGATOSPORE   DENSOSPORE   LITHOTYPE

COAL

SEAT EARTH   DULL | BRIGHT

B FLORAL FACIES, WESTPHALIAN RAISED MIRE

FERNS, LYCOPOD TREES   HERBACEOUS

(SIGILLARIA, LEPIDODENDRON)   PTERIDOSPERM   STUNTED

SPHENOPSID   LYCOPOD

CALAMITES   DULL

BRIGHT

STIGMARIA ROOTS (RARE)

EQUISITES ROOTS   SEAT EARTH

(COMMON)

C MODERN TEMPERATE RAISED MIRE

SUBAQUEOUS   SECONDARY FEN   TERTIARY RAISED

PRIMARY   STUNTED TREES

BIRCH   SPRUCE PINE

ALDER   SPHAGNUM MOSS   POND

HEATHER

HORSETAILS   PEAT LAYERS RECORD CLIMATE

D MODERN RAISED MIRE HYDROLOGY

NUTRIENTS

LAGG FEN   RAND DRY

ALWAYS WET   EXPANSE OCCASIONALLY DRY

WATER

FLOOD, RUNOFF   TABLE

AND SOIL NUTRIENTS   3   2   1   TIME

RAPID LATERAL GROWTH   SLOW   GROWTH UP

FIG. 8. Vertical sequences of coal lithotypes and miospores are interpreted in terms of a lateral vegetation zonation. This is directly analogous to vegetation zonations and peat-forming hydrology seen in modern raised-mire ecosystems. These can form in any climatic setting but are most abundant in temperate climates.

(1967) identified spore profiles from UK Carbon-iferous seams, from which a general model can be extracted (Smith 1968) in which the flora is initially a high diversity Lycospore assemblage (tree-sized lycopods) which passes up via a 'transitional' phase into the Densospore assem-blage of low diversity lycopods. These phases may then either alternate repeatedly or be interrupted by a Crassispore assemblage rich in *Sigillaria* and fern species. Smith (1968) also noted that the Lycospore phase approximately coincides with bright coals and the Densospore phase approximately coincides with dull coals. Scott (1978, 1979) also studied coal spores and

the floras in sediments beneath them. He found that seat-earth floras are radically different from the coal floras and that *Calamites* (Equisetum, horsetail) is often dominant below the coals. Thus a characteristic vertical floristic stratigraphy is present within these Carboniferous coals, which, if viewed in terms of sedimentological facies, is equivalent to a lateral progradation of floristic zones, *ie* a zonation of floras existed lateral to lake infills, with yet different floras adjacent to river channels (Gastaldo 1987). Similar spore-profile evidence has not yet been synthesized for the temperate Gondwanaland coals.

These floristic successions can be interpreted

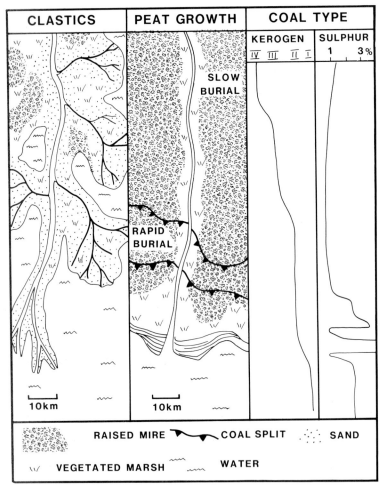

FIG. 9. Model for coal deposition overlying abandoned deltas. High sulphur vitrinite-rich bright coals occur on the faster subsiding distal delta and will form good gas source rocks, with some H$_2$S. The inland delta accumulates more durain-rich dull coals which contain more type IV kerogen.

by analogy with modern raised mires (Fig. 8) in terms of a *Calamites* flora rooted subaqueously in seat-earth ooze at a lake edge (analogous to today's ecological niche for horsetails). Mud settling from suspension gradually raised this to very near water level, whereupon Lycospore tree-sized club mosses could colonize. These may have been partly rooted in the seat earth, although *Stigmaria* root fossils are not abundant. As the peaty debris of fallen club mosses accumulated, so the vegetation became totally rooted in peat, and nutrient supply was from the peat and from rainfall. The Lycospore vegetation was rapidly buried anoxically, forming bright coal. The increasing peat thickness resulted in elevation of the peat mire above the surroundings, thus

excluding clastics and resulting in low ash contents. The elevated peat was colonized by a stunted and low diversity Densospore flora. This area was subject to rainfall flooding and also to desiccation, resulting in partly oxidized dull coal. The vegetation and lithotypes at the coal top may be symmetrical with the coal's base, suggesting gradual migration of floristic facies owing to slow drowning. If the coal top does not consist of Lycospore vegetation and bright coal, this suggests that rapid submergence occurred as a result of faulting. Many coals have a roof of subaqueous shale or are overlain by clastics eroding through a shale, supporting the inference that many mires were terminated by drowning. Close analogies occur between these ancient Carboniferous coal features and floristic features generated in modern raised mires (Smith 1968; Teichmüller & Teichmüller 1975, cited by Stach 1982, p. 9; Haszeldine 1981; McCabe 1984).

Detailed investigations of spore profiles in Gondwanaland and post-Devonian coals have not yet been undertaken, and so it is not yet certain that similar mire features were present in all coals. However, the similarity of circumstantial evidence from lithotype profiles, coal seam geometries and ash and sulphur contents suggests that some sort of raised- or blanket-mire types also formed the pre-coal peats in these cases (Fig. 9).

Modern peat analogues

A modern analogue for ancient coals must demonstrate peat-accumulating mires which are low in clastics (less than 10%), low in sulphur (less than 1%), have a lateral vegetation zonation and can form peats 10 m or more thick which will result in coals 1 m or more thick assuming a 10:1 compaction (*cf* Ryer & Langer 1980). Terminology for peats is taken from Moore (1987), where *peat* contains less than 25% ash, *mire* is any freshwater peat, *fen, carr* and *swamp* are progressively more flooded rheotrophic mires dependent on groundwater nutrients and *bog* is a mire supplied with its nutrients by rainwater.

Contemporary peat formation is a much studied subject (*cf* Moore 1984, 1987), but peat occurrences can be viewed as (i) floating mire, (ii) low-lying mire (fen or low moor) and (iii) raised mire (high moor, *ie* blanket bog or raised bog) (McCabe 1984). These equate respectively to the primary, secondary and tertiary peats of Bellamy (1972).

Floating mires rely upon intertwined roots of angiosperms such as sedges to link the vegetation mats. It is uncertain whether late Palaeozoic pre-angiosperm floras could have formed such mats. Equisites and Lycopod floras do not form them today. However, these floras are much reduced today compared with pre-angiosperm times. Where observed in groups today on lake margins they root in the mineral soil. However, *Equisetum fluviatale* has horizontal rhizomes and *Lycopodiella incundata* has extensive creeping stems, so the potential for mat forming exists. Low mires (fen, carr and swamp) are dependent upon nutrients from underlying mineral soils or incoming groundwater. These nutrients also bring in clastic matter, so that these peats have high ash contents. The 'peats' of the modern Mississippi Delta are in this category, with more than 25% ash (Kosters *et al.* 1987).

Present-day raised peats are blanket bogs and raised bogs (see below). These form low ash peats, with a laterally zoned flora in raised bogs, do not depend on the underlying soil for nutrition and can form peats 3–12 m thick which would form coals 0.3–1.2 m thick. They provide excellent analogues for ancient pre-coal peats, and are reviewed below and by McCabe (1984). The crucial factor which blanket and raised bogs possess is that they maintain their own elevated water table, in part because of the relatively poor vertical permeability of peat and the frequent rainfall. Thus no water run-off from surrounding areas affects the low diversity peat flora and no inorganic 'ash' is washed in, other than that generated botanically within the plants.

Raised-mire peats

These mires are in many ways a closed ecosystem, dependent only upon an adequate year-round rainfall, outstripping evaporation and transpiration, to supply nutrients and prevent rapid oxidative decay. These ombrogenous (rain-fed) peats can be called bogs. The peat surface at the outer edge of the mire merges with the surrounding topography or mires via a rimming fen or swamp. However, the peat slopes steeply upwards towards the bog centre, where the bog surface is flat (Fig. 8). This flat central portion forms the majority of the mire's area and prevents flooding by clastics or by surface groundwaters from outside the mire, so that the peats have ash contents of only 6.5% (Fitch 1954) down to 3%– 0.7% (Polak 1975) and 0.5%–1.5% (Styan & Bustin 1983). The E Canadian mires prograde sideways by growth of their vegetation and can move past trees in their path, surrounding and partly burying the upright trunk as they do so (Ganong 1897). It therefore seems possible for such peats to grow laterally as well as vertically.

It is important to realize that the modern raised-mire ecosystem is not restricted to the tropics but occurs most abundantly in humid temperate climates. In NW Borneo, coastal-plain peats unconformably overlie barrier coastlines and cover 670 km^2 (260 square miles) with mire centres raised 20 m (60 ft) above nearby river banks and measured peat thicknesses greater than 13 m (45 ft) (Anderson 1964; Polak 1975; Teyssen 1987). On the western Malayan peninsula, raised mires are forming in the active tidal Klang Delta (Coleman *et al.* 1970). In E Canada, Ganong (1897) made careful observations of raised *Sphagnum* mires some 600 m in diameter. In W Canada, Styan & Bustin (1983) and Bustin & Lowe (1987) studied raised mires coexisting with the Fraser River Delta. These show a range of settings from mid-delta plain to fluvial delta plain, with corresponding variations in sulphur content. In Britain, raised mires form 5 m (16 ft) peats on the Welsh estuarine coast (Wilks 1979) and well-studied examples in NW England form raised mires at least 3 m deep near estuaries, on low hills and up to 330 m (1000 ft) above sea level (Barber 1981). Similar raised mires form over many areas of cooler NW Europe (Moore 1984), northern Canada (Korpijanko & Woolnough 1977; Stanek 1977; Terasmae 1977) and northern Russia (P'yavchenko 1964; Romanov 1968). In all these areas there is a large excess of rainfall over evaporation and transpiration, and a locally flat topography. Most importantly, the rainfall is not highly seasonal but occurs all year. The exception to this generalization is the northernmost latitudes where winter botanical activity is, literally, frozen.

The plant species on the bog plateau and sides are identical with or closely related to those in the nearby blanket bog (eg Ganong 1897; Barber 1981). The raised mire itself is rimmed by a fen vegetation fed by run-off from the raised bog. In NW Borneo the raised-mire species are closely comparable with those in Miocene coals from the same area (Anderson & Muller 1975). The lateral floral zonation from the edge to the centre of the mire is produced by the changes in the abundance and dominance of various plants. The SE Asian mires are dominated by tree-sized vegetation and therefore are bog forests. The cooler temperate climate mires are *Sphagnum* dominated in high-rainfall wet areas of Britain (eg Butterburn Flow, Barber 1981) and heather dominated in dryer areas (eg Bolton Fell Moss, Barber 1981). In Canada, some of the raised mires may support spruce or birch trees which are often stunted (Ganong 1897; Stanek 1977). The *Sphagnum* species from modern raised mires have a very high cation-exchange capacity in order to extract dilute nutrients from rainwater (Barber 1981). Ancient bog floras may have been similarly endowed, and this would explain the high trace element contents (Ward 1984) in the ashes of coals.

All the modern raised mires show a lateral floral zonation (Fig. 8), often ranging from full-sized diverse trees on the mire edge to a more restricted flora on the mire slope to a wide flat plateau with a stunted low diversity flora (Ganong 1897; Anderson 1964; Barber 1981). This fits very well with the vertical floral changes recorded in ancient coals (see above). The mire plateau ('mire expanse') may have a dry surface during dry weather or climatic changes (Anderson 1964; Barber 1981; Ingram 1982), but always has a water table close beneath. Water from this surface expanse seeps laterally through the surface soil

layer (acrotelm of Ingram 1982) and down the sloping rand, which is usually the driest part of the dome. This water emerges to supply the lagg, which rims the raised mire and supports a rheotrophic fen vegetation distinct from that of the raised mire itself. The partly aerobic conditions within the acrotelm may be those which favoured oxidation to form dull coals. A highly irregular or seasonal rainfall would presumably permit desiccation, oxidation and destruction of the bog.

Workers on modern peats suggest that fen (secondary mire) evolves into raised or blanket bog (tertiary mire) (Bellamy 1972) if sufficient rainfall is available and enough time is given (Ganong 1897; Tallis 1983). The lateral lagg run-off from a raised mire presumably assists peat preservation and hence raising of the peat surface to become oligotrophic. In areas of very persistent and abundant rainfall, oligotrophic raised blanket bogs can develop very widely and are unconformable on many substrates (Taylor 1983).

Blanket-bog peats

Before the current enthusiasm for raised-bog peats sweeps us away completely (see above), it is important to recall that blanket bogs are also ombrogenous low ash peats with a water table raised above their surroundings and containing a very similar flora to that found in raised bogs. Consequently, this ecosystem is also a potential coal precursor. Temperate raised bogs today are typically small in area (a few square kilometres), and the individual tropical raised bogs of Borneo are only 250 km^2 in area (Anderson 1964; Anderson & Muller 1975). In contrast, the temperate blanket bogs in N Europe or Canada may commonly cover hundreds of square kilometres (Taylor 1983) and are very much more widespread, currently forming up to 3% of the world's land area (Clymo 1987). This should be compared with the area of 100–200 km^2 often occurring in an economic coal leaf in the British Carboniferous. The similarities of scale are attractive, and it is possible that our ancient coals may have originated from extensive ombrogenous bogs growing unconformably over a flat fluvial, deltaic or shoreline plain (Fig. 9). Whether these bogs were raised or blanket in the modern sense will be difficult to determine, but a blanket-bog analogy must remain a possibility, particularly for ancient coals formed outside the tropics. In modern situations the British blanket bogs may show a forested or scrubwood basal layer overlying the mineral substrate, followed by peat 1–7 m thick composed of *Eriophorum, Carex, Calluna* and *Sphagnum*. However, in many cases the peat is recorded as lying directly on the mineral substrate beneath (Taylor 1983). Thus these bogs could mimic the vertical floral succession found in some coals, although in these modern cases the succession was due more to climatic change than to an ecological climax (see below). These bogs are confined to parts of the world with exceptionally moist climates (*eg* rainfall of 1250 mm a^{-1} with 225 rain days per year in western Ireland and Scotland (Taylor 1983)). On areas of flat ground such as the flat Devonian lacustrine flags in Caithness, NE Scotland (similar to an ancient flat coastal plain or lake?), the blanket-bog system has large-scale run-off and drainage patterns which cause local fens. Thus the blanket bog becomes morphologically indistinguishable from raised mire (Ingram 1987).

Rates of peat accumulation

Mean rates of vertical peat accumulation in temperate-climate raised mires are 20 years cm^{-1}, based on radiocarbon dates (Barber 1981). Over short time spans, the sphagnum mosses in these mires can accumulate at 3 cm a^{-1} (Barber 1981). In the tropical mires of NW Borneo, Anderson (1964) records accumulation rates of 47–22 cm per year per 100 years (Fig. 10)—some five times faster than the British examples—although the accumulation rate seems to slow with time on this single data set. These measurements (Fig. 10) imply that 13 m of peat (1.3 m of coal?) needed a minimum of 4300 years to accumulate. Such figures must be regarded as a minimum, for climatic changes during peat growth can substantially slow the long-term accumulation rates. In addition, moss or lycopod vegetation may compact to give thinner coals (Ryer & Langer 1980; Elliott 1985) than the woody vegetation. Thus Okefenokee peats in the eastern USA have taken 9000 years to accumulate 5.9 m of peat (Cohen 1984) and European raised peats accumulate at 20 years cm^{-1} (Barber 1981). 1 m of Carboniferous coal may have needed 10 m of peat (Ryer & Langer 1980) or 20 m of Carboniferous vegetation (Elliott 1985) and so taken at least 15 000–40 000 years (say 30 000 years) to accumulate—perhaps 100 000 years if depositional diastems are assumed within the coal. The Westphalian A and B may have lasted for 4 Ma (Hess & Lippolt 1986), and about 15–25 economic seams (say 20) of 1 m or more usually occur in any Pennine coalfield (Trueman 1954). Thus pre-coal mires may have existed in any area for at least 6×10^5 years and maybe 2×10^6 years, raising the possibility that *extensive* mires may have been present for at least 6% and up to 50% of the duration of the Coal Measures. Deltaic advances between peats may have been very rapid. This inference that mires were persistent in time is strengthened by the observation that modern ombrogenous peats in tropical zones (Fig. 10) and temperate zones (Clymo 1987) show an asymptotic decrease in their rate of vertical accumulation. Thus an individual raised bog has a theoretical maximum thickness of some 22 m in the tropics (Fig. 10) and 7–15 m in temperate climates (Clymo 1987). Any bog approaching its maximum could, presumably, remain in steady-state thickness for thousands of years, until disturbed by compactional or tectonic subsidence. Upon compaction these bogs would form

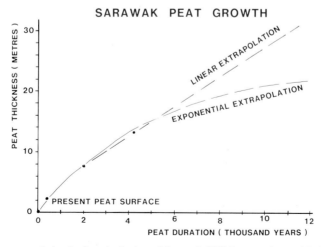

FIG. 10. Rate of peat accumulation in the raised mires of Sarawak, NW Borneo, slows with time (data from Anderson 1964).

coals perhaps 0.7–2.2 m thick. Thus in order to form any thick coals comprising several individual leaves separated by clastic bands (these are common in the UK Carboniferous), a separate time interval is needed for each pre-coal peat to grow, reach maturity, subside and be vertically succeeded by the later coal leaf.

Climate and coal accumulation

In this section I speculate that world climatic cooling may promote regular rainfall in tropical and temperate zones, leading to extensive peats and hence coals. These peats may also act as traps to remove CO_2 from the atmosphere. I have argued above that raised or blanket mires are the best modern analogue for ancient coal-forming peats, and that these mires develop in areas of reliable year-round high rainfall. The plant species on the surface of today's raised mires in Britain are controlled by the local wetness— various *Sphagnum* species prefer gradations of wetness, whereas heathers prefer drier locations. These vegetational variations have also occurred through time, linked with climatic changes in rainfall over the past 2000 years (Barber 1981, Fig. 14) to 5500 years (Aaby 1976), so that a stratigraphy of climatically controlled vegetation occurs within each British or Danish peat (Fig. 11). Periods of exceptionally heavy rain may cause the bog to burst its sides and flow out (Barber 1981).

The northern Canadian mires are closely controlled by the migration of the mixing zone between arctic and temperate air masses, which produces a reliable rainfall (Terasmae 1977). All these mires must have originated after the last ice sheet retreated at 18 000 BP. After this time, the world climate became warmer, until a cool period from 12 000–10 000 BP. Warming then occurred from 10 000 BP to reach the 'climatic optimum' at around 6000–7000 BP; since then a gradual cooling has occurred until the present day (Fig.

12). The exact dates of warming and cooling in Britain vary with latitude, altitude and local climatic factors (Taylor 1983). I have not made an exhaustive survey of the ages when mires in the British Isles were initiated, but dates given by Taylor (1983) show that only a few localized mires were developed before 8000 BP. These seem to have started to form during phases of wetter climate, and only in local ponds or lakes in valleys or estuaries. The dry climate in the period 10 000–8000 BP did not favour bog development but permitted forests of birch and pine to develop. About 8000 BP the climate became much wetter, with average rainfalls about 11% greater than today. Extensive mire development began in lowland areas all over the British Isles at about 8000 BP, with blanket bogs commencing 7500–6000 BP in most upland areas. This wetter climate, aided by human colonization, caused the demise of the pine and birch forests which do not tolerate being rooted in waterlogged peat. Periods of drier climate, often non-synchronous and only of local extent, enabled brief periods of recolonization by birch and pine to occur on dry peat surfaces.

Thus today's bogs in Canada and the British Isles find their origins in situations of regular rainfall associated with cooling of the climate on a thousands of years time-scale. Presumably the bogs of NW Europe and the USSR are similar. In the tropics, we know that extensive ombrotrophic peats are forming in Indonesia (Anderson 1983) and maybe in Brazil (Suszczynski 1984). In a geological context, we are currently experiencing a worldwide episode of peat formation, covering 3% of the Earth's surface (Clymo 1987). I suggest that the present cooling climate produces regular rainfall in wide temperate zones and in the tropics. Hence we live in a period which has the potential to deposit coals. Does this idea of coal formation in a cooling climate have any geological support?

The Tertiary brown coals of the Rhone area were thought by Teichmüller & Teichmüller (1968, Fig. 4) to have formed during an episode

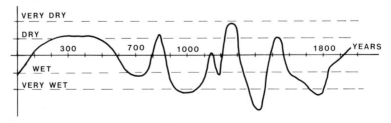

CLIMATE AT BOLTON FELL, CARLISLE SINCE 0 A.D.

FIG. 11. Once a modern raised mire has developed, an internal stratification of vegetation records changing wetness periods due to climatic changes (Barber 1981).

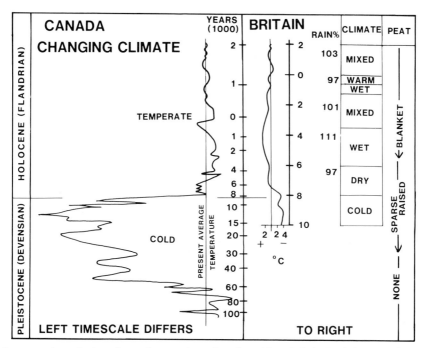

FIG. 12. Interpretations of palaeoclimate from northern Canada and from Britain. Both show a rapid warming after the last glaciation and a slow overall cooling during the past 8000 years. Widespread blanket peats started to form during the high and frequent rainfall associated with this cooling.

of cooling climatic change, with the cooling accelerating from the Eocene to Pliocene. Could this also have been a period of intense reliable rainfall?

The Permian Gondwanaland coals of South Africa, India and Australia formed above the tillites of the Dwyka glaciation (*eg* Hobday 1987; Stavrakis 1987), and so formed during and after the glaciations. Could these climatic changes have produced increased rainfall analogous to that in Canada after Pleistocene deglaciation?

Ziegler *et al.* (1987) take these climatic arguments further, to argue that coal deposits tend to be absent on a worldwide basis during periods of generally warm Earth climate (*eg* the Triassic to Palaeocene). These workers suggest that a warm hothouse Earth produces climates with a highly seasonal and unreliable rainfall. A cold icehouse Earth produces latitudinally compressed climatic zones, with only limited excursions of tropical rainfall away from the intertropical zones of airflow convergence. This type of Earth, as in the past 8000 years, produces regular rainfall in the tropics and at the mixing zone of the tropical and polar air masses N and S of 50°N. Thus, simplistically, the rainfall of a cool or cooling Earth produces peat in the tropical and temperate zones. In contrast, the rainfall of a warm or warming Earth produces rainfall and peat only in settings where local factors induce rain, such as to the E of mountain ranges or on coasts adjacent to large oceans. McCabe (1984, Fig. 7) shows that many ancient coals formed either at the palaeo-equator or at temperate latitudes N of 30°N. These are generally the zones of reliable rain on a cool Earth.

Coal and atmospheric CO_2

On a geological time-scale, major coal deposition has occurred three times: during the late Carboniferous–Permian, during the late Jurassic–early Cretaceous, and finally in the late Palaeocene–Eocene (Ronov 1976; Bestougeff 1980). These epochs represent variations in the rate of organic carbon accumulation, and therefore variations in the rate at which carbon is removed from organic cycling. Ronov (1976) has found that these same time epochs are also characterized by minimum CO_2 preserved in limestones, and are preceded 40–80 Ma earlier by large accumulations of volcanic rocks (with abundant limestone CO_3) (Fig. 13). Arrhenius (1896, in Stutzer 1940) suggested that abundant volcanism would inject CO_2 into the atmosphere. This increased CO_2

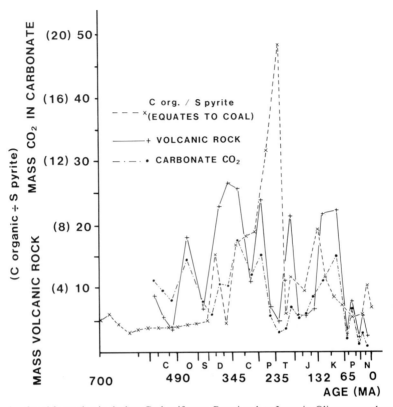

FIG. 13. Epochs of coal formation in the late Carboniferous–Permian, late Jurassic, Oligocene and present occur when limestone deposition is absent. Limestone deposition is better linked to periods of volcanism. Data from Ronov (1976) and Berner & Raiswell (1983). Thus atmospheric CO_2 may be extracted and buried by peat accumulation to buffer and maintain surface temperatures at equable levels for the biosphere (Arrhenius 1896; Lovelock 1979).

would tend to cause increased world surface temperatures via the 'greenhouse effect' of CO_2 molecules absorbing infrared solar radiation. A similar thermal effect occurs by blanketing the atmosphere with clouds of water vapour. Removal of CO_2 from the atmosphere by incorporation of carbon into aquatic carbonates or coal would tend to reduce world temperatures. The 'living' Gaian Earth may effectively be able to feed back biologically into its geosphere and biologically control and regulate temperature to maintain its own surface conditions suitable for life (Lovelock 1979; Charlson *et al.* 1987).

If the limestone 'sink' for CO_2 became insufficient in cold oceans with low CO_3 productivity, then coals formed extensively during the final stages of this atmospheric CO_2 removal and cooling during the late Carboniferous and Permian, the late Jurassic and Cretaceous, and the Palaeogene.

Conclusions

(1) Many ancient coals are interbedded between clastic sediments deposited by deltas, but coals also interbed with fluvial plains, barrier coastlines, estuaries, alluvial fans and lacustrine deposits.

(2) Local coal thickness is controlled by compaction of underlying sediment, syndepositional faulting, or deep basement structure. Thick coals often formed on the hanging walls of fault blocks, over slowly subsiding basement areas or over compactible mudstones.

(3) Ancient coals have a high sulphur content (more than 2%) if overlain within 10 m by marine sediments. However, most ancient coals have low sulphur contents (less than 2%), implying freshwater deposition.

(4) Modern peats close to fluvial channels often have high ash contents (more than 20%) and

pass laterally into fluvial facies. Ancient coals have low ash contents (less than 10%), transgressively overlie the older clastic facies and rarely pass laterally into fluvial facies. Modern mire analogies are peats formed in blanket bogs or raised ombrogenous bogs, which develop inland from the coast *after* fluvial deposition, delta progradation, barrier coastline advance or alluvial fan deposition.

(5) Coal-forming epochs relate to periods of climatic change producing regular rainfall in the tropics and N of 30°N. This has occurred either during Earth cooling or after deglacia-tion. The cooling may be an effect of self-regulating Gaian Earth cooling as the biosphere deposits atmospheric CO_2 as coal to reduce 'greenhouse' warming effects from high volcanic CO_2 contents.

ACKNOWLEDGMENTS: The accuracy of my comments, deductions and speculations have been improved after helpful comments on the typescript from H. A. P. Ingram, A. C. Scott, P. D. Moore and the forceful P. J. McCabe. Thanks are due to A. C. Scott for preprints of selected articles from his book, and to the editors M. K. G. Whateley and K. T. Pickering for sympathetic deadlines.

References

AABY, B. 1976. Cyclical climatic variations in climate over the past 5,500 yr reflected in raised bogs. *Nature* 263, 281–284.

ALLEN, P. A., HOMEWOOD, P. & WILLIAMS, G. D. 1986. Foreland basins; an introduction. *In:* ALLEN, P. A. & HOMEWOOD, P. (eds) *Foreland Basins.* Blackwell Scientific Publications, Oxford, 3–12.

ANDERSON, J. A. R. 1964. The structure and development of the peat swamps of Sarawak and Brunei. *Journal of Tropical Geography* 18, 7–16.

—— 1983. The tropical peat swamps of Western Malesia. *In:* GORE, A. J. P. (ed.) *Ecosystems of the World,* Vol 4B, *Mires, Swamp, Bog, Fen and Moor.* Elsevier, Amsterdam, 181–199.

—— & MULLER, J. 1975. Palynological study of a Holocene peat and a Miocene coal deposit from NW Borneo. *Review of Palaeobotany and Palynology* 19, 291–351.

ARRHENIUS, S. 1896. Über der Einflüss des atmosphärischen Kohlensauregehaltes auf die Temperatur der Erdoberflache. *Bihang svenska vet. Akad. Landlingar* 22, 1.

AVERETTI, P. 1984. Coal resources. *In:* ELLIOTT, M. A. (ed.) *Chemistry of Coal Utilization* (2nd edn). Wiley Interscience, New York, 55–90.

AYERS, W. B. & KAISER, W. R. 1984. Lacustrine-interdeltaic coal in the Port Union Formation (Palaeocene), Powder River Basin, Wyoming and Montana, USA. *In:* RAHMANI, R. A. & FLORES, R. M. (eds) *Sedimentology of Coal and Coal-bearing Sequences.* Blackwell Scientific Publications, Oxford, 61–184.

BALLANCE, P. F. & READING, H. G. (eds) 1980. *Sedimentation in Oblique-slip Mobile Zones.* Blackwell Scientific Publications, Oxford, 265 pp.

BARBER, K. E. 1981. *Peat Stratigraphy and Climatic Change.* Balkema, Rotterdam.

BARNSLEY, G. B. 1984. Introduction to coal. *In:* WARD, C. R. (ed.) *Coal Geology and Coal Technology.* Blackwell Scientific Publications, Oxford, 1–39.

BELLAMY, D. J. 1972. Templates of peat formation. *Proceedings of the 4th International Peat Congress, Helsinki,* Vol. 4, 7–18.

BERNER, R. A. & RAISWELL, R. 1983. Burial of organic carbon and pyrite sulfur in sediments over Phane-rozoic time: a new theory. *Geochemica et Cosmochimica Acta* 47, 855–862.

BESTOUGEFF, M. A. 1980. Summary of world coal resources and reserves. *26th International Geological Congress, Paris, Colloqium C2* 35, 353–366.

BREYER, J. A. 1987. A tidal origin for coarsening-up sequences above two Wilcox lignites in east Texas. *Journal of the Geological Society, London* 44, 463–469.

BUSTIN, R. M. & LOWE, L. E. 1987. Sulphur, low temperature ash and minor elements in humid-temperate peat of the Fraser River delta, British Columbia. *Journal of the Geological Society, London* 144, 435–450.

BUTTERWORTH, M. A. 1964. Die verteilung der Denosporites sphaerotriangularis in Westfal B der westpenninischen steinkolenfelder Englands. *Fortschritte in der Geologie von Rheinland und Westfalen* 12, 317.

CAMERON, A. R. 1971. Some petrological aspects of the Harbour coal seam, Sydney coalfield, Nova Scotia. *Geological Survey of Canada, Bulletin* 175, Department of Energy, Mines and Resources, Canada.

CASAGRANDE, D. J. 1987. Sulphur in peat and coal. *In:* SCOTT, A. C. (ed.) *Coal and Coal-bearing Strata: Recent Advances.* Geological Society of London Special Publication 32, 87–106.

CAVAROC, V. V. & FLORES, R. M. 1984. Lithological relationships of the Upper Cretaceous Gibson–Cleary stratigraphic interval: Gallup Coalfield, New Mexico, USA. *In:* RAHMANI, R. A. & FLORES, R. M. (eds) *Sedimentology of Coal and Coal-bearing Sequences.* Blackwell Scientific Publications, Oxford, 197–215.

CHARLSON, R. J., LOVELOCK, J. E., ANDREAE, M. O. & WARREN, S. G. 1987. Oceanic phytoplankton, atmospheric sulphur, cloud albedo and climate. *Nature* 326, 655–661.

CLYMO, R. S. 1987. Rainwater-fed peat as a precursor to coal. *In:* SCOTT, A. C. (ed.) *Coal and Coalbearing Strata: Recent Advances.* Geological Society of London Special Publication 32, 17–24.

COHEN, A. D. 1984. The Okefenokee swamp: a low sulphur end-member of a shoreline-related depositional model for coastal plain coals. *In:* RAHMANI,

R. A. & FLORES, R. M. (eds) *Sedimentology of Coal and Coal-bearing Sequences*. Blackwell Scientific Publications, Oxford, 231–240.

COLEMAN, J. M., GAGLIANO, S. M. & SMITH, W. G. 1970. Sedimentation in a Malaysian high tide tropical delta. *In:* MORGAN, J. P. (ed.) *Deltaic Sedimentation, Modern and Ancient*. Special Publication of the Society of Economic Paleontologists and Mineralogists **15**, 185–197.

DAVIS, A. 1984. Coal petrology and petrographic analysis. *In:* WARD, C. R. (ed.) *Coal Geology and Coal Technology*. Blackwell Scientific Publications, Oxford, 74–112.

EGGERT, D. C. 1984. The Leslie Cemetery and Francisco distributary fluvial channels in the Petersburg Formation (Pennsylvanian) of Gibson County, Indiana, USA. *In:* RAHMANI, R. A. & FLORES, R. M. (eds) *Sedimentology of Coal and Coal-bearing Sequences*. Blackwell Scientific Publications, Oxford. 309–315.

ELLIOTT, R. E. 1965. Swilleys in the Coal Measures of Nottinghamshire interpreted as palaeo-river courses. *Mercian Geologist* **1**, 133–142.

—— 1985. Quantification of peat to coal compaction stages, based especially on phenomena in the East Pennine Coalfield, England. *Proceedings of the Yorkshire Geological Society* **45**, 163–172.

FERM, J. C. 1970. Allegheny deltaic deposits. *In:* MORGAN, J. P. (ed.) *Deltaic Sedimentation, Modern and Ancient*. Special Publication of the Society of Economic Paleontologists and Mineralogists **15**, 246–255.

—— 1984. Geology of coal. *In:* WARD, C. R. (ed.) *Coal Geology and Coal Technology*. Blackwell Scientific Publications, Oxford, 151–176.

—— & STAUB, J. R. 1984. Depositional controls of mineable coal bodies. *In:* RAHMANI, R. A. & FLORES, R. M. (eds) *Sedimentology of Coal and Coal-bearing Sequences*. Blackwell Scientific Publications, Oxford, 275–289.

FIELDING, C. R. 1984a. A coal depositional model for the Durham Coal Measures of NE England. *Journal of the Geological Society, London* **141**, 919–931.

—— 1984b. Upper delta plain lacustrine and fluvio lacustrine facies from the Westphalian of the Durham coalfield, NE England. *Sedimentology* **31**, 547–567.

—— 1986. Fluvial channel and overbank deposits from the Westphalian of the Durham coalfield, NE England. *Sedimentology* **23**, 119–140.

FISK, H. N. 1960. Recent Mississippi River sedimentation and peat accumulation. *Comptes Rendus, 4ème Congrès sur l'Avancement des Etudes de Stratigraphie et de Géologie du Carbonifre, Heerlen, 1958*, Vol. 1, 187–199.

FITCH, F. H. 1954. North Borneo, mineral resources. *In:* ROE, F. W. (ed.) *Annual Report of the Geological Survey Department for 1953: British Territories in Borneo*. Government Printing Office, Kuching, Sarawak.

FLORES, R. M. 1981. Coal deposition in fluvial palaeoenvironments of the Palaeocene Tongue River Member of the Fort Union Formation, Powder River Basin. *In:* ETHRIDGE, F. G. & FLORES, R. M. (eds) *Recent and Ancient Non-marine Depositional Environments: Models for Exploration*. Special Publication of the Society of Economic Paleontologists and Mineralogists **31**, 169–190.

FRAZIER, D. E. & OSANIK, A. 1969. Recent peat deposits—Louisiana coastal plain. *In:* DAPPLES, E. C. & HOPKINS, M. E. (eds) *Environments of Coal Deposition*. Geological Society of America Special Paper **114**, 63–85.

GALLOWAY, W. E. 1981. Depositional architecture of Cenozoic Gulf Coast coastal plain fluvial systems. *In:* ETHRIDGE, F. G. & FLORES, R. M. (eds) *Recent and Ancient Non-marine Depositional Environments: Models for Exploration*. Special Publication of the Society of Economic Paleontologists and Mineralogists **31**, 127–155.

GANONG, W. F. 1897. Upon raised bogs in the province of New Brunswick. *Transactions of the Royal Society of Canada* **Section IV**, 131–163.

GASTALDO, R. A. 1987. Confirmation of Carboniferous clastic swamp communities. *Nature* **326**, 869–871.

GERSIB, G. A. & McCABE, P. J. 1981. Continental coal-bearing sediments of the Port Hood Formation (Carboniferous), Cape Linzee, Nova Scotia, Canada. *In:* ETHRIDGE, F. G. & FLORES, R. M. (eds) *Recent and Ancient Non-marine Depositional Environments: Models for Exploration*. Special Publication of the Society of Economic Paleontologists and Mineralogists **31**, 95–108.

HACQUEBARD, P. A. & DONALDSON, J. R. 1969. Carboniferous coal deposition associated with floodplain and limnic environments in Nova Scotia. *In:* DAPPLES, E. C. & HOPKINS, M. E. (eds) *Environments of Coal Deposition*. Geological Society of America Special Paper **114**, 143–191.

HASZELDINE, R. S. 1981. *Westphalian B Coalfield Sedimentology in NE England, and its Regional Setting*. Unpublished PhD thesis, Strathclyde University.

—— 1984. Muddy deltas in freshwater lakes, and tectonism in the Upper Carboniferous coalfield of NE England. *Sedimentology* **31**, 811–822.

HESS, J. C. & LIPPOLT, H. J. 1986. $^{40}Ar/^{39}Ar$ ages of tonstein and tuff sanidines: a new calibration point for the improvement of the Upper Carboniferous timescale. *Chemical Geology (Isotope Geoscience)* **59**, 143–154.

HEWARD, A. P. 1978. Alluvial fan and lacustrine sediments from the Stephanian A and B (La Magdalena, Cinera-Metallana and Sabero) coalfields, northern Spain. *Sedimentology* **25**, 451–488.

HOBDAY, D. K. 1987. Gondwana coal basins of Australia and Africa: tectonic setting, depositional systems and resources. *In:* SCOTT, A. C. (ed.) *Coal and Coal-bearing Strata: Recent Advances*. Geological Society of London Special Publication **32**, 219–234.

—— & HORNE, J. C. 1977. Tidally influenced barrier island and estuarine sedimentation in the Upper Carboniferous of southern West Virginia. *Sedimentology* **18**, 97–122.

HORNE, J. C., FERM, J. C., CARUCCIO, F. T. & BAGANZ, B. P. 1978. Depositional models in coal exploration and mine planning in the Appalachian region. *American Association of Petroleum Geologists Bulletin* **62**, 2379–2411.

HUNT, J. W. & HOBDAY, D. K. 1984. Petrographic composition and sulphur content of coals associated with alluvial fans in the Permain Sydney and Gunnedah Basins, eastern Australia. *In:* RAHMANI, R. A. & FLORES, R. M. (eds) *Sedimentology of Coal and Coal-bearing Sequences.* Blackwell Scientific Publications, Oxford, 43–60.

INGRAM, H. A. P. 1982. Size and shape in raised mire ecosystems: a geophysical model. *Nature* **297**, 300–303.

—— 1987. Ecohydrology of Scottish peatlands. *Transactions of the Royal Society of Edinburgh: Earth Sciences* **78**, 287–296.

INTERNATIONAL ENERGY AGENCY 1983. *Concise Guide to World Coalfields.* International Energy Agency, London.

KORPIJAAKKO, E. O. & WOOLNOUGH, D. F. 1977. Peatland survey and inventory. *In:* RADFORTH, N. W. & BRAWNER, C. O. (eds) *Muskeg and the Northern Environment in Canada,* University of Toronto Press, Toronto, 63–81.

KOSTERS, E. C., CHMURA, G. L. & BAILEY, A. 1987. Sedimentary and botanical factors influencing peat accumulation in the Mississippi delta. *Journal of the Geological Society, London* **144**, 423–434.

LI SITIAN, LI BAOFANG, YANG SHIGONG, HUANG JIAFU & LI ZHEN 1984. Sedimentation and tectonic evolution of late Mesozoic faulted coal basins in NE China. *In:* RAHMANI, R. A. & FLORES, R. M. (eds) *Sedimentology of Coal and Coal-bearing Sequences.* Blackwell Scientific Publications, Oxford, 387–406.

LOVE, C. G., COLEMAN, M. L. & CURTIS, C. D. 1983. Diagenetic pyrite formation and sulphur isotope fractionation associated with a Westphalian marine incursion, northern England. *Transactions of the Royal Society of Edinburgh: Earth Sciences* **74**, 165–182.

LOVELOCK, J. E. 1979. *Gaia, a New Look at Life on Earth.* Oxford University Press, Oxford, 157 pp.

MCCABE, P. J. 1984. Depositional environments of coal and coal-bearing strata. *In:* RAHMANI, R. A. & FLORES, R. M. (eds) *Sedimentology of Coal and Coal-bearing Sequences.* Blackwell Scientific Publications, Oxford, 13–42.

MOORE, P. D. 1984. *European Mires.* Academic Press, London, 367 pp.

—— 1987. Ecological and hydrological aspects of peat formation. *In:* SCOTT, A. C. (ed.) *Coal and Coal-bearing Strata: Recent Advances.* Geological Society of London Special Publication **32**, 7–16.

POLAK, W. 1975. Character and occurrence of peat deposits in the Malaysian tropics. *In:* BARTSTRA, G. & CASPARIE, W. (eds) *Modern Quaternary Research in SE Asia.* Balkema, Rotterdam, 71–81.

P'YAVCHENKO, N. I. 1964. *Peat Bogs of the Russian Steppe.* Israel Programme for Scientific Translations, Jerusalem, 156 pp.

RAYNER, D. H. 1981. *The Stratigraphy of the British Isles* (2nd edn). Cambridge University Press, Cambridge.

ROMANOV, V. V. 1968. *Hydrophysics of Bogs.* Israel Programme for Scientific Translations, Jerusalem, 299 pp.

RONOV, A. B. 1976. Global carbon geochemistry, volcanism, carbonate accumulation and life. *Geochemistry International* **13** (4), 172–195.

RYER, T. A. & LANGER, A. W. 1980. Thickness change involved in the peat-to-coal transformation for a bituminous coal of Cretaceous age in central Utah. *Journal of Sedimentary Petrology* **50**, 987–992.

SCOTT, A. C. 1978. Sedimentological and ecological control of Westphalian B plant assemblages from West Yorkshire. *Proceedings of the Yorkshire Geological Society* **41**, 461–508.

—— 1979. The ecology of Coal Measure floras from northern Britain. *Proceedings of the Geologists Association* **90**, 97–116.

—— (ed.) 1987. *Coal and Coal-bearing Strata: Recent Advances.* Geological Society of London Special Publication **32**.

SHIMOYAMA, T. 1984. Sulphur concentration in Japanese Palaeogene coal. *In:* RAHMANI, R. A. & FLORES, R. M. (eds) *Sedimentology of Coal and Coal-bearing Sequences.* Blackwell Scientific Publications, Oxford, 361–372.

SMITH, A. H. V. 1962. The palaeoecology of Carboniferous peats based on the miospores and petrography of Carboniferous coals. *Proceedings of the Yorkshire Geological Society* **33**, 423–474.

—— 1968. Seam profiles and seam characters. *In:* MURCHISON, D. & WESTOLL, T. S. (eds) *Coal and Coal-bearing Strata.* Oliver & Boyd, Edinburgh, 31–40.

—— & BUTTERWORTH, M. A. 1967. Miospores in the coal seams of the Carboniferous of Great Britain. *Special Papers in Palaeontology* **1**, Palaeontological Association, London.

SMYTH, M. 1967. Coal petrology applied to the correlation of some New South Wales Permian coals. *Proceedings of the Australasian Institute of Mining and Metallurgy* **221**, 11–17.

—— 1970. Type sequences for some Permain Australian coals. *Proceedings of the Australasian Institute of Mining and Metallurgy* **233**, 7–15.

STACH, E. 1982. *Coal Petrology* (3rd edn). Gebruder Bornträger, Berlin.

STANEK, W. 1977. Classification of muskeg. *In:* RADFORTH, N. W. & BRAWNER, C. O. (eds) *Muskeg and the Northern Environment in Canada.* University of Toronto Press, Toronto, 31–62.

STAVRAKIS, N. 1987. Karoo coals of South Africa. In: *Abstracts of the Symposium on Deltas—Sites and Traps for Fossil Fuels.* Geological Society, London.

STOPES, M. C. 1919. On the four visible ingredients in banded bituminous coals. *Proceedings of the Royal Society of London, Series B* **90B**, 470–487.

STUTZER, O. 1940. *Geology of Coal.* University of Chicago, Chicago, IL, 461 pp.

STYAN, W. B. & BUSTIN, R. M. 1983. Sedimentology of Fraser River delta peat deposits: a modern analogue for some deltaic coals. *International Journal of Coal Geology* **3**, 101–143.

SUGGATE, R. P. 1959. New Zealand coals, their geological setting and its influence on their properties. *New Zealand Department of Scientific and Industrial Research Bulletin* **134**.

SUSZCZYNSKI, E. F. 1984. The peat resources of Brazil. *Proceedings of the 7th International Peat Congress, Dublin*, Vol. 1, 468–492.

TALLIS, J. H. 1983. Changes in wetland communities. *In:* GORE, A. J. P. (ed.) *Ecosystems of the World*, Vol. 4B, *Mires, Swamp, Bog, Fen and Moor*. Elsevier, Amsterdam, Chapter 9.

TAYLOR, J. A. 1983. The peatlands of Great Britain and Ireland. *In:* GORE, A. J. P. (ed.) *Ecosystems of the world*, Vol. 4B, *Mires, Swamp, Bog, Fen and Moor*. Elsevier, Amsterdam, 1–46.

TEICHMÜLLER, M. & TEICHMÜLLER, R. 1968. Cainozoic and Mesozoic coal deposits of Germany. *In:* MURCHISON, D. & WESTOLL, T. S. (eds) *Coal and Coal-bearing Strata*. Oliver & Boyd, Edinburgh, 347–379.

TERASMAE, J. 1977. Post-glacial history of Canadian muskeg. *In:* RADFORTH, N. W. & BRAWNER, C. O. (eds) *Muskeg and the Northern Environment in Canada*. University of Toronto Press, Toronto, 9–30.

TEYSSEN, T. 1987. Coastal plain of NW Borneo. In: *Abstracts pf the Symposium on Deltas—Sites and Traps for Fossil Fuels*. Geological Society, London.

TING, F. T. C. & SPACKMAN, W. 1975. The coal lithotype and seam profile. *Comptes Rendus 7eme Congrès Internationale sur le Stratigraphie et Géologie du Carbonifre, Krefeld, 1971*, Vol. 4, 307–311.

TRINKLE, E. J. & HOWER, J. C. 1984. Petrography of the middle Pennsylvanian Upper Elkhorn no. 3 coal of eastern Kentucky, USA. *In:* RAHMANI, R. A. & FLORES, R. M. (eds) *Sedimentology of Coal and Coal-bearing Sequences*. Blackwell Scientific Publications, Oxford, 349–360.

TRUEMAN, A. 1954. *The Coalfields of Great Britain*. Edward Arnold, London, 396 pp.

WANLESS, H. R. 1970. Late Paleozoic deltas in central and eastern United States. *In:* MORGAN, J. P. (ed.) *Deltaic Sedimentation, Modern and Ancient*. Special Publication of the Society of Economic Paleontologists and Mineralogists **15**, 215–245.

WARD, C. R. 1984. Chemical analysis and classification of coal. *In:* WARD, C. R. (ed.) *Coal Geology and Coal Technology*. Blackwell Scientific Publications, Oxford, 40–73.

WEISENFLUH, G. A. & FERM, J. C. 1984. Geological controls on deposition of the Pratt seam, Black Warrior Basin, Alabama, USA. *In:* RAHMANI, R. A. & FLORES, R. M. (eds) *Sedimentology of Coal and Coal-bearing Sequences*. Blackwell Scientific Publications, Oxford, 317–330.

WILKS, P. J. 1979. Mid-Holocene sea level and sedimentation interactions in the Dovey Estuary area, Wales. *Palaeogeographica, Palaeoclimatica, Palaeoecologica* **26**, 17–26.

WILLIAMS, E. G. & KEITH, M. C. 1963. Relationship between sulphur in coals and the occurrence of marine roof beds. *Economic Geology* **58**, 720–729.

WILLS, L. J. 1954. *A Palaeogeographical Atlas of the British Isles and Adjacent Parts of Europe*. Blackie, Glasgow, 64 pp.

ZIEGLER, A. M., RAYMOND, A. L., GIERLOWSKI, T. C., HORRELL, M. A., ROWLEY, D. B. & LOTTES, A. C. 1987. Coal climate and terrestrial productivity, present and early Cretaceous compared. *In:* SCOTT, A. C. (ed.) *Coal and Coal-bearing Strata: Recent Advances*. Geological Society of London Special Publication **32**, 25–49.

R. S. HASZELDINE, Department of Applied Geology, University of Strathclyde, Glasgow G1 1XJ, UK. *Present address:* Department of Geology and Applied Geology, University of Glasow, Glasgow G12 8QQ, UK.

Deltaic coals: an ecological and palaeobotanical perspective

A. C. Scott

SUMMARY: Coals occur widely in deltaic sequences but are rarely of economic quality, being thin, split and with high ash content. Coals with brackish-water influence or overlain by marine rocks tend to have high sulphur values. Not all coals in deltaic sediments formed in swamps or marshes. Peats may form under a wide variety of conditions, and ecological and hydrological interpretations of mire types are essential and should be integrated with facies studies of accompanying clastic sediments. Botanical, physical and chemical data are needed for the interpretation of the coals.

Considerable time may elapse between delta abandonment and the onset of peat formation. Rheotrophic or flow-fed peats may form initially on abandoned delta lobes, but if peat accumulation continues under high rainfall conditions ombrotrophic low ash peats may form as raised bogs. In this case vertical and lateral zoning of the peat (coal) may occur. In any interpretation of the coals based on analogies of modern peats care must be exercised in using ecological data from Recent plant communities, as peat-forming vegetation shows significant changes from the Devonian to Recent.

Within many deltaic sequences coal seams are considered to represent the emergent phase of delta abandonment. With relation to the occurrence of coals three important questions may be addressed: (i) What is the temporal relationship between delta abandonment and the onset of peat formation? (ii) What are the biological, physical and chemical characteristics of deltaic peats and coals? (iii) To what extent are deltaic coals economic and how can their occurrence be predicted?

In this contribution modern deltaic peats are considered first. It has become clear, however, that hydrological and ecological factors in peat formation may affect several 'economic' features of the resulting coals, ie ash and sulphur content. These relationships are outlined with regard to modern peats and then with regard to ancient deltaic coals. Interpretation of ancient coals utilizes diverse data. Recent developments in the study of coals and the erection of plant ecological models are discussed. The studies used as examples are mainly from non-deltaic systems but can be used as models for the study of deltaic coals.

In addition to the occurrence of peats, brief consideration is given to the distribution of vascular land plants in deltaic sediments. The theme of this book is sites and traps of fossil fuels in deltaic sediments and environments. Whilst it is clear that some peats might eventually yield economic coal seams, the occurrence of vascular land plants must also be considered as a potential source of oil and gas.

Deltaic peats

Many geological studies of modern (and, even more, of ancient) deltas concern the clastic facies.

Even so, the occurrence of peats and organic-rich muds has been widely studied in diverse deltas from a wide range of climatic regimes. Several contrasting examples are cited here.

The occurrence of peats associated with the Mississippi Delta has long been recognized. Features of these peats have recently been discussed by Kosters et al. (1987). Many of the peats are freshwater and form planar deposits which accumulated in areas abandoned by deltaic sedimentation. These peats originated in forested swamps or under herbaceous floating mats. Peat accretion rates are generally low. A major problem concerns the relationship between delta abandonment and initiation of peat formation. The question arises whether deltaic peats are any different from those forming in other mires. Kosters et al. (1987) point out that, in their example from the Mississippi Delta, lobe abandonment and consequent coastal erosion causes saltwater intrusion into freshwater environments. The resulting phase would be of organic-rich muds rather than true peats. Because of this it has been argued by McCabe (1984) that deltaic coals would have too high an ash content to be economic. Kosters et al. (1987) claim, however, that leaching by peat solutions of inorganic constituents during early post-burial diagenesis might result in a lowering of ash content in the final coal.

Peats are found around the Gulf Coast from coastal Louisiana, associated with back-barrier environments and with the alluvial and delta plain of the Mississippi, to the Everglades mire complex in Florida (Spackman et al. 1966; Frazier et al. 1978). In coastal Louisiana, coals which overlie deltaic sediments are generally thin but inland thick peats accrue in mire complexes

From WHATELEY, M. K. G. & PICKERING, K. T. (eds), 1989, *Deltas: Sites and Traps for Fossil Fuels,*
Geological Society Special Publication No. 41, pp. 309–316.

(Frazier *et al.* 1978). The Everglades mire complex has been studied by Spackman and co-workers (Spackman *et al.* 1966; Cohen & Spackman 1972). In this work several different peat-forming plant communities have been recognized, each yielding a different peat type (Cohen & Spackman 1977; Cohen *et al.* 1987). The two major controls on peat type appear to be (i) water-table level and (ii) salinity. Saline-influenced peats tend to be dominated by the mangrove *Rhizophora* and have high ash and sulphur contents (Casagrande 1987; Cohen *et al.* 1987). The occurrence of *Rhizophora* in brackish peats has been widely reported (Baltzer 1969) and these communities are generally restricted in diversity (Cohen & Spackman 1977).

In a more tropical setting Coleman *et al.* (1970) noted the development of peats in an active delta. As was shown in the Mississippi and the Everglades, brackish marine-influenced peats were formed from a specialized flora and the ash content was high. In areas of raised bog, however, and in freshwater peats above any high tide or flood level, more diverse peats with low ash and sulphur contents could form (Coleman *et al.* 1970; McCabe 1984).

Peats have also been studied from deltas in more temperate regions. In their study of the Fraser River Delta, British Columbia, Bustin & Lowe (1987) note the wide range of peat-forming vegetation. They also note that brackish peats of the lower delta plain are characterized by high sulphur and that the highest sulphur values occur in brackish sedge peat. As in other deltas, cores of peats showed changing vegetational facies and changing ash and sulphur values, indicating that the peat was not formed under a single set of environmental conditions. This complexity is rarely considered by sedimentologists studying ancient delta sequences.

From these studies it is clear that peats may form in numerous environments from the lower delta plain through the upper delta plain into an alluvial flood plain. Those peats associated with lower-delta-plain environments tend to be planar, often split and with high ash and sulphur contents, and can rarely be considered as commercial prospects. Those peats away from marine influence in the upper delta, and even more so in alluvial-plain settings, may be thicker, domed, have a zoned vegetation and tend to have low sulphur and ash contents, and hence may be of more economic use. The realization of the distribution of these different peat types associated with distinct depositional settings has facilitated the development of predictive sedimentological models (Ferm & Staub 1984).

In all these examples the deltas are marine.

Few studies of Recent peats associated with lacustrine deltas have included petrology and chemistry as well as vegetational analysis. Some basic questions remain, most particularly the relationship of delta progradation and peat formation with the timing of peat onset and development.

In many cases plant material may accrue in deltaic sediments but may not form peats. This material, in the form of either identifiable microscopic plant remains or organic detritus, may be of academic as well as economic interest and may tell us something about the original vegetation as well as being a contributor to oil and gas. For studies of the distribution of plant material in Recent deltas the reader is referred to Spicer (1981) and Scheihing & Pfefferkorn (1984). It can be argued that peats and coals represent a significant global carbon sink. Moore (1983), however, considers that even the coals of the Carboniferous did not represent a significant carbon sink, despite their being equivalent to 1148×10^{15} g C, which is in excess of the current biomass of the Earth.

In the discussion of peats within deltaic sequences there must be some consideration of models of peat formation.

Models of peat formation

Studies of Recent mires (peat-forming systems) (Gore *et al.* 1983) have shown that there are two basic types: ombrotrophic and rheotrophic. These have been fully described and defined by Moore (1987). The two types may intergrade (Fig. 1) but have some fundamentally different features (Fig. 2).

Ombrotrophic mires are rainwater-fed peat systems. These include raised bogs and bog forests and typically have low sulphur and ash contents. The low ash values are entirely a result of their being rainwater-fed. This may lead through time to a change in vegetation type in the mire, and the peat often shows both lateral and vertical zoning. In more mature raised bogs the plants become less diverse and in some cases become stunted (*cf* McCabe 1984). These peats often show few splits and may be thick.

In contrast, rheotrophic mires are flow fed and include swamps and swamp forests. They tend to be higher in ash content but as a result more mineral nutrients are available and the vegetation is more uniform and diverse. Peats tend to be planar but show more frequent splits. Where there is marine influence the diversity of the vegetation forming the peats is less and the total sulphur values for the peats are greater.

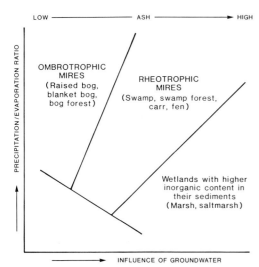

FIG. 1. Relationship between mires in terms of the relative influence of rainwater and groundwater in their hydrological input (modified from Moore 1987).

CHARACTERISTICS OF MAJOR PEAT TYPES

OMBROTROPHIC	RHEOTROPHIC	
	FRESH	BRACKISH
DOMED	PLANAR	
THICK	VARIABLE THICKNESS	
ZONED	RARELY ZONED	
LOW DIVERSITY	HIGH DIVERSITY	LOW DIVERSITY RESTRICTED FLORA
LOW ASH	HIGH ASH	VERY HIGH ASH
LOW SULPHUR	LOW TO HIGH SULPHUR	
RARE SPLITS	COMMON SPLITS	
COMMONLY ECONOMIC	VARIABLY ECONOMIC	

FIG. 2. Characteristics of major peat types (compiled from numerous sources).

Several factors influence peat formation in deltaic settings. It is also clear that the timing of delta abandonment and onset of peat formation is critical as it may affect the ash and sulphur content in particular. The use of mire terminology is essential (Moore 1987), and the interpretation that all coals in deltaic sequences are 'swamp deposits' is clearly untenable. Just as sedimentologists advocate facies analyses in clastic deltaic sediments and an understanding of process, setting *etc.*, so also the same detail is needed from the organic facies including studies of its vegetation and chemistry.

In all these examples we have considered purely peat formation. One feature of mire systems is the occurrence of fires, generally started by lightning strikes. Charcoal is common in numerous Recent peats and fossil coals (where it is called fusain), and the fires may have profound effects not only on the vegetation but also on the sedimentology (Cypert 1972; Komarek 1972; Scott & Collinson 1978; Cope & Chaloner 1985; Cohen *et al.* 1987). Fires may burn through peats, giving rise to lakes once the water has returned to its original level, and in some cases may initiate crevasse splays, termed fire splays by Cohen (1984). Layers of charcoal may commonly be found in deltaic sediments but their significance is often overlooked. Removal of vegetation cover by wildfire and the resulting change in the erosion–deposition pattern may well be indicated by charcoal accompanying an abrupt change in lithology. This aspect of delta sedimentation warrants further investigation.

Whilst investigations of peats from Recent deltaic settings are sparse, those of coals are even more restricted in their scope.

Ancient deltaic coals

The recognition of ancient deltaic coals has been complicated by the diversity of interpretations of ancient deltaic sequences. Many of these problems are reviewed elsewhere in this book. The division into lower delta plain, upper delta plain and alluvial plain has been recently discussed by Fielding (1985). It has also been pointed out that many coals may have formed in back-barrier coastal-plain settings (Cohen 1984).

It is obvious that no economic coals will be associated with pre-Devonian delta sequences (see below). Some coals are known from late Devonian deltaic sequences in the Appalachian Mountains, USA; they represent the first widespread peat-forming flora and are dominated by the prefern *Rhacophyton* (Scheckler 1986). These peats tend to be high in sulphur and associated plant fossils are permineralized by iron pyrite. By the early Carboniferous lycopods became the dominant component of peat-forming vegetation (Collinson & Scott 1987*a*). Coals from the paralic Upper Carboniferous Coal Measures of the Euramerican tropical belt are also dominated by lycopods (Phillips *et al.* 1985), although some brackish 'mangrove' coals are dominated by cordaites (Raymond & Phillips 1983). A major climatically controlled change in peat-forming vegetation has been documented from the Westphalian to the Stephanian (Phillips & Peppers 1984). This major climatic change also appears to have affected the mire types which can be correlated with this as there are major changes in ash and sulphur values in North American coals from the Westphalian to the Stephanian (Cecil *et al.* 1985).

Many of the thick Permian coals in the temperate southern hemisphere appear to be associated with alluvial-fan or alluvial-flood-plain environments, which is also the case for many Mesozoic coals (Hobday 1987). Our knowledge of Jurassic coal-forming floras is very sparse and there are no widely known detailed studies of Jurassic deltaic coals (Collinson & Scott 1987*a*). Few data are available on Cretaceous coals (but see Spicer *et al.* 1987). The problem is often that where clastic sequences are well known the coals are not, and *vice versa*.

Many Tertiary coal deposits, such as the German Miocene brown coals, have been compared with Recent peats of the SE United States, many of which are not truly deltaic (Teichmüller & Teichmüller 1968; Collinson & Scott 1987*a*). Although we have considerable data on ancient deltaic sequences, our knowledge of the associated coals is of very varied quality and even the coals of the Jurassic deltas of the northern North Sea are poorly known. There is the potential, however, of utilizing techniques from coals formed in other environments for deltaic coals and the integration of coal and clastic facies studies.

Development of plant palaeo-ecological models and methods

Several field and laboratory methods have been developed which could profitably be applied to deltaic coals.

In the field the variation in thickness and geometry of coals should be recorded. The nature of the junction with the underlying clastic sediment should also be recorded as well as the type and occurrence of plant rootlets. A study of the underlying palaeosols may yield valuable data (Retallack 1986). Coal lithotypes should also be recorded both vertically and laterally as well as the occurrence of dirt and fusain bands. Finally the occurrence of pyrite and its distribution should be noted. A complete section, or several complete sections, through the coals should be removed for laboratory study. As has been pointed out earlier, the coal may show important vertical changes reflecting changing environmental conditions and hence should not be treated as a single lithology (coal) representing a single environment (swamp).

Aspects of the biology, chemistry and physical properties of the coals can be investigated in the laboratory. Most studies consider only one of these three aspects, but a multidisciplinary approach is advocated. Physical studies include

coal petrology of polished blocks. Such analyses have proved useful in the environmental interpretation of coals (Smyth 1983) but should be combined with a botanical study (Cohen *et al.* 1987). Comments concerning the usefulness of coal petrology in this respect are also dealt with by Collinson & Scott (1987*a*).

Chemical studies may involve ash analyses including compositional studies and in addition analyses of total sulphur and forms of sulphur. It should be clear from the previous sections that these two parameters are of fundamental importance to our understanding of the original depositional environment (Casagrande 1987; Cohen *et al.* 1987).

Botanical studies, which are important, have generally received less attention than petrographic and chemical analyses. Three approaches can be used: the study of permineralized peats (revealing the structure of the constituent plants), the study of palynology (mainly miospores) and the study of total organic macerates including megaspores, cuticles and fusain (Cohen & Spackman 1972). From an analysis of plant biology it is clear that there are fundamental differences between, for example, Palaeozoic and Tertiary peat-forming vegetation (Collinson & Scott 1987*a*).

The study of Recent peat formation in deltaic environments has its limitations as a basis for the interpretation of the ancient equivalents (Collinson & Scott 1987*b*). For example, salinity tolerance appears to have evolved late in some groups of plants. Likewise the ability to colonize drier environments expanded with the evolution of seed in the late Devonian. In many instances the initiation of Recent peat formation is from aquatic vegetation. We have no evidence, however, of aquatic floating vegetation at least until the Jurassic (Collinson & Scott 1987*b*). With these facts in mind we shall consider some of the methods of studying coal-forming vegetation which can be applied to deltaic coals.

By far the most reliable method of studying the vegetation of ancient peats (coals) is to examine permineralized peat, *ie* peat which was infiltrated by mineral-rich waters precipitating silica, carbonate or pyrite. The organic walls of the plants are uncompressed and remain so during burial and the coalification process (Scott & Rex 1985). By far the most significant permineralized peats are the coal balls from the Upper Carboniferous coal seams of Euramerica. We have much ecological data for these peats, mainly from the work of Phillips and his co-workers (*eg* Phillips & Peppers 1984; DiMichele *et al.* 1985; Phillips *et al.* 1985) who have studied the plant contents of coal balls quantitatively. Whilst permineralized peats are widely known, both geographically and stratigraphically, they are relatively rare.

More commonly, vegetational analysis is derived from palynological studies. We have data on the relation of peat petrology and palynology from Recent peat-forming environments (Cohen & Spackman 1972; Styan & Bustin 1983). Detailed miospore profiles have been used to interpret changing environments through coals (Smith 1968; Fulton 1987), and it has been shown that diversity and vegetation can change both gradually and suddenly in a single seam. Heterosporous lycopods are a dominant component of the vegetation in many Palaeozoic coals. Megaspores can therefore be of additional interest (Scott & King 1981). Bartram (1987) has demonstrated that in the Upper Carboniferous Barnsley Coal there are assemblages of megaspores which indicate a change from a rheotrophic to an ombrotrophic mire (Fig. 3). The later stages of peat development were dominated by a single species of plant. Bartram (1987) also recognized a herbaceous phase which had not been recorded from the associated miospore studies.

A final method is to study macroscopic plant material macerated from the coal. Styan & Bustin (1983) have indicated the importance of identifying macroscopic constituents of the peat. Macerations of coals may yield identifiable plant cuticles and charcoal which may give data on the original peat-forming communities (Scott & Collinson 1978).

These data can be used to identify (i) the homogeneity or zonal nature of the peat-forming vegetation, (ii) the diversity of the vegetation and (iii) the occurrence of environmentally restricted forms. They can also be used to test hypotheses concerning the freshwater versus brackish-water origin of peats, the ombrotrophic or rheotrophic nature of mire systems and their temporal association with active deltaic systems. Interpretations will have to remain tentative until we have a larger data base from which to draw conclusions.

Terrestrial organic matter in deltaic sediments

Whilst thick peats are generally not known to be accruing in modern deltaic sediments, organic debris from vegetation growing on the emergent parts of the delta is often incorporated into clastic sediments (Allen *et al.* 1979; Spicer 1981; Scheihing & Pfefferkorn 1984). The importance of this organic debris is that in some cases it may be the source of oil and gas. It has been increasingly recognized that both Mesozoic and Tertiary deltaic sediments which contain terrestrial organic matter may give rise to oil (Thompson *et al.* 1985; Saxby & Shibaoka 1986). In

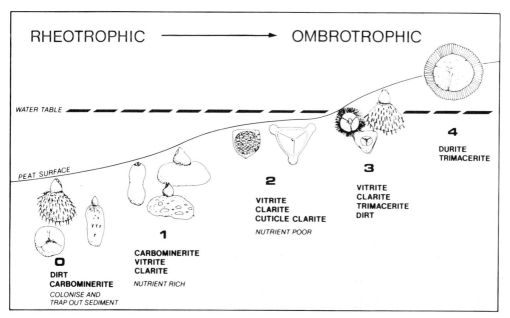

FIG. 3. A schematic diagram of an idealized uninterrupted sequence of megaspores through an Upper Carboniferous coal seam (modified from Bartam 1987). Megaspores not to scale.

addition some coals appear to be oil prone (Murchison 1987). Despite this we know little of the vegetation types that have given rise to these deposits. There is urgent need for botanical study of oil-prone coals and of detrital organic matter in deltaic sediments. A review of terrestrially derived organic matter in deltaic sediments is given by Cawley & Fleet (1987).

Conclusions

Several important conclusions can be drawn from this short review. Coals may occur associated with either lower- or upper-delta-plain settings, but in such cases they are often highly split and have high ash and sulphur values. As a consequence it is concluded that most economic coal seams do not occur in deltaic sequences.

Many deltaic coals are zoned, being initially rheotrophic and later ombrotrophic. Brackish-water rheotrophic peats (coals) have high ash and sulphur contents and have restricted and special-ized plant communities. Freshwater rheotrophic peats (coals) have a high ash content and diverse plant communities. Freshwater ombotrophic peats (coals) are usually domed, low ash and occasionally zoned with restricted plant communities. Considerable time may elapse between delta abandonment and the onset of peat formation.

Peat-forming plant communities have changed considerably through time, not least in deltaic settings. Integrated studies of coals and clastic deltaic sediments would be very rewarding. Integrated botanical, physical and chemical studies of coals will facilitate the erection of more complete sedimentological models and will aid our understanding of deltas as traps and sites for fossil fuels.

ACKNOWLEDGMENTS: I thank Professor W. G. Chaloner FRS and Dr L. Frostick for commenting on the manuscript and the symposium organizers for inviting me to make this contribution. Mr C. Hildrew kindly drafted the figures.

References

ALLEN, G. P., LAURIER, D. & THOUVENIN, J. 1979. Etude sédimentologique du Delta de la Mahakam. *Notes et Mémoires* **15**, 156 pp. Total, Compagne Française des Petroles, Paris.

BALTZER, F. 1969. Les formations végétales associées au Delta de la Dumbea. *Cahiers ORSTOM, Serie Geologie (1969)* **1**, 59–84.

BARTRAM, K. M. 1987. Lycopod succession in coals: an example from the Low Barnsley Seam (Westphalian B), Yorkshire, England. *In:* SCOTT, A. C. (ed.) *Coal and Coal-bearing Strata: Recent Advances.* Geological Society of London Special Publication **32**, 189–201.

BUSTIN, R. M. & LOWE, L. E. 1987. Sulphur, low temperature ash and minor elements in humid-temperate peat of the Fraser River Delta, British Columbia. *Journal of the Geological Society, London* **144**, 435–450.

CASAGRANDE, D. J. 1987. Sulphur in peat and coal. *In:* SCOTT, A. C. (ed.) *Coal and Coal-bearing Strata: Recent Advances.* Geological Society of London Special Publication **32**, 87–105.

CAWLEY, S. F. & FLEET, A. 1987. Deltaic petroleum source rock: a review. In: *Abstracts of the Symposium on Deltas: Sites and Traps for Fossil Fuels.* Geological Society, London.

CECIL, C. B., STANTON, R. W., NEUZIL, S. G., DULONG, F. T., RUPPERT, L. F. & PIERCE, B. S. 1985. Paleoclimate controls on late Paleozoic sedimentation and peat formation in the Central Appalachian Basin (U.S.A.). *International Journal of Coal Geology* **5**, 195–230.

COHEN, A. D. 1984. The Okefenokee Swamp: a low sulphur end-member of a shoreline-related depositional model for coastal plain coals. *In:* RAHMANI, R. A. & FLORES, R. M. (eds) *Sedimentology of Coal and Coal-bearing Sequences.* Special Publication of the International Association of Sedimentologists **7**, 231–240.

—— & SPACKMAN, W. 1972. Methods in peat petrology and their application to reconstruction of paleoenvironments. *Geological Society of America Bulletin* **83**, 129–240.

—— & —— 1977. Phytogenic organic sediments and sedimentary environments in the Everglades-mangrove complex. Part II: Origin, description and classification of the peats of South Florida. *Paleontographica B* **162**, 1–61.

——, —— & RAYMOND, R. 1987. Interpreting the characteristics of coal seams from chemical, physical, and petrographic studies of peat deposits. *In:* SCOTT, A. C. (ed.) *Coal and Coal-bearing Strata: Recent Advances.* Geological Society of London Special Publication **32**, 107–125.

COLEMAN, J. M., GAGLIANO, S. M. & SMITH, W. G. 1970. Sedimentation in a Malaysian high tide tropical delta. *In:* MORGAN, J. P. (ed.) *Deltaic Sedimentation, Modern and Ancient.* Special Publication of the Society of Economic Paleontologists and Mineralogists **15**, 185–197.

COLLINSON, M. E. & SCOTT, A. C. 1987a. Implications of vegetational change through the geological record on models for coal-forming environments. *In:* SCOTT, A. C. (ed.) *Coal and Coal-bearing Strata: Recent Advances.* Geological Society of London Special Publication **32**, 67–85.

—— & —— 1987b. Factors controlling the organisation and evolution of ancient plant communities. In:

GEE, J. H. R. & GILLER, P. S. (eds). *Organisation of Communities—Past and Present*. British Ecological Society, London, 399–420.

COPE, M. & CHALONER, W. G. 1985. Wildfire: an interaction of biological and physical processes. *In:* TIFFNEY, B. (ed.) *Geological Factors and the Evolution of Plants*. Yale University Press, New Haven, CT, 257–277.

CYPERT, E. 1972. Plant succession on burned areas in Okefenokee swamp following the fires of 1954 and 1955. *Proceedings of the 12th Annual Tall Timbers Fire Ecological Conference*. Tall Timbers Research Station, Tallahassee, FL, 199–217.

DIMICHELE, W. A. & PHILLIPS, T. L. 1985. Arborescent lycopod reproduction and paleoecology in a coal-swamp environment of late Middle Pennsylvanian age (Herrin Coal, Illinois). *Review of Palaeobotany and Palynology* **44**, 1–26.

——, —— & PEPPERS, R. A. 1985. The influence of climate and depositional environment on the distribution and evolution of Pennsylvanian coal-swamp plants. *In:* TIFFNEY, B. H. (ed.) *Geological Factors and the Evolution of Plants*. Yale University Press, New Haven, CT, 223–256.

FERM, J. C. & STAUB, J. R. 1984. Depositional controls of mineable coal bodies. *In:* RAHMANI, R. A. & FLORES, R. M. (eds) *Sedimentology of Coal and Coal-bearing Sequences*. Special Publication of the International Association of Sedimentologists 7, 275–289.

FIELDING, C. R. 1985. Coal depositional models and the distinction between alluvial and delta plain environments. *Sedimentary Geology* **42**, 41–48.

FRAZIER, D. E., OSANIK, A. & ELSIK, W. C. 1978. Environments of peat accumulation—coastal Louisiana. *Proceedings of the Gulf Coast Lignite Conference*. Texas Bureau of Economic Geology Report of Investigations **90**, 5–20.

FULTON, I. M. 1987. Genesis of the Warwickshire Thick Coal: a group of long-residence histosols. *In:* SCOTT, A. C. (ed.) *Coal and Coal-bearing Strata: Recent Advances*. Geological Society of London Special Publication **32**, 203–220.

GORE, A. J. P. (ed.) 1983. *Ecosystems of the World*, Vol. 4A, *Mires: Swamp, Bog, Fen and Moor, General Studies*. Elsevier, Amsterdam.

HOBDAY, D. K. 1987. Gondwana coal basins of Australia and Africa: tectonic setting, depositional systems and resources. *In:* SCOTT, A. C. (ed.) *Coal and Coal-bearing Strata: Recent Advances*. Geological Society of London Special Publication **32**, 221–235.

KOMAREK, E. V. 1972. Ancient fires. *Proceedings of the 12th Annual Tall Timbers Fire Ecological Conference*. Tall Timbers Research Station, Tallahassee, FL, 219–240.

KOSTERS, E. C., CHMURA, G. L. & BAILEY, A. 1987. Sedimentary and botanic factors influencing peat accumulation in the Mississippi Delta. *Journal of the Geological Society, London* **144**, 423–434.

MCCABE, P. J. 1984. Depositional environments of coal and coal-bearing strata. *In:* RAHMANI, R. A. & FLORES, R. M. (eds) *Sedimentology of Coal and Coal-bearing Sequences*. Special Publication of the

International Association of Sedimentologists **7**, 13–42.

MOORE, P. D. 1983. Plants and the palaeoatmosphere. *Journal of the Geological Society, London* **140**, 13–26.

—— 1987. Ecological and hydrological aspects of peat formation. *In:* SCOTT, A. C. (ed.) *Coal and Coal-bearing Strata: Recent Advances*. Geological Society of London Special Publication **32**, 7–15.

MURCHISON, D. G. 1987. Recent advances in organic petrology and organic geochemistry: an overview with some reference to 'oil from coal'. *In:* SCOTT, A. C. (ed.) *Coal and Coal-bearing Strata: Recent Advances*. Geological Society of London Special Publication **32**, 259–304.

PHILLIPS, T. L. & PEPPERS, R. A. 1984. Changing patterns of Pennsylvanian coal-swamp vegetation and implications of climatic control on coal occurrence. *International Journal of Coal Geology* **3**, 205–255.

——, —— & DIMICHELE, W. A. 1985. Stratigraphic and interregional changes in Pennsylvanian coal-swamp vegetation: environmental inferences. *International Journal of Coal Geology* **5**, 43–109.

RAYMOND, A. & PHILLIPS, T. L. 1983. Evidence for an Upper Carboniferous mangrove community. *In:* TEAS, H. J. (ed.) *Tasks for Vegetation Science*, Vol. 8. Junk, The Hague, 19–30.

RETALLACK, G. J. 1986. The fossil record of soils. *In:* WRIGHT, V. P. (ed.) *Paleosols: Their Recognition and Interpretation*. Blackwell Scientific Publications, Oxford, 1–57.

SAXBY, J. D. & SHIBAOKA, M. 1986. Coal and coal macerals as source rocks for oil and gas. *Applied Geochemistry* **1**, 25–36.

SCHECKLER, S. E. 1986. Geology, floristics and paleoecology of late Devonian coal swamps from Appalachian Laurentia (U.S.A.). *Annales de la Société Géologique de Belgique* **109**, 209–222.

SCHEIHING, M. H. & PFEFFERKORN, H. W. 1984. The taphonomy of land plants in the Orinoco delta: a model for the incorporation of plant parts in clastic sediments of late Carboniferous age of Euramerica. *Review of Palaeobotany and Palynology* **41**, 205–240.

SCOTT, A. C. & COLLINSON, M. E. 1978. Organic sedimentary particles: results from Scanning Electron Microscope studies of fragmentary plant material. *In:* WHALLEY, W. B. (ed.) *Scanning Electron Microscopy in the Study of Sediments*. Geoabstracts, Norwich, 137–167.

—— & KING, G. R. 1981. Megaspores and coal facies: an example from the Westphalian A of Leicestershire, England. *Review of Palaeobotany and Palynology* **34**, 107–113.

—— & REX, G. M. 1985. The formation and significance of Carboniferous coal balls. *Philosophical Transactions of the Royal Society of London, Series B* **311**, 123–137.

SMITH, A. H. V. 1968. Seam profiles and seam characters. *In:* MURCHISON, D. G. & WESTOLL, T. S. (eds) *Coal and Coal-Bearing Strata*. Oliver & Boyd, Edinburgh, 31–40.

SMYTH, M. 1983. Nature of source material for

hydrocarbons in Cooper Basin, Australia. *Journal of the American Association of Petroleum Geologists* **67**, 1422–1428.

SPACKMAN, W., DOLSEN, C. P. & RIEGEL, W. 1966. Phytogenic organic sediments and sedimentology environments in the Everglades-mangrove complex. Part I: Evidence of a transgressing sea and its effects on environments of the Shark River area of Southwestern Florida. *Palaeontolographia B* **117**, 135–152.

SPICER, R. A. 1981. *The Sorting and Deposition of Allochthonous Plant Material in a Modern Environment at Silwood Lake, Silwood Park, Berkshire, England.* US Geological Survey Professional Paper **1143**, 1–77.

——, GRANT, P. R. & PARRISH, J. T. 1987. The late Cretaceous polar coals: a sedimentological, vegetational and climatological study of the Umiat Delta, Alaska. *In: Abstracts of the Symposium on Deltas—Sites and Traps for Fossil Fuels.* Geological Society, London, PA9.

STYAN, W. B. & BUSTIN, R. M. 1983. Petrography of some Fraser Delta peat deposits: coal maceral and microlithotype precursors in temperate-climate peats. *International Journal of Coal Geology* **2**, 321–370.

TEICHMÜLLER, M. & TEICHMÜLLER, R. 1968. Cainozoic and Mesozoic coal deposits of Germany. *In:* MURCHISON, D. G. & WESTOLL, T. S. (eds) *Coal and Coal-bearing Strata.* Oliver & Boyd, Edinburgh, 347–379.

THOMPSON, S., COOPER, B. S., MORLEY. R. J. & BARNARD, P. C. 1985. Oil-generating coals. *In:* THOMAS, B. M., DORE, A. G., EGGEN, S. S., HOME, P. C. & LARSEN, R. M. (eds) *Petroleum Geochemistry in Exploration of the Norwegian Shelf.* Graham & Trotman, London, 59–73.

A. C. SCOTT, Geology Department, Royal Holloway and Bedford New College, University of London, Egham, Surrey TW20 0EX, UK.

Fan-delta–lacustrine sedimentation and coal development in th Tertiary Ombilin Basin, W Sumatra, Indonesia

M. K. G. Whateley & G. R. Jordan

SUMMARY: The Ombilin Basin is a Tertiary intermontane basin situated in the Barisan mountain range of W Sumatra. It covers an area of some 1500 km² and is known for the coal resources which have been mined in the Sawahlunto area since 1891. Seismic and borehole data indicate that the basin has been filled with approximately 4600 m of middle Eocene to early Miocene sediments. The coal developed early in the basin's history (middle Eocene). During this study the fan-delta sediments below the oldest seam, the C seam, the seam itself and the sediments immediately overlying the C seam were investigated along the western margin of the Ombilin Basin in the Sawahlunto area.

The oblique convergence of the Indian Ocean plate with the SE Asia plate created a stress field of the type that can produce a primary dextral wrench fault which in this case is named the Great Sumatran Fault Zone. The Ombilin Basin is a graben-like 'pull-apart' basin resulting from localized extension failure and related deformation. The failure and deformation are a function of irregular strike-slip movement along the Great Sumatran Fault Zone during the early Tertiary. Tertiary sedimentation began in the Eocene following graben formation. High basement relief adjacent to the newly formed basin resulted in debris flows and fan-delta formation around the margin of the central lake. Concurrent with the deposition of coarse-grained fan-delta sediments around the basin margins, fine-grained lake sediments accumulated in the basin centre.

Reduction in relief difference between the source area and the basin was associated with an expansion of the lacustrine environment. Shallowing of the lake and a humid tropical climate allowed plant colonization to take place. Peat accumulation was influenced by the underlying fan-delta lobe geometry. In interlobe areas thick sequences of low ash peat accumulated, while central lobe areas through the same interval were sites of continued sediment influx and relatively slow peat accumulation. In these areas high ash coal or carbonaceous shale are laterally equivalent to the low ash coal. A detailed investigation of the character of the coal in the C seam indicates that, while the lateral dispersion of clastic material may play some part in the variation in the ash content of the coal from one place to another, the most significant variable is the rate of accumulation of peat.

Fan deltas have been defined as alluvial fans that build out into a standing body of water from an adjacent highland (Holmes 1965; Rust 1979; Wescott & Ethridge 1980; Ricci Lucchi et al. 1981). Most modern fan deltas are located along tectonically active coastlines which are usually wave dominated and receive between 100 and 300 cm annual precipitation (Wescott & Ethridge 1980). These workers cite numerous examples of modern fan deltas. They also express the opinion that many examples of coarse clastic-wedge sequences ranging from Precambrian to Pleistocene should be re-evaluated in terms of fan-delta models, suggesting that fan deltas are more common than previously thought.

However, examples of smaller lacustrine fan deltas are not so well documented. Hyne et al. (1979) describe the recent stratigraphy of the intermontane lacustrine delta of Lake Maracaibo, Venezuela. This area receives as much as 250 cm a^{-1} rainfall, but there is little or no wave or tidal action. The original description of lacustrine fan deltas was that of Gilbert (1885), after whom 'Gilbert' deltas are named. Sneh

(1979) described the Pleistocene fan deltas along the Dead Sea Rift, which developed as a result of flash floods in an arid environment.

The sedimentology of part of the Eocene deposits which initially infilled the Ombilin Basin, Sumatra, is outlined in this study. The basin was infilled by side-fed coarse-grained fan-delta deposits and central fine-grained lacustrine deposits. The sediments accumulated in a humid tropical climate which encouraged thick coal to develop. The investigation will show how the geometry of two fan-delta lobes influenced the distribution of the overlying coal seam (Whateley 1986) and its coal quality (Norwest Resource Consultants 1987).

Location

The Ombilin Basin is a relatively small Tertiary intermontane basin situated in the Barisan mountain range of W Sumatra (Fig. 1). The basin has an area of approximately 1500 km² and is

From WHATELEY, M. K. G. & PICKERING, K. T. (eds), 1989, *Deltas: Sites and Traps for Fossil Fuels,*
Geological Society Special Publication No. 41, pp. 317–332.

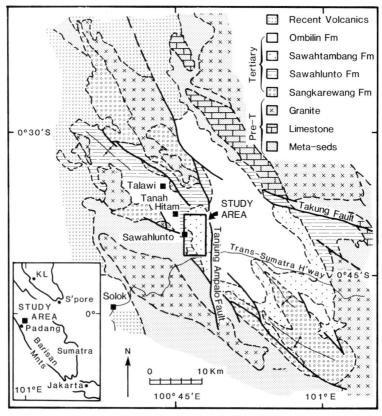

FIG. 1. Location of the study area, Ombilin Basin, showing the geological setting. The Brani Formation is included in the Sangkarewang Formation. (After Koning 1985.)

known for the coal resources which are well explored on its western side. Coal has been mined in the Sawahlunto area since 1891. The study area (Fig. 1) is approximately 11 km² and lies immediately E of Sawahlunto and some 120 km E of Padang in W Sumatra.

Methods of study

In what is essentially a subsurface investigation, some 54 borehole logs from within the study area were examined (Fig. 1; see also Fig. 11). The cores from six of the more recently drilled holes were relogged with particular attention being paid to the nature of the individual beds (such as lithology and sedimentary structure), grain size and bed contact type (eg erosive, gradational, sharp).

Four measured sections were compiled from exposures in the Waringin and Sugar areas (see Fig. 11), in the Tanah Hitam open pit and near Talawi (Fig. 1). In addition to the parameters

described above the lateral extent of beds was noted and palaeocurrent readings were taken.

Using data determined from borehole logs, structural contour maps of readily identifiable horizons (eg the top of a formation or coal seam) were constructed as well as coal-seam isopach maps and sandstone percentage maps of the interseam sediments (Norwest Resource Consultants 1987). These were interpreted in conjunction with known geology and sedimentology, and a model for the deposition of the fan-delta and coal-bearing lacustrine sediments was constructed. Studies of the coal quality data were incorporated in the development of this model.

Stratigraphy

The general geological history of the central and S Sumatra basins, including the Ombilin Basin, was discussed by de Coster (1976). He described the Tertiary sediments in these basins as mainly non-marine, consisting of a complex interrela-

TABLE 1. *Lithostratigraphy of the study area based on work by Koesoemadinata & Matasak (1981) in the Ombilin Basin*

AGE			FORMATION	THICKNESS(m)	LITHOLOGY	INTERPRETATION
PLEISTOCENE			RANAU		TUFFS	SUBAERIAL VOLCANIC DEPOSITION
TERTIARY	EARLY MIOCENE		OMBILIN	1442	GREY CALCAREOUS SHALE (MARL) WITH LIMESTONE LENSES. INTERBEDDED TUFFS IN UPPER PART.	MARINE (NERITIC)
	OLIGOCENE		SAWAHTAMBANG	625	INTERBEDDED SANDSTONE, SILTSTONE, SHALE AND COAL STRINGERS. COARSE-GRAINED SANDSTONE. BLUE/GREEN FINING UPWARD SEQUENCES OF COARSE-GRAINED SANDSTONE. FINING UPWARD SEQUENCES FROM COARSE-GRAINED SANDSTONE TO VARIEGATED MUDSTONE.	UPPER DELTA PLAIN (?) FAN DELTA DEVELOPMENT BRAIDED STREAMS { HIGH ENERGY DISCHARGE SHEET SANDS AND BASIN FILL MUDSTONE
	EOCENE		SAWAHLUNTO	0-195	INTERBEDDED COAL, SEAT EARTH, CLAYSTONE, RIPPLE CROSS-LAMINATED AND BIOTURBATED SILTSTONE AND CROSS-BEDDED SANDSTONE.	LAKE FILL SEQUENCES, PONDS AND MIRES
	PALAEOCENE ?		SANGKA-REWANG / BRANI	0-280	ALTERNATING SANDSTONE AND SHALE UNITS. DARK GREY SHALES PREDOMINATE AT BASE. THIN LATERALLY PERSISTENT SANDSTONE HORIZONS OCCUR HIGHER IN THE SEQUENCE. CONGLOMERATES, COARSE SANDSTONE WITH MATRIX SUPPORTED PEBBLES, COBBLES & BOULDERS.	FRESHWATER LACUSTRINE INFILLING DEBRIS FLOWS AND FAN DELTAS
PRE-TERTIARY BASEMENT						

tionship of lacustrine, deltaic and fluviatile environments. A detailed investigation of the stratigraphy and sedimentation of the Ombilin Basin was undertaken by Koesoemadinata & Matasak (1981).

The Tertiary sediments rest on pre-Tertiary basement rocks which are well exposed around the margin of the Ombilin Basin. The basement rocks consist predominately of Permo-Carboniferous slates, phyllites and limestones in the E, and Permo-Carboniferous and Triassic limestones, slates and volcanics and middle Cretaceous granite along the western margin (Fig. 1). The results of their field mapping enabled Koesoemadinata & Matasak (1981) to subdivide the Tertiary sediments within the Ombilin Basin into a series of formations (Table 1). They suggested that deposition commenced around the margins of the newly formed intermontane basin in the Palaeocene or Eocene (Koning 1985) with very coarse-grained alluvial fan sediments. They called this unit the Brani Formation. The Brani Formation lies with an angular unconformity upon the basement, and has been measured up to 100 m thick in the study area. It is considered that these sediments built out into the Ombilin Lake from adjacent highlands and should be considered as fan-delta sediments.

The Brani Formation interfingers laterally and basinward with the Sangkarewang Formation, which consists of lacustrine sediments deposited in the centre of the basin. These lake deposits contain Tertiary freshwater fish fossils (Musper 1929). The Sangkarewang Formation may also unconformably overstep onto the pre-Tertiary basement in places. In the study area the formation varies from a few metres to 280 m thick (Table 1), although Koning (1985) reports some 460 m in a borehole in the basin centre. Koning (1985) also reports a significant intra-Sangkarewang Formation unconformity based on vitrinite reflectance and spore colour index. This may well be the result of reactivation of a marginal fault which is common in tectonically active basins (Steel & Aasheim 1978; Ricci Lucchi *et al.* 1981).

The Sangkarewang Formation is overlain conformably and gradationally by the Eocene coal-bearing Sawahlunto Formation. The type section measured by Koesoemadinata & Matasak (1981) is 274 m thick, although Koning (1985) reports a thickness of only 170 m in the centre of the Ombilin Basin. The Sawahlunto Formation contains three main coal seams that achieve thicknesses normally considered suitable for mining. Koesoemadinata & Matasak (1981) suggest that the sediments were deposited in a fluvially dominated flood-plain system. However, as channel sands were notably rare or absent in the study area the Sawahlunto Formation sediments are considered to be lacustrine and peat deposits.

These sediments are unconformably overlain by the Sawahtambang Formation. Up to 625 m of this formation has been measured in the Ombilin Basin. The coarse-grained sands were deposited initially by high sinuosity channels with overbank fines, but later coarse-grained sands were deposited by braided river systems. The Sawahtambang Formation is overlain unconformably by the marine Ombilin Formation. These sediments consist of grey calcareous shales.

Koning (1985) confirmed the stratigraphy in the subsurface from seismic data, although he presents some significant depositional hiatuses not identified by previous workers. He recognizes up to 4600 m of Tertiary sediments in the basin, although approximately 3600 m has been eroded from the western side of the basin, leaving only the Brani, Sangkarewang, Sawahlunto and Sawahtambang Formations in the study area. For the purpose of this paper the Brani, Sangkarewang and Sawahlunto Formations, consisting of fan-delta and lacustrine sediments, are described.

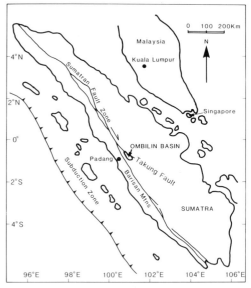

FIG. 2. Structural setting of the Ombilin Basin, W Sumatra.

Structural setting

Two major regional tectonic features, the magmatic arc and the Great Sumatran Fault Zone, dominate W Sumatra. These features extend the full length of Sumatra. The convergence of the Indian Ocean plate obliquely towards the SE Asian plate has created a dextral strike-slip fault, referred to as the Great Sumatran Fault Zone (Fig. 2). Several graben-like structures, which include the Ombilin Basin, are identified along this zone (Tjia 1970). Harding *et al.* (1985) and Koning (1985) believe that the Ombilin Basin is a graben-like pull-apart structure resulting from early Tertiary tensional tectonics related to strike-slip movement along the Great Sumatran Fault Zone. Mann *et al.* (1983) have attributed this type of formation to many similar basins.

The regional setting of the Ombilin Basin was described in terms of plate tectonics by Koesoemadinata *et al.* (1978). The local structure of the Ombilin Basin was described in more detail by Koesoemadinata & Matasak (1981). Further data from side-looking airborne radar (SAR) imagery and surface seismic studies added to the structural knowledge of the basin (Koning 1985).

The northeastern margin of the basin is delineated by the Takung Thrust Fault (Fig. 1) (Koesoemadinata & Matasak 1981). This fault trends NW–SE, parallel to the Great Sumatran Fault Zone. The Ombilin Basin is bisected by the N–S-trending Tanjung Ampalo Fault. This fault

separates the deeper eastern portion of the basin from the Sawahlunto area to the W. It is within this western portion of the basin that the coal measures are mined near the surface. The western boundary of the basin is formed by complex faulting which brings pre-Tertiary basement adjacent to the Tertiary basin sediments (Fig. 1).

Folding is developed to a minor degree. The basin axis has a NW–SE trend, and the few anticlines and synclines parallel this trend and are parallel to the marginal faults (Fig. 1). It has been demonstrated by Anadòn *et al.* (1985) that similar structures in the Ebro Basin relate to the location of basement-involved faults which have caused flexures and folds in the sedimentary cover. Harding *et al.* (1985) explain that flexures associated with wrench faults are formed predominantly by vertical displacement and that most are drag or forced folds parallel and adjacent to the fault. They suggest, however, that folds within the pull-apart basins in Sumatra owe their existence to a pre-existing basement grain. The sediments in the Sawahlunto area dip uniformly to the E at between 12° and 28°. This may reflect the regional dip of the sediments into the basin or the local effect of the Talawi syncline which forms the northwestern limb of the Ombilin Basin (Fig. 1).

Sedimentary associations

Several distinctive sedimentary associations, including fan-delta, distal-fan and lacustrine associations which are encompassed in the Eocene Brani, Sangkarewang and Sawahlunto Formations, are recognized within the coalfield.

Fan-delta association

Description

The fan-delta association is represented by the Brani Formation. These sediments lie with angular unconformity upon the pre-Tertiary basement. They are widely distributed along the basin margin (Koesoemadinata & Matasak 1981) and vary greatly in thickness as they show both lateral and basinward thinning. A structural contour map constructed for the top of the Brani–Sangkarewang Formation showed a general NW–SE strike with a large lobate (structural high) anomaly projecting NE from the western basin margin towards hole W27 (*cf* Fig. 13).

The basal fan-delta deposits consist of crudely stratified gravel to boulder conglomerate resting directly on the basement (Fig. 3). The maximum clast size is 75 cm and the clasts vary in lithology, depending upon the basement from which they were derived. In the Talawi area the angular to subrounded clasts consist predominately of quartz and quartzite with subordinate grandodiorite, sandstone, smoky quartz and slate amongst others. The sandstone matrix ranges in composition from lithic arenite to feldspathic arenite and arkose. The clasts are generally matrix supported. It is difficult to identify the basement–Brani Formation contact at first glance because of their similar mineralogy. The fan-delta deposits are mineralogically immature and texturally poorly sorted. Individual beds, 50 cm to 2 m thick (Fig. 4), are defined by changes in the abundance of clasts (there are fewer clasts towards the top of individual beds) and a change in colour. The beds are pale grey to white at the base, but towards the top the matrix becomes purple-red. No obvious cross-stratification or imbrication was seen at the outcrop near Talawi.

The individual beds show crude fining upwards, and maximum clast size and overall clast size decrease upwards in the succession until clasts are rare in coarse-grained sandstone beds (Fig. 5).

Interpretation

These coarse-grained fan-delta accumulations around the margin of the Ombilin Basin are similar to deposits described in other pull-apart basins (Heward 1978; Sneh 1979; Hempton & Dunne 1984; Link *et al.* 1985; Manspeizer 1985) and to marginal deposits described by Wescott & Ethridge (1980) and Ricchi Lucchi *et al.* (1981).

The coarse poorly sorted nature of the deposits, the lack of apparent channelling, the limited

FIG. 3. Basal contact of the Brani Formation with pre-Tertiary granite basement near Talawi, W Sumatra. Contact marked by hammer.

FIG. 4. Crude stratification in the Brani Formation defined by changes in abundance of clasts and colour changes near Talawi, W Sumatra.

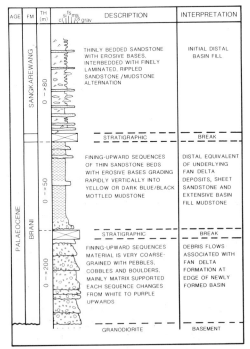

FIG. 5. Generalized composite stratigraphic section of the Brani and Sangkarewang Formations near Talawi, W Sumatra. For key see Fig. 13.

lateral and basinward extent, the presence of predominately matrix-supported clasts and the absence of sedimentary structures suggests that they are debris flow deposits resulting from rapid deposition along the flanks of the newly formed graben (Enos 1977; Heward 1978; Steel & Aasheim 1978). Deposition resulted from deceleration due to increased flow width and decrease in flow depth as floods emerged from the confines of the restricted flow area on the scarp (Bull 1964; Heward 1978; Hempton & Dunne 1984). The sediments built out into the Ombilin Lake and constructed fan deltas. The conglomerates pinch out basinward by interfingering with coarse-grained sandstone and distal-fan deposits which in turn interfinger with the finely laminated lacustrine sediments (Koesoemadinata & Matasak 1981) which show shallow-water structures.

The change in colour from pale grey to purple-red upwards in individual beds suggests emergent conditions. The fan-delta sediments were probably deposited as flood events into a lake with a raised water level. Termination of the flood event resulted in a slight lowering of the lake water level to expose the upper surface of the fan delta which allowed oxidation of the iron and thus a reddening of the sand matrix.

The decrease in maximum grain size upwards in individual sequences and upwards through the fan-delta association may indicate declining sediment supply resulting from scarp retreat and lowering of source-area topography. This possibility, in conjunction with a rapid rise in lake level, would result in an overall fining-upward sequence (Sneh 1979).

A series of late Pleistocene fan deltas which developed at the mouths of streams where they entered the precursor of the Dead Sea are described by Sneh (1979) and Manspeizer (1985). They are shown to have a lobate or delta form along the margin of the Dead Sea Rift. The largest of the fan deltas has an area of only about 6 km^2, but the majority do not exceed 2 km^2 which is comparable with the size of the fan deltas implied by the structural contour map of the top of the Brani–Sangkarewang Formation.

Distal-fan association

Description

The distal-fan association is represented in part of the Sangkarewang Formation. These sediments are characteristically fine grained, varying from shale to siltstone, with occasional medium- to coarse-grained sandstones (Fig. 6). The sandstone beds are between 1 and 2 m thick and are laterally continuous over tens of metres at outcrop. These sheet sandstones have sharp or erosive bases and individual beds fine upwards. The only structure observed in them is rare cross-lamination seen in core.

The sandstone beds are separated by shale or siltstone with thicknesses varying from a few centimetres to 10 m (Fig. 6). The sediments are light grey with occasional purple-red mottling when fresh, but become yellow to dark blue-black and mottled on weathered surfaces. They are calcareous in places. Rare interbedded carbonaceous shale occurs, but rooting is common. The sandstones and shales are normally arranged in fining-upward sequences between 2 and 15 m thick (Figs 5 and 6).

Interpretation

The laterally extensive sharp- or erosive-based sandstones which fine upwards were probably formed during sheet-flood episodes (Bull 1964) and represent the distal equivalent of the fan-delta association. Waning of the flood events gave rise to the extensive siltstone and shale sequences. Periodic lowering of the water level led to oxidation which produced the purple-red mottling. Individual flood events were separated by enough time to enable plants to take root on the sediment surface. The continued high rate of influx of fine sediment meant that plant debris could not generally accumulate in sufficient quantities to form carbonaceous shale, but did leave numerous rootlet horizons. The distal-fan deposits interfinger with both the marginal fan-delta deposits and the central lacustrine deposits.

Lacustrine association

Description

The lacustrine association is represented in the Sangkarewang and Sawahlunto Formations (Fig. 6). In the study area the dominant sediments in these formations consist of dark grey (when fresh) to brown (in outcrop), non-calcareous to calcareous, well-laminated, fissile carbonaceous shales, which are pyritic in places. The shales are finely interbedded (1–5 cm) with fine-grained light grey ripple cross-laminated micaceous sandstones (Fig. 7). In the Sawahlunto Formation these interbeds are highly bioturbated (Fig. 8) with vertical to subhorizontal burrows and trails. In some places the bioturbation is so intense that the interbeds are thoroughly mixed and the siltstone takes on a mottled dark grey appearance.

Medium-grained, dirty, light grey, poorly

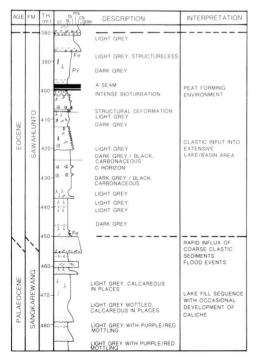

FIG. 6. Stratigraphic section of Sangkarewang and Sawahlunto Formations in hole W23, near Sawahlunto, Ombilin Basin, W Sumatra. See Fig. 11 for location of hole W23. For key see Fig. 13.

FIG. 7. Interbedded shale and fine-grained light grey ripple cross-laminated micaceous sandstone (Sawahlunto Formation) in a core from hole W23, Ombilin Basin, W Sumatra. The scale at the bottom of the core is in centimetres.

FIG. 8. Bioturbation in a core of the Sawahlunto Formation from hole W23, Ombilin Basin, W Sumatra. The scale at the bottom of the core is in centimetres.

sorted, arkosic and slightly micaceous sandstone beds up to 1 m thick occur locally towards the base of the lacustrine association. The beds have erosive bases which may be overlain by coarse-grained sandstone with pebbles. These sequences show a decrease in grain size upwards. Cross-bedding may occur at or near the base with ripple cross-lamination nearer the top. The sandstones may contain comminuted coaly fragments laid parallel to the lower bedding surface. These sandstone beds occur less frequently higher in the succession.

The shale in the Sangkarewang Formation shows crude bedding on a 1–2 m scale, outlined by colour changes. The lower sections are dark grey while the upper sections are purple-grey with distinctive layers of white calcareous material. The purple-grey coloration and the calcareous material do not appear in the lacustrine sediments of the Sawahlunto Formation.

Freshwater fish fossils, eg *Musperia radiata* (Herr) and *Scheropagus*, have been reported in

these sediments by Musper (1929). A well-preserved bird fossil, *Protoplotus beauforti*, was also discovered in these sediments (Rich & Marino-Hadiwardoyo 1977). Pond sediments are included in the lacustrine association (Fig. 9). They are composed of homogeneous light grey, silty shale containing abundant carbonized plant roots, sometimes over several metres long (*cf* Long 1981). There is a noticeable lack of bioturbation structures, which is similar to several examples given by Long (1981).

The interbedded shale–sandstone sequences grade vertically upwards and horizontally into shale, carbonaceous shale and coal in the Sawahlunto Formation (Figs 6 and 10). There is a gradual increase, in places over several metres, in the amount of carbonaceous material in the sediments beneath the oldest major seam, the C seam. The immediate floor of the C seam is identified by a grey-brown mottled clay seat earth. This is overlain gradationally by an average of 15 cm of coaly shale and 15 cm of dull coal, and this in turn is overlain by bright low ash coal.

FIG. 9. Light grey silty shale pond sediments containing carbonized plant roots in hole W22, Ombilin Basin, W Sumatra. The scale at the bottom of the core is in centimetres.

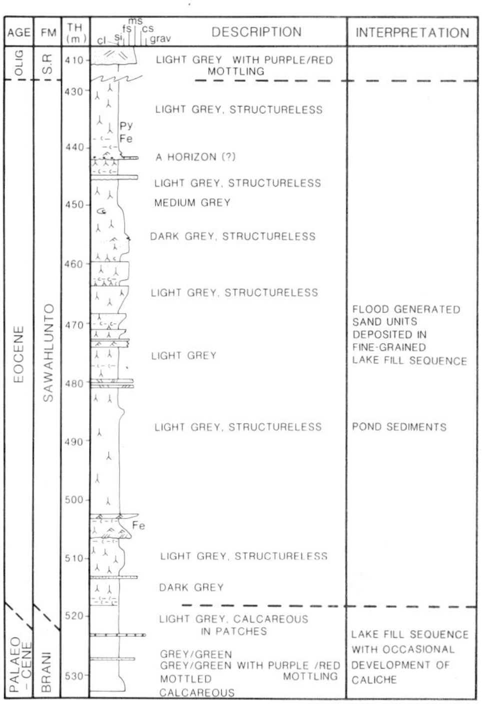

FIG. 10. Stratigraphic section of the Sawahlunto Formation in hole W22, Ombilin Basin, W Sumatra. See Fig. 11 for location of hole W22. For key see Fig. 13.

The contact of the coal with the overlying sediment is also gradational over some 20 cm.

The C seam has an average true coal thickness of 5.5 m but it is highly variable, ranging from a few centimetres to over 15 m (Fig. 11). The ash content is also variable. There appears to be an inverse correlation between seam thickness and ash content. The ash content increases rapidly to between 15% and 30% where the seam is less than 2 m thick, but becomes asymptotic at 4%–5% where the seam is greater than 4 m thick (Norwest Resource Consultants 1987). The seam has a high ash content near the western margin.

There is an average of 15 m of clastic lacustrine sediment between the C seam and the overlying B seam. The B seam has an average thickness of 1 m. Some 15 m of clastic sediment were deposited between the B seam and the A seam. The A seam has an average thickness of 1.5 m. Lacustrine sedimentation ceases a few metres above the A seam. Limited palaeocurrent data obtained from ripple cross-laminations and cross-bedding from sediments in the Sawahlunto Formation in

the Tanah Hitam open pit indicate a source in the W or SW (Fig. 12). Palaeocurrent data were collected in the Waringin and Sugar areas from the Sawahtambang Formation (Whateley 1986). These data also indicate a sediment source in the W or SW (Fig. 12).

Interpretation

Lacustrine deposits are characterized by finely interbedded shale and ripple cross-laminated siltstone, and carbonaceous material is locally abundant (Long 1981). The fine-grained freshwater lacustrine deposits in the Sawahlunto area are similar to deposits in other basins in Lake Hazar in eastern Turkey (Long 1981; Hempton & Dunne 1984) and in the Dead Sea (Manspeizer 1985). The finely interbedded, extensively bioturbated, lacustrine sediments differ from pond deposits in several ways. Pond deposits in the study area consist of homogeneous silty shale with abundant rootlets and a noticeable lack of bioturbation structures. The fine-grained lacustrine sediments accumulated in low energy environments basinward of the fan deltas and other basin-margin down-slope-directed sedi-

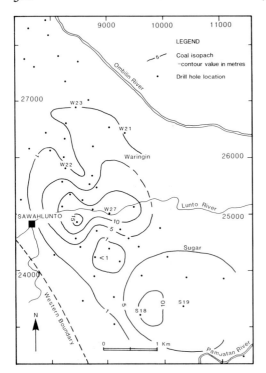

FIG. 11. Isopach map of the C seam, Ombilin Basin, W Sumatra.

ment, as well as in interlobe areas. Periodically coarser sediment (silt to fine-grained sand) entered the system either seasonally or as a result of individual monsoon storm events to give rise to the fine interbeds. The conglomerates and coarse-grained sandstone beds which were noted some way below the C seam in the relogged boreholes probably formed as sediment gravity flow deposits in front of the fan deltas (Hughes & Long 1980). Pond deposits may have collected in relatively sheltered shallow interlobe areas. These areas probably avoided the down-fan-directed sediment input to some extent, and plant colonization took place in the quiet backwater. Conditions were probably not conducive to preservation of organic material as peat, and so we see only the plant root remains in the sediments.

In more sheltered interlobe areas away from the main sediment influx, peat accumulated in considerable quantities where water-table and nutrient supply were at an optimum. Peat would not accumulate immediately adjacent to the basin margin because of the general down-slope-directed sediment input along the margin. Thus, peat accumulated slightly basinwards of the margin in interlobe areas. The C seam thins proximal to the basin margin and a rapid facies

change to carbonaceous shale is noted. This is similar to the model proposed by Heward (1978) for the Pastora coal seam in the Matallano coalfield, northern Spain. Thinning of the seam on the proximal margin may also occur owing to a postulated fluctuating water-table resulting in periodic oxidation of the peat. This is described by Ayers & Kaiser (1984) for seam 14 in the Fort Union Formation, Wyoming.

Palaeoclimatology

It has been suggested by Habicht (1979) and Frakes (1979) that Sumatra lay in an equatorial position during the Palaeogene and that a humid tropical climate existed. Calculated sea surface temperatures were about 29–30 °C during the Eocene, according to Frakes (1979, pp. 193–195).

Additional palaeoclimatic evidence is found in the study area. There are only a few purple-red coloured beds developed in the study area and these are found in the Brani Formation at the top of fining-upward sequences and locally in the Sangkarewang Formation. This suggests emergent conditions which allowed oxidation of the upper portion of the recently deposited sediments. The presence of minor calcareous layers in the Sangkarewang Formation may be attributable to the input of fine calcareous debris from the adjacent pre-Tertiary limestone highlands. This is by no means the only source of carbonate in the lacustrine environment. Carbonates can also be inorganically precipitated by evaporation or by changes in water temperature. The absence of calcareous layers in the Sawahlunto Formation suggests that the climate became humid enough to maintain a high water-table which was not conducive to the formation of carbonates in the younger lacustrine sediments. These features point to warm humid climatic conditions which were not hot or dry enough to form evaporite minerals. Evaporite minerals are not known in the study area.

The occurrence of the fossilized skeleton of the snake bird *Protoplotus beauforti* in the Sankarewang Formation is significant because its modern relative, the snake bird or darter, is restricted primarily to the wet tropics. Palynological data from the area suggest that the Palaeogene flora was dominated by tropical species (JICA 1980; Williams *et al.* 1985).

This cumulative evidence suggests that a humid tropical climate probably prevailed at the time the coal was developing. In tropical peat-forming areas today, where the annual rainfall is greater than the annual evaporation, raised mires will develop (McCabe 1984). Well-documented mod-

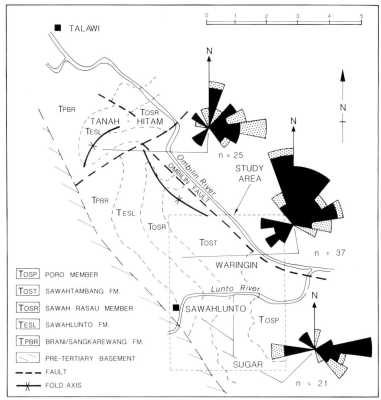

FIG. 12. Palaeocurrent data for the Sawahlunto area of the Ombilin Basin: black areas, data measured from planar and trough cross-bedding; stippled areas, data measured from ripple cross-lamination; scale bar in kilometres. Readings in the Tanah Hitam open pit were taken in the Sawahlunto Formation and those in Waringin and Sugar were taken in the Sawahtambang Formation.

ern examples are found in Malaysia and Indonesia (McCabe 1984; Cameron 1987; Cecil *et al.* 1987; Esterle & Ferm 1987) in deltaic, coastal and alluvial settings, and the raised mires are probably analogous to the Ombilin Basin coal-forming environment.

Depositional model

Exposed sections of the fan-delta and lacustrine associations are limited in the study area. Only a generalized distribution of the associations is given (Fig. 13) based on interpretations of subsurface data derived from borehole cores and structural contour or isopach maps (Fig. 11). The facies associations are shown in their true position as determined from borehole cores and are indicated on the sketch map of the distribution of depositional environments (Fig. 14).

The earliest sedimentation consisted of laterally fed coarse-grained deposits interpreted to be the product of debris flows resulting from

rapid deposition along the margins of the newly formed graben. The interpreted processes and interfingering of these sediments with fine-grained lacustrine sediments basinward appear compatible with a fan-delta origin.

The major characteristics of these fan deltas (Wescott & Ethridge 1980) can be summarized as follows.

(1) The basin margin is fault bounded with proximal fan deposits unconformably overlying basement.

(2) The deposits have a fan- or delta-shaped wedge of coarse-grained sediments.

(3) The fan-delta deposits are immature texturally and mineralogically. The conglomerates and sandstones have abundant lithic fragments reflecting the basement source rocks.

(4) There is a general fining-upward sequence caused by a rapid rise in lake level and a declining sediment supply.

(5) The fan delta was deposited subaqueously as evidence suggests that emergence took place.

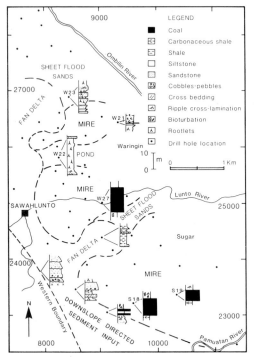

(6) The coarse-grained fan-delta deposits inter-finger with fine-grained freshwater lacustrine deposits.

(7) The fan-delta deposits vary in thickness. In this study thicknesses of some 30 m were measured; Koesoemadinata & Matasak (1981) record a maximum of 646 m in their hypostrato-type section, although Koning (1985) records no coarse-grained fan-delta sediments in the centre of the Ombilin Basin.

By analogy with recent (Link *et al.* 1985) and other Tertiary fan deltas (Sneh 1979; Hempton & Dunne 1984; Manspeizer 1985) the dimensions of the fan deltas suggest that these sediments accumulated no more than a few kilometres from the basin-margin fault scarp.

During major floods, debris flows probably passed basinward into distal-fan sheet floods, giving rise to fining-upward coarse-grained ero-sive- or sharp-based sandstone beds. These pass upwards (and laterally?) into silts and muds which accumulated by settling in ponds during interflood or minor flood episodes. These major periods of relative inactivity allowed plants to colonize and root into the substrate, resulting in the preservation of extensive carbonized rootlets.

Adjacent to and basinward of the distal-fan deposits are the lacustrine sediments within which thick coal seams are developed. The

FIG. 13. Lateral time-equivalent distribution of fan-delta, distal-fan and lacustrine deposits when the C seam peat was forming in the Ombilin Basin.

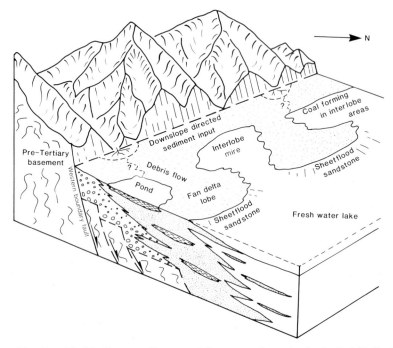

FIG. 14. Depositional model of the Eocene sediments and C seam peat formation in the Ombilin Basin, W Sumatra (modified after Heward 1978).

extensive bioturbated and ripple cross-laminated siltstone and fine-grained sandstone below and above the C seam suggests a low energy environment. The environment in which the peat formed was probably distinctly different from the environment in which the underlying and overlying sediments were deposited (McCabe 1984). The depositional environment that precedes the coal shapes the topography upon which the mire (peat-forming ecosystem of Moore (1987)) develops. The lobate form of the fan deltas resulted in interlobe topographic lows (Fig. 11) in which lacustrine sediments accumulated. Subsequent differential compaction, with the finer-grained lake sediments compacting to a greater extent than the coarser-grained fan-delta deposits, may well have perpetuated local topographic differences. Peat formation was initiated in the topographic lows during a period of low or zero sediment influx. These low-lying mires built up to near-horizontal surfaces.

Palaeocurrent evidence suggests a point source of origin for the lacustrine sediments. During a period of low sediment influx, continuous sediment supply would be restricted to the axial part of the fan delta. Lack of flood tides to impound and raise the level of the flood waters on the fan delta would not force the flood water into interlobe areas (Hyne *et al.* 1979). This would lead to relative stability in sediment transport, resulting in sediment being carried basinward down the fan delta between the peat mires. As a result the time-equivalent of the coal horizon was composed of shale or siltstone.

In tropical peat-forming areas where the annual rainfall is greater than the annual evaporation, raised mires will develop (Teichmüller & Teichmüller 1982). Raised mires have flat central areas and fairly steep convex sides. The upward growth of a peat mire is regulated by the water-table (McCabe 1984). If conditions are favourable for a rise in the water-table, such as basin subsidence or a rise in the level of the lake, peat growth will generally keep pace with this rise. Raised mires up to 13.7 m thick (Anderson 1964) in modern humid tropical mires can form adjacent to active clastic environments (Coleman *et al.* 1970), and these raised mires may have a significant influence on sedimentary processes. Smith & Smith (1980) have shown that vegetation on river banks stabilizes the environment and decreases erosion rates. If peat growth kept pace with sediment aggradation, then the sediment would be confined to a narrow belt. Sediment supplied to the basin from the fan delta by this mechanism would be confined to areas between raised mires, emphasizing the basin-like shape of the coal deposits (*cf* Read, this volume). The

overall pattern is one of a series of margin-parallel basins separated by zones of sediment influx (*cf* Ryer 1981).

There was probably a rise in the level of the lake (tectonically induced?) which drowned the peat and terminated peat growth. The interval between the C seam and the overlying B seam is composed of lacustrine deposits with varying amounts of sandstone. The map showing the percentage of sandstone between the C and B seams (Fig. 15) indicates three distinct areas of sandstone development: (i) a narrow zone in the N trending NE–SW, (ii) a central zone subparallel to the Lunto River and (iii) a southern zone trending NE–E.

The sand-size material can be found in fining-upward sheet-sand sequences or coarsening-upward sequences (Fig. 9), of which the latter are indicative of small-scale progradational deltas. The area in which the sand concentrates is significant. The two southerly sand belts overlie the thickest coal areas. This suggests that the underlying fan-delta lobe geometry is still influencing sedimentation but that the topographic lows are now sites of sediment deposition. A similar effect is seen in the N where the clastic sediment appears to follow an interlobe area.

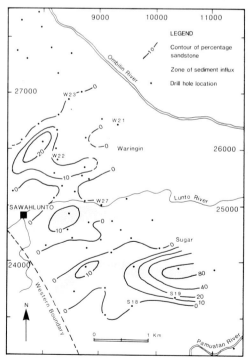

FIG. 15. Percentage sandstone between the C and B seams, Ombilin Basin, W Sumatra.

Coal quality

The depositional setting and humid tropical climate have combined to produce an unusually thick seam of coal whose thickness and ash content vary laterally. The ash content of the C seam also varies vertically. Initially, peat accumulated in the interlobe topographic lows. These areas were therefore likely to receive water-borne sediment. Samples taken at the base of the C seam have ash contents between 38% and 39%. As the peat accumulated to a near-horizontal surface clastic sediment was less likely to enter the swamp. The samples taken above the base of the C seam report ash contents between 25% and 26%.

The humid tropical climate is thought to be a major contributing factor to the subsequent development of the raised mires in the Ombilin Basin. These mires were raised above the flood levels and probably received little or no water-transported mineral matter. As a result the ash content of the upper samples from the C seam is low, between 2% and 9%. This is similar to most modern raised mires whose ash contents vary between 0.5% and 6.5% (McCabe 1984). The gradational upper contact of the coal with the overlying sediment in part reflects the oligotrophic and acidic groundwater conditions that developed as the mire became more elevated and the flora became less varied and stunted (McCabe 1987). The stunted flora on a mature raised bog may not produce sufficient organic matter to maintain the *status quo*. A degree of fluctuation of the groundwater table may have resulted in oxidation and thus dull coal at the top of the coal seam. Alternatively, and more probably, a relatively rapid rise of lake level would have drowned the mire and allowed a gradual increase in the amount of sediment entering it.

The coal thickness is reduced adjacent to the zones of sediment influx. The input of clastic sediments results in the formation of carbonaceous shale and clastic partings. The ash content of the coal in these areas rises to between 16% and 27%.

Summarizing remarks

The structural setting of the Ombilin Basin and the depositional systems in the Sawahlunto area of the Ombilin Basin are compatible with pull-apart basin-sedimentation models based on analogous modern and ancient pull-apart basin deposits. Two depositional systems developed in a humid tropical environment—coarse-grained marginal fan-delta deposits and fine-grained basinal lacustrine deposits (Hempton & Dunne 1984).

The fan-delta deposits are characterized by debris-flow-dominated conglomerates and sandstones which grade into fine-grained sediments towards the basin centre. The lacustrine deposits are characterized by finely interbedded shale and sandstone showing ripple cross-lamination, abundant bioturbation and carbonaceous material, as well as thick coals. Occasional pond deposits (Long 1981) of homogeneous siltstone with abundant carbonized roots also develop. The lacustrine environment is sometimes invaded by fining-upward sheet-flood sandstones or coarsening-upward deltaic (minor-mouth-bar) sandstones.

The newly formed pull-apart basin filled rapidly with sediment and freshwater. The rising and expanding lake led to the formation of fining-upward sequences in the fan-delta association. The lacustrine sediments lack the abundant marls, calcarenites and evaporite minerals described by Sneh (1979), Hempton & Dunne (1984) and Manspeizer (1985). In intermontane basins the lacustrine sediments are the most variable, ranging from anoxic fine clays with limited nutrient content to salts of a hypersaline lake (Hyne *et al.* 1979). In the Ombilin Basin oxygenated bioturbated carbon-rich sediments with abundant palaeobotanical and fossil evidence all point to a humid tropical climate.

Sedimentation rates slowed down which allowed thick coal to accumulate. Coal formation was influenced by fan-delta lobe geometry, with coal forming in the topographic lows in the interlobe areas (*cf* Heward 1978; Ryer 1981). The initial low-lying mires had high mineral matter content, but as the peat accumulated to form a near-horizontal surface the mineral matter decreased. Eventually raised mires developed with low mineral matter content. Around the edges of the raised mires the coal interfingers with clastic sediments and coal quality then deteriorates rapidly laterally.

Eventually the lake level rose and drowned the swamp. Clastic lacustrine sedimentation was re-established over the entire area. The fan-delta lobe geometry still influenced the pattern of sedimentation because sand-size sediment tends to lie over the thick coal where differential compaction has resulted in topographic lows which mirror the underlying interlobe geometry.

This model can be used to predict coal quality and thickness variations. These predictions are invaluable in coal exploration and mine-planning exercises. Coarse-grained transition zone deposits between the fan-delta and lacustrine sediments could provide reservoirs for oil and/or gas accumulations which originated in laterally adjacent organic-rich lacustrine sediments (Koning 1985).

ACKNOWLEDGMENTS: The project formed part of a feasibility study funded by the World Bank and being undertaken by Norwest Resource Consultants Ltd, Canada, consulting to the Indonesian Directorate General of Mines. The authors are grateful for the stimulating discussions and help given during the project by members of the project team. We would particularly like to thank Bob Morris who constructed all the earlier versions of the contour maps, Bob Sharon whose work on the geotechnical aspects of the area was used in assessing interseam sediments, and Ivan Delas who logged all the core retrieved during the project and who located all the interesting section sites. Graham Wallis and Bill Read commented on earlier versions of the manuscript and their helpful discussion is appreciated.

References

ANADÒN, P., CABRERA, L., GUIMERA, J. & SANTANCH, P. 1985. Paleogene strike-slip deformation and sedimentation along the southeastern margin of the Ebro Basin. *In:* BIDDLE, K. T. & CHRISTIE-BLICK, N. (eds) *Strike-slip Deformation, Basin Formation and Sedimentation.* Special Publication of the Society of Economic Paleontologists and Mineralogists 37, 303–318.

ANDERSON, J. A. R. 1964. The structure and development of the peat swamps of Sarawak and Brunei. *Journal of Tropical Geography* 18, 7–16.

AYERS, W. B. & KAISER, W. R. 1984. Lacustrine-interdeltaic coal in the Fort Union Formation (Palaeocene), Powder River Basin, Wyoming and Montana, USA. *In:* RAHMANI, R. A. & FLORES, R. M. (eds) *Sedimentology of Coal and Coal-bearing Sequences.* Special Publication of the International Association of Sedimentologists 7, 61–84.

BULL, W. B. 1964. *Alluvial Fans and Near Surface Subsidence in Western Fresno County, California.* US Geological Survey Professional Paper 473-A.

CAMERON, C. 1987. Factors controlling the qualitative and quantitative characteristics of peat deposits. *Annual Meeting of the Geological Society of America,* Abstract 130 930, 609.

CECIL, C. B., SUPARDI & NEUZIL, S. G. 1987. Domed peat deposits in Indonesia: a modern analog of coal formation. *Annual Meeting of the Geological Society of America,* Abstract 137 455, 615.

COLEMAN, J. M., GAGLIANO, S. M. & SMITH, W. G. 1970. Sedimentation in a Malaysian high tide tropical delta. *In: Deltaic Sedimentation, Modern and Ancient.* Special Publication of the Society of Economic Paleontologists and Mineralogists 15, 185–197.

DE COSTER, G. L. 1976. The geology of the central and south Sumatra Basins. *Proceedings of the 5th Annual Convention.* Indonesian Petroleum Association, Jakarta, 77–110.

ENOS, P. 1977. Flow regimes in debris flow. *Sedimentology* 24, 133–142.

ESTERLE, J. S. & FERM, J. C. 1987. Spatial distribution of peat types in selected deposits in Indonesia and Malaysia as analogues to ancient coal types. *Annual Meeting of the Geological Society of America,* Abstract 134 173, 657.

FOUCH, T. D. 1982. Character of ancient petroliferous lake basins of the world. *American Association of Petroleum Geologists Bulletin* 66, 1680–1681.

FRAKES, L. A. 1979. *Climates throughout Geologic Time.* Elsevier, Amsterdam.

GILBERT, G. K. 1885. The topographic features of lake shores. *Annual Report of the US Geological Survey* 5, 69–123.

HABICHT, J. K. A. 1979. Paleoclimate, paleomagnetism and continental drift. *AAPG Studies in Geology* 9, 31 pp.

HARDING, T. P., VIERBUCHEN, R. C. & CHRISTIE-BLICK, N. 1985. Structural styles, plate-tectonic settings, and hydrocarbon traps of divergent (transtensional) wrench faults. *In:* BIDDLE, K. T. & CHRISTIE-BLICK, N. (eds) *Strike-slip Deformation, Basin Formation, and Sedimentation.* Special Publication of the Society of Economic Paleontologists and Mineralogists 37, 51–78.

HEMPTON, M. R. & DUNNE, L. A. 1984. Sedimentation in pull-apart basins: active examples in eastern Turkey. *Journal of Geology* 92, 513–530.

HEWARD, A. P. 1978. Alluvial fan and lacustrine sediments from the Stephanian A and B (La Magdalena, Ciñera-Matallana and Sabero) coalfields, northern Spain. *Sedimentology* 25, 451–488.

HOLMES, A. 1965. *Principles of Physical Geology.* Ronald Press, London.

HUGHES, J. D. & LONG, D. G. F. 1980. *Geology and Coal Resource Potential of Early Tertiary Strata along Tintina Trench, Yukon Territory.* Geological Survey of Canada Paper 79-32, 22 pp.

HYNE, N. J., COOPER, W. A. & DICKEY, P. A. 1979. Stratigraphy of intermontane, lacustrine delta, Catatumbo River, Lake Maracaibo, Venezuela. *American Association of Petroleum Geologists Bulletin* 63, 2042–2057.

JICA 1980. *Pre-feasibility Study Report on the Ombilin Coal Mine Rehabilitation Project in the Republic of Indonesia (Additional Exploration).* Unpublished report, TBO, 49 pp.

KOESOEMADINATA, R. P. & MATASAK, T. 1981. Stratigraphy and sedimentation, Ombilin Basin, Central Sumatra. *Proceedings of the 10th Annual Conference.* Indonesian Petroleum Association, Jakarta, 217–249.

——, HARDJONO, SUMADIRDJA, H. & USNA, I. 1978. Tertiary coal basins of Indonesia, UN ESCAP. *CCOP Technical Bulletin* 12, 43–86.

KONING, T. 1985. Petroleum geology of the Ombilin intermontane Basin, West Sumatra. *Proceedings of the 14th Annual Conference.* Indonesian Petroleum Association, Jakarta, 117–137.

LINK, M. H., ROBERTS, M. T. & NEWTON, M. S. 1985. Walker Lake Basin Nevada: an example of late Tertiary (?) to Recent sedimentation in a basin adjacent to an active strike-slip fault. *In:* BIDDLE, K. T. & CHRISTIE-BLICK, N. (eds) *Strike-slip Deformation, Basin Formation and Sedimentation.* Special Publication of the Society of Economic Paleontologists and Mineralogists **37**, 105–125.

LONG, D. G. F. 1981. Dextral strike slip faults in the Canadian Cordillera and depositional environments of related fresh-water intermontane coal basins. *In:* MIALL, D. (ed.) *Sedimentation and Tectonics in Alluvial Basins.* Geological Association of Canada Special Paper **23**, 153–186.

MANN, P., HEMPTON, M. R., BRADLEY, D. C. & BURKE, K. 1983. Development of pull-apart basins. *Journal of Geology* **91**, 529–554.

MANSPEIZER, W. 1985. The Dead Sea Rift: impact of climate and tectonism on Pleistocene and Holocene sedimentation. *In:* BIDDLE, K. T. & CHRISTIE-BLICK, N. (eds) *Strike-slip Deformation, Basin Formation and Sedimentation.* Special Publication of the Society of Economic Paleontologists and Mineralogists **37**, 143–158.

MCCABE, P. J. 1984. Depositional environments of coal and coal-bearing strata. *In:* RAHMANI, R. A. & FLORES, R. M. (eds) *Sedimentology of Coal and Coal-bearing Strata.* Special Publication of the International Association of Sedimentologists **7**, 13–42.

—— 1987. Facies studies of coal and coal-bearing strata. *In:* SCOTT, A. C. (ed.) *Coal and Coal-bearing Strata: Recent Advances.* Geological Society of London Special Publication **32**, 51–56.

MOORE, P. D. 1987. Ecological and hydrological aspects of peat formation. *In:* SCOTT, A. C. (ed.) *Coal and Coal-bearing Strata: Recent Advances.* Geological Society of London Special Publication **32**, 7–15.

MUSPER, K. A. F. R. 1929. Beknopt verslag over de uitkomsten van nieuwe geologische onderzoekingen in de Padangsche Bovenlanden, *Jaarb. Mijnw. Ned. Indie* **58**, 303–313. *In:* KOESOEMADINATA, R. P. & MATASAK, T. 1981. Stratigraphy and Sedimentation, Ombilin Basin, Central Sumatra (West Sumatra Province). *Proceedings of the 10th Annual Convention.* Indonesian Petroleum Association, Jakarta, 217–249.

NORWEST RESOURCE CONSULTANTS LTD 1987. *The Final Report on the Mine Feasibility and Transportation Study of the Ombilin II Coal Project, Sumatra.*

Unpublished report, Indonesian Directorate General of Mines (5 volumes).

RICCI LUCCHI, F., COLELLA, A., ORI, G. G., OGLIANI, M. L. & COLALONGO, M. L. 1981. Pliocene fan deltas of the Intra-Apenninic Basin, Bologna. *IAS 2nd European Meeting, Excursion Guidebook, Excursion 4*, 78–162.

RICH, P. V. & MARINO-HADIWARDOYO, H. R. 1977. *Protoplotus Beauforti:* The world's oldest member of the bird family Anhingidae. *Geosurvey Newsletter* **9**, Direktorat Geologi, Bandung.

RUST, B. R. 1979. Coarse alluvial deposits. *In:* WALKER, R. G. (ed.) *Facies Models.* Geoscience Canada Reprint Series **1**, 9–21.

RYER, T. A. 1981. Deltaic coals of Ferron Sandstone Member of Mancos Shale: predictive model for Cretaceous coal-bearing strata of western interior. *American Association of Petroleum Geologists Bulletin* **65**, 2323–2340.

SMITH, D. G. & SMITH, N. D. 1980. Sedimentation in anastomosed river systems: examples from alluvial valleys near Banff, Alberta. *Journal of Sedimentary Petrology* **50**, 157–164.

SNEH, A. 1979. Late Pleistocene fan-deltas along the Dead Sea rift. *Journal of Sedimentary Petrology* **49**, 541–552.

STEEL, R. & AASHEIM, S. M. 1978. Alluvial sand deposition in a rapidly subsiding basin (Devonian, Norway). *In:* MIALL, A. D. (ed.) *Fluvial Sedimentology.* Memoir of the Canadian Society of Petroleum Geologists **5**, 385–412.

TEICHMÜLLER, M. & TEICHMÜLLER, R. 1982. The geological basis of coal formation. *In:* STACH, E. (ed.) *Stach's Textbook of Coal Petrography* (3rd edn). Gebruder Borntraeger, Berlin, 5–86.

TJIA, H. D. 1970. Nature of displacements along the Semangko Fault Zone, Sumatra. *Journal of Tropical Geography* **30**, 63–67.

WESCOTT, W. A. & ETHRIDGE, F. G. 1980. Fan delta sedimentology and tectonic setting, Yallahs fan delta, southeast Jamaica. *American Association of Petroleum Geologists Bulletin* **64** (3), 374–399.

WHATELEY, M. K. G. 1986. *The Depositional History of the Sediments in the Ombilin II Project Area, with Reference to Exploration and Mining Applications of the Depositional Model.* Unpublished report, Norwest Resource Consultants Ltd, 70 pp.

WILLIAMS, H. H., KELLEY, P. A., JANKS, J. S. & CHRISTENSEN, R. M. 1985. The Paleogene rift basin source rocks of central Sumatra. *Proceedings of the 14th Annual Convention.* Indonesian Petroleum Association, Jakarta, 57–90.

M. K. G. WHATELEY, Geology Department, University of Leicester, University Road, Leicester LE1 7RH, UK.

G. R. JORDAN, Norwest Resource Consultants Ltd, Calgary, Canada.

The influence of basin subsidence and depositional environment on regional patterns of coal thickness within the Namurian fluvio-deltaic sedimentary fill of the Kincardine Basin, Scotland

W. A. Read

SUMMARY: Quantitative studies of a varied spectrum of Namurian fluvio-deltaic sediments deposited in Scotland's most rapidly subsiding Namurian basin, the Kincardine Basin, have revealed the following three simple regional geometrical patterns of coal thickness: (i) basin patterns, (ii) inverse basin patterns and (iii) wedge patterns. These characterize both the overall thickness of coal within a given stratigraphical interval and the thickness of some individual coal seams. The patterns become clearer once local sedimentary and tectonic effects have been filtered out by trend-surface analysis.

In a basin pattern coal thickness is closely related to the net subsidence of the basin. In contrast, in an inverse basin pattern coal-forming peat was selectively preserved on the flanks of the basin and vegetation growth was inhibited in the centre by repeated flooding and channel erosion. A wedge pattern shows coal thickness increasing fairly regularly right across the basin, so that it is directly linked to net subsidence on only one flank. Such a pattern can be linked to saline-water incursions from one direction or to optimal peat growth in a locality sheltered from repeated flooding. Similar coal-thickness patterns seem to be present in some other basins, of various ages, scattered over the world.

Despite the recent increase in interest in basin analysis and the increasing recognition of the importance of coal as an energy resource and a source of methane, relatively few attempts have been made to analyse the overall regional patterns that characterize coal as a bulk lithological constituent of the sedimentary fill of differentially subsiding basins, and it is significant that McCabe (1987) has recently stressed the need for more detailed studies of coal thickness. Most papers on regional patterns of coal thickness have concentrated on only one part of a sedimentary basin and on detailed studies of single seams, or groups of closely adjacent seams, particularly in relation to penecontemporaneous channels, delta lobes and crevasse splays (eg Houseknecht & Iannacchionne 1982; Ferm & Staub 1984; Hughes 1984; Guion 1987) or to sheet-sand platforms (Levey 1985). Of the few more general papers, those of Donaldson et al. (1985) and Fielding (1987) are of particular interest. The former concentrates on the effects of depositional systems and climates, whereas the latter suggests that tectonic subsidence was the dominant control on coal distribution.

A particularly varied spectrum of Namurian coal-bearing deltaic and fluvial deposits has been investigated in the Kincardine Basin, E of Stirling in the Midland Valley of Scotland. The results suggest that geometrically simple patterns of total coal thickness are present within relatively thick stratigraphical intervals that extend over the whole of this differentially subsiding basin. These patterns become clearer when local sedimentary and tectonic effects are filtered out by trend-surface analysis. Three patterns of coal thickness can be identified. In the first, which is termed a basin pattern (Read & Dean 1967), the total thickness of coal is closely correlated with the net tectonic subsidence of the basin, so that the thickest coal lies near the axis of the basin. Such a pattern would obviously support Fielding's (1987) hypothesis that subsidence is the dominant control.

In the second pattern, which is here termed an inverse basin pattern, coal is thinnest towards the centre of the basin and thickest on its flanks. Such a pattern runs counter to Fielding's hypothesis. The third pattern, termed a wedge pattern (Read & Dean 1967), shows the total thickness of coal increasing fairly regularly across the basin from one flank to the other. This pattern does not fit Fielding's hypothesis either. The same three simple thickness patterns characterize both the total thickness of coal as a lithological constituent of a basin fill and the thicknesses of some individual coal seams. Similar patterns of coal thickness seem to be present in some other sedimentary basins, and the present paper is intended to stimulate research into finding out how widely these patterns occur. The recent rapid growth and sophistication of computer-operated geological data bases should make this research both feasible and relatively cheap.

The growth and, more critically, the preservation of potentially coal-forming vegetation depends on the complex interaction of many factors, including tectonic and compactional subsidence,

From WHATELEY, M. K. G. & PICKERING, K. T. (eds), 1989, *Deltas: Sites and Traps for Fossil Fuels,*
Geological Society Special Publication No. 41, pp. 333–344.

depositional setting, climate, water-table, and regional and local botanical–ecological factors, including nutrition (see Teichmüller & Thompson 1958; Hacquebard & Donaldson 1969; Scott 1977, 1980; McCabe 1984, 1987; Cecil et al. 1985; Donaldson et al. 1985). Kosters & Bailey (1983) have found that modern deltaic peats are particularly affected by depositional setting, by the balance between subsidence and sedimentation and by marine inundation, all three of which are closely interrelated. Unfortunately, because of the inherent limitations in the available data, it is not always easy to interpret patterns of coal thickness in terms of all these factors or to assess their relative importance.

The present study is an attempt to select and synthesize some of the results of earlier quantitative basin analysis studies, based on bulk lithological variations within the Namurian succession in the Kincardine Basin, and to assess these results in terms of the coal-thickness patterns described from other sedimentary basins, in so far as this is possible. As stated earlier, the results suggest that tectonic subsidence is not always the dominant control.

Area and stratigraphical intervals studied

The area studied lies in the northern part of the Midland Valley of Scotland (Fig. 1). The Silesian tectonic setting of the Midland Valley has recently been discussed in some detail by Read (1988), who has attempted to assess the current hypotheses and has concluded that a modified version of Dewey's (1982) model accords most closely with the evidence currently available. This model postulates that during the Silesian the Midland Valley was undergoing thermal subsidence, following an earlier phase of lithospheric stretching and crustal rifting. Right-lateral strike-slip was superimposed upon this thermal subsidence. The Kincardine Basin, which abuts against the major Ochil Fault, may itself be a product of this strike-slip (Read 1988). During the Namurian it subsided more rapidly than any other part of the Midland Valley. Fortunately, this basin is sufficiently small and intensively explored for virtually the whole of its Silesian sedimentary fill to have been investigated in detail, and in a uniform manner, by a single small team of geologists from the British Geological Survey and British Coal who have made detailed logs of fully cored boreholes. For the present study, records of 80 unfaulted borehole and shaft sections have been selected, but these are supported by records from many other boreholes and shafts and from extensive coal and fireclay workings. The sites of the selected sections are shown in Fig. 1. Their names and National Grid references are listed in the following papers: Interval A, Read & Dean

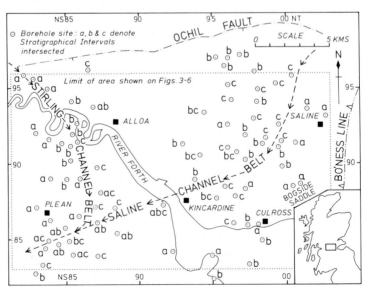

FIG. 1. Sketch map of the area studied, showing the positions of the principal settlements (■), lines of major structures and persistent channel belts active during the Namurian, and sites of borehole and shaft sections used in this study. For the names and National Grid references of these sections see the following references: Interval A, Read & Dean (1967, Fig. 1); Interval B, Upper Hirst Coal and Plean No. 1 Coal, Read et al. (1971, Fig. 1); Interval C, Read & Dean (1982, Fig. 2).

(1967); Interval B, Read *et al.* (1971); Interval C, Read & Dean (1982).

The three stratigraphical intervals described in this paper have already been the subjects of earlier detailed quantitative studies of variations in bulk lithology and lithological sequence. Figure 2 shows their positions relative to each other and to the major lithostratigraphical and chronostratigraphical divisions within the Scottish Namurian. Each interval is bounded by laterally extensive and easily recognized marine horizons.

Interval A, which was sampled by 33 boreholes and shafts (Read & Dean 1967, 1968; Read 1976), lies in the upper part of the Limestone Coal Group between the major marine transgressions marked by the Black Metals and the Index Limestone, and is of Pendleian (E_1) age. This interval is of Coal Measures facies, but was subject to more definite marine influences than the Westphalian Coal Measures of Britain and western Europe. Much of Interval A is composed of upward-coarsening delta-lobe and crevasse-splay sequences (Read 1969*a*), but fluvially influenced sequences are also present (*eg* Read 1961). The peats which formed the relatively thick and laterally persistent coals that characterize this interval accumulated largely under freshwater conditions (*cf* Scott 1978).

Interval B, which has been sampled by 37 boreholes (Read *et al.* 1971; Read & Dean 1972), lies within the upper part of the Upper Limestone Group between the thick and extensively worked Upper Hirst Coal and the subeconomic but laterally persistent Plean No. 1 Coal. Most of this interval was deposited during and immediately after a major widespread, possibly glacial-eustatic, transgression in Arnsbergian (E_2) times, when the sea level rose rapidly (Read 1988) and the Kincardine Basin acted as an effective sediment trap for sands and silts. Much of the succession consists of small delta lobes which were built out repeatedly from two or more sources into fairly deep, generally brackish, water and were commonly colonized by vegetation. Finally, these lobes, together with related channels and crevasse splays, completely filled the basin and formed a stable platform of sandy sediment on which peat accumulated over the whole area.

Interval C, which has been sampled by 36 boreholes (Read 1969*b*; 1981; 1984; Read & Dean 1982), lies in the lower part of the Passage Group between No. 2 Marine Band and the Netherwood Marine Band, and is of Arnsbergian (E_2) to Kinderscoutian (R_1) age. This interval was laid down during a major regression, and is dominated by upward-coarsening erosive-based channel-fill and overbank sequences laid down

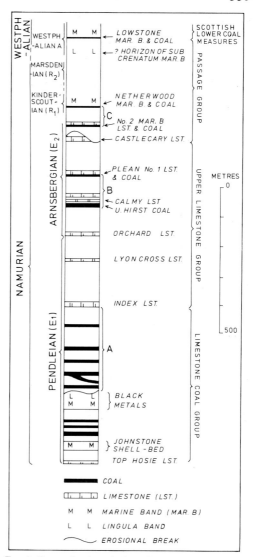

FIG. 2. Generalized vertical section of the Namurian and basal Westphalian in the study area, together with the relevant chronostratigraphical and lithostratigraphical subdivisions and major marker horizons. The limits of Intervals A, B and C are indicated by braces immediately to the right of column.

by meandering rivers. Coal-forming peats were rare and were probably related to infrequent general rises in water-table.

Data sets from the three intervals have also already been used as part of the input for wider quantitative studies of autocyclic processes in relation to net subsidence (Read & Dean 1975, 1976), which provide further background support to the present study.

Methods

The well-tried technique of trend-surface analysis (Miller 1956; Krumbein 1959; Harbaugh & Merriam 1968) has been used to filter out purely local sedimentary and tectonic features and thus clarify the regional sedimentation patterns. Third-degree polynomial surfaces have been chosen in preference to those of a lower or higher order as they provide a convenient level of separation between regional and local factors, as well as fitting the regional subsidence patterns in the Kincardine Basin remarkably well (Read & Dean 1967). The percentage of the total sums of squares satisfied by polynomial trend surfaces of increasing order also provides a useful cross-check on how well these surfaces satisfy the observed data and on the underlying regional sedimentation patterns (Table 1).

The total thicknesses of coal within each interval, plus the total thickness of strata within that interval, have been extracted from each of the chosen borehole records and used to construct firstly a set of contour plots and secondly a set of third-degree polynomial trend surfaces (Figs 3–5). Thickness data for the two individual coal seams that bound Interval B, namely the Upper Hirst Coal and Plean No. 1 Coal, have been treated in the same way (Fig. 6) for comparative purposes. Earlier studies, using principal component analysis, have already demonstrated the

TABLE 1. *Percentage of total variation (calculated as the total sum of squares of the relevant variable) satisfied by least-squares-fit first-degree, second-degree and third-degree polynomial trend surfaces*

Interval or seam	Degree of trend surface	Percentage of total variation satisfied (%)	
		Total thickness strata	Total thickness coal
A	1	34.4	61.5
	2	77.6	70.2
	3	89.6	81.7
B	1	36.0	4.8
	2	59.1	59.4
	3	71.5	80.3
C	1	8.4	17.0
	2	74.8	46.3
	3	76.9	71.1
Plean No. 1 Coal	1	—	0.5
	2	—	61.1
	3	—	79.3
Upper Hirst Coal	1	—	78.4
	2	—	85.3
	3	—	86.9

See Harbaugh & Merriam (1968) for method.

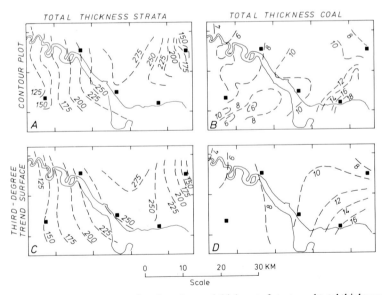

FIG. 3. Contour plots and third-degree trend surfaces for total thickness of strata and total thickness of coal in Interval A, between the Black Metals and the Index Limestone, within the Limestone Coal Group. For sites of sections used, see Read & Dean (1967, Fig. 1). The positions of the principal settlements shown in Fig. 1 are indicated (■) and contours are in metres in this diagram and in Figs 4–6.

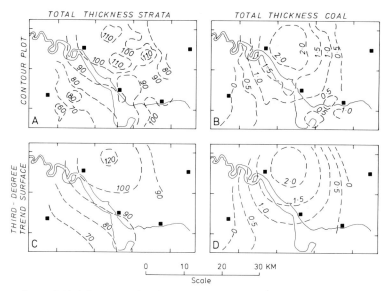

FIG. 4. Contour plots and third-degree trend surfaces for total thickness of strata and total thickness of coal in Interval B, between the Upper Hirst Coal and the top of Plean No. 1 Coal, within the Upper Limestone Group. For sites of sections used, see Read *et al.* (1971, Fig. 1).

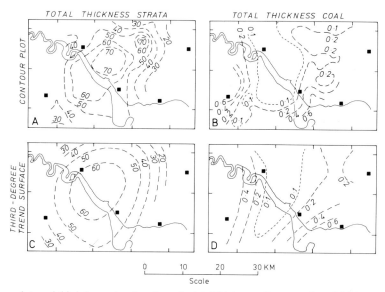

FIG. 5. Contour plots and third-degree trend surfaces for total thickness of strata and total thickness of coal within Interval C, between the base of No. 2 Marine Band and the top of the Netherwood Coal, within the Passage Group. For sites of sections used, see Read & Dean (1982, Fig. 2).

relationships between the total thickness of strata and the total thickness of coal in each interval and a number of other lithological variables (Read & Dean 1968, 1972, 1982). These studies have provided a further quantitative cross-check on the results of trend-surface analysis.

Results

Whereas the contour plot and third-degree trend surface for the total thickness of strata in Interval A (Figs 3*A*, *C*) both bring out the regional pattern

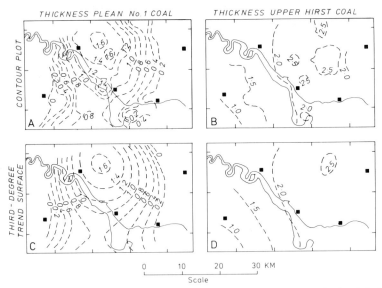

FIG. 6. Contour plots and third-degree trend surfaces for the thickness of Plean No. 1 Coal and the Upper Hirst Coal at the top and bottom respectively of Interval B. For sites of sections used, see Read *et al.* (1971, Fig. 1).

of a fairly simple asymmetrical basin, the contour plot for the total thickness of coal within this interval (Fig. 3*B*) is more complex and less obvious. This is because the simple underlying regional pattern of coal thickness is confused by the local effects of channel belts and structural features. However, the third-degree trend surface for this variable (Fig. 3*D*), which shows a fairly steady increase in the total thickness of coal from WNW to ESE across the basin, satisfies 81.7% of the total variation and brings out the simple underlying wedge pattern. This pattern is confirmed by the fact that a planar first-degree trend surface satisfies no less than 61.5% of the total variation (Table 1).

The pattern for the total thickness of strata in Interval B, which is considerably thinner than Interval A, indicates a more symmetrical basin with an axis trending NW–SE. This comes out in both the contour plot and the third-degree trend surface (Figs 4*A*, *C*). However, in contrast with Interval A, both the contour plot and the third-degree trend surface for total thickness of coal (Figs 4*B*, *D*) also show a similar symmetrical basin pattern. Significantly, a planar first-degree trend surface satisfies only 4.8% of the total variation for the total thickness of coal within this interval.

In fluvially dominated Interval C the contour plot for total thickness of strata (Fig. 5*A*) brings out the effects of a NE–SW Caledonoid graben which was active in the basement below the

Kincardine Basin during Passage Group times (see Read & Dean 1982), but these effects are smoothed by the third-degree trend surface (Fig. 5*C*) which shows a fairly symmetrical basin but nevertheless satisfies 76.9% of the total variation (Table 1). Both the contour plot (Fig. 5*B*) and the third-degree trend surface (Fig. 5*D*) for the total thickness of coal within this interval show an inverse basin pattern with coal thickest on the flanks of the basin and thinnest towards its axis. A planar first-degree trend surface for this variable satisfies only 17.0% of the total variation.

The contour plots and third-degree trend surfaces for the Upper Hirst Coal (Figs 6*B*, *D*) show a wedge pattern and those for Plean No. 1 Coal (Figs 6*A*, *B*) show a basin pattern. The proportion of the total variation satisfied by planar first-degree trend surfaces for the two seams accordingly varies greatly, with no less than 78.4% being satisfied for the Upper Hirst Coal as against only 0.5% for Plean No. 1 Coal.

In addition to the figures quoted above, the values shown in Table 1 confirm the conclusions reached by visual comparison of the third-degree trend surfaces shown in Figs 3–6. Thus only low proportions of the total variation in basin patterns and inverse basin patterns of coal thickness are satisfied by planar first-degree surfaces, and there are large increases in the proportions satisfied by curved second- and third-degree surfaces. In contrast, in wedge patterns comparatively high proportions are satisfied by planar first-degree surfaces and the increases in the proportions satisfied by the higher-degree polynomials are much less.

Discussion

Basin-pattern coals

Basin patterns for the total thicknesses of coal within given stratigraphical intervals strongly suggest that net subsidence, particularly tectonic subsidence of the basin, is the dominant underlying controlling factor. Interval B typifies such a relationship, as the geographical patterns for the total thickness of coal within the interval (Figs 4B, D) and for Plean No. 1 Coal at the top of the interval (Figs 6A, C) both closely resemble the pattern for total thickness of strata (Figs 4A, C) and hence reflect the net subsidence of the basin.

Such a pattern suggests that, during at least part of the interval, a balance existed over the whole basin between the factors that encouraged peat accumulation and those that inhibited it, with accumulation being limited by the relatively slow rate of subsidence on the flanks of the basin yet without vegetation having been drowned out by flooding that resulted, either directly or indirectly, from the greater subsidence in the centre of the basin. This is somewhat unexpected as this interval was particularly subject to strong marine influences, especially towards the centre of the basin, although a strongly autocyclic relationship is known to exist between the number of delta lobes that were colonized by vegetation and the net subsidence of the basin (Read & Dean 1972, 1976). Possibly, the total thickness of coal within Interval B is unduly dominated by the basin pattern of Plean No. 1 Coal at the very top, which grew under more stable conditions than most of the earlier seams within this interval (Read *et al.* 1971).

Inverse basin-pattern coals

An inverse basin pattern of total coal thickness obviously indicates that subsidence was not the direct underlying control. Such a pattern could arise if vegetation was selectively drowned out by flooding or choked out by repeated crevasse splays (*cf* Kosters & Bailey 1983) towards the centre of the basin, or if peats were repeatedly removed by erosive channels. All these processes probably operated during the deposition of Interval C when meander belts persistently tended to occupy the most rapidly subsiding parts of the Kincardine Basin (Read & Dean 1982). In contrast, overbank environments, in which peat

could accumulate more readily, prevailed towards the flanks.

Wedge-pattern coals

Wedge patterns of coal thickness are rather more difficult to explain in the sedimentary and tectonic setting of the Kincardine Basin than are the other two patterns. One possible explanation is that vegetation growth was more frequently inhibited by saline groundwater conditions towards the W than towards the E. Marine influences were certainly more prevalent towards the W during the deposition of Interval A and the Upper Hirst Coal. Many of the roof shales that overlie the coals in Interval A contain an essentially brackish-water fauna (Forsyth & Read 1962), and a more definitely marine fauna lies only a short distance above the Upper Hirst Coal (Read *et al.* 1971).

Another explanation, more closely related to the local structural and depositional setting, is that peat tended to accumulate most rapidly in the general vicinity of a local WSW-trending slowly subsiding structural platform in the basement, known as the Bogside Saddle (Fig. 1) (Read *et al.* 1971, Fig. 1), that lay towards the SE of the Kincardine Basin between a persistent channel belt, known as the Saline Channel Belt (Fig. 1), that ran from NE to SW diagonally across the basin (Read 1976) and a N–S monoclinal flexure, known as the Bo'ness Line, which bounded the E flank of the basin and probably formed the western margin of a major complex area of low subsidence (Read 1988).

A possible reconstruction is as follows. The Bo'ness High to the E shielded the platform from any major westward-flowing channels and thus limited the input of clastic sediments from this direction (Read 1988). On the NW of the platform the Saline Channel Belt was constrained by increased tectonic subsidence to follow the line of a WSW-trending graben in the basement (Read & Dean 1967). Reduced subsidence over the Bogside Saddle would also have tended to protect any vegetation growing above it from flooding and crevassing from the Saline Channel Belt and periodic incursions of brackish water that entered the Kincardine Basin from the W (Forsyth & Read 1962). It is known that splits in coal seams die out and separate coal seams come together in parts of the Limestone Coal Group within this general area (*eg* Read 1961). Tectonic subsidence was nevertheless probably sufficient to allow thick deposits of peat to accumulate here, probably as a succession of ombrotrophic raised mires (Moore 1987) analogous to those presently forming on coastal plains in parts of

Malaysia and Indonesia (Coleman *et al.* 1970; Anderson 1983; Cecil *et al.* 1985). Such a situation would be compatible with the coal-thickness patterns shown in Figs 3*B*, *D*. A similar situation seems to have existed during the deposition of the Upper Hirst Coal (Read *et al.* 1971). Other areas towards the centre and W of the basin were probably more subject to repeated flooding and crevassing which, together with periodically more saline groundwater conditions, would have tended to inhibit vegetation growth and the formation of raised mires.

A tendency for the Upper Hirst Coal to thin in the extreme E of the basin, as seen in Fig. 6*B*, suggests that peat accumulation may have been inhibited by low subsidence towards the Bo'ness Line and indicates some slight tendency towards a basin pattern of coal thickness.

Compaction effects

If compaction of the original peat were to be allowed for, this would not greatly change the overall regional patterns but only their amplitude, particularly as most of the coals have a relatively low ash content.

The various compactional stages in the coalification process are known to be associated with a reduction in moisture content and associated changes in rheological properties (Elliott 1985), but these are not particularly easy to monitor because of factors like the rapid initial compaction of thick peats under the load of penecontemporaneous channel fills (Thiadens & Haites 1944; Britten *et al.* 1975). Nevertheless, it is interesting to note how the sedimentary fill of the Kincardine Basin during Interval A times would become more symmetrical if the coal were decompacted.

Total coal thicknesses would need to be increased by a factor of about 3.7 if compaction by biogenic accumulation, forest loading and penecontemporaneous sediment loading were allowed for (Elliott 1985, Table 1), and this factor would need to be increased to between 11 and 30 if the further effects of loading by the geological column and coalification proper were added (Ryer & Langer 1980, Table 1).

Comparison with other areas

With a few notable exceptions (*eg* Donaldson *et al.* 1985; Fielding 1987), relatively little has been published to date on coal as a bulk lithological constituent of complete basin fills, although there are many recent papers that describe the geometry of single coals, or groups of closely associated seams, within parts of basins. One common

pattern of coal thickness which emerges from these papers can be termed a lenticular shoreline-parallel pattern. This generally consists of pods of thick coal which lie in a belt that runs subparallel to a sea or lake shore. These pods pinch out both basinwards and landwards and are commonly separated laterally by a framework of penecontemporaneous channels which are associated with belts of thin coal and seam splitting, or even gaps in the seams. Such lenticular shoreline-parallel patterns are commonly associated with a line of deltas that prograded out into relatively deep standing water (Ryer 1981, 1984; Ayers & Kaiser 1984; Donaldson *et al.* 1985, Fig. 6; Budai & Cummings 1987). Some individual coals may be both thick and laterally extensive, particularly if they are underlain by a sheet-sandstone platform (Levey 1985) or grew as peats on top of an extensive filled-in lagoon (Flores *et al.* 1984). Variants of a lenticular shoreline-parallel pattern can be produced by lines of peat-capped minor delta lobes or crevasse splays which have prograded out from major fluvial channels into persistent flood-basin lakes (Flores & Hanley 1984, Fig. 5), or by coal-forming peats which have accumulated on the distal portions of, and interlobe areas between, a line of fan deltas that prograded out into a lake (Heward 1978; Whateley & Jordan, this volume). Similar patterns are also found in coastal belts of peat which either fringe the landward shores of lagoons (Snedden & Kersey 1981) or have accumulated in hollows between earlier beach ridges or barriers (Staub & Cohen 1979; Cohen 1984; Flores *et al.* 1984). Some peats with lenticular shoreline-parallel thickness patterns may have formed well back from the contemporary coastline if there was a hiatus between the peat and the underlying clastic sediments (*eg* McCabe 1987, Figs 7 and 8).

Where a stacked sequence of linearly aligned prograding deltas has been preserved, the effects of pods of thick coal separated by thinner shoreline-normal belts of thinner coal tend to be smoothed out, so that a lenticular shoreline-parallel pattern of total coal thickness tends to emerge (see Donaldson *et al.* 1985, Figs 3–6). A lenticular shoreline-parallel pattern of coal thickness represents a fourth simple regional pattern, in addition to basin, inverse basin and wedge patterns, but significantly, if only the basinward side of a lenticular shoreline-parallel pattern were to be preserved, it would closely correspond to a wedge pattern of coal thickness.

Some suggestions of the three simple patterns of coal thickness found in the Kincardine Basin, *ie* basin patterns, inverse basin patterns and wedge patterns, can be traced in the literature on other basins, although these suggestions are not always easy to decipher, particularly as so few workers show the isopachs of the stratigraphical interval within which the coals lie, as well as the isopachs of coals themselves. A few examples, of varying degrees of certainty, are listed below.

Basin-pattern coals

An unequivocal example of a basin pattern for total thickness of coal is seen in the small Westfield Basin, which lies within the Midland Valley only about 25 km E of the Kincardine Basin, and in part of the Passage Group succession which lies approximately between the top of Interval C and the base of the Westphalian. Here unusually thick coals accumulated in a rapidly subsiding local basin which was associated with right-lateral strike-slip along the Ochil Fault and was sheltered by highs from streams bringing coarse-grained clastics (Francis 1961; Brand *et al.* 1980; Read 1988).

Other basin-like patterns of coal thickness have been described from the Middle Pennsylvanian of the Appalachians, where the basin was markedly asymmetrical (Donaldson *et al.* 1985, Fig. 5), from the Palaeocene Ravenscraig Formation, where this was deposited in salt-solution troughs superimposed on broader tectonic basins within the northern part of the major Williston Basin, Saskatchewan (Broughton 1980), from the Palaeocene Fort Union Formation of the Sand Wash Basin in NW Colorado and Wyoming (Beaumont 1979), and from the Oligocene and Miocene lignite-bearing deposits of the Lower Rhine Basin (Quitzow 1966; Hager 1968, Figs 2 and 3; Gliese & Hager 1978).

Inverse basin-pattern coals

One possible example of an inverse basin pattern of coal thickness can be found in the Eocene Upper Eastern View Type Coals of the Gippsland Basin of Victoria, Australia, which accumulated selectively in areas of slower subsidence around the margins of the 'offshore' Gippsland Basin (Smith 1982). A somewhat similar pattern could also have been produced if fan deltas bordered by peats (Heward 1978; Whateley & Jordan, this volume) had been built out into a central lake from both sides of, or all sides of, a fault-bounded basin. Subsequent erosion might not allow such a pattern to be resolved into separate lenticular shoreline-parallel elements.

Wedge-pattern coals

A good example of a wedge-like pattern of coal thickness within a single basin is found in the Middle Pennsylvanian Fire Clay (Hazard No. 4) Coal of the East Kentucky Syncline which caps a lobate delta system that prograded westwards into the sea from the eastern flanks of the basin. Here thin belts of high ash coal, which are associated with penecontemporaneous channels, tend to divide the coal into a series of thick pods (Haney *et al.* 1986). Somewhat similar examples of wedge-like patterns of coal thickness have been described by Hamilton (1985, Fig. 12) from part of his lower-delta-plain facies in the Late Permian Lower Black Jack Formation of the Gunnedah Basin, New South Wales, and by Casshyap (1975, Figs 8*a*, *c*) from the Westphalian A Lower and Upper Bochumer Formations of the southern Ruhr Coalfield. In contrast, the total coal thickness in the intervening Middle Bochumer Formation corresponds more closely to a lenticular shoreline-parallel pattern with pods of thick coal that taper out both basinwards and landwards, thus supporting the idea of a link between some wedge patterns and lenticular shoreline-parallel patterns of coal thickness.

McCabe (1987, Fig. 6) has illustrated another example of a wedge-like pattern of total coal thickness from part of the Maastrichtian Horseshoe Canyon Formation in central Alberta. This is also associated with a lenticular shoreline-parallel pattern of coal thickness in a succeeding formation (McCabe 1987, Fig. 8). It is postulated that the coal-forming peats accumulated some 40–60 km to the landward of the contemporary coast.

Conclusions

Quantitative basin analysis of a varied spectrum of Namurian fluvio-deltaic sediments that were deposited in the differentially subsiding Kincardine Basin in the northern part of the Midland Valley of Scotland has revealed the following three simple (regional) geometrical patterns of coal thickness: (i) basin patterns, (ii) inverse basin patterns and (iii) wedge patterns. These characterize both the total thicknesses of coal within particular stratigraphical intervals, once the effects of local sedimentary and tectonic features have been filtered out, and the thickness patterns of some individual coal seams.

Basin patterns of coal thickness have been strongly influenced by net subsidence, and dominantly by the tectonic subsidence of the basin. Such a pattern suggests that, although relatively low subsidence may have inhibited peat accumulation on the flanks of the basin, vegetation growth in the centre of the basin was not unduly inhibited by being drowned out by flooding, choked out by crevassing or inhibited by repeated incursions of brackish water. The reverse was probably true for inverse basin patterns in which peat accumulated more readily on the stable flanks of the basin rather than in the centre where vegetation growth was repeatedly interrupted by the emplacement of fluvial meander belts.

Wedge patterns, in which the thickness of coal is not obviously related to basin subsidence, are rather less easy to explain. In these the total thickness of coal increases fairly regularly across the basin. Such a pattern might be explicable in terms of a line of deltas that prograded basinwards from one flank of the basin out into standing water, but, in the present examples in the Kincardine Basin, it is more likely to be related to optimal peat accumulation and repeated ombrotrophic raised-mire formation in an area on one flank of the basin that was underlain by a

slowly subsiding platform and was partly sheltered from both freshwater flooding and periodic brackish-water incursions.

ACKNOWLEDGMENTS: The author must acknowledge his indebtedness to his former colleagues in the British Geological Survey and British Coal who so carefully logged the cored boreholes on which this study is based. He is particularly grateful to his friend and co-researcher Mr J. M. Dean, who shared with him the logging of many of these boreholes and who was subsequently jointly responsible for many of the original quantitative basin analysis studies of the Silesian deposits in the Midland Valley of Scotland. He would also like to thank his many academic and Geological Survey friends for many fruitful and stimulating discussions on this topic, particularly Professor E. S. Belt of Amherst College, Amherst, MA, who drew his attention to several important North American references.

References

ANDERSON, J. A. R. 1983. The structure and development of the peat swamps of western Malesia. In: GORE, A. J. P. (ed.) Ecosystems of the World, Vol. 4B, Mires: Swamp, Bog, Fen and Moor. Elsevier, Amsterdam, 181–199.

AYERS, W. B. & KAISER, W. R. 1984. Lacustrine–interdeltaic coal in the Fort Union Formation (Paleocene), Powder River Basin, Wyoming and Montana, U.S.A. In: RAHMANI, R. A. & FLORES, R. M. (eds) Sedimentology of Coal and Coal-bearing Sequences. Special Publication of the International Association of Sedimentologists 7, 61–84.

BEAUMONT, E. A. 1979. Depositional environments of Fort Union (Tertiary, Northwest Colorado) and their relation to coal. American Association of Petroleum Geologists Bulletin 63, 194–217.

BRAND, P. J., ARMSTRONG, M. & WILSON, R. B. 1980. The Carboniferous strata at the Westfield Opencast Site, Fife, Scotland. Report of the Institute of Geological Sciences 79/11.

BRITTEN, R. A., SMYTH, M., BENNETT, A. J. R. & SHIBAOKA, M. 1975. Environmental interpretations of Gondwana Coal Measure sequences in the Sydney Basin of New South Wales. In: CAMPBELL, K. S. W. (ed.) Gondwana Geology. Australian National University Press, Canberra, 233–247.

BROUGHTON, P. L. 1980. Origin of Coal Basins by Salt Solution Tectonics in Western Canada. Unpublished PhD thesis, University of Cambridge.

BUDAI, C. M. & CUMMINGS, M. L. 1987. A depositional model for the Antelope Coalfield, Powder River Basin, Wyoming. Journal of Sedimentary Petrology 57, 30–38.

CASSHYAP, S. M. 1975. Lithofacies analysis and palaeogeography of Bochumer Formation (Westfal A2), Ruhrgebiet. Sonderdruck aus der Geologischen Rundschau 64, 610–640.

CECIL, C. B., STANTON, R. W., NEUZIL, S. G., DULONG,

F. T., RUPPERT, L. F. & PIERCE, B. S. 1985. Paleoclimate controls on Late Paleozoic sedimentation and peat formation in the Central Appalachian Basin (USA). International Journal of Coal Geology 5, 195–230.

COHEN, A. D. 1984. The Okefenokee Swamp: a low sulphur end-member of a shoreline-related depositional model. In: RAHMANI, R. A. & FLORES, R. M. (eds) Sedimentology of Coal and Coal-bearing Sequences. Special Publication of the International Association of Sedimentologists 7, 231–240.

COLEMAN, J. M., GAGLIANO, S. M. & SMITH, W. G. 1970. Sedimentation in a Malaysian high tide tropical delta. In: MORGAN, J. P. (ed.) Deltaic Sedimentation, Modern and Ancient. Special Publication of the Society of Economic Paleontologists and Mineralogists 7, 185–197.

DEWEY, J. F. 1982. Plate tectonics and the evolution of the British Isles. Journal of the Geological Society, London 139, 371–412.

DONALDSON, A. C., RENTON, J. J. & PRESLEY, M. W. 1985. Pennsylvanian depo-systems and paleoclimates of the Appalachians. International Journal of Coal Geology 5, 167–193.

ELLIOTT, R. E. 1985. Quantification of peat to coal compaction stages, based on phenomena in the East Pennine Coalfield, England. Proceedings of the Yorkshire Geological Society 45, 163–172.

FERM, J. C. & STAUB, J. R. 1984. Depositional controls of mineable coal bodies. In: RAHMANI, R. A. & FLORES, R. M. (eds) Sedimentology of Coal and Coal-bearing Sequences. Special Publication of the International Association of Sedimentologists 7, 275–289.

FIELDING, C. R. 1987. Coal depositional models for deltaic and alluvial plain sequences. Geology 15, 661–664.

FLORES, R. M. & HANLEY, J. H. 1984. Anastomosed

and associated coal-bearing fluvial deposits: Upper Tongue River Member, Paleocene Fort Union Formation, northern Powder River Basin, Wyoming, U.S.A. *In:* RAHMANI, R. A. & FLORES, R. M. (eds) *Sedimentology of Coal and Coal-bearing Sequences.* Special Publication of the International Association of Sedimentologists **7**, 85–103.

——, BLANCHARD, L. F., SANCHEZ, J. D., MARLEY, W. E. & MULDOON, W. J. 1984. Paleogeographic controls of coal accumulation, Cretaceous Blackhawk Formation and Star Point Sandstone, Wasatch Plateau, Utah. *Bulletin of the Geological Society of America* **95**, 540–550.

FORSYTH, I. H. & READ, W. A. 1962. The correlation of the Limestone Coal Group above the Kilsyth Coking Coal in the Glasgow–Stirling Region. *Bulletin of the Geological Survey of Great Britain* **19**, 29–52.

FRANCIS, E. H. 1961. Economic geology of the Fife Coalfields, Area II (2nd edn). *Memoir of the Geological Survey of Great Britain.*

GLIESE, J. & HAGER, H. 1978. On brown coal resources in the Lower Rhine Embayment (West Germany). *Geologie en Mijnbouw* **57**, 517–525.

GUION, P. D. 1987. The influence of a palaeochannel on seam thickness in the Coal Measures of Derbyshire, England. *International Journal of Coal Geology* **7**, 269–299.

HACQUEBARD, P. A. & DONALDSON, J. R. 1969. Carboniferous coal deposition associated with floodplain and limnic environments in Nova Scotia. *In:* DAPPLES, E. C. & HOPKINS, M. E. (eds) *Environments of Coal Deposition.* Geological Society of America Special Paper **114**, 143–192.

HAGER, H. 1986. Zur Gleichstellung und Genese der Flöze im rheinischen Braunkohlenrevier. *Fortschritte in der Geologie von Rheinland und Westfalen* **16**, 73–84.

HAMILTON, D. S. 1985. Deltaic depositional systems, coal distribution and quality and petroleum potential, Permian Gunnedah Basin, N.S.W., Australia. *Sedimentary Geology* **45**, 35–75.

HANEY, D. C., COBB, J. C., CHESNUT, D. R. & CURRENS, J. C. 1986. Structural controls on environments of deposition, coal quality, and resources in the Appalachian Basin in Kentucky. *Compte Rendu 10ème Congrès International de Stratigraphie et de Géologie du Carbonifère, Madrid, 1985,* Vol. 2, 69–78.

HARBAUGH, J. W. & MERRIAM, D. F. 1968. *Computer Applications in Stratigraphic Analysis.* Wiley, New York.

HEWARD, A. P. 1978. Alluvial fan and lacustrine sediments from the Stephanian A and B (La Magdalena, Cinera-Metellana and Sabero) coalfields, northern Spain. *Sedimentology* **25**, 451–488.

HOUSEKNECHT, D. W. & IANNACCHIONE, A. T. 1982. Anticipating facies-related coal mining problems in the Hartshorne Formation, Arkoma Basin. *American Association of Petroleum Geologists Bulletin* **66**, 923–946.

HUGHES, J. D. 1984. Geology and depositional setting of the Late Cretaceous, Upper Bearpaw and Lower Horseshoe Canyon Formations in the Dodds-Round Hill Coalfield of Central Alberta—a computer-based study of closely-spaced exploration data. *Geological Survey of Canada Bulletin* **361.**

KOSTERS, E. C. & BAILEY, A. 1983. Characteristics of peat deposits in the Mississippi River delta plain. *Transactions of the Gulf Coast Association of Geological Societies* **33**, 311–325.

KRUMBEIN, W. C. 1959. Trend surface analysis of contour-like maps with irregular control-point spacing. *Journal of Geophysical Research* **64**, 823–824.

LEVEY, R. A. 1985. Depositional models for understanding geometry of Cretaceous coals: major coal seams, Rock Springs Formation, Green River Basin, Wyoming. *American Association of Petroleum Geologists Bulletin* **69**, 1359–1380.

MCCABE, P. J. 1984. Depositional environments of coal and coal-bearing strata. *In:* RAHMANI, R. A. & FLORES, R. M. (eds) *Sedimentology of Coal and Coal-bearing Strata.* Special Publication of the International Association of Sedimentologists **7**, 13–42.

—— 1987. Facies studies of coal and coal-bearing strata. *In:* SCOTT, A. C. (ed.) *Coal and Coal-bearing Strata: Recent Advances.* Geological Society of London Special Publication **32**, 51–66.

MILLER, R. L. 1956. Trend surfaces: their application to analysis and description of environments of sedimentation. *Journal of Geology* **64**, 425–446.

MOORE, P. D. 1987. Ecological and hydrological aspects of peat formation. *In:* SCOTT, A. C. (ed.) *Coal and Coal-bearing Strata: Recent Advances.* Geological Society of London Special Publication **32**, 7–15.

QUITZOW, H. W. (ed.) 1966. *Survey of Geology and Mining in the Lower Rhine Brown Coal District.* State Geological Survey of Nordrhein-Westfalen, Krefeld and Rheinische Braunkolenwerke Aktiengesellschaft, Cologne, Krefeld.

READ, W. A. 1961. Aberrant cyclic sedimentation in the Limestone Coal Group of the Stirling Coalfield. *Transactions of the Edinburgh Geological Society* **18**, 271–292.

—— 1969a. Analysis and simulation of Namurian sediments in Central Scotland using a Markov-process model. *Journal of the International Association for Mathematical Geology* **1**, 199–219.

—— 1969b. Fluviatile deposits in Namurian rocks of Central Scotland. *Geological Magazine* **106**, 331–347.

—— 1976. An assessment of some quantitative methods of comparing lithological succession data. *In:* MERRIAM, D. F. (ed.) *Quantitative Techniques for the Analysis of Sediments.* Pergamon, Oxford, 35–51.

—— 1981. Facies breaks in the Scottish Passage Group and their possible correlation with the Mississippian–Pennsylvanian Hiatus. *Scottish Journal of Geology* **17**, 295–300.

—— 1984. Geographical variation in the numbers of meandering river cycles in a Scottish Namurian Basin. *Compte Rendu 9eme Congrès International de Stratigraphie et de Géologie du Carbonifère, Urbana, Illinois, 1979,* Vol. 3, 589–598.

—— 1988. Controls on Silesian sedimentation in the

Midland Valley of Scotland. *In:* BESLY, B. M. & KELLING, G. (eds) *Sedimentation in a Synorogenic Basin Complex: the Upper Carboniferous of Northwest Europe.* Blackie, Glasgow, 222–241.

—— & DEAN, J. M. 1967. A quantitative study of a sequence of coal-bearing cycles in the Namurian of Central Scotland, 1. *Sedimentology* **9**, 137–156.

—— & —— 1968. A quantitative study of a sequence of coal-bearing cycles in the Namurian of Central Scotland, 2. *Sedimentology* **10**, 121–136.

—— & —— 1972. Principal component analysis of lithological variables from some Namurian (E_2) paralic sediments in Central Scotland. *Bulletin of the Geological Survey of Great Britain* **40**, 83–89.

—— & —— 1975. A statistical relation between net subsidence and number of cycles in Upper Carboniferous paralic facies successions in Great Britain. *Compte Rendu 7eme Congrès International de Stratigraphie et de Géologie du Carbonifère, Krefeld*, Vol. 4, 153–159.

—— & —— 1976. Cycles and subsidence: their relationships in different sedimentary and tectonic environments in the Scottish Carboniferous. *Sedimentology* **23**, 107–120.

—— & —— 1982. Quantitative relationships between numbers of fluvial cycles, bulk lithological composition and net subsidence in a Scottish Namurian basin. *Sedimentology* **29**, 181–200.

——, —— & COLE, A. J. 1971. Some Namurian (E_2) paralic sediments in Central Scotland: an investigation of depositional environments and facies changes using iterative-fit trend-surface analysis. *Journal of the Geological Society, London* **127**, 137–176.

RYER, T. A. 1981. Deltaic coals of Ferron Sandstone Member of Mancos Shale—predictive model for Cretaceous coal-bearing strata of Western Interior. *American Association of Petroleum Geologists Bulletin* **65**, 2323–2340.

—— 1984. Transgressive–regressive cycles and the occurrence of coal in some Upper Cretaceous strata of Utah, U.S.A. *In:* RAHMANI, R. A. & FLORES, R. M. (eds) *Sedimentology of Coal and Coal-bearing Sequences.* Special Publication of the

International Association of Sedimentologists **7**, 217–227.

—— & LANGER, A. W. 1980. Thickness change involved in the peat-to-coal transformation for a bituminous coal of Cretaceous age in Central Utah. *Journal of Sedimentary Petrology* **50**, 987–992.

SCOTT, A. C. 1977. A review of the ecology of Upper Carboniferous plant assemblages with new data from Strathclyde. *Palaeontology* **20**, 447–473.

—— 1978. Sedimentological and ecological control of Westphalian B plant assemblages from West Yorkshire. *Proceedings of the Yorkshire Geological Society* **41**, 461–508.

—— 1980. The ecology of some Upper Palaeozoic floras. *In:* PANCHEN, A. L. (ed.) *The Terrestrial Environment and the Origin of Land Vertebrates.* Academic Press, New York, 87–115.

SMITH, G. C. 1982. A review of the Tertiary–Cretaceous tectonic history of the Gippsland Basin and its controls on coal measure sedimentation. *In:* MALLET, C. W. (ed.) *Coal Resources, Origin, Exploration and Utilization in Australia.* Geological Society of Australia, Melbourne, 1–38.

SNEDDEN, J. W. & KERSEY, D. G. 1981. Origin of the San Miguel lignite deposit and associated lithofacies, Jackson Group, South Texas. *American Association of Petroleum Geologists Bulletin* **65**, 1099–1109.

STAUB, J. R. & COHEN, A. D. 1979. The Snuggeldy Swamp of South Carolina: a back barrier estuarine coal forming environment. *Journal of Sedimentary Petrology* **49**, 133–144.

TEICHMÜLLER, M. & THOMPSON, P. W. 1958. Vergleichende mikroscopische und chemische Untersuchungen der witchtigsten Fazies-Typen im Hauptflöz der niederrheinischen Braunkohle. *Fortschritte in der Geologie von Rheinland und Westfalen* **2**, 573–598.

THIADENS, A. A. & HAITES, T. B. 1944. Splits and washouts in the Netherland Coal Measures. *Mededeelingen van de Geologische Stichting Maestricht, Series C2*, **1**.

W. A. READ, Department of Geology, University of Leicester, Leicester LE1 7RH, UK.

Index